Probability and Statistical Symbols Used in this Book	Meaning	Probability and Statistical Symbols Used in this Book	Meaning
$EOL(A_j)$	Expected opportunity loss of act A_j (12.4)	L	...west observation (3.2)
EVPI	Expected Value of Perfect Information (12.5)	L_{Md}	The lower limit of the class containing the median (3.10)
$E(X)$	Expected value of the random variable X; i.e., the expected value of the probability distribution of X (5.2)	MA	Moving average figures in seasonal variation analysis (13.5)
e	Specified sampling error in a determination of sample size (7.4)	Md	Median (3.10)
F	F ratio; the ratio of the between-column variance to the between-row variance (9.2)	MS_c	Between-column mean square (9.2)
		MS_r	Between-row mean square (9.2)
$F(df_1, df_2)$	F ratio in which df_1 and df_2 are the numbers of degrees of freedom for the numerator and denominator, respectively (9.2)	μ	Arithmetic mean of a probability distribution (5.4)
		μ	Arithmetic mean of a population (3.8)
f	Number of observations (frequency) in a class interval of a frequency distribution (3.8)	$\mu_{\bar{x}}$	Mean of the sampling distribution of the mean; it is equal to the mean of the population (6.5)
F_{Md}	Frequency of the class containing the median (3.10)	$\mu_{\bar{p}}$	Mean of the sampling distribution of a proportion (7.2)
f_p	Frequencies in classes preceding the one containing the median (3.10)	$\mu_{\bar{p}_1 - \bar{p}_2}$	Mean of the sampling distribution of the difference between two proportions (8.3)
f_t	Theoretical or expected frequency in a χ^2 test (9.1)	μ_r	Mean of the sampling distribution of the number of runs r (11.4)
H	Value of the highest observation (3.2)	μ_U	Mean of the sampling distribution of U in the Mann-Whitney U test (rank sum test) (11.3)
H_0	Null hypothesis; basic hypothesis that is being tested (8.1)	$\mu_{\bar{x}_1 - \bar{x}_2}$	Mean of the sampling distribution of the difference between two sample means (8.3)
H_1	Alternative hypothesis; rejection of the null hypothesis H_0 implies tentative acceptance of the alternative hypothesis H_1 (8.1)	N	Number of observations in a population (3.8)
I	Effect of the irregular factor in time series analysis (13.4)	n	Number of observations in a sample (3.8)
i	Size of a class interval in a frequency distribution (3.2)	$\binom{n}{x}$	Number of combinations of n objects taken x at a time (5.3)
		$P(A)$	Probability of the event A (4.1)
k	Number of classes in a frequency distribution (3.2)	$P(A_1 \text{ and } A_2)$	Joint probability of the events A_1 and A_2 (4.2)
		$P(A_1 \text{ or } A_2)$	Probability of the occurrence of at least one of the events A_1 and A_2 (4.2)

Basic Statistics: A Modern Approach

DATE DUE

Basic Statistics:
A Modern Approach

Morris Hamburg

The Wharton School
University of Pennsylvania

Under the general editorship of
Martin K. Starr
Columbia University

Harcourt Brace Jovanovich, Inc.
New York   Chicago   San Francisco   Atlanta

Preface

The major objective of *Basic Statistics: A Modern Approach* is to present the fundamental concepts and methods of statistics in a clear and straightforward way as an aid to students in the development of critical judgment and decision-making ability using quantitative tools. The book is designed for a first course in statistics primarily for students of business and public administration, but also for students in the social sciences and liberal arts.

In their future lives, most readers of this book will be primarily consumers of statistical analysis rather than producers. Hence, emphasis is placed on understanding methods and conclusions rather than on computational routine. The reasoning and logic underlying techniques and concepts and the interpretation and use of statistical results are highlighted.

The book assumes that the basic philosophy, concepts, and methods of statistical analysis can be grasped by any intelligent reader with a modest mathematical background. In fact, the book employs mathematics at about the lowest level consistent with an accurate and understandable statement of the concepts and methods of modern statistical analysis. The emphasis is on helping the reader to appreciate the power of statistical thinking and the scope and versatility of the methods, without bogging him down in mathematical formalities. These formalities often prevent the student with limited mathematical training from seeing a fascinating forest because he or she is concentrating on the incomprehensible individual trees.

This book is based on *Statistical Analysis for Decision Making* (HBJ, 1970) but differs from it in a number of ways. Not only is this book shorter—it is appropriate for use in a one-semester or two-quarter course—it is also more informal and intuitive in approach. The topics included fall under the general headings of descriptive statistics, probability, statistical inference, and statistical decision theory. The two chapters on Bayesian decision theory and nonparametric statistics are included to introduce the student to newer methods of statistical thinking and to expand his horizons beyond the more traditional material.

Many students approach statistics, or any quantitative subject, with uneasiness and a feeling that the subject is unrelated to their needs. However, to function as an intelligent citizen, a person must know how to make sensible inferences and decisions and must be able to evaluate inferences and decisions made by others. That is what statistics is all about! In this book I have tried to impart the feeling that statistics is a useful and exciting field that deals with a scientific method for obtaining, processing, and using knowledge to draw inferences and to make decisions in a world filled with uncertainty.

The structure of the book allows the instructor a great deal of flexibility in designing a course. The fundamentals of classical statistics are presented in the first ten chapters. Chapters 11, 12, 13, and 14 present self-contained discussions of nonparametric statistics, Bayesian decision analysis, time series, and index numbers, respectively. Selection of sections within chapters has been facilitated by the inclusion of exercises at the ends of individual sections rather than at the ends of chapters.

A glossary of symbols, keyed to the section in which a symbol is first introduced, is given on the endpapers of the book. Answers to even-numbered problems are included in the back of the book. Worked-out solutions to all problems are given in the *Solutions Manual*. A *Study Guide* provides supplementary exercises, worked-out problems, and explanations.

I gratefully acknowledge the assistance of the many individuals and organizations, too numerous to mention specifically, who have contributed to this book. I am deeply indebted to Professors Kenneth R. Eberhard, Chabot College; Lawrence L. Lapin, California State University, San Jose; and Andrea Lubov, Eastern Washington State College, for their careful reviews of the manuscript and for their many insightful comments, criticisms, and helpful suggestions. My sincere gratitude is extended to my colleagues in the Statistics and Operations Research Department of the Wharton School, University of Pennsylvania, who originated many of the ideas in the examples and exercises. Many thanks go to David Ostwald and Steven Epstein for checking the arithmetic, to Jonathan Anolik for general research assistance, and to Kenneth Johnson for help in preparing the *Solutions Manual*. Special mention goes to Sylvia Balis for her splendid assistance in secretarial and typing work.

I am indebted to the Literary Executor of the late Sir Ronald A. Fisher, F.R.S., to Dr. Frank Yates, F.R.S., and to Longman Group Ltd., London, for permission to reprint Tables III and IV from their book *Statistical Tables for Biological, Agricultural and Medical Research*. I am also indebted to the other authors and publishers whose generous permissions to reprint tables and excerpts from tables are acknowledged at the appropriate places.

This book is dedicated to my wife, June, and my children, Neil and Barbara. No formal statement is adequate to express my appreciation to them for their cheerful help, encouragement, and understanding.

Morris Hamburg

Contents

Basic Statistics: A Modern Approach

> *"Statistical thinking will one day be as necessary for efficient citizenship as the ability to read and write."*

H. G. WELLS

Introduction 1

INTRODUCTION   1.1

What do the following questions have in common?

Personal questions: Whom should I marry? What field should I choose as my life's work? Which automobile should I buy? What clothes should I wear today?

A private business corporation's questions: Should we buy or lease this building? Should this new product be placed on the market? In which of these capital investment proposals should we participate? Should we invest now or wait a year? Or two?

A national government's questions: Should price and wage controls be instituted? Should the central bank's discount rate be lowered? Should a proposed new crime prevention program be adopted? Should a national health insurance program be initiated? If so, which one of three competing programs should be selected?

All these questions require decisions to be made under conditions of uncertainty. In most cases, the benefits and costs associated with these decisions can only be roughly estimated, and there is uncertainty about the outcomes that will affect these benefits and costs. The required choices must be based on incomplete information. Nevertheless, choices must be made. Even a failure to make a decision constitutes a choice—one that may have net benefits far less or far more desirable than those that flow from an explicit decision.

In its modern interpretation statistics is a body of theory and methodology for drawing inferences and making decisions under conditions of uncertainty.

1

Statistics also deals with how these inferences may be extended beyond the particular set of data examined and how rational decisions may be based on appropriate analyses of such data. From this interpretation, it would seem that the field of statistics has much to contribute to the answering of some of the questions posed earlier, and indeed it does. The raw material of statistics is statistical data, or numbers that represent counts and measurements of objects. The theory and methodology of statistics aid in determining what data should be compiled, and how they should be collected, analyzed, interpreted, and presented in order to make the best inferences and decisions.

However, you may protest, "The field of statistics is not going to help me answer some of the earlier questions such as whom should I marry, or what clothes should I wear today."

We agree that for some decisions a careful scientific approach based on quantitative data does not seem very appropriate. Many decisions require a substantial component of intuitive judgment. Nonetheless, in many areas of human activity, statistical analysis provides a solid foundation for decision making. Statistical concepts and methods such as those discussed in this book bring a logical, objective, and systematic approach to decision making in business, governmental, social, and scientific problems. They are not meant to replace intuition and common sense judgments. On the contrary, they assist in structuring a problem and in bringing the application of judgment to it.

As an example of the last statement in the preceding paragraph, let us return to the question, "Whom should I marry?" You may feel that this question is the least likely of those posed to be subject to a meaningful solution by statistical reasoning. However, let us consider the following line of thought. If you marry a (the?) "right" person, you have made no error. Also, if you do not marry someone you should not have, again you have made a correct decision. On the other hand, if you fail to marry someone with whom you would have been happy, you have made an error. Or, if you marry someone with whom your future life falls far from a state of connubial bliss, again you have made an error. After studying the subject of hypothesis testing, you will recognize such mistakes as Type I and Type II errors. Which error is more serious? If you are considering the possibility of marrying someone, which is the best course of action, to marry or not to marry? Which action has associated with it the higher expected "payoff"? Although the problem and questions are raised here tongue in cheek and without the expectation of a serious answer, they illustrate a way in which you may begin to structure the problem and apply judgment to it. We cannot dispute that one's romantic sensibilities may be offended by any attempt to solve this sort of problem by the suggested method of analysis. Our purpose is merely to indicate that even so unlikely a problem for quantitative analysis may be structured in a framework that can assist decision making. It is currently recognized that the field of statistics aids in providing the basis for arriving at rational decisions regarding a tremendous variety of matters relating to business, public affairs, science, and other fields. In many of these matters, such quantitative analysis has been found to be not only applicable but helpful.

As indicated in the preceding discussion, statistical concepts and methods are applied in many areas of human activity. In the sciences, the applications range from the design and analysis of experiments to the testing of new and competing hypotheses. In industry, statistics makes its contributions in short- and long-range planning and decision making as well as in day-to-day operational decision making and control. Many firms use statistical methods to analyze patterns of change and to forecast future activity for a firm, an industry, and the economy as a whole. Such forecasts often provide the foundation for corporate planning and control, with areas such as purchasing, production, and inventory control dependent upon short-range forecasts, and capital investment and long-term development decisions dependent upon longer range forecasts. In addition to forecasts, areas such as production control, inventory control, and quality control often employ statistical methods on a standard basis as well.

Statistical concepts are also applied in areas such as auditing, the determination of manpower requirements, personnel selection, marketing research, advertising, industrial experimentation, credit risk selection, financial analysis, distribution analysis, and research and development.

In public administration and in the social sciences, statistical methods are widely employed, particularly in the analysis of political and social problems. Quantitative studies have been made of poverty, population, sex behavior, voting patterns, vehicular accidents, educational matters, and public health matters such as the relationship between smoking and various types of disease. Sample surveys are heavily relied upon in such studies to obtain the statistical data concerning the human activities and conditions involved. A variety of statistical techniques has been and is currently being used to collect, analyze, and present the data from such studies, and for planning and decision-making purposes.

During the past few decades a large number of quantitative techniques and procedures designed to aid and improve managerial decision making have been developed. The field of statistics has provided many of the fruitful ideas and techniques in this development. Currently, applications of statistics pervade almost every area of activity of the business firm. Reliance on the use of statistical methods has increased, and the methods themselves have become more sophisticated. Statistical methods are an integral part of the general development of more rational and quantitative approaches to the solution of business problems. Characteristic of this development has been the growing use of scientific decision-making tools using mathematical models. These models are mathematical formulas or equations that state the relationship among the important factors or variables in a problem or system. A concurrent development has been the increasing use of modern computers, that is, electronic data processing equipment—a development that has added momentum to the spread of the use of statistical techniques. Quantitative analyses that would not have been attempted a generation or two ago because they would have been too costly and time consuming are now being carried out on massive bodies of data. Statistical analysis plays an increasingly important role in many areas of human activity. It is extremely useful for communicating information, for drawing conclusions and inferences from data, and for the guidance of rational planning and decision making.

1.3 DESCRIPTIVE AND INFERENTIAL STATISTICS

The most widely known statistical methods are those that summarize numerical data in terms of averages and other measures for purposes of description. For example, after grading an examination taken by a particular class, a teacher may calculate an average grade to summarize the overall performance of the class. If the teacher is only interested in a description of the performance of the specified class on that particular test and not in further *generalization,* he is dealing with a problem of *descriptive statistics.* Presentations of data in the form of graphs and tables also fall under the heading of *descriptive statistics.*

However, much of modern statistics is concerned with the theory and methodology for the drawing of *inferences* that extend beyond the particular set of data examined. For example, suppose a government agency draws a sample of 5000 families at random in a given city and, on the basis of collected data, classifies a certain percentage of these families as living in poverty. (Note that poverty is relative and the agency would first have to define the term.) Suppose the agency then goes farther and on the basis of calculations from the collected data draws an *inference* about the percentage of *all* families in the city that are living in poverty. The problem of estimating the percentage for all families in the city is one of *statistical inference.*

The agency would also be engaged in a problem of *inference* if it attempted to *test a hypothesis,* say, that at least ten percent of all families in the city were living in poverty. The conclusion drawn from the test would involve a generalization beyond the data for the sample of 5000 families. The mathematical *theory of probability* provides the basis for the generalizations from the sample of data studied to the inferences about all families in the city.

Let us note a few points concerning the preceding examples. An inference was desired concerning the percentage "poverty families" represented of all families in the city. However, since it would have been too expensive and too time-consuming to obtain the relevant data for every family in the city, only a sample of 5000 families was drawn. All families in the city, or more generally, the totality of the elements about which the inference is desired is referred to in statistics as the *universe* or *population.* The 5000 families, which represent a collection of only some of the universe elements, as we have seen, are referred to as a *sample.* In statistics, *sample data* are observed in order to make *inferences* or *decisions* about the *populations* from which samples are drawn.

In recent years, there has been an extension of statistical inference, with emphasis placed on the problem of decision making under conditions of uncertainty. This formulation has come to be known as *statistical decision theory* or *Bayesian decision theory*. The latter term is often used to emphasize the role that a theorem derived by the Reverend Thomas Bayes[1] plays in this theory. Statistical decision theory addresses itself to the problem of making rational selections among alternative courses of action when information is incomplete and uncertain. As an example, let us assume that the problem of the preceding section was extended from drawing an inference about the percentage of poverty families in a city to making a decision based on these findings. For example, the alternative courses of action might be for the government either to grant or not to grant financial aid to the city. The various possible values for the percentage of poverty families may be viewed as events or outcomes that affect the achievement of the decision maker's objectives. The "payoffs" or net benefits of the two alternative actions would be the consequences that flow from the action selected and the event that occurs. Statistical decision theory presents the principles and methods for making the best decisions under specified conditions. Although a comprehensive treatment of Bayesian decision theory is beyond the scope of this book, an introduction to that subject is given in Chapter 12.

APPLICATIONS OF STATISTICS   1.5

As implied by the earlier discussion, statistical methods are versatile and have a wide scope of applicability. Indeed, the range of useful applications is so extensive that it can only be partially touched on here. However, in order to suggest the power of statistical methodology and reasoning, a few examples will be discussed in this section.

A central concept in the theories of John Maynard Keynes, the British economist (1883–1946), is the consumption function, that is, the relationship between consumer expenditures and income. Many statistical studies have been carried out, usually in the form of regression analyses (discussed in Chapter 10), to obtain quantitative measures of this relationship. These and other measures established by statistical studies aid in assessing the effects of various fiscal measures and proposed governmental monetary actions.

More generally, statistical methods are applied to relevant numerical data to assess past trends and current status and to project future economic activity. These methods provide measures of human and physical resources, economic growth, well-being, and potential. They are essential tools for appraising the performance and for analyzing the structure of an economy.

[1] The Reverend Thomas Bayes (1702–1761) was an English Presbyterian minister and mathematician.

At the level of the individual business firm statistical methods are used in innumerable ways for planning, decision making, and control. For example, they are used in connection with the forecasting of general economic activity and the corresponding activity for the industry of which the firm is a part and for the firm itself. In marketing research, statistical methods have been used to measure the relationships between the demand for a company's products and the socioeconomic and demographic characteristics of the consumers of the product, such as income, savings, market value of home, family size, and family composition. On the basis of the relationships observed, the company can choose the groups of consumers toward which it is most profitable to direct its marketing efforts. Also, firms often employ sample surveys to determine the most effective methods of promotion of their products. Such surveys aid in the evaluation of promotional methods such as television advertising, direct mail promotion, and advertising in periodicals.

Statistical techniques have had far-ranging and very effective applications in the quality control of manufactured products. They are used to maintain the average level of manufacturing processes within tolerable limits and are also used to measure and control the variability of these processes. In this type of application, statistical methods are used to differentiate between variations attributable to chance causes and those too great to be considered a result of chance. The latter type of variation can be analyzed and remedied. Often applications of these statistical quality control methods have resulted in substantial improvements in product quality and in lower costs because of reduction in rework and spoilage.

Many applications have been made of the use of sampling techniques in which information is obtained at far less cost and effort than if complete counts were made. Furthermore, in many cases, the data obtained have proven to be more accurate than corresponding complete counts. For example, throughout American as well as many foreign industries, sampling inspection is used for acceptance and rejection of incoming and outgoing products. Relatively small samples are randomly drawn from incoming shipments from suppliers or from outgoing shipments of lots produced to determine whether to accept or reject these shipments and lots. Also, the U.S. Bureau of the Census has often employed samples of individuals and business firms to estimate certain population figures. These sampling procedures have generally been found to be superior to complete counts because more rigorous measurement and inspection techniques can usually be employed on the smaller numbers of items in the samples. In the case of sample surveys of human populations, it is often feasible to avoid the high proportions of nonresponse that occur when attempts are made to cover an entire population. The reason is that in sample surveys, as opposed to complete counts, it is generally more practicable and less expensive to employ detailed follow-up techniques to obtain the required information.

The foregoing illustrations are primarily in the areas of economics and business, and even these merely scratch the surface of the variety of applications of statistical analysis in those fields. The following questions illustrate other areas in which statistical methods have been useful.

What is the relationship between educational performance of students and the social and economic characteristics of families of these students? What is the relationship between obesity and heart disease? What are the attitudes of the American population on nationalized health insurance? How does a crackdown on speeding affect automobile accident rates? What are the relationships between the genetic characteristics of parents and their offspring? How effective is the Salk vaccine? What is the relationship between population growth and various measures of the quality of life (again, "quality" is relative and must be strictly defined)? Do some people really possess powers of extra-sensory perception far in excess of those of most people?

The usefulness of statistical methods has been amply demonstrated. However, there are many possibilities for error in the carrying out and interpretation of statistical studies and many potential pitfalls and limitations. A healthy skepticism about the results of any statistical investigation is essential. Although it is rather difficult to elaborate on these errors, pitfalls, and limitations before presenting a discussion of statistical methods, a few of these problems are indicated in the next section. To appreciate fully appropriate applications of statistical methods, it is important to have an understanding of possible misuses.

ERRORS IN STATISTICAL STUDIES   1.6

Bias

Many statistical studies lead to incorrect conclusions because of bias in the collection of the data. Bias as used here refers to any persistent or systematic error. In ordinary usage, the term "bias" often refers simply to prejudice. Bias may be present in statistical data for reasons of prejudice expressed in responses to questions in statistical surveys, but it may also arise from reasons entirely unrelated to prejudice. Examples of bias arising from prejudiced responses have occurred in attitude surveys concerning race relations and other sensitive topics. For instance, in one study it was found that when blacks were asked whether the army is unfair to blacks, 35 percent said yes to black interviewers, whereas only 11 percent said yes to white interviewers.[2]

On the other hand, a classic case of bias or systematic error related not to prejudice but rather to the collection procedures in a sample survey is that of the *Literary Digest* prediction of the presidential election of 1936. During the election campaign between Franklin D. Roosevelt and Alfred M. Landon, the *Literary Digest* magazine sent questionnaire ballots to about ten million persons whose names appeared in telephone directories and automobile registration lists. Over two million ballots were returned by the respondents. On the basis of these replies, the *Literary Digest* erroneously predicted that Landon would be the next president of the U.S. The reasons why the conclusion based

[2] Survey carried out by the National Opinion Research Center, reported in the University of Chicago Magazine, April 1952.

on this survey was unrelated to the actual outcome are rather clear. In 1936, during the Great Depression, the presidential vote was cast largely along economic lines. Voters who did not own telephones or automobiles were not included in the sample. The majority of these people, who represented a lower economic level than possessors of telephones and automobiles, voted for Roosevelt, the Democratic candidate. A second reason for the false conclusion stemmed from the nonresponse group, which represented about four-fifths of those polled. Typically, individuals of higher educational and economic status are more likely to respond to voluntary questionnaires than those with lower educational and economic status. Therefore, the nonresponse group doubtless contained a higher percentage of this lower status group than did the group that responded to the questionnaire. Again, this factor added a bias due to under-representation of Democratic votes. In summary, the sample used for prediction purposes contained a greater proportion of persons of higher socioeconomic status than were present in the relevant population, namely, those who cast votes on election day. Since this factor of socioeconomic status was related to the way people voted, a systematic overstatement of the Republican vote was present in the sample data. It is of interest that the sampling method used by the *Literary Digest* had been successfully employed in four prior presidential elections to predict the winning candidate.

A number of methodological lessons can be learned from this example. First, it is dangerous to sample a population that differs considerably from the target population, that is, the one about which inferences are desired. As we have seen, the population sampled in this case consisted of owners of telephones and automobiles, whereas an inference was desired about the population that appeared at the polls on election day. Second, procedures must be established to deal with the problem of nonresponse in statistical surveys. Clearly, even if the proper target universe is sampled, the problem of nonresponse still must be properly handled. Third, if a biased sampling procedure is used, simply increasing the size of sample will not result in greater accuracy. Note that if the same sampling procedure were used as in the preceding example, doubling or quadrupling the size of the sample would probably not have given more satisfactory results. The same problem of systematic error would still be present. It may be noted that the sample size of over two million was much larger than is possible in most investigations, yet the conclusions were sadly in error.

Another example of bias occurred in the results of a survey of the alumni of a university. One of the items of information requested was gross earned income of the respondents. When the data were compiled, the average incomes of every class were far in excess of reasonable expectations based on various types of external data and other evidence. A number of reasons may account for the upward bias clearly present in these average income figures. First, if misreporting of income did occur, it probably was in an upward direction because of a desire on the part of respondents to appear financially successful. Second, even if we assume no misreporting of incomes, a bias probably occurred because of selective response. That is, if we assume that the graduates who had lower incomes were less likely to respond than those with higher incomes, then an average of the reported incomes would be higher than the correct average for all alumni.

On the other hand, when the Internal Revenue Service assembles data on income based on the figures reported on income tax forms, it is doubtful that averages derived from these data are similarly biased in an upward direction. Since the reporting requirement is mandatory, selective response does not operate. Furthermore, it seems reasonable to assume that there is no systematic pattern of overstatement of gross incomes to the Internal Revenue Service. Indeed, it seems reasonable to suppose that there is a downward bias in these data in the aggregate. Thus, we have the interesting situation of the same type of data being gathered by two different agencies, one set being biased in an upward direction, the other downward. Therefore, it is clear that in using statistical information, informed critical judgment must be exercised to extract meaningful inferences. This judgment must include such practical considerations as methods of data collection and auspices under which studies are conducted. Questions should be raised such as "Who collected the data and how?"; "Was there anything about the way the data were collected that would tend to produce biased figures?"; "Did the person who reported the conclusions have a vested interest in the results of the study?"; and "If yes, did this vested interest seem to affect the way the data were collected and the way the results were reported?"

An amusing example of a case in which biased results were obtained because of a vested interest on the part of respondents is that of a survey attempted by a church in New York City during the depression of the 1930's. The church ran a kitchen where it was customary for people to line up in order to obtain a free bowl of soup. One day the church decided it would like to get some information about the people in the soup lines. The question of religion was asked. The first man in line responded, "Baptist", and received his bowl of soup. The message that "Baptist is O.K." quickly passed down the line. The data collected that day gave a strong indication of bias in the direction of an over-representation of Baptists.

The type of selective response suggested in the alumni income illustration virtually always occurs in mail questionnaires. That is, just as in the sample where those of higher economic status tended to be more apt to respond, those who have the strongest opinions, who have a greater interest in the results of the investigation, who may derive some benefit from the results, or who have a loyalty to express tend to have higher response rates than others.

Improper inferences

As indicated earlier, much of statistical reasoning involves inferences about populations from data observed in samples. These inferences may be characterized as inductive in nature, that is, they represent reasoning from the particular (the sample) to the general (the population). Many misuses of statistics involve erroneous inferences or generalizations. Only a few of the reasons will be touched on here.

A survey on shoppers' attitudes was conducted by stationing an interviewer on a particular street corner in the central business area of a large city. He asked questions of individuals who passed by and recorded the responses of those who were willing to reply. Without concerning ourselves with the conclusions drawn from this study, we can see the major difficulty arising from the sampling method employed. It is unclear to what specific population any inferences drawn from the sample pertain. First, we can again note the uncertain effect of selective response. However, even if everyone who passed responded, we still would have difficulty in defining the population to which the sample results generalized. As it happened, the street corner was in an affluent shopping district. Quite different patterns of response might have been observed had the questioning been carried out a few blocks away in a less affluent shopping area. Even if some interviewing had been done on a number of street corners in shopping areas of different levels of affluence, serious difficulties would be encountered in trying to aggregate properly the results obtained. As we shall see in subsequent discussions, particularly in Chapters 2, 4, and 5, the population about which inferences are desired must be specifically defined, and probability methods of sampling should be employed. In this example, a sample was obtained, but the corresponding population was unclear. However, even if we could specify what the population was, using such terms as persons passing by the specified street corners within certain time intervals, this in all likelihood is not the population about which generalizations are really desired.

Another example of the same type of difficulty is that of a study reported in a newspaper concerning the current credit-granting practices of ten "large banks" in a certain city. Vague generalizations concerning credit-granting practices were stated. No definition of the term "large bank" was given nor was there any indication of how the sample of banks (presumably "commercial banks") was drawn. If we assume that the banks were not drawn by a probability sampling method, that is, a method in which the chances of inclusion of every element in the population are known, the ten banks constitute what is referred to in statistics as a "judgment sample" or "chunk sample." A basic difficulty with such samples is that there is no methodology for specifying the precision of inferences drawn. On the other hand, the methodology for specifying the precision of estimates from "probability samples" is clear-cut and is a major topic of this book.

Sometimes improper generalizations from sample results occur because of invalid comparisons. During the 1930s a study was made of the incidence of tuberculosis among men and women in the garment industry. Samples were drawn (assume "probability samples") of men and women currently at work. The data indicated that men had a higher incidence of tuberculosis than women. The conclusion was drawn that men in the garment trade were more susceptible to tuberculosis than were women.

The invalidity of the comparison made in this case and of the conclusion drawn becomes clear when a few points are noted. The average age of the women was much lower than that of the men in the samples compared. This was because many women left the industry after getting married, whereas the men tended to make the work in this industry their permanent occupations. Hence, if a longer duration of employment in the industry meant a greater exposure to the hazard of tuberculosis, then the men had been exposed, on the average, much longer than the women. Furthermore, the following subtle factor was operative. Assume that a woman had left the industry prior to the study and had later contracted tuberculosis and had the disease at the time of the study. Since she was no longer in the industry, she would not have been included as a female tuberculosis case. Another factor not possible to detect from the reported results of the study may also have been operative. Suppose a higher proportion of women than men held office positions as opposed to factory jobs. If the exposure to the hazard of contracting tuberculosis was greater in factory work than in office work, it would follow that a higher incidence of tuberculosis would be found in men than in women in this industry. A preferable methodology in this case would have involved comparing matched pairs of women and men, who were performing the same work and had been employed in the industry about the same length of time. Care would have to be exercised that the matched pairs did not differ in any other characteristics related to the risk of contracting tuberculosis.

Finally, another point may be noted concerning this illustration. The conclusion that men in the garment industry were more susceptible to tuberculosis than women was drawn on the basis that the data appeared to be consistent with that hypothesis. However, as we have seen, with some analytical reasoning, we find that the data are consistent with other hypotheses. In many somewhat more subtle misuses of statistics, seemingly valid generalizations are drawn. However, on closer examination, it may be found that other quite different generalizations are equally tenable.

Concluding causation from correlations

Numerous examples have been given to demonstrate the pitfalls in attempting to conclude that because two factors are correlated, one of these factors caused the variations in the other. For example, the Cuban National Commission for the Propaganda and Defense of Havana Tobacco once noted that since the discovery of the tobacco plant the human life span has doubled. One surely would be unjustified in inferring from this statement that increased usage of tobacco over the years has *caused* this increase in length of life. Indeed, suppose one had data that showed an increase in crime rates over the same period. Should one then conclude that increased use of tobacco *causes* increases in crime? Clearly not. The point is that many factors, other than the use of tobacco, that may have affected length of life and crime rates have been present over the same time periods. It may actually be very difficult and perhaps even impossible to make valid statements from such statistical data concerning cause-effect relationships. However, it suffices for our present purposes merely to indicate that many potential booby traps await one who attempts to draw cause and effect conclusions from statistical data concerning correlated factors.

Misuses of ratios

Many misuses of statistics involve technical errors, sometimes of a very simple nature, and often of a more sophisticated nature. Only a few of the simpler errors will be noted here. The errors that follow involve misuses of ratios. As indicated in Chapter 2, ratios may be expressed as fractions, decimals, or percentages.

A fairly common error is to interpret a change of percentage points as though it were a percentage change. For example, assume a price index whose base period is 1959. Hence, the value of the index in 1959 is 100. Suppose the index is at 150 in 1965 and moves to 200 in 1971. What was the percentage change in prices from 1965 to 1971? This type of change has often been incorrectly referred to as a 50 percent increase in prices. The index has increased by 50 *percentage points,* but the percentage increase in the level of prices from 1965 to 1971 was $(200 - 150)/150 = 33\frac{1}{3}\%$. The percentage increase from 1965 to 1971 must be computed using the 1965 figure as a base. The type of error indicated here has often been made in newspaper reports concerning month-to-month changes in the U.S. Bureau of Labor Statistics Consumer Price Index.

Many misuses of ratios involve vague or unstated bases on which percentage changes are computed. Unfortunately, we are all to familiar with advertisements of the type "Our tires give 50 percent more wear," "Users of our toothpaste have 30 percent fewer cavities." More than what? Less than what? Since the bases are not given, no meaningful inferences can be drawn.

Some misuses occur because of a failure to distinguish correctly between a percentage relative and a percentage rate of change. For example, the sales of a corporation were $10,000,000 in 1960 and $30,000,000 in 1970. The corporation reported that its sales had increased 300 percent from 1960 to 1970. Of course, the percentage increase was only $(\$30,000,000 - \$10,000,000)/\$10,000,000 = 200$ percent. The level of sales in 1970 relative to 1960 was $\$30,000,000/\$10,000,000 = 300$ percent.

Sometimes a misuse occurs when there is a failure to use a ratio when one is needed. For example, the statement was made that in a certain community automobile drivers under 25 years old were involved in more accidents than drivers over 65. This comparison of the *numbers* of accidents was entirely misleading. Not only were there more younger drivers, but also the younger drivers tended to drive more miles than the older persons. Hence, what was needed for a more meaningful comparison was a comparison of accident rates for the two groups. These accident rates might be computed for each group in terms of numbers of accidents divided by total numbers of miles driven. Of course, even if such data were available, many problems remain concerning the comparison of the rates for the two groups, particularly if there were wide differences in the general conditions under which persons in the two age groups drive.

Another common misuse of ratios involves calculations of percentages using an incorrect base. A furniture company advertised a sale of a sofa at a 200 percent discount. Was the company giving the sofa away free (a 100 percent discount) and giving the customer some money in addition? No, the company was not engaged in any such philanthropic endeavor! The original price of the sofa was $300 and the reduced price was $100. The company had calculated the 200 percent reduction by dividing the discount of $200 by the new price of $100. Of course, the decrease of $200 in price should have been divided by the original price of $300, yielding a 66$\frac{2}{3}$ percent reduction.

Sometimes, percentages are computed on bases that are so small that distorted inferences result. A fund-raising organization reported a 200 percent increase in the number of corporate donors from a certain industry to a charity drive, whereas the increase from other industries averaged only 10 percent. For the industry in question, the number of donors had increased from one to three. The arithmetic was correct, but percentages computed from such small bases tend to be misleading. This is one reason, for example, why recession or depression years, where prices are unusually low, are ordinarily not used as base periods.

The misuses indicated above represent only a small number of the possible misapplications of statistics and are not necessarily the most important categories of such misuses. However, they are presented to alert you to the advisability of skeptically questioning every stage of a statistical investigation from the design of the study, through the collection and analysis of data to the presentation of conclusions. It is said that Disraeli once stated "there are three kinds of lies: lies, damned lies, and statistics." Just as it is important for a policeman and detective to know the methods of operations of those who violate the law, so it is essential to understand misuses of statistics in order to apply statistical methods in valid and fair-minded ways, and to be able to evaluate applications of these methods by others. Many types of misuses of statistics are indicated along with the discussion of statistical analysis in the pages that follow.

MARK TWAIN

2 Statistical Investigations and Data

2.1 FORMULATION OF THE PROBLEM

A statistical investigation arises from the need to solve some sort of a problem. These problems may be classified in a variety of ways. For example, problems can be classified as (1) choices among alternative courses of action, and (2) informational reporting. The problems of managerial decision making typically fall in category 1. For example, an industrial corporation wishes to choose a location for a plant from among alternative sites. A financial vice-president might wish to decide among alternative methods of financing planned increases in production facilities. An advertising manager must choose media in which he will advertise from among many possible choices. Under heading 2, many illustrations can be given of statistical data collected merely for reporting purposes. For example, a trade association may report on the characteristics of its member companies. A research organization may publish data on the relationship between children's achievement in school and the socioeconomic characteristics of their parents. An economist may report on the distribution of family incomes in a particular city.

Throughout our lives we are involved in answering questions and solving problems. As we noted in the last chapter, for many of these questions and problems, careful, detailed investigations are simply inappropriate. For example, to answer the questions "What clothes should I wear today?", or "What type of transportation should I take to get to a friend's house?", does not require painstaking, objective, scientific investigation. However, in this discussion, we are concerned with the investigation of problems that do require careful planning and objective, scientific approaches to arrive at meaningful solutions. Many of these problems arise as rather vague, original questions. These questions must be translated into a series of other questions, which then form the basis of the investigation. In most carefully planned investigations the problem will be defined and redefined many times. The purposes and importance of an investigation will determine the type of study to be conducted. In all studies it is critical to state as precisely as possible the purposes and objectives of the investigation. All subsequent analysis and interpretation depend upon these objectives, and only by spelling them out very carefully can we know the questions to be answered by the inquiry.

DESIGN OF THE INVESTIGATION 2.2

Observational studies

We are interested in those types of investigations that may be referred to as "controlled inquiries." In most statistical investigations in business and economics, it is not possible to manipulate people and events in the direct way that a physical scientist manipulates his experimental materials. For example, if we want to investigate the effect of income on a person's expenditure pattern, it would not be feasible for us to vary this individual's income. On the other hand, we can observe the expenditure patterns of people who fall in different income groups, and therefore we can make statistical generalizations about how expenditures vary with differences in income. This would be an example of an *observational study.* In this type of study the analyst essentially examines historical relationships that exist among variables of interest. By observing the important and relevant properties of the group under investigation, the study can be carried out in a controlled manner. For example, if we are interested in how family expenditures vary with family income and color, we can record data on family expenditures, family incomes and color, and then tabulate data on expenditures by income and color classifications, such as white, black. In this way, if we observe the differences in family expenditures for white and black families within the same income group, we have, in effect, "controlled" for the factor of income. That is, since the families observed are in the same income group, income cannot account for the differences in the expenditures observed.

If observational data represent historical relationships, it may be particularly difficult to ferret out causes and effects. For example, suppose we observe past data on the advertising expenses and sales of a particular company. Also, let us assume, that both of these series have been increasing over time. It may be quite incorrect to assume that it is the changes in advertising expenditures that have caused sales to increase. If a company's practice in the past had been to budget 3% of last year's sales for advertising expense, one may state that advertising expenses depend upon sales with a one-year lag. In this situation sales might be increasing quite independently of changes in advertising expenses. Thus, one certainly would not be justified in concluding that changes in advertising expenses cause changes in sales. The point may also be made that many factors, other than advertising, may have influenced changes in sales. If data were not available on these other factors, it would not be possible to determine cause and effect relationships from these past observational data. The specific difficulty in attempting to derive cause-effect relationships in mathematical terms from historical data is that the pertinent environmental factors will not ordinarily have been controlled nor have remained stable.

Direct experimentation studies

The use of direct experimentation studies is becoming more prevalent in fields other than the physical sciences, where they have been traditionally employed. In such studies, the investigator directly controls or manipulates factors that affect a variable of interest. For example, a marketing experimenter may vary the amounts of direct mail exposure of a particular consumer audience. He may also use different types of periodical advertising and observe the effects upon some experimental group. Various combinations of these direct mail exposures and periodical advertising, as well as other types of promotional expenditures, such as the size of the sales force, may be used. Thus the investigator may be able to observe from his experiment that high levels of periodical advertising produce high sales effects only if there is a high concentration of sales force activity. These scientifically controlled experiments for generating statistical data can be very efficiently used to reduce the effect of uncontrolled variations. The real importance of this type of planning or design is that it gives greater assurance that the statistical investigation will yield valid and useful results.

Ideal research design

An important concept of a statistical inquiry is that of the ideal research design. The investigators should think through what the ideal research experiment would be without reference to the limitations of data available, or data that can be feasibly collected. Then, if compromises must be made because of the practicalities of the real world situation, the investigator will at least be completely aware of the specific compromises and expedients employed. As an example, suppose we wanted to answer the question of whether women or men are better automobile drivers. Clearly, it would be incorrect simply to obtain past data on the accident rates of men versus women. First of all, men drive under quite different conditions than do women. For example, driving may constitute a large proportion of the work that many men do. On the other hand, the conditions under which women drive often differ considerably from those for men with respect to exposure to accident hazards. Many other reasons may be indicated for differences in accident rates between men and women apart from the essential driving ability of these two groups. Thus, as a first approximation to the ideal research design, perhaps we would like to have data for quite homogeneous groups of men and women, for example, women and men of essentially the same age, driving under essentially the same driving conditions, using the same types of automobiles. It may not be within the resources of a particular statistical investigation to gather data of this sort. However, once the ideal data required for a meaningful answer to the question have been thought through, the limitations of other somewhat more practical sets of data become apparent.

As noted earlier, you will probably play the role of a consumer of statistical analysis far more frequently than that of a producer. Hence, in your role of consumer, when you are considering a study carried out by others, it is often extremely useful to attempt to think through what might have constituted the ideal research design. It may then be much more feasible to evaluate critically the methodology and the findings of the study.

CONSTRUCTION OF METHODOLOGY   2.3

An important phase of a statistical investigation is the construction of the conceptual or *mathematical model* to be used. A model is simply a representation of some aspect of the real world. Mechanical models are very profitably used in industry. For example, airplane models may be tested in a wind tunnel. Ship models may be tested in experimental water basins. Experiments may be carried out by varying certain factors and observing the effect of these variations on the mechanical models employed. Thus, we can manipulate and experiment upon the models and draw corresponding inferences about their real world counterparts. The advantages of this procedure are obvious when compared with attempting to manipulate an experiment using the real world counterparts, such as actual airplanes or ships. In statistical investigations, models are often used to state in mathematical terms the relationships among the relevant variables. These models are conceptual abstractions that attempt to describe, to predict, and often to control real world phenomena. For example, the law of gravity describes and predicts the relationship between the distance an object falls and the time elapsed. This conceptual model describes the relationship between time and distance if an object is dropped in a vacuum. Such models can be tested by physical experimentation.

In well-designed statistical investigations, the nature of the model or models to be employed should be carefully thought through in the planning phases of the study. In fact, the nature of these models provides the conceptual framework that dictates the type of statistical data to be collected. Let us consider a few simple examples. Suppose a market research group wants to investigate the relationship between expenditures for a particular product and income and several other socioeconomic variables. The investigators may want to use a mathematical model such as a regression equation (to be discussed later in this book), which states in mathematical form the relationship among the above variables. When the investigators determine the variables most logically related to the expenditures for the product, they also determine the types of data that will have to be collected in order to construct their model.

Even in the case of relatively simple informational reporting, there is a conceptual model involved. For example, suppose an agency wishes to determine the percentage of unemployed in a given community. Also, assume that the agency must gather the data by means of a sample survey of the labor force in this community. The ratio "proportion unemployed" is itself a model. It states a mathematical relationship between the numerator (number of persons unemployed) and the denominator (total number of persons in the labor force). The agency may wish to go further and state the range within which it is highly confident that the true percentage unemployed falls. In such a situation, as we shall see later when we study estimation of population values, there is an implicit model, which is the probability distribution of a sample proportion. Obviously, the nature of the data to be observed and, furthermore, the nature of the analysis to be carried out will flow from the type of conceptual model used in the investigation.

2.4   SOME FUNDAMENTAL CONCEPTS

Statistical universe

In the problem formulation stage, it is necessary to define very carefully the relevant *statistical universe* of observations. The universe or *population* consists of the total collection of items or elements that fall within the scope of a statistical investigation. The purpose of defining a statistical population is to provide very explicit limits for the data collection process and for the inferences and conclusions that may be drawn from the study. The items or elements that comprise the population may be individuals, families, employees, schools, corporations, and so forth. Time and space limitations must be specified, and it should be clear whether or not a particular element falls within or outside the universe.

In survey work, a listing of all the elements in the population is referred to as the *frame,* or *sampling frame.* A *census* is a survey that attempts to include every element in the universe. The word "attempts" is used here because often in surveys of very large populations, despite every effort to do so, complete coverage may not be effected. Thus, for example, the Bureau of the Census readily admits that its national "censuses" of population invariably result in a lower than actual count. Although strictly speaking, any partial enumeration of a population constitutes a *sample,* the term "census" is used as defined here. In most practical applications, it is not feasible to attempt a complete enumeration of a population, and, therefore, typically, only a sample of items is drawn. If the population is well defined in space and time, the problem of selecting a sample of elements from it is considerably simplified.

Let us illustrate some of the above ideas by means of a simple example.

Suppose we draw a sample of 1000 families in a large city to estimate the arithmetic mean family income of all families in the city. The aggregate of all families in the city constitutes the universe and each family is an element of the universe. The income of the family is a characteristic of the unit. A listing of all families in the city would comprise a frame. If instead of drawing the sample of 1000 families, an attempt had been made to include all families in the city, a census would have been conducted. The definition of the universe would have to specify the geographic boundaries that constitute the city and also the time period for which income would be observed. The terms "family" and "income" would also have to be rigidly specified. Of course, the precise definitions of all of these concepts would depend upon the underlying purposes of the investigation.

The terms *universe* and *sample* are relative. An aggregate of elements that constitutes a population for one purpose may merely be a sample for another. Thus, if one wants to determine the average weight of students in a particular classroom, the students in that room would represent the population. However, if one were to use the average weight of these students as an estimate of the corresponding average for all students in the school, the students in the one room would be a sample of the larger population. The sample might not be a good one from a variety of viewpoints, but it is, nevertheless, a sample.

If the number of elements in the population is fixed, that is, if it is possible to count them and come to an end, the population is said to be finite. Such universes may range from a very small to a very large number of elements. For example, a small population might consist of three balls in an urn; a large population might be the retail transactions that occur in a large city during a one-year period. A point of interest concerning these two examples is that the balls in an urn represent a fixed and unchanging population, whereas the retail transactions illustrate a dynamic population, which might differ considerably over time and space.

Infinite populations

An *infinite population* is composed of an infinitely large number of items. Usually, such populations are conceptual constructs in which data are generated by processes that may be thought of as repeating indefinitely, such as the rolling of dice and the repeated measurement of weight of an object. Sometimes, the population that is sampled is finite, but so large that it makes little practical difference if it is considered infinite. For example, suppose that a population consists of 1,000,000 manufactured articles and a sample of 100 of these articles is drawn and tested. Frequently, in such situations where a finite population is very large relative to sample size, the simplest procedure is to treat the population as infinite. The important point here is that since a finite population is depleted by sampling without replacement whereas an infinite population is inexhaustible, if the depletion causes the population to change only slightly, it may be simpler for computational purposes to consider the population infinite.

Target populations

Another useful concept is the *target population,* or the universe about which inferences are desired. Sometimes in statistical work, it is impractical or perhaps impossible to draw a sample directly from this *target population,* but it is possible to obtain a sample from a very closely related one. That is, the list of elements that constitutes the *frame* that is sampled may be related to but definitely different from the list of elements that comprises the target population. For example, suppose it is desired to predict the winner in a forthcoming municipal election by means of a polling technique. The target population is the collection of individuals who will cast votes on election day. However, since the specific individuals who will show up at the polls on election day are unknown, it is not possible to draw a sample directly from this population. It may be possible to draw a sample from a closely related population, such as the eligible voting population. In this case, the list of eligible voters constitutes the sampling frame. The percentage of the eligible voting population that would vote for a given candidate may differ from the corresponding figure for the election day population. Furthermore, the percentage of the eligible voter population that would vote for a given candidate will probably change as the election day approaches. Thus, we have a situation in which the population that can be sampled changes over time and differs from that about which inferences are to be made. The fact that many voters do not decide on a candidate until very shortly before the election must be taken into account. There have been several incorrect predictions of national election results because researchers have stopped sampling too far in advance of election day.

In many statistical investigations, the target population coincides with a population that can be sampled. However, in any situation where one must sample a past statistical universe and yet make estimates for a future universe, the above mentioned problem of inference about the target universe is present.

EXERCISES

1. A sample survey was conducted in a certain city to determine sentiment concerning the construction of a stadium that would cost $10,000,000 to be paid for by an increase in the personal intangible property tax. Mail questionnaires were sent to 10,000 persons whose names appeared in the city's telephone book and a certain number responded.
   (a) Identify the statistical universe from which the sample was drawn.
   (b) What would you consider to be the relevant "target universe," that is, the universe that ideally should have been sampled? Comment briefly on the discrepancy between these two statistical universes.

2. Indicate whether you think a sample or census would be advisable in each of the following situations. Give a brief reason for your choice in each case.
   (a) A professor wishes to determine whether the students in one of his classes would prefer an open- or closed-book final examination.
   (b) A nationally distributed magazine wishes to determine the socioeconomic characteristics of its subscribers.

(c) A manufacturer of toaster ovens wants to determine customer satisfaction with the ovens they have purchased.

(d) A trade association of 25 firms wishes to obtain current data on numbers of employees in these firms.

3. The school board of Lower Fenwick wished to ascertain voter opinion concerning a special assessment to permit the expansion of school services. Lower Fenwick is an industrial community on the fringe of a metropolitan area and has a population of 25,000. There are 5000 pupils enrolled in the public schools of the community. The board selected a random sample of these children and sent questionnaires to their parents.

(a) Identify the statistical universe from which the sample was drawn.

(b) Is the sample chosen a random sample of parents in Lower Fenwick? Of parents of public school children in Lower Fenwick? Why or why not?

(c) If you had been asked to assist the board, would you have approved the universe it studied? Defend your position.

4. The personnel director of a large manufacturing concern wished to ascertain employee opinion with respect to the annual Christmas party. The party had been limited to employees in the past, but the director thought husbands and wives of employees might be included for the next party. He selected a random sample of 100 persons who had attended the party the previous year and asked their opinions. Eighty wanted the party to be "employees only," but 20 requested that husbands and wives be included. Based on the results the director decided to continue the present practice. Briefly evaluate the procedure employed by the director, indicating any possible sources of bias.

5. Give an example of a situation where the population of interest would be
(a) an infinite population,
(b) considered an infinite population yet in reality is a finite population,
(c) a target population.

6. If one is interested in the percentage of consumers in the New York City area who would wear a certain type of man's suit at varying prices, what is the population of interest? Is this a fixed and unchanging population or a dynamic population? Explain your answer.

Control groups

Probably the most familiar setting involving the concept of a control group is the situation in which an experimental group is given some type of treatment. In order to determine the effect of the treatment, another group is included in the experiment and not given the treatment. These "no-treatment" cases are known as the "control group." The effect of the treatment can then be determined by comparing the relevant measures between the "treatment" and "control" groups. For example, in testing the effectiveness of an inoculation against a particular disease, the inoculation may be administered to a group of school children (the treatment group) and not given to another group of school children (the control group).[1] The effectiveness of the inoculation can then be determined by comparing the incidence of the relevant disease between the two groups. As a point in the design of the experiment, it may be noted that there

[1] Difficult ethical questions arise in cases that involve human experimentation. If the inoculation is indeed effective, clearly it should not be withheld from anyone who wants it. In cases of new treatments whose effectiveness is highly questionable, yet human experimentation appears necessary, the treatment group often is composed entirely of volunteers. Ethical questions still exist as regards the volunteers.

should be no systematic difference between the two groups at the outset that would make one group more susceptible to the disease than the other. Therefore, as indicated in an illustration in Chapter 1, the experiment may be designed with so-called "matched-pairs," where pairs of persons having similar characteristics are drawn into the experiment. (This is extremely difficult—who is exactly like someone else?) For example, if age and health are felt to have some effect on incidence of the disease, pairs of school children may be drawn who are similar with respect to these characteristics. The treatment is given to one child of a pair and not to the other. Since the children are of similar ages and have the same health backgrounds, these factors cannot explain the fact that one child contracts the disease, whereas the other does not. In the language of experimental design, age and general health conditions may be said to have been "designed out" of the experiment. Numerous other techniques are employed in experimental design to ensure that treatment effects can be meaningfully discerned.

The concept of a control group is important in many statistical investigations in the areas of business and economics. In fact, in many instances, the results of an investigation may be uninterpretable unless one or more suitable control groups have been included in the study. It is a sad fact that often after statistical investigations are completed at considerable expense it is found that because of faulty design and inadequate planning, the results cannot be meaningfully interpreted or the data collected are inappropriate for testing the hypotheses in question. It is of paramount importance that during the planning stage, the investigators project themselves to the completion of the study. They should ask the question, "If the collected data show thus-and-so, what conclusions can we reach?" This simple yet critical procedure will often highlight difficulties connected with the study design.

The example that follows illustrates the use of control groups in statistical studies. Suppose a mail-order firm decided to conduct a study to determine the characteristics of its "high-volume" customers. Its purpose is to determine the distinguishing characteristics of these heavy purchasers in order to direct future campaigns to noncustomers who possess similar attributes. Assume that the firm decides to study all of its high-volume customers. At the conclusion of the investigation it will be able to make statements such as, "The income of high-volume customers is so-many dollars." Or, it may calculate that X% of these heavy purchasers have a certain characteristic. However, such population figures will be virtually useless unless the company has an appropriate control group against which to compare these figures. It is important that the company be able to isolate the distinguishing characteristics of high-volume purchasers. Thus, in studying its customers, the company should have separated them into two groups: "high volume" and "non-high volume." Now, if it studied both groups, it would be in a position to determine the properties in which the two groups differ. Thus, returning to the earlier statements, if the company found that the high-volume and non-high-volume customers had the *same* mean incomes and that in both groups X% possessed a certain characteristic, it could not use these properties to distinguish between the two groups. The properties

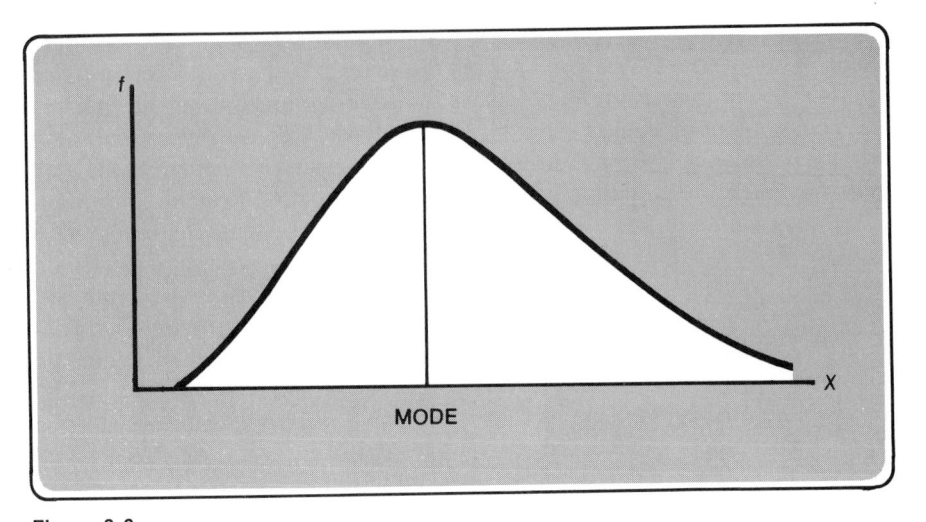

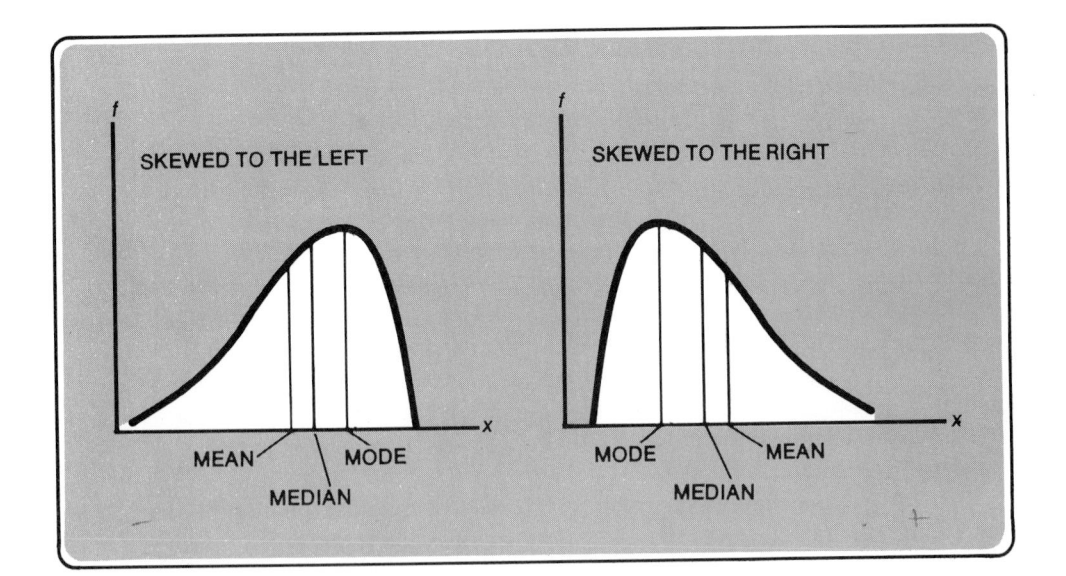

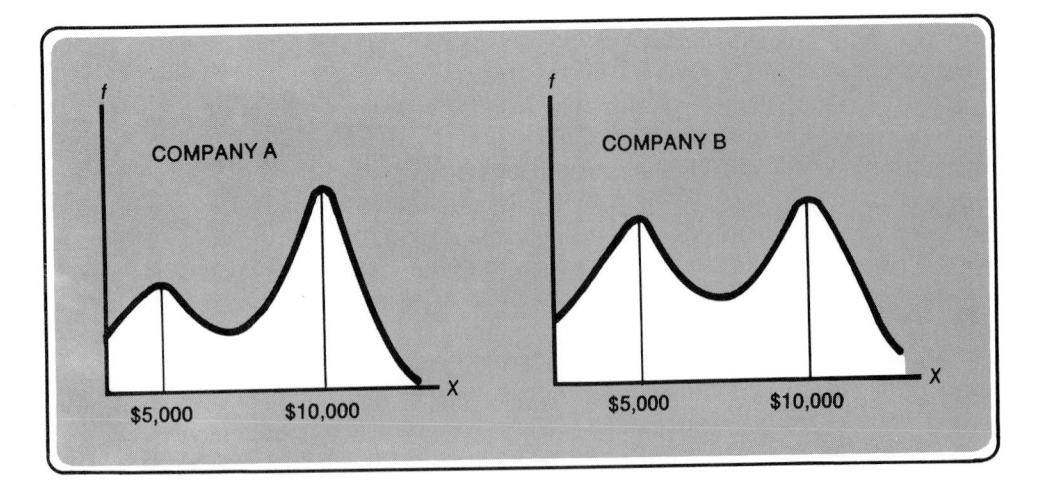

(3.9)
$$s = \sqrt{\frac{\Sigma(X - \bar{X})^2}{n - 1}}$$

> *"The record of a month's roulette playing at Monte Carlo can afford us material for discussing the foundations of knowledge."*

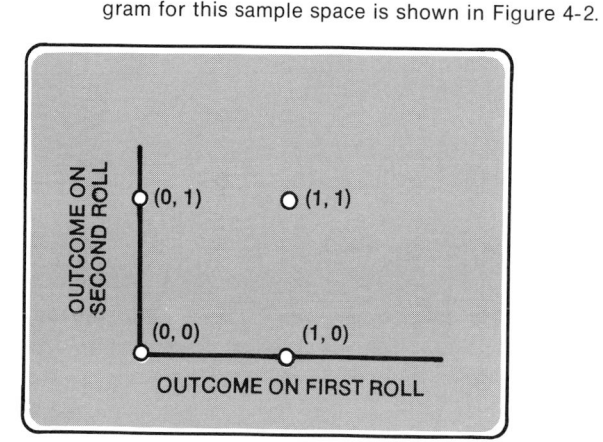

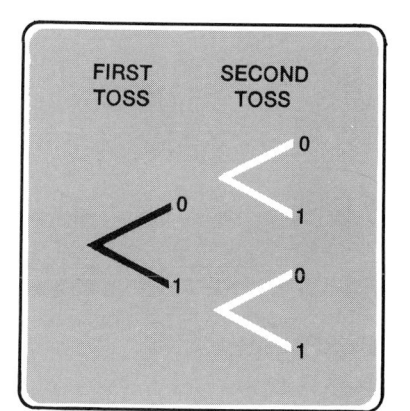

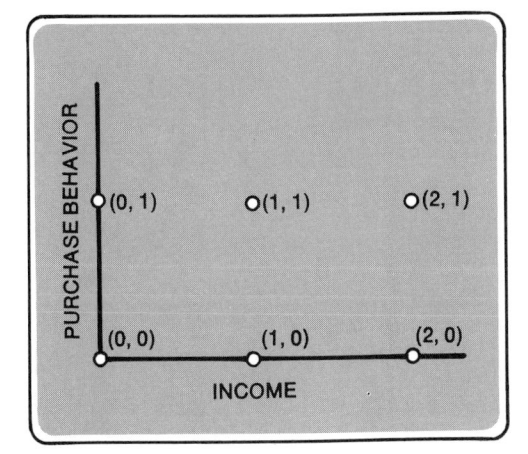

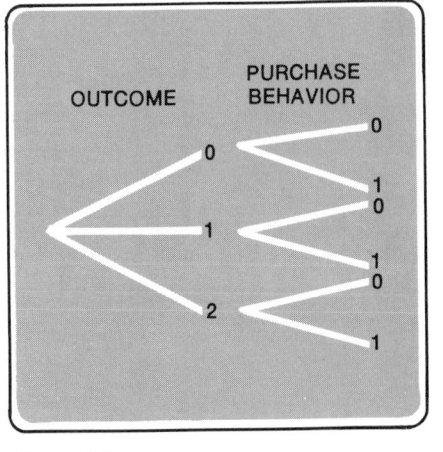

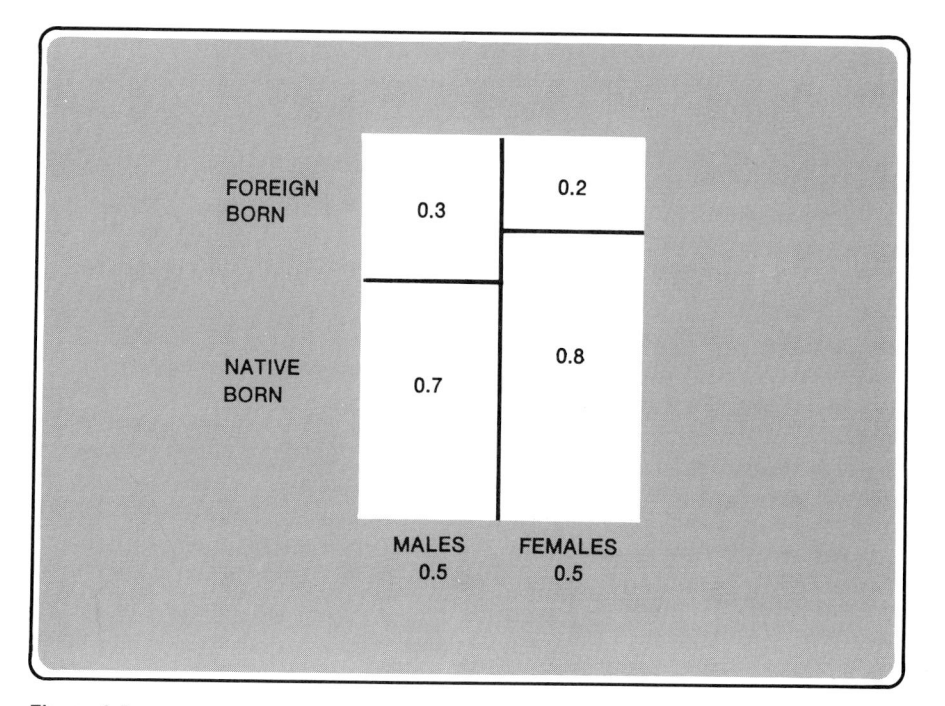

> *"If six monkeys were set before six typewriters it would be a long time before they produced by mere chance all the written books in the British Museum; but it would not be an infinitely long time." (Excerpt from an address to the British Association for the Advancement of Science — sometimes referred to as the British Ass.)*
>
> *. . . Thank you, thank you, men of science!*
> *Thank you, thank you, British Ass!*
> *I for long have placed reliance*
> *On the tidbits that you pass.*
> *And this season's nicest chunk is*
> *Just to sit and think of those*
> *Six imperishable monkeys*
> *Typing in eternal rows!*

LUCIO, *Manchester Guardian*

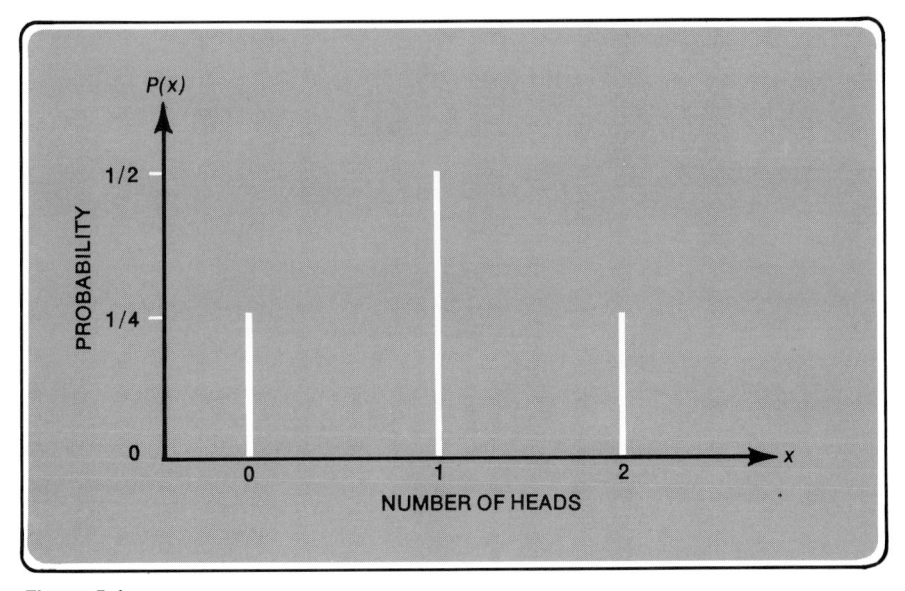

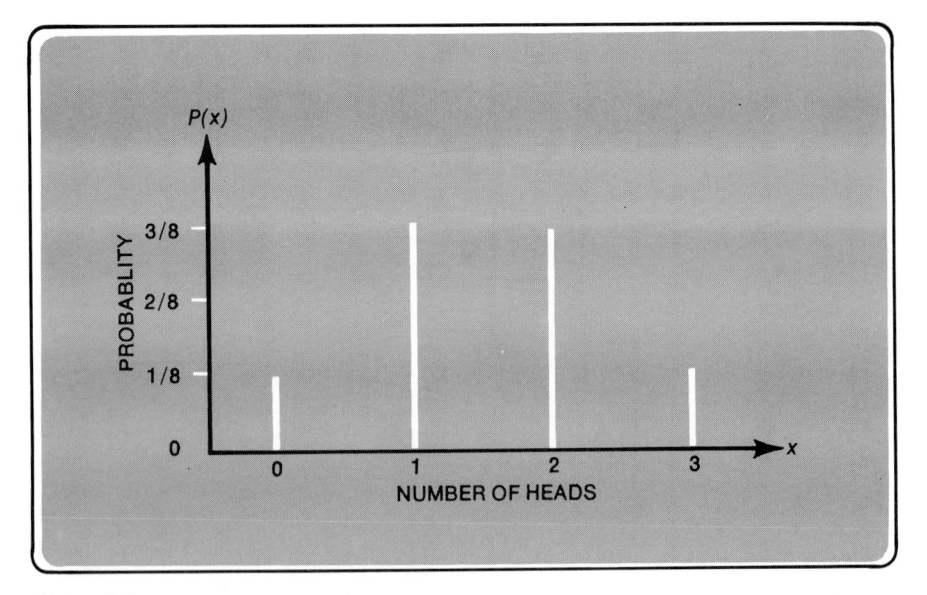

$$\underset{\substack{n-x \\ \text{failures}}}{FFF \cdots F} \quad \underset{\substack{x \\ \text{successes}}}{SSS \cdots S}$$

$$P\left(\begin{array}{c}\text{at least}\\\text{one six}\end{array}\right) = \frac{10}{36} + \frac{1}{36} = \frac{11}{36}$$

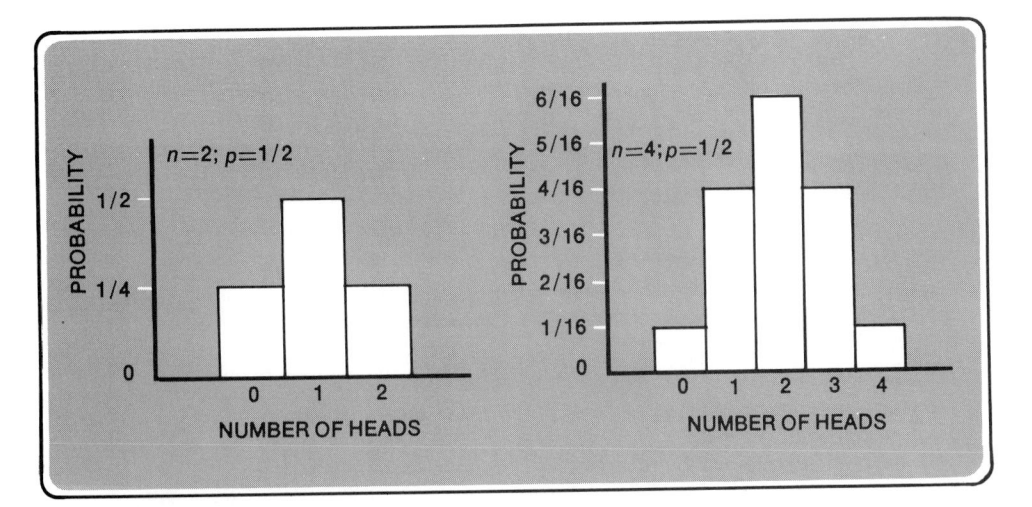

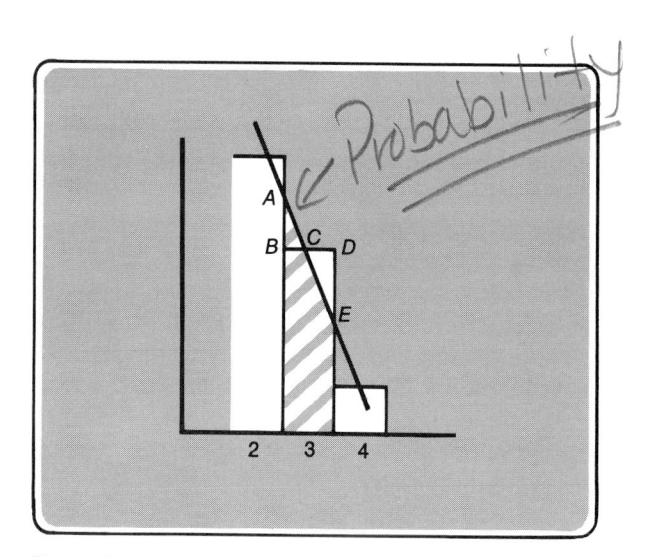

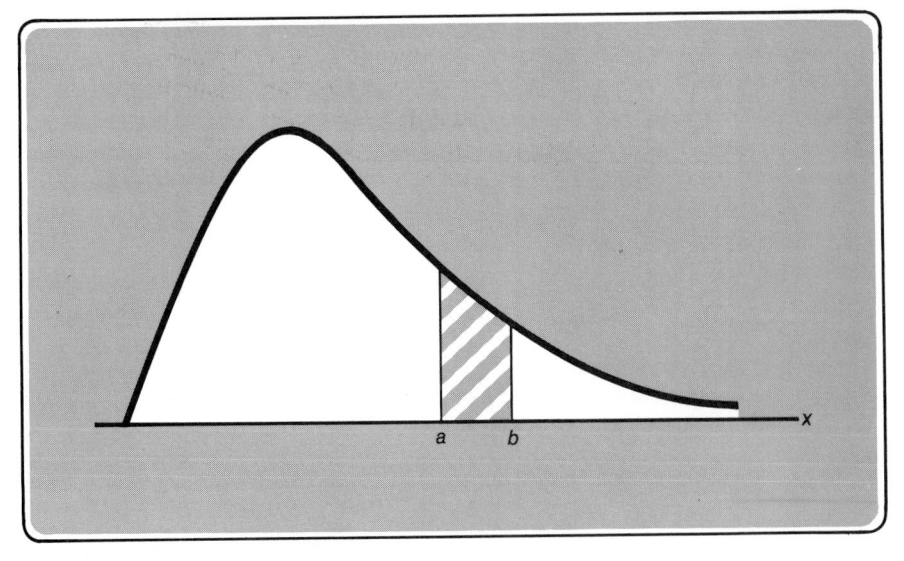

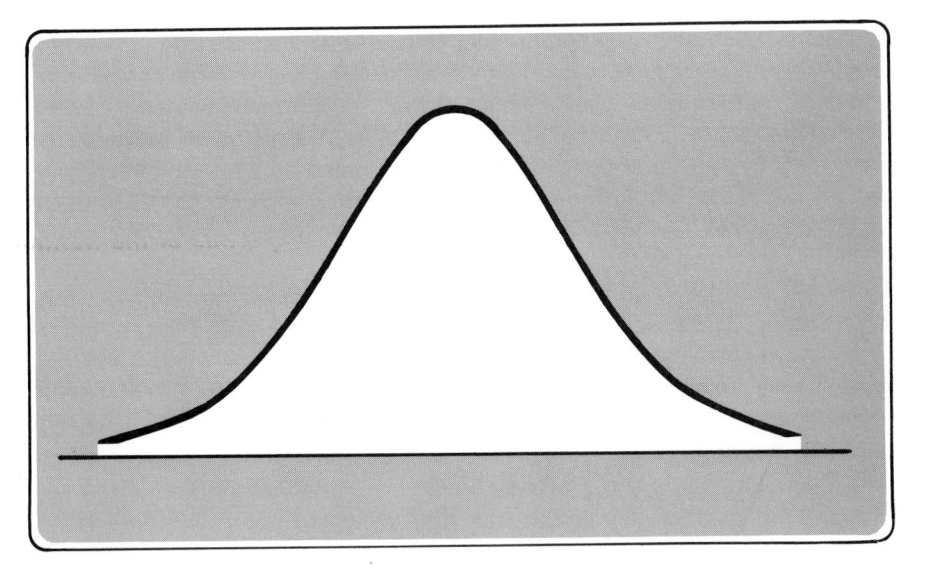

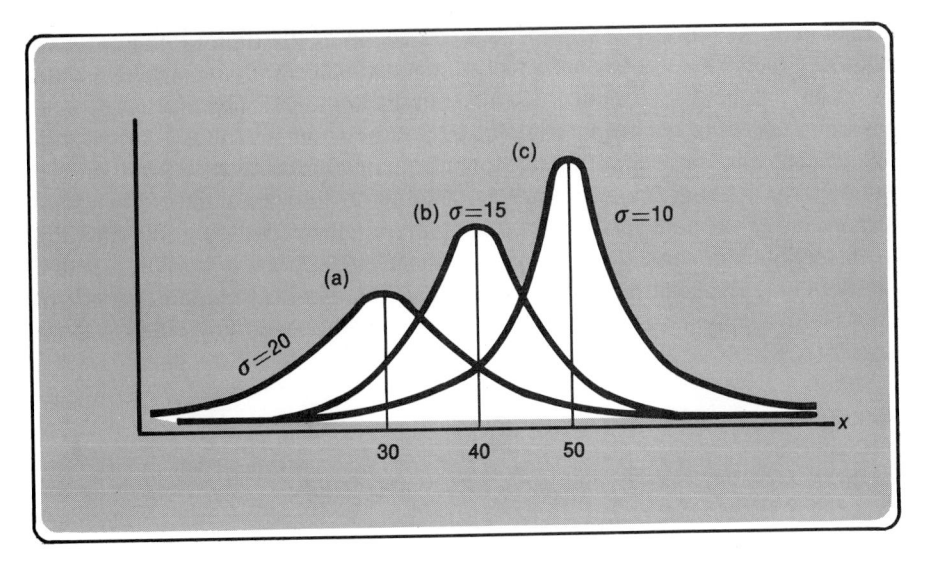

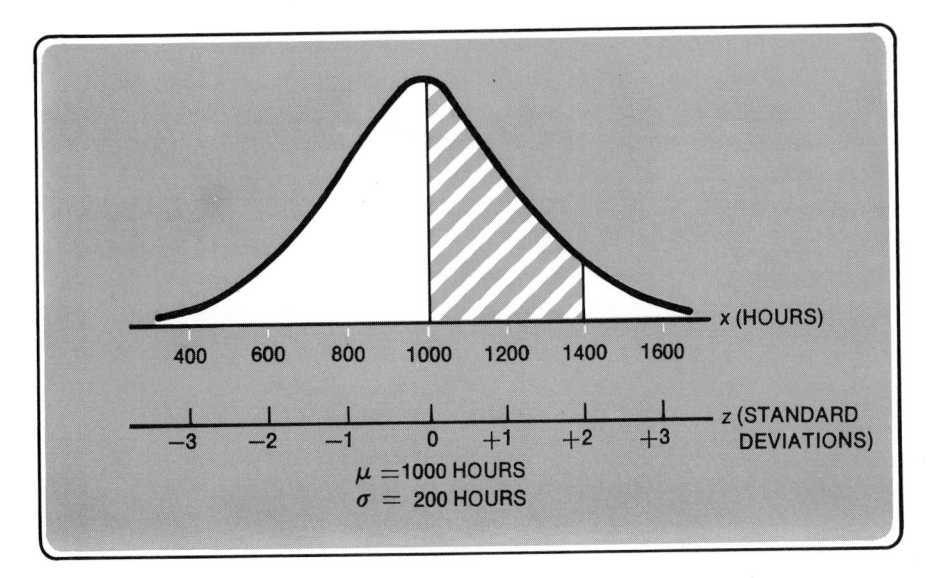

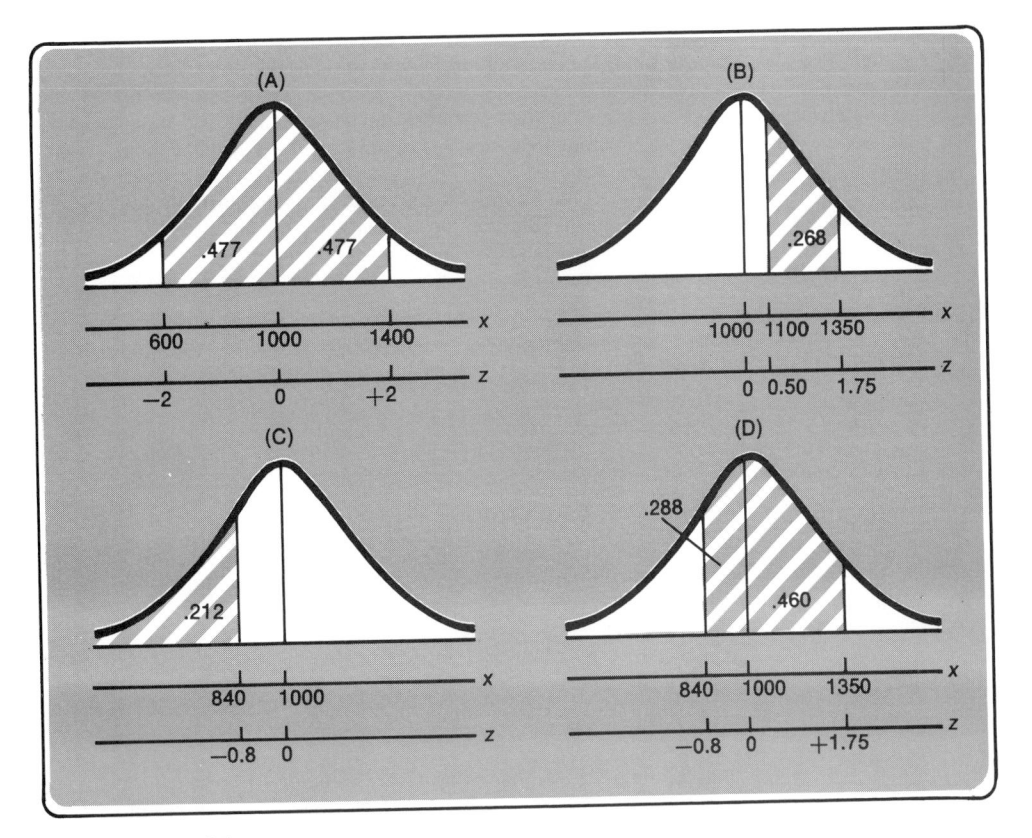

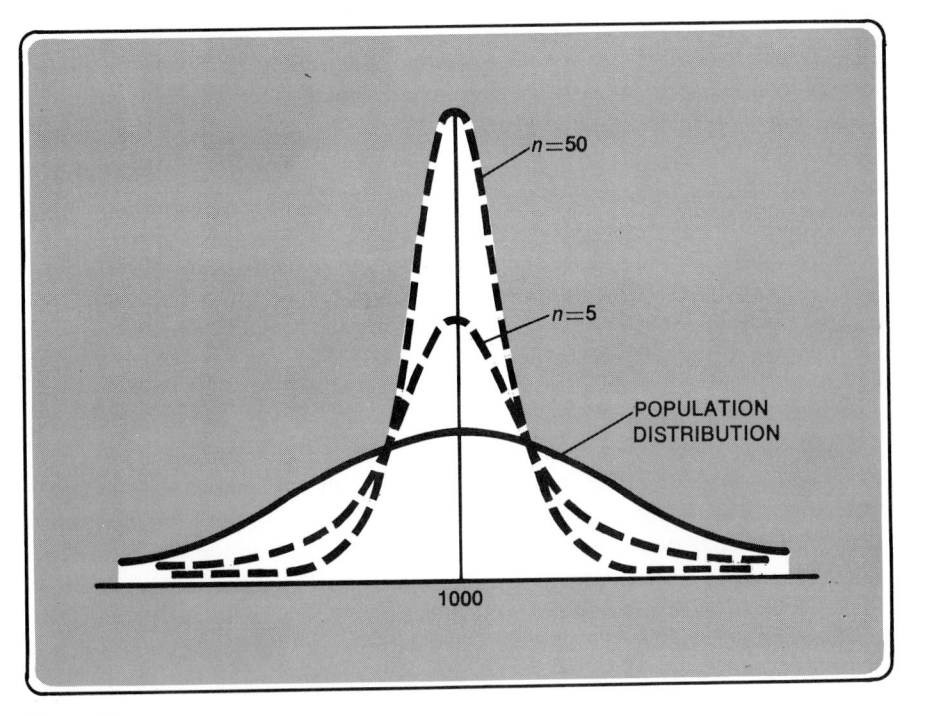

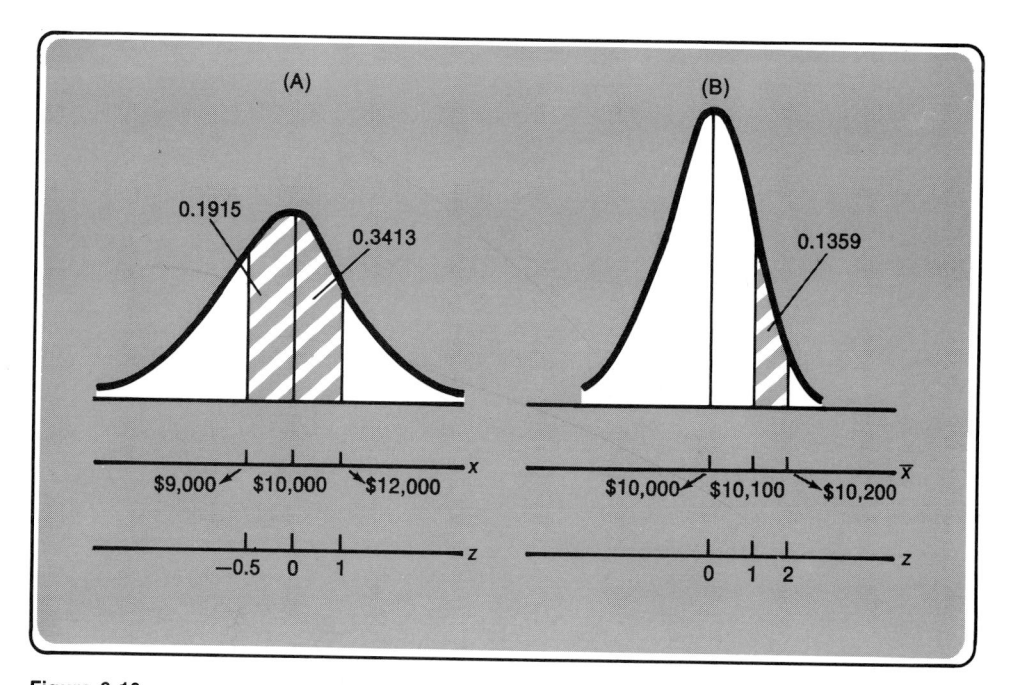

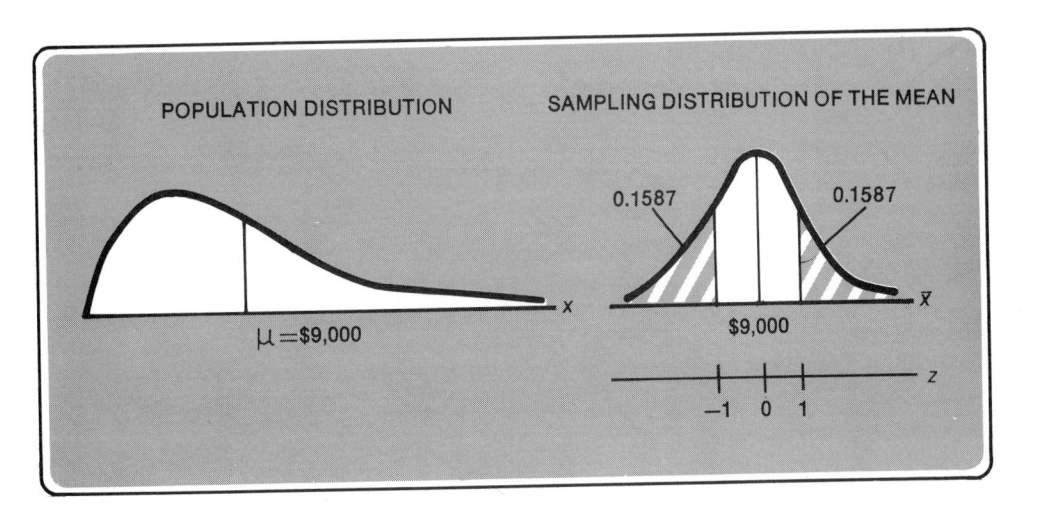

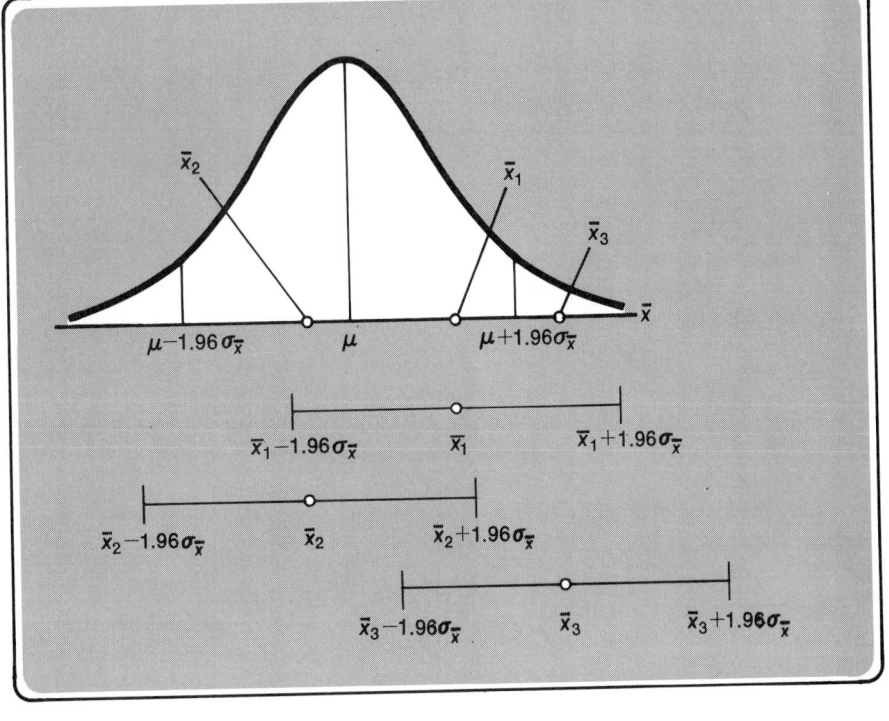

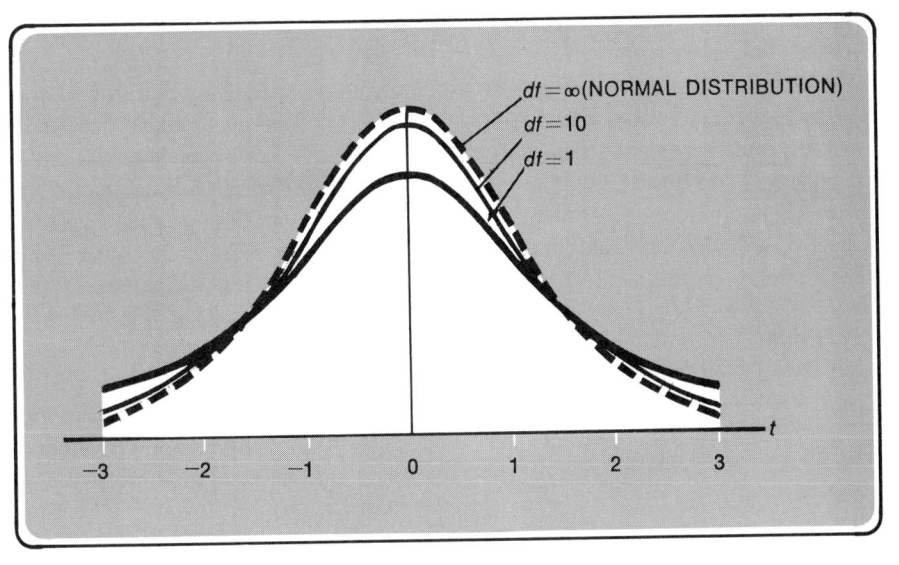

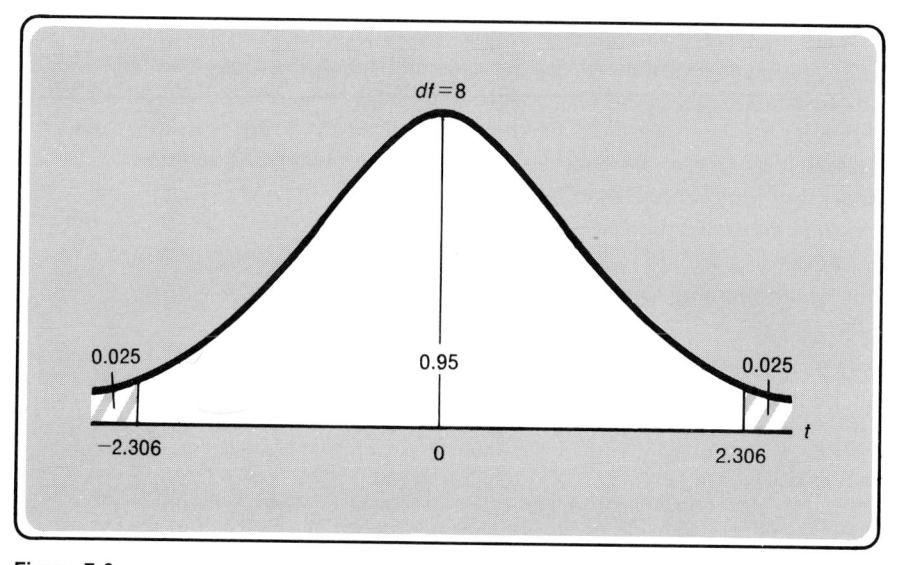

that differed most between the two groups would obviously be the most useful ones for spelling out the distinguishing characteristics of high-volume purchasers. In summary, the firm could have used the non-high-volume customers as a control group against which to compare the properties of the high-volume group, which in the terminology used earlier would represent the "treatment group."

A comment in the form of a warning is pertinent in the above illustration. Care must be used in the selection of the properties of the two groups to be observed. These properties should bear some logical relationship to the characteristic of high- versus non-high-volume purchases. Otherwise, the properties may be spurious indicators of the distinguishing characteristics between the two groups. For example, income level would be logically related to purchasing volume. If the high-volume purchaser group had a substantially higher income than the non-high-volume group, then income would evidently be a possible distinguishing characteristic. On the other hand, suppose at the time of the survey the high-volume purchaser group happened to have a higher percentage of persons who wore black shoes than did the non-high purchaser group. This characteristic of shoe color would *not* seem to be logically related to volume of purchases. Hence, we would not be surprised if the relationship between shoe color and volume of purchases disappeared in subsequent investigations or even reversed itself.

A couple of comments may be made on the construction of control groups. If the treatment group is symbolized as A and the control group as B, then an alternative control group to the one used would have been the treatment and control groups combined, or A + B. Thus, in the above example, if the relevant data had been available for the "high-volume" and "non-high-volume" customers combined, or for all customers, this group could have constituted the control. For example, let us assume for simplicity that high-volume and non-high-volume customers were equal in number. Suppose that 90% of high-volume customers possessed characteristic X, whereas only 50% of non-high-volume customers had this characteristic. The same information would be given by the statements that 90% of high-volume customers possessed characteristic X, while 70% of all customers possessed this characteristic. The 70% figure is, of course, the weighted mean of 90% and 50%. With the knowledge of equal numbers of persons in the high-volume and non-high-volume groups, it can be inferred that 50% of non-high-volume customers had the property in question. This point is of importance, because sometimes historical data may be available for an entire group A + B, whereas available resources may permit a study only of the treatment group A or what is more usual only a sample of this group.

EXERCISES

1. Indicate the control groups that you think should be used in each of the following situations. Briefly defend your choices.
   (a) It is desired to try a new method of teaching sixth-grade arithmetic in a certain city.
   (b) A medical researcher has data on the families of 180 children whom he is treating

for juvenile rheumatoid arthritis. He believes that he observes a large number of mongoloid children in these families.

(c) A production manager wishes to determine the characteristics of workers in his plant who are "unusually efficient" as measured by the percentage of defective articles they have produced.

2. Explain the way that the concept of a control group should be used by criticizing the following situations:

(a) Of the persons in the U.S. who contracted polio in a certain year, only 6% were from families with incomes over $10,000 per year. Therefore, it was concluded that families in that income group had less to fear from polio than those in lower income groups.

(b) In a certain university, only 20% of those failing to graduate were women. This suggests that women are better students than men in this university.

(c) A manufacturer of a preparation for treating the common cold claimed that 95% of cold sufferers who used his product were free of their colds within a one-week period.

3. What is the difference between an observational study and a "controlled inquiry"?

4. Discuss the need for and advantages of having a control group in a study, and give an example of a study in which use of a control group would be of value.

5. In each of the following situations, state whether a control group would be of use, and if so, what the control group would be.

(a) You are interested in investigating the accident rates in low-income urban areas and you have data on numbers of accidents occurring in low-income urban areas, numbers of cars registered in low-income areas, actual area of low-income areas, and other similar information.

(b) You are interested in evaluating the effectiveness of a new safety lighting program to be installed in your plant.

Types of errors

The concept of error is a central one throughout all statistical work. Wherever we have measurement, inference, or decision making, the possibility of error is present. In this section, we shall concern ourselves with errors of measurement. The problems of errors of inference and decision making are treated in subsequent chapters.

It is useful to distinguish two different types of errors that may be present in statistical measurements, namely, *systematic errors* and *random errors*. Systematic errors, as the term implies, cause a measurement to be incorrect in some systematic way. If observations have arisen from a sample drawn from a statistical universe, systematic errors persist even when the sample size is increased. They are errors involved in the procedures of a statistical investigation and may occur in the planning stages, or during or after the collection process. As noted in Chapter 1, another term conventionally employed for systematic error is *bias*. Among the causes of bias are faulty design of a questionnaire such as misleading or ambiguous questions, systematic mistakes in planning and carrying out the collection and processing of the data, nonresponse and refusals by respondents to provide information, and too great a discrepancy between the sampling frame and the target universe. As a generalization, these errors may be viewed as arising primarily from inaccuracies or deficiencies in the measuring instrument.

On the other hand, *random errors* or *sampling errors* may be viewed as arising from the operation of a large number of uncontrolled factors, conveniently subsumed under the term "chance." As an example of this type of error, if repeated random samples of the same size are drawn from a statistical universe (replacing each sample after it is drawn), a particular statistic, such as an arithmetic mean, will differ from sample to sample, even if the same definitions and procedures are used. These sample means tend to distribute themselves below and above the "true" population parameter (arithmetic mean), with small deviations between the statistic and the parameter occurring relatively frequently and large deviations occurring relatively infrequently. The word "true" has quotation marks around it, because it refers to the figure that would have been obtained through equal complete coverage of the universe, that is, a complete census using the same definitions and procedures as had been used in the samples. The difference between the mean of a particular sample and the population mean is said to be a *random error* or a *sampling error,* as it is termed in later chapters. All the factors that could explain why the sample mean differed from the population mean are unknown, but we can conveniently lump them together and refer to the difference as a random or chance error. A random error is one that arises from differences between the outcomes of trials (or samples) and the corresponding universe values using the same measurement procedures and instruments. The sizes of the differences are indications of reliability or precision. Random errors decrease on the average as sample size is increased. It is precisely for this reason that we prefer a larger sample of observations to a smaller one, all other things being equal. That is, since sampling errors are on the average smaller for larger samples, the results are more reliable or more precise.

Systematic and random errors may occur in experiments where the variables are manipulated by the investigator, or in survey work where the elements of a population are observed without any explicit attempt to directly manipulate the variables involved. A few examples will be given here of how bias or systematic error may be present in a statistical investigation. The problems of how random errors are measured and what constitute suitable models for the description of such errors represent central topics of statistical methods and are discussed extensively later in this chapter and in Chapters 6 through 9.

Systematic error — biased measurements

The possible presence of biased measurements in an experimental situation may be illustrated by a simple example. Suppose that a group of individuals measured the length of a 36-inch table top using the same yardstick. Let us further assume that the yardstick, although calibrated as though it were 36 inches long, was in fact 35 inches long, and this fact was unknown to the individuals making the measurements. There would then be a systematic error of one inch present in each of the measurements, and a statistic such as an arithmetic mean of the readings would reflect this bias. In this type of situation, the systematic error could be detected if another, correctly calibrated, yardstick were used as a standard against which to test the incorrect one. This is an important methodological point.

Often, systematic error can be discovered through the use of an independent measuring instrument. Even if the independent instrument is inaccurate, a comparison of the two measuring instruments may give clues as to where to begin the search for sources of bias. The variation among the individual measurements using the incorrect yardstick would be a measure of what is usually referred to as "experimental error," that is, differences among individual observations that are not attributable to specific causes of variation. It may be noted that the observations may have been very precise, in the sense that each person's measurement was very close to that of every other person. Thus, the random error would be small, and there would be good repeatability because in repeating the experiment, each measurement would be close to preceding measurements. These random or chance errors may be assumed to be compensating in that some observations would tend to be too large and some too small. Since the table top is 36 inches long, the measurements would tend to cluster around a value about one inch greater than the true length of the table top. In summary, we have a model in which each individual measurement may be viewed as the sum of three components, (1) the true value, (2) systematic error, and (3) experimental error. This relationship is stated in equation form as

$$\text{(2.1)} \qquad \frac{\text{Individual}}{\text{Measurement}} = \frac{\text{True}}{\text{Value}} + \frac{\text{Systematic}}{\text{Error}} + \frac{\text{Experimental}}{\text{Error}}$$

EXERCISES

1. Distinguish clearly the difference between systematic errors and random errors. Explain which error will decrease with a larger sample size, which will not, and why. Which error can and should be eliminated?

2. State whether each of the following errors should be considered random, systematic, or both, and why:
   (a) In a study which attempted to estimate the percentage of students who smoke, the first 100 students who entered the student lounge, the only area in the building where students are permitted to smoke, were asked if they smoked. The study resulted in an overestimate of the true percentage.
   (b) In a study to estimate the average life of a certain type of vacuum tube, five tubes were purchased from five different stores, in five different wholesale sales regions. The average life of the five tubes tested was shorter than the "true" average life.
   (c) In a study to determine the true weight of a process which fills one pound cans, 50 cans were selected randomly and weighed on a scale that measured .1 ounce too heavy. The process in fact filled the cans on the average with one pound but the 50 cans averaged 1.06 pounds.
   (d) One thousand questionnaires were sent out asking the respondent to rate the community services of police, fire, garbage, and so forth as bad, adequate, or

good. Of the 87 responses, a majority rated the services either bad or good. Yet the majority of the people in the community considered the services adequate.

3. On a weekday in 1973, a telephone survey of households made between the hours of 9 A.M. and 4 P.M. showed that 80% of the persons reached favored having a woman as a Vice President of the United States. Do you think that both sampling and systematic errors are present in the results of this study? Explain.

4. A magazine with 200,000 subscribers sent a mail questionnaire to a random sample of 10,000 of these persons inquiring about their attitudes toward the magazine's editorial policy. About 1500 respondents mailed back completed questionnaires. Do you think that both sampling and systematic errors are present in the results of this study? Explain.

COLLECTION OF STATISTICAL DATA   2.5

Data sources

After formulation of the problem in a statistical investigation, and after adequate attention has been given to matters such as the ideal research design, the statistical universe to be investigated, and the methodology to be employed, it becomes necessary to obtain the required data. Although frequently the investigator may have to collect his own data, he may be able to find a portion of the data or perhaps all of them in published or unpublished form. Sources include the internal records of business firms and surveys conducted by governmental agencies and nongovernmental organizations such as trade associations and private research companies. Furthermore, it is essential for the analyst to review earlier work on the problem under investigation. Again, he may find that some of the desired information has already been collected, and he may obtain useful ideas for his own study.

The most highly organized and extensive systems for the collection and dissemination of statistical data are those of the federal government. State and local governments and the United Nations publish large numbers of series of statistical data. Also, there are many nongovernmental sources of data such as private agencies, professional organizations, and trade associations.

It is important to note that data collected from any of the above sources, including the federal government, should never be accepted at face value, but should be scrutinized with care. The investigator must be alert for changes in definition, in the nature of the statistical units employed, in the coverage and scope of series, for revisions of series that may have originally been issued in preliminary form, and for errors of various types. To the extent possible, data obtained from one source should be checked against analogous figures from other sources. Where data from alternative sources differ, reconciliations are often possible in terms of adjustments for differences in definition, coverage, and so forth. Above all, it is essential to proceed with caution and with a critical, questioning attitude when using data collected by others, whether under governmental or nongovernmental auspices.

Direct collection of data

There are a variety of methods of collection of original data, including the planning and execution of surveys, and only a very brief discussion of them is feasible at this point. The statistical universe under investigation may be either a human or nonhuman population. Some problems are similar for these two types of populations such as the question whether a census or a sample should be conducted and the requirement of careful editing of the data for errors. However, there are generally many more problems involved in collecting of high-quality data from human populations, so we will confine the discussion to that case.

Data are generally obtained from people by means of telephone interviews, personal interviews, mail questionnaires, or some combination of these. Each method has its advantages and disadvantages. Ordinarily, much higher response rates are obtainable from interviews, both telephone and personal, than from mail questionnaires. Many persons almost automatically discard mail questionnaires, whereas they are more likely to reply to a direct interview. As previously indicated (Section 1.6), mail questionnaires generally result in highly selective responses, which tend to produce biased data. Furthermore, more accurate replies can usually be obtained from interviews because of the possibility of communication between the interviewer and respondent. If a question is misunderstood, the interviewer can explain it. (However, this also may introduce bias depending on the interviewer's explanation.) If the respondent gives a reply that is inconsistent with previous answers, the interviewer can ask him to resolve the inconsistency.

On the other hand, mail questionnaires are often used rather than personal interviews because of economy and practicality. It is ordinarily far less expensive to send a questionnaire to a potential respondent than to conduct a personal interview. If the population is geographically widely dispersed, it may be both more economical and more feasible to use questionnaires than personal interviews. However, since the response rate on questionnaires is generally so much lower than that for personal interviews, the cost per completed return may be surprisingly close for the two data collection methods. For all methods, it is good practice to use follow-up techniques to obtain data from nonrespondents.

The construction of questionnaires or schedules, as they are often called, is a detailed field of study in itself. There are many pitfalls in the wording of questions. Experience has demonstrated that differences in the wording of questions can give rise to important differences in responses received. As an example, so-called "leading questions," that is, those that tend to evoke particular types of response should be avoided. A classic leading question is that of the prosecuting attorney when he asks the defendant on the witness stand, "Have you stopped beating your wife?" If the accused answers "Yes," the impression is created that he was indeed engaged in the nefarious practice, but claims that he has ceased. If he answers "No," he admits a past and present engagement in the wife-beating activity. In a market research study, a question such as, "Do you prefer XYZ toothpaste?" is apt to produce a higher proportion of such preferences than a question such as "Which toothpaste do you prefer?" Many examples can be given of somewhat more subtly worded leading questions that also tend to lead to seriously biased replies.

So many possible difficulties can arise in connection with the construction of a questionnaire that it is invariably useful to employ a small pilot study or pretest. In such a study, the questionnaire is given to a small number of respondents from the relevant population before carrying out the intended survey. The difficulties encountered are useful in redesigning the questionnaire.

In most studies, after the data are collected, the schedules should be subjected to an *editing* procedure. This process includes deletion of obviously incorrect information, reconciliation of inconsistent answers, and so forth. Frequently, it is necessary to query the respondent in order to resolve the problems experienced. Then some or all of the data may have to be encoded for tabulation purposes. Usually this coding takes the form of assigning numbers to the various classifications, both qualitative or quantitative. For example, male may be encoded as zero, and female as one. Income below $1000 may be zero, from $1000 to $1999 one, from $2000 to $2999 two, and so forth. Depending on the size of the study, tabulations and calculations may then be carried out by hand, calculating equipment, or computers. Today, virtually all of the editing, coding, and computational procedures can be carried out by modern electronic computers. Indeed, computers have radically changed many of the processing aspects of statistics. Computers can be used as aids in all of the phases of the direct collection of data, in tabular and graphic presentations, and in carrying out practically every type of statistical analysis discussed in this text. Modern electronic data processing equipment plays an increasingly important role in every aspect of statistics particularly because it can store and retrieve large masses of data and perform incredibly rapid calculations.

RATIOS 2.6

After statistical data have been collected, often only relatively simple methods may be required to indicate the existence of situations of importance or to trigger appropriate courses of action. Sometimes only the calculation of certain key ratios is necessary. For example, an executive in charge of production may have his attention focused on a particular department by a sharp decline in its production rate. Or, a financial institution may decide that a loan application should not be approved because the applicant has an excessively high ratio of debt to net worth. Of course, more detailed inquiry is frequently required. The calculation of ratios may represent an essential part of the analysis, whether the analysis is simple or complex in nature.

A ratio is simply a statement of comparison between two numbers. This statement may take several forms, such as a fraction, a decimal, or a percentage. For example, if a company has $10,000,000 in total assets, $2,000,000 of which are in the form of current assets, the relationship between current assets and total assets may be expressed as $2,000,000/$10,000,000 = 2/10, 0.20, or 20 percent. When the units of measure in the numerator and denominator of a ratio differ, the fraction is often referred to as a "rate." Hence, if the total of a day's production of a group of workers is divided by the number of workers, the result, say, of 50 items per worker per day may be referred to as a production rate. Other examples of rates include accidents per mile, per capita gross national product, population per square mile, and crop production per acre. However, since all rates are ratios, the distinction between these two terms is frequently not made.

The basic applications of ratios or rates are in making comparisons or in combining these measures in some way. Although the calculation of a ratio or rate requires nothing more than elementary arithmetic, the comparison and combination of these measures is the source of many errors.

Comparison of ratios

In constructing ratios it is important that the numerator and denominator be clearly and directly related. For example, in computing an unemployment rate, the ratio of the number unemployed to the total population would not be a very meaningful figure. The problem is that the denominator "total population" includes many persons such as infants, retired persons, and people in institutions such as mental hospitals who are not members of the working population. Hence, it is the usual practice to "refine" the ratio, that is, to exclude from the numerator and denominator irrelevant components. Thus, the denominator in an unemployment rate usually refers to the labor force, that is, the working population, those ready, willing, and able to work. The numerator then includes only those in the labor force who are unemployed.

Many misuses of ratios occur because of comparisons of ratios that should have been refined or that have been refined to different degrees. For example, suppose the unemployment rates of two towns are compared where one town is the location of the county mental institution. Assume the population of the mental institution represents about half of the population of the town in which it is located. If unemployment rates for the two towns are computed using the ratio of unemployed to total population including the mental patients, clearly no meaningful inference can be drawn concerning the relative severity of the unemployment problem for the two towns.

As another example, suppose that a comparison were made of the so-called "crude" death rates of two cities for a particular year. The crude death rate is the number of deaths divided by the number of persons alive at the beginning of the year. If the first city had a much higher proportion of people in older age groups than the second, as, for example, would occur if the first city were a "retirement community," it would tend to have a much higher crude death rate than the second. This would happen even if the death rates for each age group, say in ten-year intervals, were exactly the same for the two cities. It is clear that many types of misleading inferences might be drawn from these crude death rates. Hence, the usual procedure in comparing death rates is to compute so-called "standardized rates." For example, if the rates are standardized by age, an age distribution for a "standard population" is assumed. This standard age distribution might be the age distribution of the population of the United States as a whole. The standardized death rate for each city is then computed by multiplying the death rate for each age group by the proportion of the standard population falling in the age group and totaling these products for all age groups. The resulting figure for each city represents the death rate that would have prevailed had the age distribution of the city's population been that of the standard population. A comparison can then be made of the death rates for the two cities unaffected by the factor of age. The procedure of standardizing of ratios may be summarized as breaking the original ratio into several component parts and calculating a weighted average of the several part ratios using an appropriate set of weights.

Of course, death rates can be standardized for other pertinent factors such as race and sex, and in the field of vital statistics, these adjustments are conventionally made. The standardizing of ratios technique is of methodological interest since it represents a means of enabling a comparison of one factor (mortality), while assuring that this comparison is not disturbed by differences in other factors (age, race, sex, etc.).

Combining of ratios

In the preceding illustration the ratios were combined as well as compared. That is, the death rates for the individual age groups were combined into a single overall death rate for each city. Often, ratios are appropriately combined by using the denominators of the original ratios as weights. For example, suppose a company consists of two divisions and the net profit-to-sales percentages for a particular year were: Division A, 4%; Division B, 6%. Further suppose the sales of the two divisions for the year were $1,000,000 and $4,000,000. What is the net profit to sales percentage for the two divisions combined, that is, for the company as a whole? The answer is obtained by weighting, that is, multiplying the individual percentages by sales, the denominators of the original ratios, adding these products and dividing the sum by total sales. This procedure, which represents an application of the weighted arithmetic mean, discussed in Chapter 3, is illustrated in Table 2-1.

An alternative procedure to weighting the profit to sales ratios by the actual values of the denominators of these ratios would be to weight by a percentage breakdown of the denominators, in this case, a percentage breakdown of total sales. For example, in the computation shown in Table 2-1, if instead of weights of $1,000,000 and $4,000,000, weights of 20% and 80% had been applied, the same answer of 5.6% would have resulted. Shifts in the pattern of weights applied to a set of ratios sometimes represent the key to a correct analysis of a given situation.

TABLE 2-1

Calculation of the Average Profit to Sales Ratio for Two Divisions Combined

DIVISION	NET PROFIT TO SALES RATIO	SALES	NET PROFIT
A	0.04	$1,000,000	$ 40,000
B	0.06	4,000,000	240,000
		$5,000,000	$280,000

Combined Ratio $= \dfrac{\$280,000}{\$5,000,000} = .056 = 5.6\%$

EXERCISES

1. In 1972 the crude death rate in St. Petersburg, Florida, was greater than the corresponding death rate in Chicago, Illinois. It was concluded, therefore, that Chicago is a healthier place to live than St. Petersburg. Comment briefly on this conclusion as a misuse of ratios, taking into account that St. Petersburg is a favorite retirement community.

2. Assume that data on family savings and income were obtained from a panel of families in 1967 and 1972. The average savings-income ratios for the families in each of two income groups are as follows:

FAMILY INCOME	SAVINGS-INCOME RATIO 1967	1972
Under $10,000	3%	2%
$10,000 and over	30%	25%

(a) What additional information must be known before the average savings-income ratio for all families combined can be calculated for each year?

(b) Demonstrate clearly how it would or would not be possible for the savings ratio for all families combined to have increased from 1967 to 1972. Assume any additional information that you need.

Frequency Distributions and Summary Measures 3

Suppose data had been collected on the ages of all individuals in the United States, and that you were required to describe these approximately 200 million figures in some generally useful manner. How might you go about it, assuming that adequate resources for processing the data were available? Since it would be very difficult to see important characteristics of the data by merely listing them, you probably would try to group the age figures into classes. For example, you might set up classes of zero and under five years, five and under ten, and so forth. You could tabulate the number of persons in each class, and if you divided these numbers by the total population you would have the proportions of the population in each class. You would then find it relatively easy to summarize the general characteristics of the age distribution of the population. If you compared similar distributions, say, for the years 1915 and 1970, a number of important features would be observable without any further statistical analysis. For example, the range of ages in both distributions would be clear at a glance. The higher proportions of young persons under the age of 20 and older persons over the age of 65 in 1970 than in 1915 would stand out. Also, smaller percentages of persons in the age categories 30 to 40 years would be observed for 1970, reflecting the decline in births during the 1930–1940 decade. In summary, through the simple device of grouping the age figures into classes and recording the frequency of occurrence in these classes, some of the underlying characteristics of the nation's age composition for each year emerge from the data. Generalizations about age patterns thus become easier to make.

You might also be inclined to continue your description of the age distribution by calculating one or more types of averages (either of the original or of the grouped data). For example, you might be interested in computing an "average age" for the population in 1915 and in 1970 to determine whether this average had increased or decreased. Furthermore, if you wished to obtain a more exact notion of the fact that in the later period there were heavier concentrations of persons in the younger and older age groups, you might attempt to construct a measure of how the ages were spread around an average age. Because of the heavier concentrations of persons in the younger and older age groups in 1970 than in 1915, the measure of spread or dispersion around the average would tend to be larger in 1970 than in the earlier period.

The types of statistical techniques that might be used to summarize and describe the characteristics of the age figures constitute the subject matter of this chapter. The table into which the data are grouped is referred to as a "frequency distribution." The average or averages that can be computed are measures of "central tendency" or "central location" of the data, and the measure of spread around the average is a measure of "dispersion." These and other techniques that group, summarize, and describe data are referred to as "descriptive statistics." If the data treated by "descriptive statistics" represent a sample of a larger group or population, as noted in Section 1.3, inferences may be desired about this larger group. The subject of statistical inference is discussed in subsequent chapters.

The methods to be discussed in this chapter are useful for describing patterns of variation in data. Variation is a basic fact of life. As individuals, we differ in age, sex, height, weight, and intelligence, in the quantities of the world's goods we possess, in the amount of our good or bad luck, and in myriad other characteristics. In the business world variations are observed in the articles produced by manufacturing processes, in the yields of the economic factors of production, in production costs, financial costs, marketing costs, and so forth. Such variations occur both in data observed at a point in time as well as in data occurring over a period of time.

The term "cross-sectional data" refers to data observed at a point in time, whereas "time series data" pertains to sets of figures that vary over a period of time. Frequency distribution analysis is concerned with cross-sectional data. In particular, such analysis deals with data where the order in which the observations were recorded is of no importance, as for example, the ages of the present members of the labor force in the United States, the present wage distribution of employees in the automobile industry, or the distribution of U.S. corporations by net worth on a given date. On the other hand, if we recorded quality control data of a manufactured product we would ordinarily be very much concerned with the order in which the articles were produced. For example, if a sudden run of defective articles were produced, we would be interested in knowing when this occurred and what was the general time pattern of production of defective and good articles. Similarly, in the study of economic growth, we might be interested in the variation over time of such data as real income per person or real gross national product per person. General methods of time series analysis are treated in Chapter 13.

As indicated earlier, when we are confronted by large masses of ungrouped data, that is, listings of individual figures, it is very difficult to generalize about the salient information they contain. However, if a frequency distribution of the figures is formed, many of these features become readily discernible.

A frequency distribution or frquency table simply records data grouped into classes and the numbers of cases that fall in each class. The numbers in each class are referred to as "frequencies," hence the term "frequency distribution." When the numbers of items are expressed by their proportion in each class, the table is usually referred to as a "relative frequency distribution," or simply a "percentage distribution."

How the classes of a frequency distribution are stated depends on the nature of the data. In all cases, we deal with objects having characteristics that can be counted or measured. Thus, individuals have characteristics such as color, nationality, sex, and religion, and counts can be made of the numbers of persons that fall in each of the relevant categories. For example, if a classification by sex is used, the frequency distribution of sex for residents of Euphoriaville on January 1, 1974 may be shown as in Table 3-1.

TABLE 3-1
Distribution of Residents of Euphoriaville
on January 1, 1974 Classified by Sex

SEX	NUMBER OF PERSONS
Male	56,427
Female	60,511
TOTAL	116,938

Characteristics such as color, nationality, sex, and religion that can be expressed in qualitative classifications or categories are often referred to as "attributes" or "discrete variables." When dealing with such cases, we make counts of the number of persons or objects that fall into each attribute classification. It is always possible to encode the attribute classifications to make them numerical. Thus, for example, in the preceding illustration, "male" could have been denoted 0 and "female" 1. In certain cases the data seem to fall naturally into simple numerical classifications. For example, families may be grouped according to number of children; the classes would run 0, 1, 2, and so forth.

Graphs are often useful for presenting the salient features of a set of statistical data as contrasted with statistical tables, which show much more specific detail. Data for qualitative characteristics or discrete variables can be presented graphically in terms of simple bar charts. Figure 3-1 gives a bar chart representation of the data given in Table 3-1.

Sometimes, lines are used rather than bars to represent frequency of occurrence for discrete variables, as shown, for example, in the graph of the binomial distribution given in Figure 6-1 on page 125.

To obtain "continuous variables," or "continuous data," that is, data that can assume any value in a given range, numerical measurements are performed rather than counts. When large numbers of measurements are made, it is convenient to use intervals or groupings of values and to tabulate the numbers or frequency of occurrence in each class. With this procedure, a few problems have to be resolved concerning the number of class intervals, the size of these intervals, and the manner in which class limits should be stated.

3.2 CONSTRUCTION OF A FREQUENCY DISTRIBUTION

We will illustrate the method of constructing a frequency distribution by considering the figures in Table 3-2, which represent the earnings of 100 semi-skilled employees of the Beco Company for a particular week in 1973. Although the data have been arrayed from lowest earnings to highest, it is difficult to discern patterns in the ungrouped figures. However, when a frequency distribution is constructed, the nature of the data clearly emerges.

The decisions about the number and size of the classes in a frequency distribution are essentially arbitrary ones. However, these two choices are clearly interrelated. Frequency distributions generally are constructed with from five to twenty classes. When class intervals are of equal sizes, comparisons of classes are made easier and subsequent calculations from the distribution are simplified. However, this is not always a practical procedure. For example, with data on the annual incomes of families, in order to show the detail for the portion of the frequency distribution where the majority of incomes lie, class intervals of $1000 or $2000 may be used up to about $10,000; then intervals of $5000 may be used up to $25,000 and a final class of $25,000 and over. It is clear that maintaining equal size classes of, say, $1000 throughout the entire range of income would result in too many classes. On the other hand, if much larger class intervals were used, too many families would be lumped together in the first one or two classes, and we would lose the information concerning how these incomes were distributed. The use of unequal class intervals and an open-ended interval for the highest class provides a simple way out of the dilemma. An open-ended class interval is one that contains only one specific limit and an "open" or unspecified value at the upper or lower end, as for example *$25,000 and over* or *110 pounds and under*. The use of unequal class sizes and open-ended intervals generally becomes necessary when most of the data are concentrated within a certain range, when

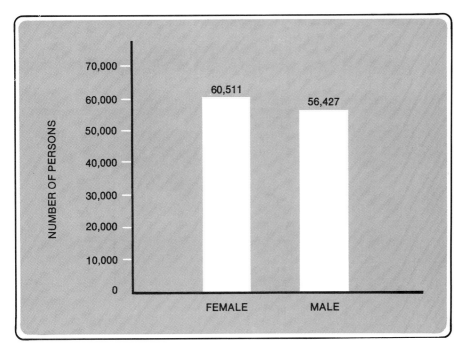

Figure 3-1
Bar chart for the number of residents of Euphoriaville on January 1, 1974 classified by sex.

		TABLE 3-2		
	Earnings of 100 Semi-skilled Employees of the			
	Beco Company for One Week in 1973.			
$160.03	$181.92	$192.03	$197.03	$209.12
163.24	181.98	192.69	197.94	209.12
167.39	182.17	192.83	198.28	209.62
168.50	182.55	192.99	199.60	209.74
171.10	182.68	193.89	200.03	210.01
171.50	183.91	193.94	200.65	210.92
172.27	184.03	194.07	201.74	211.12
173.87	184.88	194.23	202.58	211.97
174.20	185.22	194.23	203.95	214.08
175.10	185.73	194.35	204.33	214.92
175.16	185.96	194.88	204.91	214.96
176.82	187.22	195.01	204.97	215.89
177.18	188.31	195.26	205.04	215.97
178.22	188.63	195.26	205.09	216.22
178.43	189.21	195.26	205.82	218.68
178.92	189.36	195.87	206.94	219.87
179.04	190.12	195.92	207.86	222.37
179.17	191.35	196.08	208.27	224.84
180.05	191.37	196.92	208.35	226.98
181.79	191.49	196.99	208.39	229.65

gaps appear in which relatively few items are observed, and finally when there are a very few extremely large or extremely small values. Open-ended intervals are sometimes also used to retain confidentiality of information. For example, the identity of the small number of individuals or companies in the highest class may be general knowledge, and stating an upper limit for the class might be considered excessively revealing.

Let us assume that for the list of earnings figures shown in Table 3-2 we would like to set up a frequency distribution with seven classes and that we want them to be of equal size. A simple formula to obtain an estimate of the appropriate size interval is

$$i = \frac{H - L}{k}$$

where i = the size of the class interval
H = the value of the highest item
L = the value of the lowest item.
k = the number of classes

This formula for class interval size simply divides the total range of the data (that is, the difference between the values of the highest and lowest observations) by the number of classes. The resultant figure indicates how large the class interval would have to be in order to cover the entire range of the data in the desired number of classes. Some considerations involved in determining an appropriate number of classes are discussed in Section 3.4.

In the case of the weekly earnings data,

$$i = \frac{(\$229.65 - \$160.03)}{7} = \$9.95$$

Since it is desirable to have convenient sizes for class intervals, the $9.95 figure may be rounded to $10.00 and the distribution may be tentatively set up on that basis. The frequency distribution shown in Table 3-3 results from a tally of the number of items that fall in each $10.00 class interval.

Table 3-3
Frequency Distribution of Earnings of 100 Semi-skilled Employees
of the Beco Company for One Week in 1973.

WEEKLY EARNINGS	NUMBER OF EMPLOYEES
$160.00 and under 170.00	4
170.00 and under 180.00	14
180.00 and under 190.00	18
190.00 and under 200.00	28
200.00 and under 210.00	20
210.00 and under 220.00	12
220.00 and under 230.00	4
TOTAL	100

Some important features of these data are immediately discernible from the frequency distribution. The approximate value of the range, or the difference between the values of the highest and lowest items, is revealed. Of course, since the identity of the individual items is lost in the grouping process, we cannot tell from the frequency table alone what the exact values of the highest and lowest items are. Also, the frequency distribution gives at a glance some notion of how the elements are clustered. For example, more of the weekly earnings figures fall in the $190.00 and under 200.00 interval than in any other single class. When the frequencies in the classes immediately preceding and following the $190.00 and under 200.00 grouping are added to the 28 in that interval, a total of 66, or 2/3 of the 100 employees, are accounted for. Furthermore, the distribution shows how the data are spread or dispersed throughout the range from the lowest to the highest value. We can quickly determine whether the items are bunched near the center of the distribution or spread rather evenly throughout. Also, we can see whether the frequencies fall away rather symmetrically on either side of a class near the center of the distribution or whether there is a decided lack of such symmetry. We will now consider various statistical measures for describing these characteristics in a more exact manner, but much information can be gained by simply studying the distribution itself. There is no single perfect frequency distribution for a given set of data. Several alternative distributions with different class interval sizes and different highest and lowest values may be equally appropriate.

CLASS LIMITS   3.3

The way in which class limits of a frequency distribution are stated depends upon the nature of the data. Figures on ages provide a good illustration of this point. Suppose that ages were rounded according to the *last completed year,* that is, ages were recorded as of the last birthday. Then a clear and unambiguous way of stating the class limits is as follows: 15 and under 20, 20 and under 25, etc.

Consider the first class interval, "15 and under 20." Since ages have been recorded as of the last birthday, this class encompasses individuals who have reached at least their fifteenth birthday but not their twentieth birthday. If a person is 19.999 years of age, that is, a fraction of a day away from his twentieth birthday, he falls into the first class. However, upon attaining his twentieth birthday, he falls in the second class, "20 and under 25." Thus, these class intervals are five years in size. The midpoints of the classes, that is, values located halfway between the class limits, are, respectively, 17.5, 22.5, 27.5, 32.5, and 37.5. These values are used in computations of statistical measures for the distribution. Note that with class limits established and stated as above, the "stated limits" are, in fact, the true boundaries or "real limits" of the classes. Of course, there are other ways of wording the limits such as "at least 15 but under 20," or "15 to but not including 20."

Suppose, on the other hand, that age data were rounded to the nearest birthday. According to a widely accepted convention, the class limits would be stated as follows: 15–19, 20–24, etc. Even though the stated limits in each class are only four years apart, it is important to realize that the size of these class intervals is still five years. For example, since the ages are given as of the nearest birthday, everyone between 14.5 and 19.5 years of age falls in the class "15–19." Thus, when data recorded to the nearest unit are grouped into frequency distribution classes, the lower real limit or lower boundary of any given class lies one-half unit below the lower stated limit and the upper real limit or upper boundary lies one-half unit above the upper stated limit. The midpoints of the class intervals may be obtained by averaging the lower and upper real limits, or the lower and upper stated class limits. For example, the midpoint of the class "15–19" is 17, which is the same figure obtained by averaging 14.5 and 19.5.

In summary, when raw data are rounded to the *last* unit, the stated class limits and real class limits are identical. When raw data are rounded to the *nearest* unit, the respective real limits are one-half unit removed from the corresponding stated limits. With both types of data, the midpoints of classes are halfway between the stated limits, or, equivalently, halfway between the real limits.

Unfortunately, conventions are not universally observed. Often, one must use a frequency distribution constructed by others, and the nature of the raw data may not be clearly indicated. Needless to say, standards of good practice oblige the producer of the frequency distribution to indicate the nature of the underlying data.

Of course, class intervals should always be mutually exclusive and the class each item falls into should always be clear. If class limits are stated as 30–40, 40–50, etc., for example, it is not clear whether 40 belongs to the first class or the second.

3.4 OTHER CONSIDERATIONS IN CONSTRUCTING FREQUENCY DISTRIBUTIONS

A number of other points should be taken into account in the construction of a frequency distribution. If the data are such that there are concentrations of particular values, it is desirable that these values be at the midpoints of class intervals. For example, assume that data are collected on the amounts of the lunch checks in a student cafeteria. Suppose these checks predominantly occur in multiples of five cents, although not exclusively so. If class intervals are set up as $0.70–0.74, $0.75–0.79, etc., a preponderance of items would be concentrated at the lower limits. In calculating certain statistical measures from the frequency distribution, the assumption is made that the midpoints of classes are average (arithmetic mean) values of the items in these classes. If, in fact, most of the items lie at the lower limits of the respective classes, a systematic error will be introduced by this assumption, since the actual averages within classes will typically fall below the midpoints.

Another factor to be considered in constructing a frequency distribution is the desirability of having a relatively smooth progression of frequencies. Many frequency distributions of business and economic data are characterized by having one class that contains more items than any other single class, and a more or less gradual dropping off of frequencies on either side of this class. Table 3-3 is an example of such a distribution. As indicated in Section 3.2, the distribution may not be at all symmetrical. However, erratic increases and decreases of frequencies from class to class tend to obscure the overall pattern. Erratic progressions of frequencies often arise from the use of class intervals that are too small. Increasing the size of class intervals usually results in a smoother progression of frequencies. But, wider classes reveal less detail than narrower classes. Thus, a compromise must be made in the construction of every frequency distribution. At one extreme, if we use class interval sizes of one unit each, every item of raw data is assigned to a separate class; at the other extreme, if we use only one class interval as wide as the range of the data, all items fall in the single class. Within the limits of these considerations, some freedom exists for the choice of an appropriate class interval size.

A final point to be considered in setting up a frequency distribution is that the numerical values of statistical measures computable from the grouped data should be close to the analogous values calculated from the ungrouped raw data. Since the user of the frequency distribution ordinarily does not have the raw data available, he makes his computations solely from the distribution. Clearly, if the values calculated from the frequency distribution, such as the arithmetic mean and standard deviation discussed later in this chapter, depart considerably from the corresponding measures computed from the ungrouped raw data, assuming no computational errors, distortion of the statistical characteristics of the data has been introduced by the grouping process.

GRAPHIC PRESENTATION OF FREQUENCY DISTRIBUTIONS 3.5

The use of graphs for displaying frequency distributions will be illustrated in terms of the data on weekly earnings shown in Table 3-3. One method is to represent the frequency of each class by a rectangle or bar. Such a chart is generally referred to as a histogram. A histogram for the frequency table given in Table 3-3 is shown in Figure 3-2. In agreement with the usual convention, values of the variable are depicted on the horizontal axis and frequencies of occurrence are shown on the vertical axis.

An alternative method for the graphic presentation of a frequency distribution is the frequency polygon. In this type of graph, the frequency of each class is represented by a dot at the appropriate height plotted opposite the midpoint of each class. The dots are joined by line segments. Thus, a many-sided figure,

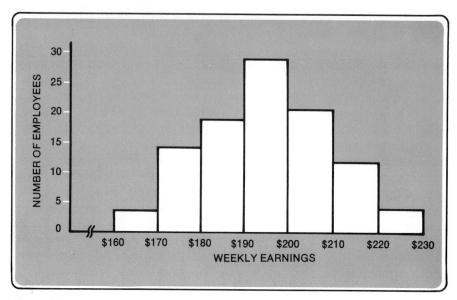

Figure 3-2
Histogram of frequency distribution of earnings of 100 semi-skilled employees of the Beco Company for one week in 1973.

or polygon, is formed. Stated differently, a frequency polygon is the line graph obtained by joining the midpoints of the tops of the bars in a histogram. By convention, the polygon is closed at both ends of the distribution by line segments, drawn, respectively, from the dot representing the frequency in the

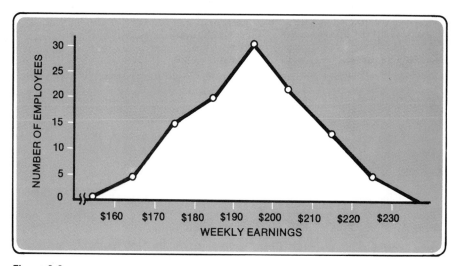

Figure 3-3
Frequency polygon for the distribution of earnings of 100 semi-skilled employees of the Beco Company for one week in 1973.

lowest class to a point on the horizontal axis one-half a class interval below the lower limit of the first class, and from the dot representing the frequency in the highest class to a point one-half a class interval above the upper limit of the last class. A frequency polygon for the distribution given in Table 3-3 is shown in Figure 3-3. It is important to realize that the line segments are drawn for convenience in reading the graph. That is, the only significant points are the plotted frequencies for the given midpoints. Interpolation for intermediate values between such points would be meaningless. Often the midpoints of classes are shown on the horizontal axis directly below the plotted points rather than the class limits shown in Figure 3-3.

If the class sizes in a frequency distribution were gradually reduced and the number of items were increased, the frequency polygon would approach a smooth curve more and more closely. Thus, as a limiting case, the variable of interest may be viewed as continuous rather than as occurring in discrete classes, and the polygon would assume the shape of a smooth curve. The frequency curve approached by the polygon for the weekly earnings would thus appear as shown in Figure 3-4.

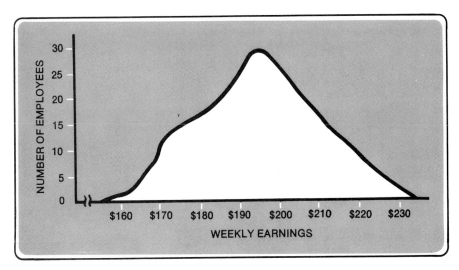

Figure 3-4
Frequency curve for the distribution of earnings of 100 semi-skilled employees of the Beco Company for one week in 1973.

CUMULATIVE FREQUENCY DISTRIBUTIONS   3.6

Sometimes interest centers in the number of cases that lie below or above specified values rather than within intervals as shown in a frequency distribution. In such situations, it is convenient to use a *cumulative* frequency distribution rather than the usual frequency distribution. A so-called "less than" cumulative distribution is shown for the weekly earnings distribution shown in Table 3-4. The cumulative numbers of employees with earnings less than the lower class limits of $160.00, $170.00, and so forth are given. Thus, there were no employees with weekly earnings of less than $160.00, four employees with earnings of less than $170.00, $4 + 14 = 18$ employees with less than $180.00, and so forth.

The graph of a cumulative frequency distribution is referred to as an *ogive* (pronounced "ojive"). The ogive for the cumulative distribution shown in Table 3-4 is given in Figure 3-5. The plotted points represent the number of employees having less than the earnings shown on the horizontal axis. The vertical coordinate of the last point represents the sum of the frequencies, in this case, 100. The S-shaped configuration depicted in Figure 3-5 is quite typical of the appearance of a "less than" ogive. A "more than" ogive for the daily sales distribution would have class limits reading "more than $160.00," and so forth. In this case a reversed S-shaped figure would have been obtained, sloping downward from the upper left to the lower right on the graph.

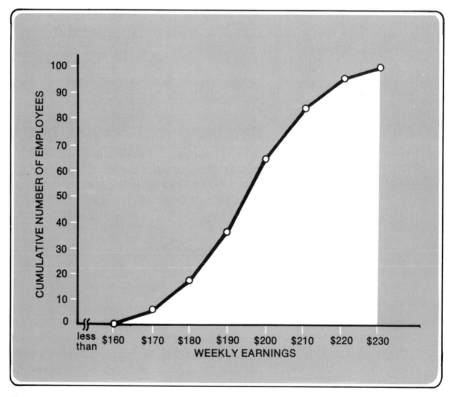

Figure 3-5
Ogive for the distribution of earnings of 100 semi-skilled employees of the Beco Company for one week in 1973.

TABLE 3-4

Cumulative Frequency Distribution of Earnings of 100 Semi-skilled
Employees of the Beco Company for One Week in 1973

WEEKLY EARNINGS	NUMBER OF EMPLOYEES
Less than $160.00	0
Less than 170.00	4
Less than 180.00	18
Less than 190.00	36
Less than 200.00	64
Less than 210.00	84
Less than 220.00	96
Less than 230.00	100

EXERCISES

1. The commissions paid to nineteen Metroperidock insurance salesmen for a one-week
period were

$125.00	$138.50	$216.74	$ 82.75	$140.40
169.95	212.55	158.00	162.63	190.15
175.00	101.82	205.01	160.00	228.00
182.05	238.09	171.13	198.50	

Construct
(a) a frequency table, assuming the data were recorded to the last cent,
(b) a histogram,
(c) a frequency polygon.

2. Daily sales of a small retail establishment are given below:

$ 97.60	$132.67
102.65	145.68
141.02	175.92
174.68	106.34
92.06	125.27
172.21	127.72
83.77	137.66
104.01	83.66
102.44	116.70
156.52	136.79
149.59	129.99
136.97	125.41
124.31	124.17
123.23	91.70
118.94	128.29

(a) Construct a frequency distribution for sales during the period, assuming the data
were recorded to the last cent. Use about five classes.
(b) Sketch a frequency polygon showing the distribution in part (a).

3. The ages measured to the last birthday of the employees of Smith, Inc. are as follows:

22	21	26	42
28	39	20	32
35	45	49	28
31	49	42	30
31	39	37	36
33	47	38	48

Set up a frequency table having six classes and specify the midpoint of each class.

4. Compute the cumulative frequency distribution for the distribution given below.

ANALYSIS OF ORDINARY LIFE INSURANCE

POLICY SIZE	NUMBER OF ORDINARY LIFE INSURANCE POLICIES IN FORCE PER 1000 POLICIES, IN A CERTAIN MONTH
Under $1000	31
$1000–2499	181
2500–4999	108
5000–9999	213
10,000 or more	467

5. Comment critically on the following systems of designating class intervals:

(a) 83–102   (b) 83–under 102
    102–121   103–under 121
    etc.   122–under 141
                    etc.

6. Blah Beer Company has two production lines working each day. The number of cases produced per eight-hour shift for a week during June 1973 was
   Line 1: 6921, 8205, 6658, 6835, 7830, 6935, 7563, 6503, 6777, 6250, 6000, 6935, 7000, 7012, 5530
   Line 2: 5655, 6138, 7891, 7951, 6666, 7891, 7589, 7002, 8131, 7662, 7555, 8495, 7832, 7477, 6886
   Using six classes and "and under" upper class limits,
   (a) Construct and graph the frequency distribution for the number of cases produced per eight-hour shift.
   (b) Construct and graph the frequency distribution for each production line separately.
   (c) Which answer, that in part (a) or part (b), gives a better picture of the distribution of production of cases of beer at the Blah Beer Company? Explain your answer.

3.7   DESCRIPTIVE MEASURES FOR FREQUENCY DISTRIBUTIONS

As indicated in Section 3.2, once a frequency distribution is constructed from a set of figures, certain features of the data become readily apparent. For most purposes, however, it is necessary to have a more exact description of these characteristics than can be ascertained by a casual glance at the distribution. Thus, analytical measures are usually computed to describe such characteristics as the *central tendency, dispersion,* and *skewness* of the data. These measures constitute summary descriptions of the frequency distribution, which is itself a summarization of the set of original data.

Averages are the measures used to describe the characteristic of *central tendency* or *location* of data. One such measure is the arithmetic mean. This is doubtless the most familiar average; in fact, in common usage, it is often referred to as "the average." For example, in ordinary conversation or print, we encounter such terms as "average income," "average growth rate," "average profit rate," "average man," etc. Actually, several different types of averages or measures of central tendency are implied in these terms. In this section, we will consider only the most commonly employed and most generally useful averages. Averages attempt to convey in summary form the notion of "central location" or the "middle property" of a set of data. As we shall see, the type of average to be employed depends on the purpose of the application and the nature of the data being summarized.

Dispersion refers to the spread or variability in a set of data. One method of measuring this variability is in terms of the difference between the values of selected items in a distribution, such as the difference between the values of the highest and lowest items. Another more comprehensive method is in terms of some average of the deviations of all the items from an average. Dispersion is, of course, a very important characteristic of data, in that interest frequently centers as much upon the uniformity in a set of data as upon its central tendency.

Skewness refers to the symmetry or lack of symmetry in the shape of a frequency distribution. This characteristic is of particular importance in judging the typicality of certain measures of central tendency.

We begin the discussion of averages or measures of location by considering the most familiar one, the arithmetic mean.

THE ARITHMETIC MEAN 3.8

Probably the most widely used and most generally understood way of describing the central tendency or central location of a set of data is the average known as the *arithmetic mean*. The arithmetic mean, or simply the *mean,* is the total of the values of a set of observations divided by their number. For example, if $X_1, X_2, \ldots, X_n$ represent the values of n items or observations, the arithmetic mean of these items, denoted $\overline{X}$, is defined as

$$\overline{X} = \frac{X_1 + X_2 + \cdots + X_n}{n} = \frac{\sum_{i=1}^{n} X_i}{n}$$

For simplicity, subscript notation such as that given above will not be used in this book. (However, before continuing, you should turn to Appendix B and work out the examples given there.) Thus when the subscripts are dropped, the formula becomes

(3.1)
$$\overline{X} = \frac{\Sigma X}{n}$$

where the symbol Σ (Greek capital "sigma") means "the sum of."

For example, suppose that an accounting department established accounts receivable in the following amounts during a one-hour period: $600, $350, $275, $430, and $520. The arithmetic mean of the amounts of the accounts receivable is

$$\overline{X} = \frac{\$600 + \$350 + \$275 + \$430 + \$520}{5} = \frac{\$2175}{5} = \$435$$

The mean of $435 may be thought of as the size of each account receivable that would have been set up if the total of the five accounts was $2175 and all the accounts were the same size. That is, the mean is the value each item would have if they were all identical and the total value and number of items remained unchanged.

Symbolism

A brief note on symbolism is appropriate at this point. In keeping with standard statistical practice, we will use the symbol $\overline{X}$ here and throughout the book to denote the mean of a *sample* of observations. We will denote the number of observations in the sample by a lower case letter n. A value such as $\overline{X}$, that is, a number computed from sample data, is referred to as a *statistic*. A statistic may be used as an estimate of an analogous population measure, known as a *parameter*. Thus, the sample mean $\overline{X}$ is a statistic that may be thought of as an estimate of the mean of the population from which the sample was drawn. The population mean is a parameter. It is conventional to denote population parameters by Greek letters and sample statistics by Roman letters. In keeping with this convention, we will denote a population mean by the Greek letter μ ("mu").

If the population mean were calculated from the data collected from a census, that is, from the entire N population observations, then

(3.2)
$$\mu = \frac{X_1 + X_2 + \cdots + X_N}{N} = \frac{\Sigma X}{N}$$

In practice, the population mean, μ, is not ordinarily calculated because it is usually neither feasible nor advisable to accomplish a complete enumeration of the population.

Grouped data

When data have been grouped into the form of a frequency distribution, the arithmetic mean can be computed by a generalization of the definition for the mean of ungrouped data. As given in equation (3.1), the formula for the mean of a set of ungrouped data is $\overline{X} = \Sigma X/n$. However, in grouped data, since the identity of the individual items has been lost, an estimate must be made of the total of the values of the observations, ΣX. This estimate is obtained by multiplying the midpoint of each class in the distribution by the frequency of that class and summing over all classes. In symbols, if X denotes the midpoint of a class and f the frequency, the arithmetic mean of a frequency distribution may be computed from the following formula, known as the direct method:

(3.3)
$$\overline{X} = \frac{\Sigma fX}{n}$$

The computation of $\overline{X}$ for the frequency distribution of weekly earnings data shown in Table 3-3 is given in Table 3-5. The mean earnings figure of $194.80 calculated from the frequency distribution is very close to the corresponding mean of $194.91 for the ungrouped original data given in Table 3-2. The small difference in these two figures illustrates the slight loss in accuracy involved in calculating statistical measures from frequency distributions rather than from ungrouped data. Furthermore, when there is a large number of observations, the calculations from original data are far more tedious than from frequency distributions.

TABLE 3-5
Calculation of the Arithmetic Mean for Grouped Data
by the Direct Method: Weekly Earnings Data

WEEKLY EARNINGS	NUMBER OF EMPLOYEES f	MIDPOINTS X	fX
$160.00 and under $170.00	4	$165.00	$660
170.00 and under 180.00	14	175.00	2450
180.00 and under 190.00	18	185.00	3330
190.00 and under 200.00	28	195.00	5460
200.00 and under 210.00	20	205.00	4100
210.00 and under 220.00	12	215.00	2580
220.00 and under 230.00	4	225.00	900
	$n = \Sigma f = 100$		$19,480

$$\overline{X} = \frac{\Sigma fX}{n} = \frac{\$19,480}{100} = \$194.80$$

Shortcut formulas are often useful for calculating the arithmetic mean and other measures for frequency distributions. One such shortcut formula, known as the step-deviation method, is explained in Appendix C.

for juvenile rheumatoid arthritis. He believes that he observes a large number of mongoloid children in these families.

   (c) A production manager wishes to determine the characteristics of workers in his plant who are "unusually efficient" as measured by the percentage of defective articles they have produced.

2. Explain the way that the concept of a control group should be used by criticizing the following situations:
   (a) Of the persons in the U.S. who contracted polio in a certain year, only 6% were from families with incomes over $10,000 per year. Therefore, it was concluded that families in that income group had less to fear from polio than those in lower income groups.
   (b) In a certain university, only 20% of those failing to graduate were women. This suggests that women are better students than men in this university.
   (c) A manufacturer of a preparation for treating the common cold claimed that 95% of cold sufferers who used his product were free of their colds within a one-week period.

3. What is the difference between an observational study and a "controlled inquiry"?

4. Discuss the need for and advantages of having a control group in a study, and give an example of a study in which use of a control group would be of value.

5. In each of the following situations, state whether a control group would be of use, and if so, what the control group would be.
   (a) You are interested in investigating the accident rates in low-income urban areas and you have data on numbers of accidents occurring in low-income urban areas, numbers of cars registered in low-income areas, actual area of low-income areas, and other similar information.
   (b) You are interested in evaluating the effectiveness of a new safety lighting program to be installed in your plant.

Types of errors

The concept of error is a central one throughout all statistical work. Wherever we have measurement, inference, or decision making, the possibility of error is present. In this section, we shall concern ourselves with errors of measurement. The problems of errors of inference and decision making are treated in subsequent chapters.

It is useful to distinguish two different types of errors that may be present in statistical measurements, namely, *systematic errors* and *random errors*. Systematic errors, as the term implies, cause a measurement to be incorrect in some systematic way. If observations have arisen from a sample drawn from a statistical universe, systematic errors persist even when the sample size is increased. They are errors involved in the procedures of a statistical investigation and may occur in the planning stages, or during or after the collection process. As noted in Chapter 1, another term conventionally employed for systematic error is *bias*. Among the causes of bias are faulty design of a questionnaire such as misleading or ambiguous questions, systematic mistakes in planning and carrying out the collection and processing of the data, nonresponse and refusals by respondents to provide information, and too great a discrepancy between the sampling frame and the target universe. As a generalization, these that differed most between the two groups would obviously be the most useful ones for spelling out the distinguishing characteristics of high-volume purchasers. In summary, the firm could have used the non-high-volume customers as a control group against which to compare the properties of the high-volume group, which in the terminology used earlier would represent the "treatment group."

A comment in the form of a warning is pertinent in the above illustration. Care must be used in the selection of the properties of the two groups to be observed. These properties should bear some logical relationship to the characteristic of high- versus non-high-volume purchases. Otherwise, the properties may be spurious indicators of the distinguishing characteristics between the two groups. For example, income level would be logically related to purchasing volume. If the high-volume purchaser group had a substantially higher income than the non-high-volume group, then income would evidently be a possible distinguishing characteristic. On the other hand, suppose at the time of the survey the high-volume purchaser group happened to have a higher percentage of persons who wore black shoes than did the non-high purchaser group. This characteristic of shoe color would *not* seem to be logically related to volume of purchases. Hence, we would not be surprised if the relationship between shoe color and volume of purchases disappeared in subsequent investigations or even reversed itself.

A couple of comments may be made on the construction of control groups. If the treatment group is symbolized as A and the control group as B, then an alternative control group to the one used would have been the treatment and control groups combined, or $A + B$. Thus, in the above example, if the relevant data had been available for the "high-volume" and "non-high-volume" customers combined, or for all customers, this group could have constituted the control. For example, let us assume for simplicity that high-volume and non-high-volume customers were equal in number. Suppose that 90% of high-volume customers possessed characteristic X, whereas only 50% of non-high-volume customers had this characteristic. The same information would be given by the statements that 90% of high-volume customers possessed characteristic X, while 70% of all customers possessed this characteristic. The 70% figure is, of course, the weighted mean of 90% and 50%. With the knowledge of equal numbers of persons in the high-volume and non-high-volume groups, it can be inferred that 50% of non-high-volume customers had the property in question. This point is of importance, because sometimes historical data may be available for an entire group $A + B$, whereas available resources may permit a study only of the treatment group A or what is more usual only a sample of this group.

EXERCISES

1. Indicate the control groups that you think should be used in each of the following situations. Briefly defend your choices.
   (a) It is desired to try a new method of teaching sixth-grade arithmetic in a certain city.
   (b) A medical researcher has data on the families of 180 children whom he is treating

phenomenon, St. Petersburg has a much higher proportion of persons in the "65 and older" age group than does Los Angeles. If we think of the crude death rate of a city as the weighted mean of the death rates for component age groupings of the population, the high death rates of the older age groups get relatively greater weight in the case of St. Petersburg than in Los Angeles, and the lower death rates of the younger age groupings get correspondingly less weight.

EXERCISES

1. In reference to Exercise 1 on page 45, calculate
   (a) the arithmetic mean from the original data;
   (b) the arithmetic mean using the frequency table.
   (c) Which is the true arithmetic mean value? Explain your answer.

2. The Center City Bank reports bad debt ratios (dollar losses to total dollar credit extended) of .04 for personal loans and .02 for industrial loans in 1973. For the same year, the Neighborhood Bank reports bad debt ratios of .05 for personal loans and .03 for industrial loans. Can one conclude from this that Center City's overall bad debt ratio is less than Neighborhood's? Justify your response.

3. Consider the following data:

Percentage of Civilian Labor Force Unemployed in a Certain Area

COUNTY	% UNEMPLOYED	CIVILIAN LABOR FORCE
A	3.6	114,395
B	3.8	214,758
C	2.5	206,324
D	6.5	843,160

   (a) What is the unweighted average of the percent unemployed per county?
   (b) What is the weighted average of the percent unemployed for the four counties combined?
   (c) Explain the reason for the difference in the figures obtained in parts (a) and (b).

THE MEDIAN   3.10

The *median* is another well-known and widely used average. It has the connotation of the "middlemost" or "most central" value of a set of numbers. For ungrouped data, the median is defined simply as the value of the central item when the data are arrayed by size. If there is an odd number of observations, the median is directly ascertainable. On the other hand, if there is an even number of items, there are two central values. Then, by convention, a value halfway between these two central observations is designated the median.

For example, suppose that a test in five new small economy cars of the same brand yielded the following numbers of miles per gallon of gasoline:

27, 29, 30, 32, and 33. The median number of miles per gallon would be 30. If another car were tested, and the number of miles per gallon obtained was 34, the array would now read: 27, 29, 30, 32, 33, 34. The median would be a value halfway between 30 and 32, or 31.

Another way of viewing the median is as a value below and above which lie an equal number of items, when these items have been arrayed from the smallest to the largest value. Thus, in the illustration involving five observations, two lie above the median and two below. In the example involving six observations, three fall above the median and three fall below. Of course, in an array with an even number of items, any value lying between the two central items may, strictly speaking, be referred to as a median. However, as indicated earlier, the convention is to use the midpoint between the two central items. When values are tied at the center of a set of observations, there may be no value such that equal numbers of items lie above and below it. Nevertheless, the central value, as defined in the preceding paragraph, is still designated as the median. For example, in the following array: 52, 60, 60, 60, 60, 61, 62, the number 60 is the median, although unequal numbers of items lie above and below this value.

Since the identity of the original observations is not retained in a frequency distribution, the median, of necessity, is an estimated value. Because the data in a frequency distribution are arranged in order of magnitude, frequencies can be cumulated to determine the class in which the median observation falls. It is then necessary to make some assumption about the way in which observations are distributed in that class. Conventionally it is assumed that observations are equally spaced or evenly distributed throughout the class containing the median. The value of the median is then established by interpolating. For example, consider the distribution of weekly wages of semi-skilled workers previously given in Table 3-3. That distribution is shown again in Table 3-6. First, we explain the calculation of the median without the use of symbols. Then we generalize the procedure by stating it as a formula.

TABLE 3-6
Calculation of the Median for a Frequency Distribution:
Weekly Wage Data

WEEKLY WAGES	NUMBER OF WORKERS f	
$160.00 and under $170.00	4	
170.00 and under 180.00	14	
180.00 and under 190.00	18	
		$\Sigma f_p = 36$
190.00 and under 200.00	28	
200.00 and under 210.00	20	
210.00 and under 220.00	12	
220.00 and under 230.00	4	
	100	

In the distribution shown in Table 3-6, since there are 100 weekly wages represented, the median lies between the 50th and 51st figures. Since 36 wage figures occur prior to the class "$190.00 and under $200.00," it is clear that the median is contained in that class. Assuming that the 28 wage figures are evenly distributed between $190.00 and $200.00, we can determine the median observation by interpolating 14/28 of the distance through this $10.00 class. In summary, the median is calculated by adding 14/28 or 1/2 of $10.00 to the $190.00 lower limit of the class containing the median. That is,

$$Md = \$190.00 + \left(\frac{50 - 36}{28}\right)(\$10) = \$190 + \left(\frac{1}{2}\right)(\$10) = \$195.00$$

Thus, the formula for calculating the median of a frequency distribution is

(3.5)
$$Md = L_{Md} + \left(\frac{n/2 - \Sigma f_p}{f_{Md}}\right)(i)$$

where Md = the median

L_{Md} = the (real) lower limit of the class containing the median

n = the total number of observations in the distribution

Σf_p = the sum of the frequencies in classes preceding the one containing the median

f_{Md} = the frequency of the class containing the median

i = the size of the class interval

CHARACTERISTICS AND USES OF THE ARITHMETIC MEAN AND MEDIAN
3.11

The preceding sections have concentrated on the mechanics of calculating means and medians for ungrouped and grouped data. We now turn to the characteristics and uses of these averages.

The arithmetic mean is doubtless the most widely used and most familiar measure of central tendency. It has the advantage of being a rigidly defined mathematical value and, therefore, can be manipulated algebraically. For example, the means of two related distributions can be combined by suitable weighting. Also if two of the quantities in the formula $\bar{X} = \Sigma X/n$ are known, the third can be obtained directly. Because of such mathematical properties, the arithmetic mean is used more often in advanced statistical techniques than any of the other averages.

The arithmetic mean is also important in connection with statistical inference as well as for descriptive purposes. For example, the arithmetic mean of a random sample of observations may be used to estimate the value of the corresponding arithmetic mean of the population from which the sample was drawn. Thus, the mean family income of a sample of families in the city of New York may be used as an estimate of the mean income of all families in that city. In this type of estimation, the mean is a more reliable estimator than other averages, such as the median or mode. That is, the mean is less affected by sampling fluctuations than are the other measures of central tendency and it tends to estimate the corresponding population figures more closely.

Another useful interpretation of the mean is as an estimated value for any item in a distribution. The sum of the deviations of a set of observations from the mean is equal to zero. Therefore, if the mean of the set is used as an estimate of the value of any observation picked at random, and this procedure is repeated, then on the average, the mean amount of error, taking into account the sign of the error or deviation, is zero.

A disadvantage of the mean is its tendency to be distorted by extreme values at either end of a distribution. In general, it is pulled in the direction of these extremes. Another disadvantage is that the mean cannot be determined from a frequency distribution with an open-ended interval such as "$25,000 and over," since the midpoint of such an interval is unknown. However, if the order of magnitude of the figures in the open-ended interval is known, assumptions can be made that will permit the calculation of the mean. What is needed is an estimate of the total value (fX) of observations in the open-ended interval. Therefore, an arithmetic mean for these items may be estimated and multiplied by the number of items in the interval to give an estimate of the total value. For example, to find the mean in the illustration with an interval of family incomes of $25,000 and over, an estimate is obtained of the total incomes falling in that interval.

The median is also a very useful measure of central tendency. Its relative freedom from distortion by skewness in a distribution makes it a particularly desirable average for descriptive purposes. Thus, it is often used to convey the idea of a "typical" observation. It is primarily affected by the number rather than the size of observations. This can be seen if we consider an array in which the median has been determined. Now assume that the largest item is increased, say, one hundredfold. The median would be unchanged. The arithmetic mean would, of course, be pulled toward the large extreme item.

The median may be interpreted as a "best estimate" value in a somewhat different sense than the mean. As previously indicated, the mean is a best estimate in the sense that if observations are repeatedly drawn at random from the distribution and the mean is used to estimate each value, the mean amount of error or deviation is zero. This follows from the fact that the algebraic sum of the deviations from the mean of a distribution is equal to zero. On the other hand, if the same experiment of repeated estimation is used with the *median* being the estimated value, the average (mean) amount of absolute error (that is, disregarding the sign of the error) is a *minimum*. In other words, the average deviation of the observations from the median is less than from any other value in the distribution. Thus, in an estimation situation, if one wants to minimize the average absolute amount of error and the sign of the error is not particularly important, the median is preferable to the arithmetic mean.

Another merit of the median is that it can ordinarily be computed for open-ended distributions. This is true because the value of the median is determined solely from the interval in which it falls. It would virtually never fall into the open-ended interval, since that interval would have to contain more than one-half the frequencies to include the median. It is extremely improbable that such a frequency distribution would ever be constructed. Therefore, for practical purposes, the median can be computed for open-ended distributions.

The major disadvantage of the median is that it is an average of position and hence not a mathematical concept suitable for further algebraic treatment. For example, if one knows the medians of two distributions, there is no algebraic way of combining these two figures to obtain the median of the combined distributions.

Also, the median tends to be a rather unstable value if the number of items is small. Furthermore, as observed earlier, the median is a less reliable average than the mean for estimation purposes, since it is more affected by sampling variations.

THE MODE 3.12

Another average that is conceptually very useful but often not explicitly calculated is the *mode.* In French, to be "à la mode" is to be in fashion. The mode as a statistical average is the observation that occurs with the greatest frequency and thus is the most "fashionable" value. The mode is rarely determined for ungrouped data. The reason is that quite accidentally or haphazardly an item might occur more often than any other, yet lie at the lower or upper end of the array of observations and be a very unrepresentative figure. Therefore, determination of the mode is generally not even attempted for ungrouped data.

When data are grouped into a frequency distribution, it is not possible to specify the observation that occurs most frequently since the identity of the individual items is lost. However, we can determine the so-called "modal class," or the class that contains more observations than any other. Of course, class intervals should be of the same size when this determination is made. As a practical matter, most analysts of business and economic data do not usually proceed any further than the specification of a modal class in the measurement of a mode for an empirical frequency distribution. When the location of the modal class is considered along with the arithmetic mean and median, much useful information is generally conveyed not only about central tendency but also about the skewness of a frequency distribution.

Several formulas have been developed for determining the location of the mode within the modal class. These usually involve assumptions about the use of frequencies in the classes preceding and following the modal classes as weighting factors that tend to pull the mode up or down from the midpoint of the modal class. We shall not present any of these formulas here. For our pur-

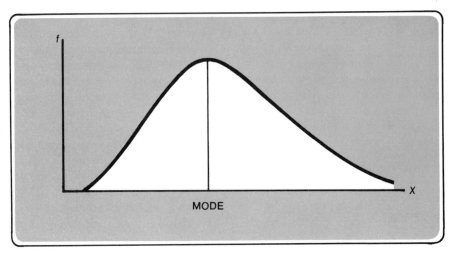

Figure 3-6
The location of the mode.[a]

[a]In this and subsequent graphs, the X on the horizontal axis denotes values of the observations and the f on the vertical axis denotes frequency of occurrence.

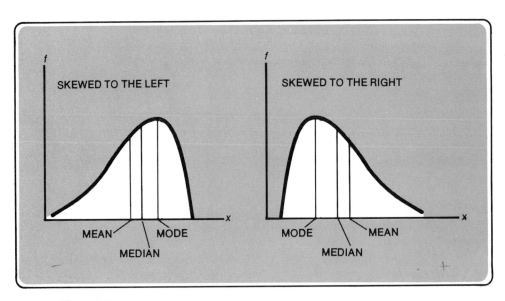

Figure 3-7
Skewed distributions depicting typical positions of averages.

poses, the midpoint of the modal class may be taken as an estimate of the mode. It is useful to consider the meaning of the mode. Let us visualize the frequency polygon of a distribution and then the frequency curve approached as a limiting case by gradually reducing class size. In the limiting situation, the variable under study may be considered as continuous rather than as occurring in discrete classes. The mode may then be thought of as the value on the horizontal axis lying below the maximum point on the frequency curve (see Figure 3-6).

The mode of a frequency distribution has the connotation of a typical or representative value. It specifies a location in the distribution at which there is maximum clustering. In this sense, it serves as a standard against which to judge the representativeness or typicality of other averages. If a frequency distribution is symmetrical, the mode, median, and mean coincide. As noted earlier, extreme values in a distribution pull the arithmetic mean in the direction of these extremes. Stated somewhat differently, in a skewed distribution, the mean is pulled away from the mode toward the extreme values. The median also tends to be pulled away from the mode in the direction of skewness, but is not affected as much as the mean. If the mean exceeds the mode,[1] a distribution is said to have "positive skewness" or to be "skewed to the right"; if the mean is less than the mode, the terms "negative skewness" and "skewed to the left" are used. The order in which the averages fall in these types of distributions is shown in Figure 3-7.

Many distributions of economic data in the United States are skewed to the right. Examples include the distributions of incomes of individuals, savings of individuals, corporate assets, sizes of farms, and company sales within many industries. In many of these instances, the arithmetic mean is pulled so far from the mode as to be a very unrepresentative figure.

Multimodal distributions

If more than one mode appears, the frequency distribution is referred to as "multimodal"; if there are two modes it is referred to as "bimodal." Extreme care must be exercised in analyzing such distributions. For example, consider a situation in which you want to compare the arithmetic mean wages of workers in two different companies. Assume that the mean calculated for Company A exceeds that of Company B. If you conclude from this finding that workers in Company A earn higher wages, on the average, than those in Company B, without recognizing the fact that the wage distribution for each of these companies is bimodal, you may make serious errors of inference. To illustrate the principle involved, let us assume that the mean annual wage for unskilled workers is $5000 and for skilled workers is $10,000 at each of these companies. Let us also assume that the individual distributions of wages of unskilled and skilled workers are symmetrical and that there are the same total number of workers in each company. Further, let us assume that these companies only

---

[1] Sometimes the median, rather than the mode, is used for this comparison.

have workers in the aforementioned two skill classifications. However, suppose 75% of the Company A workers are skilled, whereas only 50% of the Company B workers are skilled. Figure 3-8 shows the frequency curves of the distributions of annual wages at the two companies. Clearly, the mean annual wage of workers in Company A exceeds that at Company B. This is true simply because there is a higher percentage of skilled workers at Company A. However, if you were ignorant of this fact, you might be tempted to infer reasons why workers at Company A earn more than those at Company B. The fact of the matter is that unskilled workers at both companies earn the same wages, on the average. The same holds true for the skilled workers. What is required here is to separate two wage distributions at each company, one for skilled workers and one for unskilled. A comparison of the mean wages of unskilled workers at the two companies would reveal their equality—similarly, for skilled workers.

The principle involved here is one of homogeneity of the basic data. The fact that a wage distribution is bimodal suggests that two different "cause systems" are present, and that two distinct distributions should be recognized. The data on wages may be said to be nonhomogeneous with respect to skill level. Other bimodal distributions might, for example, include the merging of height data for men and women or merging data on dimensions of products from two different suppliers.

If a basis for separating a bimodal distribution into two distributions cannot be found, then extreme care must be used in describing the data. In cases such as that shown for Company B in Figure 3-8, where the heights of the two modes are about equal, the arithmetic mean and median will probably fall between the modes and will not be representative of the large concentrations of values lying at the modes below and above these averages.

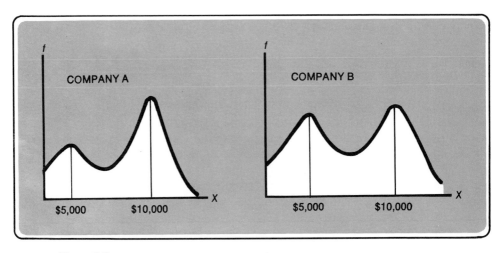

Figure 3-8
Bimodal frequency distributions: annual wages of workers in two companies.

EXAMPLE 3-1

The following distribution gives the dollar cost per unit of output for 200 plants in the same industry:

DOLLAR COST PER UNIT OF OUTPUT	NUMBER OF PLANTS
$1.00 and under $1.02	6
1.02 and under 1.04	26
1.04 and under 1.06	52
1.06 and under 1.08	58
1.08 and under 1.10	39
1.10 and under 1.12	15
1.12 and under 1.14	3
1.14 and under 1.16	1
TOTAL	200

(a) Calculate the arithmetic mean of the distribution.
(b) Calculate the median of the distribution.
(c) Is the answer in (a) the same as you would have obtained had you computed the following ratio?

$$\frac{\text{Total dollar cost for the 200 plants}}{\text{Total number of units of output of the 200 plants}}$$

(d) Would you be willing to say that 50% of the units produced cost less than your answer to part (b)? Explain.
(e) Suppose that the last class had read "$1.14 and over." What effect, if any, would this have had on your calculation of the arithmetic mean and of the median? Briefly justify your answers.

SOLUTION

(a)
$$\overline{X} = \frac{\$213.20}{200} = \$1.066$$

(b) Median class = $1.06 and under $1.08

$$\text{Median} = \$1.06 + \left(\frac{100 - 84}{58}\right)(\$.02) = \$1.066$$

(c) No. The ratio in (c) is conceptually a weighted mean of 200 cost per unit output ratios. The mean calculated in (a) is unweighted. It is analogous to adding up 200 cost per unit output ratios and dividing by 200 plants.
(d) No. We have no way of telling how many units each plant produces. For example, the six plants whose costs are under $1.02 per unit might produce (say) 60 percent of the total number of units in the industry. However, we can say that the grouped data procedure indicates that 50% of the *plants* had a per unit cost less than the median figure of $1.066.
(e) The open-end interval would have no effect on the median, which is concerned only with the order of the frequencies. The mean, however, which is concerned with the actual values, would have to be recalculated. A mean value for the items in the open interval would have to be assumed and multiplied by the frequency to obtain an estimate of the *fX* value for that class.

EXAMPLE 3-2

What is the proper average(s), if any, to use in the following situations? Briefly justify your answers.

    (a) In determining which members of a class are in the upper half with respect to their overall grade averages.

    (b) In determining the average death rate for six cities combined.

    (c) In a profit-sharing plan in which a firm wishes to find the average amount each worker is to receive to ensure equal distribution.

    (d) In a frequency distribution in which the first class has no lower limit and you are unwilling to estimate the midpoint of this class.

    (e) In determining how high to make a bridge (not a drawbridge). The distribution of the heights of boats expected to pass under the bridge is known and is skewed to the left.

    (f) In determining a typical wage figure for use in later arbitration for a company that employs 100 men, several of whom are highly paid specialists.

    (g) In determining the average annual percentage rate of net profit to sales of a company over a ten-year period.

SOLUTION

    (a) The median, since (generally) one-half of the grade averages will fall above and below this figure.

    (b) The weighted arithmetic mean. The death rates of the six cities would be weighted by the population figures of these cities.

    (c) The arithmetic mean. Divide the total profits to be shared (ΣX) by the number of workers (n).

    (d) The median. Of course, this assumes that the number of observations in the first class does not exceed one-half the total frequencies. However, such a frequency distribution would indeed be unusual, if not, inappropriate.

    (e) None of the averages would be appropriate, since they would result in a large number of boats being unable to pass under the bridge. The bridge should be high enough to allow all the expected traffic to pass under it.

    (f) The median. Because of the tendency of the arithmetic mean to be distorted by a few extreme items, it would tend to be a less typical figure.

    (g) The weighted arithmetic mean. The profit rates would be weighted by the sales figures for each year.

EXERCISES

    1. The number of resignations received by a certain firm per month during 1973 was

$$8, 3, 5, 3, 4, 3, 1, 0, 3, 4, 0, 7$$

Calculate and interpret the arithmetic mean, mode, and median.

    2. Compute the mean and median of both the raw data and the frequency distribution in Exercise 2 on page 45.

3. Explain or criticize the following statement.

The frequency distributions of family income, size of business, and wages of skilled employees all tend to be skewed to the right.

4. The following data represent chainstore prices paid by farmers for two products during a certain month:

COMPOSITION ROOFING PRICE RANGE DOLLARS/90-POUND ROLL	NUMBER OF REPORTS	DOUGLAS AND INLAND FIRS, 2 × 4'S STANDARD OR BETTER PRICE RANGE DOLLARS/THOUSAND POUND-FOOT	NUMBER OF REPORTS
$2.35 and under $2.75	1	76–85	2
2.75 and under 3.15	6	86–95	4
3.15 and under 3.55	33	96–105	8
3.55 and under 3.95	51	106–115	7
3.95 and under 4.35	121	116–125	28
4.35 and under 4.75	50	126–135	48
4.75 and under 5.15	44	136–145	57
5.15 and under 5.55	13	146–155	71
5.55 and under 5.95	5	156–165	45
		166–175	20
		176–185	1

(a) Graph the frequency distribution for each product.
(b) Indicate the median and modal class for each distribution.
(c) Calculate and graph the cumulative distribution for each frequency distribution.
(d) Compare the skewness of the two distributions.

5. The following table presents a frequency distribution of the gross income of males in a certain city in 1973.

INCOME	NUMBER OF MALES
$1999 or less	7035
2000–3999	6642
4000–5999	6823
6000–7999	8775
8000–9999	10,642
10,000–14,999	9410
15,000–19,999	5312
20,000–24,999	5165
25,000 and over	2542

(a) Why do you think open-ended intervals are used in this distribution?
(b) Which is the modal class?
(c) Which is the median class?

6. The following are the earnings for all employees of Company A for a certain week:

EARNINGS FOR WEEK	NUMBER OF EMPLOYEES WITH GIVEN EARNINGS
$ 87.50 and under $ 95.00	2
95.00 and under 102.50	7
102.50 and under 110.00	9
110.00 and under 117.50	14
117.50 and under 125.00	10
125.00 and under 132.50	6
132.50 and under 140.00	2
	50

(a) Compute the arithmetic mean of the distribution.
(b) Is the answer in (a) the same as you would have obtained had you calculated the following ratio?

$$\frac{\text{Total earnings of all employees of Company A for the week}}{\text{Total number of employees of Company A for the week}}$$

Why or why not?
(c) Compute the median of the distribution.
(d) In which direction are these data skewed?
(e) The comptroller of the company stated that the total payroll for the week was $5675.18. Do you have any reason to doubt this statement? Support your position very briefly.
(f) Would you say that the arithmetic mean that you computed in (a) provides a satisfactory description of the typical earnings of these 50 employees for the given week? Why or why not?

3.13 DISPERSION—DISTANCE MEASURES

Central tendency, as measured by the various averages already discussed, is an important descriptive characteristic of statistical data. However, although two sets of data may have similar averages, they may differ considerably with respect to the spread or dispersion of the individual observations. Measures of dispersion describe this variation in numerical observations.

There are two types of measures of dispersion. The first, which may be referred to as *distance measures,* describes the spread of data in terms of the distance between the values of selected observations. The very simplest of such measures is the range or the difference between the values of the highest and lowest items. For example, if the loans extended by a bank to five corporate customers are $82,000, $125,000, $140,000, $212,000, and $245,000, the range of these loans is $245,000 − $82,000 = $163,000. Such a measure of dispersion may be useful for obtaining a rough notion of the spread in a set of data, but it is certainly inadequate for most analytical purposes. A disadvantage of the range is that it describes dispersion in terms of only two selected values in a set of observations. Thus, it ignores the nature of the variation among all other observations. Furthermore, the two numbers used, the highest and the lowest, are extreme rather than typical values.

Other distance measures of dispersion employ more typical values. For example, the *interquartile range* is the difference between the *third quartile* and the *first quartile* values. The third quartile is a figure such that three-quarters of the observations lie below it; the first quartile is a figure such that one-quarter of the observations lie below it. Thus, the distance between these two numbers measures the spread between the values that bound the middle 50% of the values in a distribution. However, a main disadvantage of such a measure is again that it does not describe the variation *among* the items between the middle 50% (nor among the lower and upper one-fourth of the values). The method of calculating quartile values will not be explicitly discussed here, but for frequency distribution data, its calculation proceeds in a manner completely analogous to that of the median, which is itself, the *second quartile* value. That is, two-quarters of the observations in a distribution lie below the median, and two-quarters lie above it. Quartiles are special cases of general measures known as *fractiles,* which refer to values that exceed specified fractions of the data. Thus, the ninth decile exceeds 9/10 of the items, the ninety-ninth percentile exceeds 99/100 of the items in a distribution, etc. Clearly, many arbitrary distance measures of dispersion could be developed, but they are infrequently used in practical applications.

DISPERSION—AVERAGE DEVIATION METHODS   3.14

The most comprehensive descriptions of dispersion are those in terms of the *average deviation* from some measure of central tendency. The most important of such measures are the variance and the standard deviation.

The variance of the observations in a population, denoted σ^2 (Greek lower case "sigma"), is the arithmetic mean of the squared deviations from the population mean. In symbols, if $X_1, X_2, \ldots, X_N$ represent the values of the N observations in a population, and μ is the arithmetic mean of these values, the population variance is defined by

(3.6)
$$\sigma^2 = \frac{(X_1 - \mu)^2 + (X_2 - \mu)^2 + \cdots + (X_N - \mu)^2}{N}$$
$$= \frac{\Sigma(X - \mu)^2}{N}$$

As usual in the simplified form on the right-hand side of formula (3.5), subscripts have been dropped.

Although the variance measures the extent of variation in the values of a set of observations, it is in units of squared deviations or squares of the original numbers. In order to obtain a measure of *dispersion* in terms of the units of the original data, the square root of the variance is taken. The resulting measure is known as the standard deviation. Thus, the *standard deviation* of a population is given by

(3.7)
$$\sigma = \sqrt{\frac{\Sigma(X - \mu)^2}{N}}$$

By convention, the positive square root is used. The standard deviation is a measure of the spread in a set of observations. If all the values in a population were identical, each deviation from the mean would be zero. In such a completely uniform distribution, the standard deviation would be equal to zero, its minimum value. On the other hand, as items are dispersed more and more widely from the mean, the standard deviation becomes larger and larger.

If we now consider the corresponding measures for a sample of n observations, it would seem logical to substitute the sample mean $\overline{X}$ for the population mean μ and the sample number of observations n for the population number N in formulas (3.6) and (3.7). However, it can be shown that when the sample variance and standard deviation are defined with $n - 1$ divisors better estimates are obtained of the corresponding population parameters. Hence, in keeping with modern usage, we define the sample variance and sample standard deviation, respectively, as in formulas (3.8) and (3.9).

SAMPLE VARIANCE

(3.8)
$$s^2 = \frac{\Sigma(X - \overline{X})^2}{n - 1}$$

and

SAMPLE STANDARD DEVIATION

(3.9)
$$s = \sqrt{\frac{\Sigma(X - \overline{X})^2}{n - 1}}$$

The term s^2 is usually referred to verbally as simply the "sample variance" or by the longer term "estimator of the population variance," and similarly for the sample standard deviation.

A brief justification will be given here for the division by $n - 1$. It can be shown mathematically that for an infinite population when the sample variance is defined with an $n - 1$ divisor as in formula (3.8) it is a so-called "unbiased estimator" of the population parameter, σ^2. This means that if all possible samples of size n were drawn from a given population and the variances of these samples were averaged (arithmetic mean), this average would be equal to the population variance, σ^2. Thus, when the sample variance is defined with an $n - 1$ divisor, on the average, it correctly estimates the population variance. Of course, for large samples, the difference in results obtained by using an n rather than $n - 1$ divisor would tend to be very slight, but for small samples the difference can be rather substantial.

We will now concentrate on methods of computing the standard deviation, both for ungrouped and grouped data. Then we discuss how this measure of dispersion is used. However, the major uses of the standard deviation are in connection with sampling theory and statistical inference, which are discussed in subsequent chapters.

The same illustrative accounts receivable data used in the case of the arithmetic mean will be employed in the calculation of the standard deviation.

In Section 3.8, the following ungrouped data were given as the accounts receivable established by an accounting department during a one-hour period: $600, $350, $275, $430, and $520. The arithmetic mean was $435. The calculation of the standard deviation using the defining formula (3.9) is illustrated in Table 3-7.

The resulting standard deviation of $129.71 is an absolute measure of dispersion, which means that it is stated in the units of the original data. Whether this is a great deal or only a small amount of dispersion cannot be immediately determined. This sort of judgment is based on the particular type of data analyzed, for example, in this case, accounts receivable data. Furthermore, as we shall see in Section 3.15, relative measures of dispersion are preferable to absolute measures for comparative purposes.

TABLE 3-7

Calculation of the Standard Deviation for Ungrouped Data
by the Direct Method: Accounts Receivable Data

AMOUNT OF ACCOUNTS RECEIVABLE X	DEVIATION FROM MEAN $X - \overline{X}$	SQUARED DEVIATIONS $(X - \overline{X})^2$
$600	$600 $-$ $435 = $165	27,225
350	350 $-$ 435 = $-$85	7225
275	275 $-$ 435 = $-$160	25,600
430	430 $-$ 435 = $-$5	25
520	520 $-$ 435 = 85	7225
$\overline{X}$ = $435	$0	67,300

$$s = \sqrt{\frac{\Sigma(X - \overline{X})^2}{n - 1}} = \sqrt{\frac{67,300}{4}} = \sqrt{16,825} = \$129.71$$

In the calculation of the standard deviation for data grouped into frequency distributions, it is merely necessary to adjust the foregoing formulas to take account of this grouping. The defining formula (3.9) generalizes to

(3.10)
$$s = \sqrt{\frac{\Sigma f (X - \overline{X})^2}{n - 1}}$$

where, as usual for grouped data, X represents the midpoint of a class, f the frequency in a class, $\overline{X}$ the arithmetic mean, and n the total number of observations. This calculation is illustrated in Table 3-8 for the frequency distribution of weekly earnings previously given in Tables 3-3 and 3-5. As shown in Table 3-8, the standard deviation is equal to $14.70.

Calculation of the sample standard deviation by the direct definitional formula can be tedious, particularly if the class midpoints and frequencies contain several digits and the arithmetic mean is not a round number. A shortcut method of calculation often useful in such cases is given in Appendix C.

TABLE 3-8

Calculation of the Standard Deviation for Grouped
Data by the Direct Method: Weekly Earnings Data.

WEEKLY EARNINGS	NUMBER OF EMPLOYEES f	MIDPOINTS X	DEVIATION $X - \bar{X}$	$(X - \bar{X})^2$	$f(X - \bar{X})^2$
$160.00 and under 170.00	4	$165	−29.80	888.04	3552.16
170.00 and under 180.00	14	175	−19.80	392.04	5488.56
180.00 and under 190.00	18	185	−9.80	96.04	1728.72
190.00 and under 200.00	28	195	0.20	.04	1.12
200.00 and under 210.00	20	205	10.20	104.04	2080.80
210.00 and under 220.00	12	215	20.20	408.04	4896.48
220.00 and under 230.00	4	225	30.20	912.04	3648.16
	100				21,396.00

$\bar{X} = 194.80$

$$s = \sqrt{\frac{\Sigma f(X - \bar{X})^2}{n - 1}} = \sqrt{\frac{21396}{99}} = \sqrt{216.1212}$$

$s = \$14.70$

Uses of the standard deviation

The standard deviation of a frequency distribution is very useful in describing the general characteristics of the data. For example, in the so-called "normal distribution" (a bell-shaped curve), which is discussed extensively in Chapters 5, 6, and 7, the standard deviation is used in conjunction with the mean to indicate the percentage of items that fall within specified ranges. Hence, if a population is in the form of a normal distribution the following relationships apply:

$\mu \pm \sigma$ includes 68.3% of all of the items
$\mu \pm 2\sigma$ includes 95.5% of all of the items
$\mu \pm 3\sigma$ includes 99.7% of all of the items

For example, if a production process is known to produce items that have a mean length of $\mu = 10$ inches and a standard deviation of 1 inch, then we can infer that 68.3% of the items have lengths between $10 - 1 = 9$ inches and $10 + 1 = 11$ inches. About 95.5% have lengths between $10 - 2 = 8$ and $10 + 2 = 12$ inches, and 99.7% have lengths between $10 - 3 = 7$ and $10 + 3 = 13$ inches. Thus, a range of $\mu \pm 3\sigma$ includes virtually all the items in a normal distribution. As discussed in Chapter 4, the normal distribution is perfectly symmetrical. If the departure from a symmetrical distribution is not too great, the rough generalization that virtually all the items are included within a range from 3σ below the mean to 3σ above the mean still holds.

The standard deviation is also useful in describing how far individual items in a distribution depart from the mean of the distribution. Suppose the population of students who took a certain aptitude test displayed a mean score of $\mu = 100$ with a standard deviation of $\sigma = 20$. If a certain student scored 80 on the examination, his score can be described as lying one standard deviation below the mean. The terminology usually employed is that his *standard score* is -1, that is, if his examination score is denoted X, then

$$\frac{X - \mu}{\sigma} = \frac{80 - 100}{20} = -1$$

The standard score of an observation is simply the number of standard deviations the observation lies below or above the mean of the distribution. Hence, in the example, the student's score deviates from the mean by -20 units, which is equal to -1 in terms of units of standard deviations away from the mean. If standard scores are computed from sample rather than universe data, the formula $(X - \overline{X})/s$ would be used instead.

Comparisons can thus be made for items in distributions that differ in order of magnitude or in the units employed. For example, if a student scored 120 on an examination in which the mean was $\mu = 150$ and $\sigma = 30$, his standard score would be $(120 - 150)/30 = -1$. This standard score would place the 120 at the same number of standard deviations below the mean as the 80 in the preceding example. Analogously, we could compare standard scores in a distribution of wages with comparable figures in a distribution of length of employment service, etc.

The standard deviation is doubtless the most widely used measure of dispersion, and considerable use is made of it in later chapters of this text.

RELATIVE DISPERSION – COEFFICIENT OF VARIATION   3.15

Although the standard score discussed earlier is useful for determining how far an *item* lies from the mean of a set of data, we often are interested in comparing the dispersion of *an entire set of data* with the dispersion of another set. As observed earlier, the standard deviation is an absolute measure of dispersion, whereas a relative measure is required for comparative purposes. This is essential whenever the sets of data to be compared are expressed in different units, or even when the data are in the same units but are of different orders of magnitude. Such a relative measure is obtained by expressing the standard deviation as a percentage of the arithmetic mean. The resulting figure, referred to as the coefficient of variation CV, is defined symbolically in Equation (3.11).

(3.11)
$$CV = \frac{s}{\overline{X}}$$

Thus, for the frequency distribution of weekly earnings data of semi-skilled workers the standard deviation is $14.70 with a mean of $194.80. The coefficient of variation is

$$CV = \frac{\$14.70}{\$194.80} = 7.5\%$$

Let us assume that the corresponding figures for the earnings of a group of highly skilled workers revealed a standard deviation of $18 with an arithmetic mean of $300. The coefficient of variation for this set of data is $CV = \$18/300 = 6\%$. Therefore, the earnings of the highly skilled group were relatively more uniform, or stated differently, displayed relatively less variation than did the earnings of the semi-skilled group. It may be noted that the earnings of the highly skilled group had the larger standard deviation, but because of the higher average weekly earnings, relative dispersion was less.

Both absolute and relative measures of dispersion are widely used in practical sampling problems. To give just one example, a question frequently arises about the sample size required to yield an estimate of a universe parameter with a specified degree of precision. For example, a finance company may want to know how large a random sample of its loans it must study in order to estimate the average dollar size of all its loans. If the company wants this estimate within a specified number of *dollars,* an absolute measure of dispersion is appropriate. On the other hand, if the company wants the estimate to be within a specified *percentage* of the true average figure, a relative measure of dispersion would be used.

3.16   ERRORS OF PREDICTIONS

In this chapter, we have discussed descriptive measures for empirical frequency distributions, with emphasis on measures of central tendency and dispersion. Some interesting relationships between these two types of measures are observable when certain problems of prediction are considered. Suppose we want to guess or "predict" the value of an observation picked at random from a frequency distribution. Let us refer to the penalty of an incorrect prediction as the "cost of error." If there were a *fixed* cost of error on each prediction, no matter what the size of the error, we should guess the mode as the value of the random observation. This would give us the highest probability of guessing the *exact value* of the unknown observation. Assuming repeated trials of this prediction experiment we would thus minimize the average (arithmetic mean) cost of error.

Suppose, on the other hand, the cost of error varies directly with the size of error regardless of its sign, that is, whether the actual observation is above or below the predicted value. In this case, we would want a prediction that minimizes the average *absolute error.* The median would be the "best guess," since it minimizes average absolute deviations. The mean deviation about the median would be a measure of this minimum cost of error.

Finally, suppose the cost of error varies according to the square of the error. For example, an error of two units costs four times as much as an error of one unit. In this situation, the mean should be the predicted value, since the average of the squared deviations about it is less than around any other figure. Here the variance, which may be interpreted as the average cost of error per observation, would represent a measure of this minimum error. Another point previously observed for the mean is that the average amount of error, taking account of sign, would be zero.

A practical business application of these ideas is in the determination of the optimum size of inventory to be maintained. Let us assume a situation in which the cost of overstocking a unit (cost of overage) is equal to the cost of being short one unit (cost of underage). Further, it may be assumed that the cost of error varies directly with the absolute amount of error. For example, having two units in excess of demand costs twice as much as one unit. In this situation, the optimum stocking level is the median of the frequency distribution of numbers of units demanded.

PROBLEMS OF INTERPRETATION   3.17

Many of the most common misinterpretations and misuses of statistics involve measures and concepts such as those discussed in this chapter: averages, dispersion, and skewness. Sometimes misleading interpretations are drawn from the use of averages that are not "typical" or "representative." Reference was made in Section 3.11 to the possible distortion of the typicality of the arithmetic mean because of the presence of extreme items at one end of a distribution. An interesting example of this distortion effect occurred in the case of a survey conducted by a popular periodical. One of the purposes of the survey was to determine the current status of persons who had graduated from college during the early Depression years. Among those included were the graduates of Princeton University for three successive years during the early 1930's. The results of the Princeton survey indicated that the arithmetic mean income of the respondents in the class that graduated in the second year was far higher than the corresponding mean income for the first- and third-year classes. The analysts attempted to rationalize this result in various ways. However, a re-examination of the data yielded a very simple explanation, which precluded potential misinterpretations. It turned out that one of the graduates of the second-year class was a member of one of the wealthiest families in the United States and was an heir to an immense fortune. His very large income exerted an obvious upward pull on the mean income of his class, making it an unrepresentative average.

Misinterpretations of averages often arise because of a neglect to take dispersion into account. Prospective college students are sometimes discouraged when they observe the mean scholastic aptitude test scores of classes admitted to colleges or universities in which they are interested. Admissions officers have commented that students sometimes erroneously assume they will not be admitted to a school because their test scores are somewhat below the published mean scores for that school. Of course, such students fail to take into account dispersion around this average. Assuming a roughly symmetrical distribution, about one-half of the admitted students on whom the published means were based had test scores that fell below that average.

Because of the shape of the underlying frequency distribution, sometimes no average will be typical. In Section 3.12, reference was made to bimodal distributions of wages of workers. Arithmetic means or medians for such distributions tend to fall somewhere between the two modes. Hence, they are not typical of the groups characterized by either of the modes. Of course, as indicated in Section 3.12, the solution when non-homogeneous data are present is to separate the distinct distributions. However, sometimes U-shaped frequency distributions are encountered where the separation into different distributions is not warranted. In such distributions frequencies are concentrated at both low and high values of the variable under consideration. For example, suppose the test scores of a mathematics class yield grades that are either very high, say in the 90's, or very low, say in the 60's. Means or medians, which might be about 75, would clearly be unrepresentative of the concentrations at either end of the distribution. When averages are presented for such distributions, without some indication of the nature of the underlying data, it is quite easy for misinterpretations to occur.

EXERCISES

1. The closing prices of two common stocks traded on the American Stock Exchange for a week in September 1973 were

	HIGHFLY	STABIL
Monday	$28	$28
Tuesday	34	26
Wednesday	18	22
Thursday	20	24
Friday	25	25

(a) Compare the two stocks simply on the basis of measures of central tendency.
(b) Compute the standard deviation for each of the two stocks. What information do the standard deviations give concerning the price movements of the two stocks?

2. The standard deviation of sales for the past five years for Company A was $1405 and for Company B $18,580. Can we conclude that sales are more stable for Company A than B?

3. The following is a distribution of the weights of 100 draftees who reported to an army camp for basic training.

WEIGHTS (IN POUNDS)	FREQUENCY
120 and under 140	14
140 and under 160	20
160 and under 180	36
180 and under 200	18
200 and under 220	8
220 and under 240	4
	100

(a) Compute the standard deviation and the arithmetic mean of the distribution of weights.
(b) Can one compare the variability in the weights of the draftees with the variability in their heights? If "yes," how? If "no," why not?

4. The following is a distribution of lifetimes, in hours, of 100 vacuum tubes.

LIFETIME (HOURS)	FREQUENCY
100 and under 200	12
200 and under 300	28
300 and under 400	20
400 and under 500	18
500 and under 600	14
600 and under 700	8
	100

(a) Compute the mean and coefficient of variation for the distribution.
(b) Compute and *interpret* the median.
(c) Suppose that the last class had read "600 and over." What effect, if any, would this have had on your calculations in (a) and (b)?

4 *Introduction to Probability*

4.1 THE MEANING OF PROBABILITY

The development of a mathematical theory of probability began during the seventeenth century when the French nobleman Antoine Gombauld, known as the Chevalier de Méré, raised certain questions about games of chance. Specifically, he was puzzled about the probability of obtaining two sixes at least once in twenty-four rolls of a pair of dice. (This is a problem you should have little difficulty solving after reading this chapter.) De Méré posed the question to Blaise Pascal, a young French mathematician, who solved the problem. Subsequently, Pascal discussed this and other puzzlers raised by de Méré with the famous French mathematician, Pierre de Fermat. In the course of their correspondence, the mathematical theory of probability was born.

The several different methods of measuring probabilities represent different conceptual approaches and reveal some of the current intellectual controversy concerning the foundations of probability theory. In this chapter we discuss three conceptual approaches: *classical* probability, *relative frequency of occurrence,* and *subjective* probability. Regardless of the definition of probability used, the same mathematical rules apply in performing the calculations (that is, measures of probability are always added or multiplied under the same general circumstances).

Classical probability

Since probability theory had its origin in games of chance, it is not surprising that the first method developed for measuring probabilities was particularly appropriate for gambling situations. A so-called *classical* concept of probability defines the probability of an event as follows: If there are *a* possible outcomes favorable to the occurrence of an event *A*, and *b* possible outcomes unfavorable to the occurrence of *A*, and all are equally likely and mutually exclusive, then the probability that *A* will occur, denoted *P(A)*, is

$$P(A) = \frac{a}{a+b} = \frac{\text{number of outcomes favorable to occurrence of } A}{\text{total number of possible outcomes}}$$

Thus, if a fair coin with two faces, denoted head and tail, is tossed into the air in such a way that it spins end over end many times, the probability that it will fall with the head uppermost is $P(\text{Head}) = 1/(1 + 1) = \frac{1}{2}$. In this case, there is one outcome favorable to the occurrence of the event "head" and one outcome unfavorable. The extremely unlikely situation that the coin will stand on end is defined out of the problem; that is, it is not classified as an outcome for the purpose of the probability calculation.

The equation above can also be used to determine the probability that a certain face will show when a true die is rolled. A die is a small cube with a number of dots on each of its faces denoting 1, 2, 3, 4, 5, or 6, respectively. A "true" die is uniform in shape and density and is therefore equally likely to show any of the six numbers on its uppermost face when rolled. The probability of obtaining a "one" if such a die is rolled is $P(1) = 1/(1 + 5) = \frac{1}{6}$. Here, there is one outcome favorable to the event "one" and five unfavorable. Untrue dice, which are not uniform in density, are said to be "loaded"; such dice are outside the scope of the examples discussed in this text, and we hope they will remain outside your experience as well.

Some of the terms used in the classical concept of probability require further explanation. The *event* whose probability is sought consists of one or more possible outcomes of the given activity of tossing a coin, rolling a die, or drawing a card. These activities are known in modern terminology as *experiments,* a term referring to processes that result in different possible outcomes or observations. The term *equally likely* in referring to possible outcomes is undefined and is considered an intuitive foundation concept. Two or more outcomes are said to be *mutually exclusive* if when one of the outcomes occurs, the others cannot. Thus, the appearances of a "one" and a "two" on a die are mutually exclusive events, since if a "one" results, a "two" cannot. The results of an experiment are conceived of as a complete or exhaustive set of mutually exclusive outcomes.

These classical probability measures have two very interesting characteristics. First, the objects referred to as *fair* coins, *true* dice, or *fair* decks of cards are abstractions in the sense that no real world object exactly possesses the features postulated. For example, in order to be a *fair* coin, thus equally likely to fall "head" or "tail," the object would have to be a perfectly flat, homogeneous disk—an unlikely object. Second, in order to determine the probabilities in the above examples, no coins had to be tossed, no dice rolled, nor cards shuffled. That is, no experimental data had to be collected; the probability calculations were based entirely upon logical prior reasoning.

In the context of this definition of probability, if it is *impossible* for an event *A* to occur, the probability of that event is said to be zero. For example, if the event *A* is the appearance of a seven when a single die is rolled, then $P(A) = 0$. A probability of one is assigned to an event that is *certain* to occur. Thus, if the event *A* pertains to the appearance of any one of the numbers 1, 2, 3, 4, 5, or 6 on a single roll of a die, then $P(A) = 1$. According to the classical definition, as well as all others, the probability of an event is a number lying between zero and one, and the sum of the probabilities that the event will occur and that it will not occur is equal to one.

Relative frequency of occurrence

Although the classical concept of probability is useful for solving problems involving games of chance, serious difficulties occur with a wide range of other types of problems. For example, it is inadequate for answering such questions as: What are the probabilities that (a) a black male American, aged 30, will die within the next year; (b) a consumer in a certain metropolitan area will purchase a company's product during the next month; (c) a production process used by a particular firm will produce a defective item? In none of these situations is it feasible to establish a set of complete and mutually exclusive outcomes, each of which is equally likely to occur. For example, in (a), only two occurrences are possible: the individual will either live or die during the ensuing year. The likelihood that he will die is, of course, much smaller than the likelihood that he will live. How much smaller? This type of question requires reference to data. The probability that a 30-year-old black male American will live through the next year is greater than the corresponding probability that a 30-year-old, black male inhabitant of India will survive the year. However, how much greater is it and precisely what do these probabilities mean?

We know that the life insurance industry establishes mortality rates by observing how many of a sample, of, say, 100,000 black American males, aged 30, die within a one-year period. In this instance, the number of deaths divided by 100,000 is the *relative frequency of occurrence* of death for the 100,000 individuals studied. It may also be viewed as an estimate of the *probability* of death for Americans in the given color-sex-age group. This relative frequency of occurrence concept can also be illustrated by a simple coin tossing example.

Suppose you are asked to toss a coin known to be biased, that is, it is not a fair coin. You are not told whether it is more likely that a head or a tail will be obtained if the coin is tossed. However, you are asked to determine the probability of the appearance of a head by means of many tosses of the coin. Assume that 10,000 tosses of the coin result in 7000 heads and 3000 tails. Another way of stating the results is that the relative frequency of occurrence of heads is 7000/10,000 or 0.70. It certainly seems reasonable to assign a probability of 0.70 to the appearance of a head with this particular coin. On the other hand, if the coin had been tossed only three times and one head resulted, you would have little confidence in assigning a probability of $\frac{1}{3}$ to the occurrence of a head.

As a working definition, probability in the relative frequency approach may be interpreted as the *proportion of times that an event occurs in the long run under uniform or stable conditions.* As a practical matter, past relative frequencies of occurrence are often used as probabilities. Hence, in the mortality illustration, if 800 of the 100,000 individuals of the given group died during the year, the relative frequency of death or probability of death is said to be 800/100,000 for individuals in the color-sex-nationality-age group.

Subjective probability

The *subjective or personalistic concept of probability* is a relatively recent development.[1] Its application to statistical problems has occurred virtually entirely in the post World War II period. According to this concept, the probability of an event is the *degree of belief or degree of confidence placed in the occurrence of an event by a particular individual based on the evidence available to him.* This evidence may consist of relative frequency of occurrence data and any other quantitative or nonquantitative information. If the individual believes it is unlikely an event will occur, he assigns a probability close to zero to its occurrence; if he believes it is very likely an event will occur, he assigns it a probability close to one. Thus, for example, in a consumer survey, an individual may assign a probability of $\frac{1}{2}$ to the event that he will purchase an automobile during the next year. An industrial purchaser may assert a probability of $\frac{4}{5}$ that a future incoming shipment will have 2% or less defective items.

Subjective probabilities should be assigned on the basis of all objective and subjective evidence currently available. These probabilities should reflect the decision maker's current degree of belief. Reasonable persons might arrive at different probability assessments because of differences in experience, attitudes, values, etc. Furthermore, in general, these probability assignments may be made for events that will occur only once, in situations where neither classical probabilities nor relative frequencies appear to be appropriate.

This approach is thus a very broad and flexible one, permitting probability assignments to events for which there may be no objective data, or for which there may be a combination of objective and subjective data. These events may occur only once and may lie entirely in the future. However, the assignments of these probabilities must be consistent. For example, if the purchaser assigns a probability of $\frac{4}{5}$ to the event that a shipment will have 2% or less defective items, then a probability of $\frac{1}{5}$ must be assigned to the event that a shipment will have more than 2% defective items. In this book we accept the concept of subjective or personal probability as a reasonable and useful one, particularly in the context of situations in business decision making.

[1] The concept was first introduced in 1926 by Frank Ramsey who presented a formal theory of personal probability in F. P. Ramsey, *The Foundation of Mathematics and Other Logical Essays* (London: Kegan Paul; New York: Harcourt Brace Jovanovich, 1931). The theory was developed primarily by de Finetti, Koopman, I. J. Good, and L. J. Savage.

Sample spaces and experiments

The concept of an *experiment* is a central one in probability and statistics. In this connection, an experiment simply refers to any process of measurement or observation of different outcomes. The collection or totality of the possible outcomes of an experiment is referred to as its *sample space.* The experiment may be real or conceptual. Thus, the collections of outcomes of the experiments of tossing a coin once, twice, or any number of times are all sample spaces. The objects that comprise the sample space are referred to as *elements* of the sample space. The elements are usually enclosed within braces, and the symbol S is conventionally used to denote a sample space.

In the illustration involving the single toss of a coin, there are two possible outcomes (tail $= T$, head $= H$) and, thus,

$$S = \{T, H\}$$

If the coin is tossed twice, there are four possibilities, and

$$S = \{(T, T), (T, H), (H, T), (H, H)\}$$

In these examples, a physical experiment may actually be performed, or we may easily conceive of the possibility of such an experiment. In the first case, the experiment consists of *one trial,* a single toss of the coin; in the second case, the experiment contains *two trials,* the two tosses of the coin.

In other situations, although no sequence of repetitive trials is involved, we may conceive of a set of outcomes as an experiment. These outcomes may simply be the result of an observational process and need not bear any resemblance to a laboratory experiment. It suffices that the outcomes be well defined. Thus, we may think of each of the following two-way classifications as constituting sample spaces:

0	1
Customer was granted credit.	Customer was not granted credit.
Employee elected a stock purchase plan.	Employee did not elect a stock purchase plan.
The merger will take place.	The merger will not take place.
The company uses direct mail advertising.	The company does not use direct mail advertising.

The elements in these two-element sample spaces may be designated zero and one, respectively, as indicated by the column headings. Therefore, each of the four illustrative sample spaces may be conveniently symbolized as

$$S = \{0, 1\}$$

There are at least two methods of graphically depicting sample spaces: (1) graphs using the conventional rectangular coordinate system and (2) tree diagrams. These methods are most feasible for sample spaces with relatively small numbers of sample points. Tree diagrams are more manageable because of the obvious graphic difficulties encountered by the coordinate system method beyond three dimensions. These methods are illustrated in Examples 4-1 and 4-2.

EXAMPLE 4-1

Depict graphically the sample space generated by the experiment of tossing a coin twice.

SOLUTION

The sample space was earlier designated as

$$S = \{(T, T), (T, H), (H, T), (H, H)\}$$

Let $T = 0$ and $H = 1$. The sample space may now be written

$$S = \{(0, 0), (0, 1), (1, 0), (1, 1)\}$$

and graphed in two-dimensional coordinates as shown in Figure 4-1. A tree diagram for this sample space is shown in Figure 4-2.

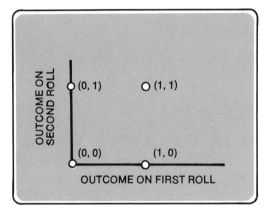

Figure 4-1
Graph for coin-tossing experiment.

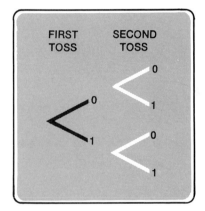

Figure 4-2
Tree diagram for coin-tossing experiment.

EXAMPLE 4-2

A market research firm studied the differences among consumers by income groups in a certain city in terms of their purchase of or failure to purchase a given product during a one-month period. The income groups used were low, middle, and high. Consumers were classified as (a) failed to purchase or (b) purchased the product at least once. Show this situation graphically.

SOLUTION

Let 0 stand for "low" income, 1 for "middle" income, and 2 for "high" income. For purchase behavior let 0 stand for "failed to purchase" and 1 stand for "purchased at least once." The sample space may be expressed as

$$S = \{(0, 0), (1, 0), (2, 0), (0, 1), (1, 1), (2, 1)\}$$

The graph is given in Figure 4-3, and the tree diagram in Figure 4-4. To illustrate

THE MEANING OF PROBABILITY   **77**

how a tree diagram is read, let us consider the element (0, 1). This element is depicted in the tree by starting at the left-hand side and following the uppermost branch to the "0," then continuing from this fork down the branch leading to a "1." This (0, 1) element denotes a consumer classified as "low" for income and "purchased at least once" for buying behavior.

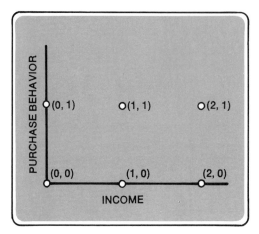

Figure 4-3
Graph for classification of consumers by income and purchase behavior.

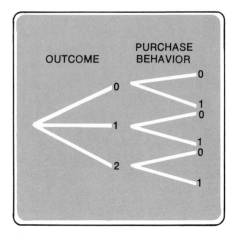

Figure 4-4
Tree diagram for classification of consumers by income and purchase behavior.

Events

The term "event" as used in ordinary conversation usually has no ambiguity about its meaning. However, since the concept of an event is fundamental to probability theory, it requires explicit definition. Once a sample space S has been specified, an event may be defined as a collection of elements each of which is also an element of S. An *elementary event* is a single possible outcome of an experimental trial. It is thus an event that cannot be further subdivided into a combination of other events.

For example, a single roll of a die constitutes an experiment. The sample space generated is

$$S = \{1, 2, 3, 4, 5, 6\}$$

We may define an event E, say, as the "appearance of a 2 or a 3," which may be expressed as

$$E = \{2, 3\}$$

If either a 2 or a 3 appears on the upper face of the die when it is rolled, the event E is said to have "occurred." Note that E is not an elementary event, since it can be subdivided into the two elementary events {2} and {3}.

Two events are of particular importance, the *complement* of an event and the *certain event*. The complement of an event A in the sample space S is the collection of elements that are not in A. We will use the symbol $\bar{A}$, read "A bar" for the complement of A. Thus, in the experiment of rolling a die one time, the complement of the event that a 1 appears on the uppermost face is the event that a 2, 3, 4, 5, or 6 appears.

When a sample space *S* has been defined, *S* itself is referred to as the *certain event* or the *sure event,* since in any single trial of the experiment that generates *S*, one or another of its elements must occur. Hence, in the experiment of rolling a die one time the certain event is that either a 1, 2, 3, 4, 5, or 6 appears.

Two events A_1 and A_2 are said to be *mutually exclusive events* if when one of these events occurs, the other cannot. Thus, in the roll of a die one time, the event that a 1 appears is mutually exclusive of the event that a 2 appears. On the other hand, the appearance of a 1 and the appearance of an odd number (1, 3, or 5) are not mutually exclusive events, since 1 is an odd number.

EXERCISES

1. If 0 represents a nonresponse to a mailed questionnaire, and 1 represents a response, depict the set of outcomes representing responses to three out of four questionnaires.

2. A certain manufacturing process produces parachutes. A worker tests each parachute as it is produced and continues testing until he finds one defective. Describe the sample space of possible outcomes for the testing process.

3. An investment company is planning to add two new stocks to its portfolio. Its research group recommends five stocks, Ranox, BIM, Bavo, Park Mining, and Goldflight. Describe the sample space representing the possible choices.

4. An econometric model predicts whether the GNP will increase, decrease, or remain the same the following year. Let *X* represent "the model's prediction, coded" and *Y* the "actual movement of GNP, coded." Graph the possible outcomes of *X* and *Y*.

5. A special electronic part is ordered by a firm in Youngstown, Ohio, from a firm in London. The London firm can fly the part to either New York, Philadelphia, or Chicago. Once it reaches one of these cities, it can be sent by train or truck to Youngstown. Use a tree diagram to find all possible shipping routes.

6. Draw a tree diagram depicting the possible outcomes resulting from flipping a coin three times.

7. Which of the following pairs of events are mutually exclusive?
   (a) (1) Park New common stock closes higher on a given day; (2) Park New common stock closes lower on the same day.
   (b) In a shipment of two relays, (1) exactly one is defective, (2) two are defective.
   (c) Three people, *A, B, C* apply for two job openings. (1) *A* is hired, (2) *B* is hired.
   (d) On two rolls of a die, (1) a six occurs, (2) the sum of the two faces is five.
   (e) On two rolls of a die, (1) a four occurs, (2) the sum of the two faces is six.

Probability and sample spaces

Probabilities may be thought of as numbers assigned to points in a sample space. These probabilities must have the following characteristics:

1. They are numbers equal to or greater than zero.
2. They must add up to one.

EXAMPLE 4-3

Let us consider again the conceptual experiment of rolling a true die one time. The sample space of elementary events is

$$S = \{1, 2, 3, 4, 5, 6\}$$

If probability assignments are made to these sample elements according to the classical definition, we have the following:

EVENTS	PROBABILITY OF EVENTS
A_i	$P(A_i)$
$A_1 : 1$	$\frac{1}{6}$
$A_2 : 2$	$\frac{1}{6}$
$A_3 : 3$	$\frac{1}{6}$
$A_4 : 4$	$\frac{1}{6}$
$A_5 : 5$	$\frac{1}{6}$
$A_6 : 6$	$\frac{1}{6}$
	1

A brief comment about the notation in Example 4-3 is in order. There are six events namely A_1, denoting the appearance of a one, $A_2, \ldots, A_6$, with the corresponding probabilities $P(A_1)$, $P(A_2)$, $\ldots$, $P(A_6)$. The first characteristic given immediately preceding the example is satisfied since each probability is equal to or greater than zero, specifically, each $P(A_i) = \frac{1}{6}$. The second condition is met since the probabilities add up to one. It should be noted that the events in a sample space need not be elementary events. Thus, if the event A_1 refers to the appearance of a 1 or 2, A_2 refers to a 3 or 4, and A_3 refers to a 5 or 6, then $P(A_1) = \frac{1}{3}$, $P(A_2) = \frac{1}{3}$, and $P(A_3) = \frac{1}{3}$. However, the events defining the sample space must be an exhaustive set of mutually exclusive events. It may be noted that the sum of the probabilities of occurrence of any event A and its complement $\bar{A}$ is equal to one, i.e., $P(A) + P(\bar{A}) = 1$.

4.2 ELEMENTARY PROBABILITY RULES

In most applications of probability theory, we are interested in combining probabilities of events that are related in some meaningful way. In this section, we discuss two of the fundamental ways of combining probabilities: *addition* and *multiplication*.

Before considering the combining of probabilities by addition, we will define a new term. The symbol $P(A_1 \text{ and } A_2)$ is used to denote the probability that both events A_1 and A_2 will occur. That is, $P(A \text{ and } B)$ refers to the probability of the joint occurrence of A and B. Hence, for example, if A_1 denotes the occurrence of a 6 on the first roll and A_2 the occurrence of a 6 on the second roll, then $P(A_1$ and $A_2)$ refers to the probability of obtaining a 6 on the first roll *and* a 6 on the second roll.

We now state the general addition rule for any two events A_1 and A_2 in a sample space S.

Addition rule

(4.1) $$P(A_1 \text{ or } A_2) = P(A_1) + P(A_2) - P(A_1 \text{ and } A_2)$$

It follows as a corollary of the addition rule that if A_1 and A_2 are mutually exclusive events, that is, they cannot both occur,

$$P(A_1 \text{ and } A_2) = 0$$

and therefore

(4.2) $$P(A_1 \text{ or } A_2) = P(A_1) + P(A_2)$$

The following is an application of equation (4.2) for the addition of probabilities of mutually exclusive events in terms of Example 4-3. The probability that either a 1 or a 2 will occur on a single roll of a die is $P(A_1 \text{ or } A_2) = P(A_1) + P(A_2) = \frac{1}{6} + \frac{1}{6} = \frac{1}{3}$. In Example (4-4) we consider an application of equation (4.1) for two events that are not mutually exclusive. Also in that example, an explanation is given of the inclusive meaning of the word "or" as used in $P(A_1 \text{ or } A_2)$.

EXAMPLE 4-4

What is the probability of obtaining a six on either the first or second roll of a die, or both? Another way of wording this question is, "What is the probability of obtaining a six at least once in two rolls of a die?"

SOLUTION

Let A_1 denote the appearance of a six on the first roll and A_2 the appearance of a six on the second roll. The question asks for the value of $P(A_1 \text{ or } A_2)$. A comment on this point is appropriate. The word "or" in probability theory is always inclusive in meaning. That is, the symbol $P(A_1 \text{ or } A_2)$ means the probability that either A_1 or A_2 occurs or that they both occur. Hence, the "or" has the meaning of the legal "and/or." For example, in this problem, a success is obtained if either A_1 occurs or A_2 occurs or both A_1 and A_2 occur.

Consider the sample space of equally likely elements listed below in columns. The sample space has 36 elements. The first and second numbers in each element represent the outcomes on the first and second rolls, respectively.

1,1	2,1	3,1	4,1	5,1	6,1
1,2	2,2	3,2	4,2	5,2	6,2
1,3	2,3	3,3	4,3	5,3	6,3
1,4	2,4	3,4	4,4	5,4	6,4
1,5	2,5	3,5	4,5	5,5	6,5
1,6	2,6	3,6	4,6	5,6	6,6

The probability that a six will appear on both the first and second rolls is $\frac{1}{36}$, i.e., $P(A_1 \text{ and } A_2) = \frac{1}{36}$. The probability that a six will appear on the first roll is $P(A_1) = \frac{1}{6}$; and on the second roll, $P(A_2) = \frac{1}{6}$. Hence, applying the addition rule, we have

$$P(A_1 \text{ or } A_2) = P(A_1) + P(A_2) - P(A_1 \text{ and } A_2)$$

$$= \frac{1}{6} + \frac{1}{6} - \frac{1}{36}$$

$$= \frac{11}{36}$$

It may be noted that eleven points in this sample space represent the event "six at least once in two trials." Therefore, the same result could have been obtained by using the classical concept of probability. The ratio of outcomes favorable to the occurrence of "six at least once" to the total number of outcomes is $\frac{11}{36}$.

The addition rule can, of course, be extended to more than two events. The generalization for the case where n events are not mutually exclusive will not be given here. If the n events $A_1, A_2, \ldots, A_n$ are mutually exclusive, then

(4.3) $P(A_1 \text{ or } A_2 \text{ or } \ldots A_n) = P(A_1) + P(A_2) + \cdots + P(A_n)$

The addition rule is applicable whenever we are interested in the probability that any one of several events will occur. On the other hand, in many applications, we are interested in the probability of the *joint occurrence* of two or more events. These events may occur simultaneously or successively. For example, we may be interested in computing the probability of obtaining two kings if two cards are drawn simultaneously from a well-shuffled deck of cards. Or, we may be interested in the probability of the appearance of two kings if two cards are drawn, one after the other from the deck. Just as the probability that any one of several events will occur is obtained by adding up appropriate probabilities, the probability of the joint occurrence of two or more events is obtained by *multiplying* appropriate probabilities.

Conditional probability

In considering joint probabilities, it is necessary to introduce the concept of *conditional probability*. The meaning of any probability depends upon the set of elements to which the discussion is limited, that is, to some sample space S. In that sense, the probability is conditional upon the definition of S. For example, if we are interested in the probability that a manufacturing firm will have 100 employees or more, that figure will depend on whether the sample space consists of the manufacturing firms in an industry, a city, a state, etc. Thus, every probability statement may be viewed as a conditional probability. However, the term "conditional probability" is usually reserved for probability statements about a reduced sample space, within a given sample space S. To illustrate this concept, consider the data shown in Table 4-1, which may be

TABLE 4-1

Ten Persons Classified by Age and Sex: A_1, the Set
of Males; and A_2, the Set of Persons Under 40 Years of Age

SEX	AGE UNDER 40 (A_2)	40 AND OVER	TOTAL
Males (A_1)	2	3	5
Females	3	2	5
TOTAL	5	5	10

thought of as depicting a sample space of ten equally likely elements, classified by age and sex. Let A_1 denote "male" and A_2 "under 40 years of age." If a person is selected at random from this group, the probability that the individual is a male is

$$P(A_1) = \frac{5}{10}$$

On the other hand, suppose we are interested in the probability that an individual is under 40 years of age, given that he is a male. In this case, the relevant sample space is restricted to the number of sample elements lying within the event A_1, "male," and we now want to know the proportion of such points which possess the property "under 40 years of age." This conditional probability is denoted $P(A_2|A_1)$ and is read "the probability of A_2, given A_1." The required probability is

$$P(A_2|A_1) = \frac{2}{5}$$

The reduced sample space consists of the five elements lying within A_1, representing the event "male." The two elements lying in the cell that is the intersection of A_1 and A_2 represent the number of those males, who also possess the property of being under 40 years of age. Thus, the *conditional probability*, $P(A_2|A_1)$, is given by the ratio of the number of elements in the cell A_1 and A_2 to the number of points in the reduced sample space A_1. Probabilities such as $P(A_1)$ and $P(A_2)$ are usually referred to as *"unconditional probabilities."* This nomenclature will be used to differentiate clearly between probabilities such as $P(A_1)$ and $P(A_2|A_1)$, although as indicated at the outset of the discussion of conditional probabilities, all probabilities are "conditional" or "depend upon" the sample space within which they are defined.

We can now state the multiplication rule for two events A_1 and A_2.

Multiplication rule

(4.4)
$$P(A_1 \text{ and } A_2) = P(A_1)\, P(A_2|A_1)$$

Let us again consider the data in Table 4-1 to illustrate the application of this rule. If a person is selected at random from this group, the probability that this individual is both male and under 40 years of age is

$$P(A_1 \text{ and } A_2) = \frac{2}{10}$$

Hence, we note that this figure represents the product of $P(A_1)$, the probability of a male, and $P(A_2|A_1)$, the probability that the individual is under 40, given that he is a male. That is,

$$\frac{2}{10} = \left(\frac{5}{10}\right)\left(\frac{2}{5}\right)$$

Conditional probabilities are often computed by means of this multiplication rule. Solving for $P(A_2|A_1)$, we obtain

(4.5) $$P(A_2|A_1) = \frac{P(A_1 \text{ and } A_2)}{P(A_1)} \qquad \text{where } P(A_1) \neq 0$$

The statement that $P(A_1) \neq 0$ is added in order to rule out the possibility of division by zero.

Interchanging the designations A_1 and A_2 gives

$$P(A_1|A_2) = \frac{P(A_1 \text{ and } A_2)}{P(A_2)} \qquad \text{where } P(A_2) \neq 0$$

If $P(A_2|A_1) = P(A_2)$, then the multiplication rule becomes

(4.6) $$P(A_1 \text{ and } A_2) = P(A_1)P(A_2)$$

and events A_1 and A_2 are said to be *independent events*. If two events A_1 and A_2 are independent, then knowing that one of the events has occurred does not affect the probability that the other will occur. Thus, if A_1 and A_2 are independent, then $P(A_2|A_1) = P(A_2)$ and $P(A_1|A_2) = P(A_1)$. If the events A_1 and A_2 are not independent, they are said to be *dependent* or *conditional* events.

EXAMPLE 4-5

As a simple illustration of these ideas, let us consider Example 4-5. What is the probability of obtaining two kings in drawing two cards from a shuffled deck of cards?

SOLUTION

The first thing to note is that this is a poorly defined question. We must specify how the two cards are to be drawn from the deck. That is, are the two cards drawn simultaneously? If not, if the cards are to be drawn successively, is the first card replaced prior to drawing the second?

Let us assume that the two cards are drawn successively without replacing the first card and denote the events "king on first card" and "king on second card" by A_1 and A_2, respectively. It is clear that these are dependent events. That is, the probability of obtaining a king on the second draw is dependent upon whether a king was obtained on the first. The required joint probability for the successive occurrence of A_1 and A_2 is then $P(A_1 \text{ and } A_2) = P(A_1)P(A_2|A_1)$. If we consider the draw of the first card, since there are 52 cards in the deck, of which four are kings, $P(A_1) = 4/52$. Given that a king was obtained on the first draw, the probability of drawing a king on the second draw is $P(A_2|A_1) = 3/51$. This follows since only 51 cards remain, three of which are kings, given that a king was obtained on the first draw. Therefore the required probability is

$$P(A_1 \text{ and } A_2) = \frac{4}{52} \cdot \frac{3}{51} = \frac{1}{221}$$

What is the probability of obtaining two kings if the two cards are drawn simultaneously from the deck? The answer to this question is exactly the same as in the preceding computation. Here we think of one of the two cards as the "first" and the other as the "second" although they have been drawn together. A_1 is again used to denote the event "king on first card" and A_2 "king on second card."

Now let us assume the two cards were drawn successively, but that the first card was replaced in the deck, and the deck was reshuffled prior to drawing the second card. Since "king on first card" and "king on second card" are now independent events, their joint probability is given by the formula $P(A_1 \text{ and } A_2) = P(A_1)P(A_2)$. Therefore,

$$P(A_1 \text{ and } A_2) = \frac{4}{52} \cdot \frac{4}{52} = \frac{1}{169}$$

In this case, of course, the composition of the deck is the same on both drawings.

This simple example has interesting implications for sampling theory. If the deck of cards is viewed as a statistical universe, we have drawn a sample of two items at random from this universe (1) *without replacement* and (2) *with replacement* of the sampled elements. In human populations, sampling is usually carried out without replacement. That is, after the necessary data are obtained, an individual drawn into the sample is usually not replaced (either conceptually or actually) prior to the drawing of another individual. Sometimes it is convenient, for simplicity, to calculate probabilities in a "sampling without replacement" situation as though "replacement" had taken place.

The generalization of the multiplication rule is straightforward. The joint probability of the occurrence of n events, $A_1, A_2, \ldots, A_n$, may be expressed as

(4.7) $P(A_1 \text{ and } A_2 \text{ and } \ldots \text{ and } A_n) = P(A_1)P(A_2|A_1)P(A_3|A_2 \text{ and } A_1)$
$$\ldots P(A_n|A_{n-1} \text{ and } \ldots \text{ and } A_1)$$

This notation means that the joint probability of the n events is given by the product of the probability that the first event A_1 has occurred; the conditional probability of the second event A_2, given that A_1 has occurred; the conditional probability of the third event A_3, given that both A_2 and A_1 have occurred; etc. Of course, the n events can be numbered arbitrarily; therefore any one of them may be the first event, any of the remaining $n - 1$ may be second, and so forth.

The joint probability of the n events, $A_1, A_2, \ldots, A_n$, in the special case where these events are independent, is

(4.8) $P(A_1 \text{ and } A_2 \text{ and } \ldots \text{ and } A_n) = P(A_1)P(A_2)P(A_3) \ldots P(A_n)$

As Examples 4-6 through 4-8 show, it is possible to solve a large variety of probability problems using only the addition and multiplication rules, either separately or together.

EXAMPLE 4-6

A fair coin is tossed twice. What is the probability of obtaining exactly one head?

SOLUTION

One way of arriving at the solution is through the combined use of the addition and multiplication theorems. Denote the appearance of a head by H and a tail by T. The event "exactly one head" in two trials may occur by obtaining a head on the first trial followed by a tail on the second or a tail followed by a head. These two events are mutually exclusive. Thus, by the addition rule,

$$P(\text{exactly one head}) = P((H \text{ and } T) \text{ or } (T \text{ and } H)) = P(H \text{ and } T) + P(T \text{ and } H)$$

The appearance of a head on the first toss and a tail on the second are independent events, as are a tail on the first toss and a head on the second. Thus, by the multiplication rule,

$$P(H \text{ and } T) = P(H)P(T) = \frac{1}{2} \times \frac{1}{2} = \frac{1}{4}$$

and

$$P(T \text{ and } H) = P(T)P(H) = \frac{1}{2} \times \frac{1}{2} = \frac{1}{4}$$

Hence,

$$P(\text{exactly one head}) = \frac{1}{4} + \frac{1}{4} = \frac{1}{2}$$

Another way of obtaining the solution is to consider the sample space of equiprobable points.

$$S = \{(T, T), (T,H), (H,T), (H,H)\}$$

Since two of the four sample points represent the occurrence of exactly one head, we have

$$P(\text{exactly one head}) = \frac{2}{4} = \frac{1}{2}$$

EXAMPLE 4-7

A batch of transistors contains 10% defectives. Three transistors are drawn at random from the batch one at a time, each being replaced prior to the next draw. What is the probability of obtaining at least one defective transistor?

SOLUTION

Let D and G represent the appearance of a defective and a good transistor, respectively, when a transistor is drawn from the batch. Then, since the sampling is done with replacement, on any draw of one transistor

$$P(D) = 0.1 \qquad \text{and} \qquad P(G) = 0.9$$

Let E denote the event "at least one defective in a sample of three transistors." Then $\bar{E}$, the complement of E, denotes the event "zero defectives in a sample of three transistors." Since the event $\bar{E}$ represents the successive occurrence of three good transistors, and since the appearances of good transistors on the first, second, and third drawings are independent events, we obtain by the multiplication rule

$$P(\bar{E}) = P(G \text{ and } G \text{ and } G) = P(G)P(G)P(G) = (0.9)(0.9)(0.9) = 0.729$$

Therefore,

$$P(E) = 1 - P(\bar{E}) = 0.271$$

EXAMPLE 4-8

The following table refers to the 2500 employees of the Johnson Company, classified by sex and by opinion on a proposal to emphasize fringe benefits rather than wage increases in an impending contract discussion

SEX	IN FAVOR	OPINION NEUTRAL	OPPOSED	TOTAL
Male	900	200	400	1500
Female	300	100	600	1000
TOTAL	1200	300	1000	2500

(a) Calculate the probability that an employee selected from this group will be
   (1) a female opposed to the proposal.
   (2) neutral.
   (3) opposed to the proposal, given that the employee selected is a female.
   (4) either a male or opposed to the proposal.
(b) Are opinion and sex independent for these employees?

SOLUTION

We use the following representation of events

A_1: Male B_1: In favor
A_2: Female B_2: Neutral
              B_3: Opposed

In (a), we have

(1) $P(A_2 \text{ and } B_3) = 600/2500 = 0.24$
(2) $P(B_2) = 300/2500 = 0.12$
(3) $P(B_3|A_2) = \dfrac{P(B_3 \text{ and } A_2)}{P(A_2)} = \dfrac{600/2500}{1000/2500} = 0.60$
(4) $P(A_1 \text{ or } B_3) = P(A_1) + P(B_3) - P(A_1 \text{ and } B_3)$

$$= \frac{1500}{2500} + \frac{1000}{2500} - \frac{400}{2500}$$

$$= \frac{2100}{2500} = 0.84$$

In (b), in order for opinion and sex to be statistically independent, the joint probability of the intersection of each pair of A events and B events would have to be equal to the product of the respective unconditional probabilities. That is, the following equalities would have to hold:

$$P(A_1 \text{ and } B_1) = P(A_1)P(B_1) \qquad P(A_2 \text{ and } B_1) = P(A_2)P(B_1)$$
$$P(A_1 \text{ and } B_2) = P(A_1)P(B_2) \qquad P(A_2 \text{ and } B_2) = P(A_2)P(B_2)$$
$$P(A_1 \text{ and } B_3) = P(A_1)P(B_3) \qquad P(A_2 \text{ and } B_3) = P(A_2)P(B_3)$$

Clearly, these equalities do not hold, as, for example,

$$P(A_1 \text{ and } B_1) \neq P(A_1)P(B_1)$$
$$\frac{900}{2500} \neq \frac{1500}{2500} \times \frac{1200}{2500}$$

Another way of viewing the problem is that each conditional probability would have to be equal to the corresponding unconditional probability. Thus, the following equalities would have to hold:

$$P(A_1|B_1) = P(A_1) \qquad P(A_1|B_2) = P(A_1) \qquad P(A_1|B_3) = P(A_1)$$
$$P(A_2|B_1) = P(A_2) \qquad P(A_2|B_2) = P(A_2) \qquad P(A_2|B_3) = P(A_2)$$

These equalities do not hold, as, for example,

$$P(A_1|B_1) \neq P(A_1)$$
$$\frac{900}{1200} \neq \frac{1500}{2500}$$

The nature of the dependence (lack of independence) can be summarized briefly as follows: The proportion of males declines as we move from favorable to opposed opinions. This type of relationship is sometimes referred to in the following terms: There is a direct (inverse) relationship between the proportion of males (females) and favorableness of opinion.

EXERCISES

1. Let $P(A) = 0.5$; $P(B) = 0.4$; $P(A \text{ and } B) = 0.2$
   (a) Are A and B mutually exclusive events? Why?
   (b) Are A and B independent events? Why?

2. In a group of 20 adults there are eight males and nine Republicans. Further, there are five male Republicans. If a random selection is made, what is the probability of selecting a female who is not a Republican?

3. A certain family has three children. Male and female children are equally probable and successive sexes are independent. Let M stand for male, F for female.
   (a) List all elements in the sample space.
   (b) What are the probabilities of the ordered events MMM and MFM?
   (c) Let A be the event that both sexes appear. Find $P(A)$.
   (d) Let B be the event that there is at most one girl. Find $P(B)$.
   (e) Prove that A and B are independent.

4. Events A and B have the following probability structure:
   $$P(A \text{ and } B) = 1/6$$
   $$P(\bar{A} \text{ and } B) = 2/9$$
   $$P(\bar{A} \text{ and } B) = 1/3$$
   (a) What is the probability of $\bar{A}$ and $\bar{B}$?
   (b) Are A and B independent events?

5. If the probability that Company *A* will buy Company *B* is 0.6, what are the odds that Company *A* will not buy Company *B*?

6. In roulette there are 38 slots in which a ball may land. There are numbers 0, 00, and 1 through 36. The odd numbers are red, even numbers are black, and the zeroes are green. A ball is thrown randomly into a slot.
(a) What is the probability that it is red?
(b) What is the probability it is number 27?
(c) What is the probability it is either red or the number 27?
(d) What are the odds in favor of black?
(e) What are the odds in favor of the number 27?
(f) If you play black an infinite number of times what fraction of times will you win? What fraction of times will you lose?

7. An investment firm purchases three stocks for one-week trading purposes. It assesses the probability that the stocks will increase in value over the week as 0.9, 0.7, and 0.6, respectively. What is the probability that all three stocks will increase, assuming that the movements of these stocks are independent? Is this a reasonable assumption?

8. The probability that a life insurance salesman following up a magazine lead will make a sale is 0.3. A salesman has two leads on a certain day. Assuming independence, what is the probability that
(a) he will sell both?
(b) he will sell exactly one policy?
(c) he will sell at least one policy?

9. There are two major reasons for classifying a bottle of soda defective; the filler (a machine) either overfills or underfills the bottle. Two percent of the time it underfills and 1% of the time it overfills. What is the probability that a bottle will be rejected because of the filler?

10. A firm has five engineering positions to fill and is trying to recruit recent graduates for the positions. In the past, 40% of the college students who were offered similar positions have turned them down. The firm offers positions to six graduates. Is the firm justified in doing so? Explain, assuming independence.

11. A national franchising company is interviewing prospective buyers in the Tulleytown area. The probability that an interviewee will buy the franchise is 0.1. Assuming independence, what is the probability the firm will have to interview more than five people before making a sale?

12. A census of a company's 500 employees in regard to a certain proposal showed 125 of its 150 white-collar workers in favor of the proposal, and a total of 125 workers opposed to the proposal. (All the workers can be classified as either white-collar or blue-collar.)
(a) What is the probability that an employee selected at random will be a blue-collar worker opposed to the proposal?
(b) What is the probability that an employee picked at random will be in favor of the proposal?
(c) What is the probability that if a blue-collar worker is chosen, he will be in favor of the proposal?
(d) Are job type (blue- or white-collar) and opinion on the proposal independent? Prove your answer and give a short statement as to the implication of this finding.

13. The following information pertains to new-car dealers in the United States:

TYPE OF DEALERSHIP	NORTH	SOUTH	REGION OF DEALER MIDWEST	WEST	TOTAL
Admiral Motors	155	50	135	110	450
Bord Motors	90	65	40	90	285
Shysler Motors	50	50	30	85	215
U.S. Motors	35	35	15	65	150
TOTAL	330	200	220	350	1100

If a name is selected randomly from the American Automobile Dealer's Association list of all United States dealers handling the four American manufacturers' brands, what is the probability that
(a) it is a Southern Admiral Motors dealer?
(b) it is a Southern dealer?
(c) it is an Admiral Motors dealer?
(d) it is a Southern dealer if he is known to be an Admiral Motors dealer?
(e) Are type of dealership and region independent?

14. A certain proposal was put forth by a company's management to all of its sales representatives in different sales regions. Questionnaires were sent to each salesman and the results were

OPINION	EAST	REGION MIDDLE WEST	PACIFIC COAST	TOTAL
Opposed	20	15	15	50
Not opposed	80	85	285	450
TOTAL	100	100	300	500

(a) What is the probability that a questionnaire selected at random is that of an Eastern salesman opposed to the proposal?
(b) What is the probability that a questionnaire selected at random is that of a Midwestern salesman?
(c) If a questionnaire is selected from the group that responded unfavorably to the proposal, what is the probability that the salesman comes from the Pacific Coast region?
(d) Are the salesman's regional district and his opinion on the proposal independent? If yes, prove it. If no, specify what the numbers in the cells of the table would have been had the two factors been independent.

4.3   BAYES' THEOREM

The Reverend Thomas Bayes (1702–1761), an English Presbyterian minister and mathematician, considered the question of how one might make inferences from observed sample data about the populations that gave rise to these data. His motivation came from his desire to prove the existence of God by examining the sample evidence of the world about him. Mathematicians had previously concentrated on the problem of deducing the consequences of specified hypotheses. Bayes was interested in the inverse problem of drawing conclusions about hypotheses from observations of consequences. He derived a theorem that calculated probabilities of "causes" based on the observed "effects." The theorem may also be thought of as a means of revising prior probabilities of events based on the observation of additional information. In the period since World War II, a body of knowledge, known as *Bayesian decision theory,* has been developed to solve problems involving decision making under uncertainty.

The structure of the following problem illustrates the nature of the probability calculation made by Bayes' theorem.

Assume that a room contains 100 people, who possess the following characteristics:

$\frac{1}{2}$ are males; of the males, $\frac{3}{10}$ are foreign born and $\frac{7}{10}$ are native born
$\frac{1}{2}$ are females; of the females $\frac{2}{10}$ are foreign born and $\frac{8}{10}$ are native born

Figure 4-5 depicts graphically the above situation. Note that the numbers (0.5) below "males" and "females" are unconditional probabilities, whereas the probabilities given inside the bars are conditional probabilities. For example, the 0.3 in the upper part of the first bar is the conditional probability that the individual is foreign born given that he is a male.

Let us consider two different probability questions. Suppose that a person is selected at random from the entire group. What is the probability that the in-

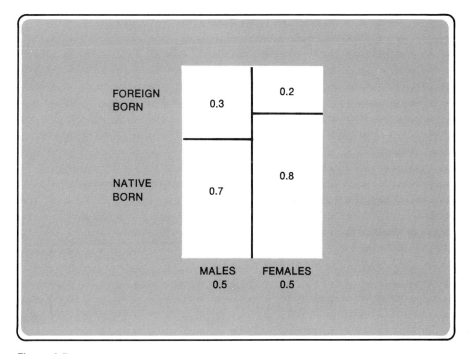

Figure 4-5
Unconditional and conditional probabilities.

dividual is a male? Since one-half of the 100 people are males, and it is equally likely that any individual would be selected, we assign a probability of 0.5 to the event "male." That is, the unconditional probability P(male) equals 0.5.

Now, let us think about a different problem. Suppose a person is selected at random from the group, and we are given the *additional information* that the individual is foreign born. What assessment would we now make of the probability that the individual is male, *given* that the individual is foreign born? That is, what is the conditional probability P(male|foreign born)?

It is important to note the nature of these probability questions. The first probability, P(male), may be thought of as a prior probability in the sense that it is assigned prior to the observation of any experimental information. The second probability, P(male|foreign born), may be thought of as a "posterior probability" or a "revised probability" in the sense that it is assigned after the observation of experimental or additional information. The posterior probability is the type of probability computed by Bayes' theorem. We will examine the logic in such calculations before stating the theorem.

Our prior probability assignment to the event "male" is 0.5. How should we revise this probability assignment if we are given the additional information that the individual is foreign born? We can reason as follows. Of all the people in the room $0.5 \times 0.3 = 0.15$ are males *and* foreign born, whereas $0.5 \times 0.2 = 0.10$ are females *and* foreign born. Hence, the probability that this foreign born individual is male is equal to

$$\frac{0.15}{(0.15 + 0.10)} = 0.60$$

The logic of the above calculations can be expressed algebraically by Bayes' theorem, which is simply a formula for computing a conditional probability. Let A_1 denote the event "male"; A_2, "female"; and B, "foreign born." Then Bayes' theorem is given by

(4.9)
$$P(A_1|B) = \frac{P(A_1)\,P(B|A_1)}{P(A_1)\,P(B|A_1) + P(A_2)\,P(B|A_2)}$$

For example, in terms of the previous illustration we have

$$P(A_1|B) = \frac{(0.5)(0.3)}{(0.5)(0.3) + (0.5)(0.2)} = 0.60$$

As we have seen, Bayes' theorem weights prior information with experimental evidence. The manner in which it does this may be seen by laying out the calculations in a form such as Table 4-2.

If there are more than two basic events (A_i), then correspondingly additional terms appear in the denominator of equation (4.9).

The first column of Table 4-2 gives the basic events of interest. The second column shows the prior probability assignments to these basic events, "male" and "female." The third column shows the conditional probabilities of the additional information given the basic events. As noted in the caption, such conditional probabilities are referred to as "likelihoods." In the illustrative

TABLE 4-2

Bayes' Theorem Calculations for Illustrative Problem

EVENTS A_i	PRIOR PROBABILITIES $P(A_i)$	LIKELIHOODS $(P(B\|A_i)$	JOINT PROBABILITIES $P(A_i)\,P(B\|A_i)$	REVISED PROBABILITIES $P(A_i\|B)$
A_1: Male	0.5	0.3	0.15	0.15/0.25 = 0.60
A_2: Female	0.5	0.2	0.10	0.10/0.25 = 0.40
			0.25	1.00

problem these are, respectively, P(foreign born|male) and P(foreign born|female). The fourth column gives the joint probabilities of the basic events and the additional information. When these probabilities are divided by their total (in this case, 0.25), the results are the revised probabilities shown in the last column. The first probability shown in the last column (0.15/0.25 = 0.60) is the one required in the illustrative problem.

In modern Bayesian decision theory, subjective prior probability assignments are made in many applications. It is argued that it is meaningful to assign prior probabilities concerning hypotheses based upon degree of belief. Bayes' theorem is then viewed as a means of revising these probability assignments. In business applications, this has meant that executives' intuitions, subjective judgments, and present quantitative knowledge are captured in the form of prior probabilities; these figures undergo revision as relevant empirical data are collected. This procedure seems sensible and fruitful for a wide variety of applications. Examples 4-10 through 4-12 suggest some of the many possible types of applications of this very interesting theorem.

EXAMPLE 4-10

A man regularly plays darts in the recreation room of his home, observed by his eight-year-old son. The father has a history of making bulls-eyes 30% of the time. The son, who habitually sits injudiciously close to the dart board, reports to his father whether or not bulls-eyes have been made. However, as is often the case with eight year olds, the son is an imperfect observer. He reports correctly 90% of the time. On a particular occasion, the father throws a dart at the board. The son reports that a bulls-eye was made. What is the probability that a bulls-eye was indeed scored?

SOLUTION

Let A_1 be the event "bulls-eye scored" and A_2 "bulls-eye not scored." Let B represent "son reports that a bulls-eye has been scored." Then,

$$P(A_1|B) = \frac{(0.3)(0.9)}{(0.3)(0.9) + (0.7)(0.1)} = 0.79$$

This is illustrative of a class of problems involving an information system that transmits uncertain knowledge. That is, the son may be thought of as an information system of 90% "reliability." A more colorful way of expressing it is that the son has "error in him" or is a "noisy" information system (where the term "noise" refers to random error).

EXAMPLE 4-11

Assume that 1% of the inhabitants of a country suffer from a certain disease. A new diagnostic test is developed that gives a positive indication 97% of the time when an individual has this disease and a negative indication 95% of the time when the disease is absent. An individual is selected at random, is given the test, and reacts positively. What is the probability that he has the disease?

SOLUTION

Let A_1 represent "has the disease" and A_2 "does not have the disease." Let B represent a positive indication. Then, by Bayes' theorem,

$$P(A_1|B) = \frac{(0.01)(0.97)}{(0.01)(0.97) + (0.99)(0.05)} = 0.16$$

This is a surprisingly low probability, which doubtless runs counter to intuitive feelings based on the 97% and 95% figures given. If the "cost" incurred is high when a person is informed that he has a disease (e.g., such as cancer) on the basis of such a test when in fact he does not, it would appear that action based solely on conditional probabilities such as the above 97 and 95% might be seriously misleading and costly.

EXAMPLE 4-12

A corporation uses a "selling aptitude test" to aid it in the selection of its sales force. Past experience has shown that only 65% of all persons applying for a sales position achieved a classification of "satisfactory" in actual selling, whereas the remainder were classified "unsatisfactory." Of those classified as "satisfactory," 80% had scored a passing grade on the aptitude test. Only 30% of those classified "unsatisfactory" had passed the test. On the basis of this information, what is the probability that a candidate would be a "satisfactory" salesperson given a passing grade on the aptitude test?

SOLUTION

If A_1 stands for a "satisfactory" classification as a salesperson and B stands for "passes the test," then the probability that a candidate would be a "satisfactory" salesperson, given a passing grade on the aptitude test, is

$$P(A_1|B) = \frac{(0.65)(0.80)}{(0.65)(0.80) + (0.35)(0.30)} = 0.83$$

Thus, the tests are of some value in screening candidates. Assuming no change in the type of candidates applying for the selling positions, the probability that a random applicant would be satisfactory is 65%. On the other hand, if the company only accepts an applicant who passes the test, this probability increases to 0.83.

1. A firm is contemplating changing the packaging sizes of its product, eliminating the three-ounce size and offering at a slightly higher price a four-ounce size. The marketing manager feels the probability that this change will increase profits is 70%. The change is tried in a limited test area and results in reduced profits. The probability that this result would occur even if the change would actually increase profits nationally is .4, whereas if it would not increase profits nationally, the probability is .8. What should be the manager's revised probability of the profitability of the change?

2. An investor feels the probability that a certain stock will go up in value during the next month is 0.7. Value-Dow, an investment advisory firm predicts the stock will not go up over the period. Value-Dow over time has proven to be correct 80% of the time. What probability should the investor assign to the stock going up in light of Value-Dow's prediction?

3. Twenty percent of the items produced by a certain machine are defective. The company hires an inspector to check each item before shipment. The probability the inspector will incorrectly ship an item is 0.3. If an item is shipped, what is the probability that it is not defective?

4. TEC Exploration Company is involved in a mining exploration in Northern Canada. The chief engineer originally feels that there is a 50-50 chance that a significant mineral find will occur. A first test drilling is completed and the results are favorable. The probability that the test drilling would give misleading results is 0.3. What should be the engineer's revised probability that a significant mineral find will occur?

5 *Probability Distributions*

5.1 PROBABILITY DISTRIBUTIONS

In Chapter 3, we saw how frequency distributions constitute a convenient device for summarizing the variations in observed data. In this chapter, we deal with *theoretical frequency distributions* or *probability distributions* that analogously describe how outcomes may be *expected* to vary. The concept of the probability distribution is a central one in statistical inference and decision making under uncertainty.

Examples 5-1 and 5-2 introduce the idea of a probability distribution.

EXAMPLE 5-1

Let us consider the experiment of tossing a fair coin two times. We will concentrate on the number of heads obtained on the two tosses and their respective probabilities of occurrence. The elements of the sample space and the number of heads corresponding to each are listed in the first two columns of Table 5-1.

TABLE 5-1

Sample Space, Number of Heads, and Probabilities
in the Two Tosses of a Coin Experiment

ELEMENTS OF THE SAMPLE SPACE	NUMBER OF HEADS	PROBABILITY
T,T	0	$\frac{1}{4}$
T,H	1	$\frac{1}{4}$
H,T	1	$\frac{1}{4}$
H,H	2	$\frac{1}{4}$

As indicated, the possible numbers of heads obtained in two tosses of a coin are 0, 1, and 2. We are interested in the probabilities associated with these events. A probability of $\frac{1}{4}$ is assigned to each point in the sample space as shown in the third column of Table 5-1.

The last two columns of Table 5-1 can be more compactly shown by adding up the probabilities associated with each distinct number of heads. The resulting distribution is known as a *"probability function"* or *"probability distribution."* (In keeping with general practice, we will usually use the term "probability distribution.") The probability distribution for "number of heads" in the two tosses of a coin experiment is shown in Table 5-2.

TABLE 5-2

Probability Distribution of Number
of Heads in the Two Tosses
of a Coin Experiment

NUMBER OF HEADS x	PROBABILITY P(x)
0	$\frac{1}{4}$
1	$\frac{1}{2}$
2	$\frac{1}{4}$

As noted in the column headings of the table, the various values that the number of heads can take on are symbolized as x and the associated probabilities are denoted $P(x)$.

When a probability distribution is graphed, it is conventional to display the values of the variable of interest on the horizontal axis and their probabilities on the vertical scale. The graph of the probability distribution for the two tosses of a coin experiment is shown in Figure 5-1.

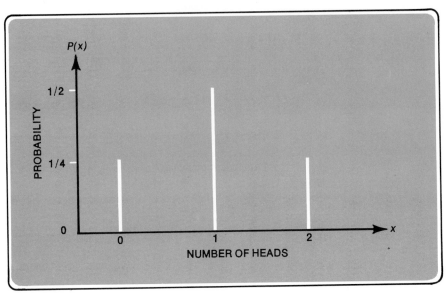

Figure 5-1
Graph of the probability distribution of number of heads obtained in two tosses of a coin.

As an exercise, verify that the experiment of tossing a fair coin three times will result in the probability distribution and graph shown in Table 5-3 and Figure 5-2, respectively.

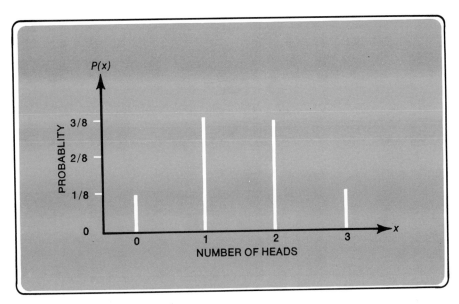

Figure 5-2
Graph of the probability distribution of number of heads obtained in three tosses of a coin.

TABLE 5-3
Probability Distribution of Number
of Heads in the Three Tosses
of a Coin Experiment

NUMBER OF HEADS x	PROBABILITY P(x)
0	$\frac{1}{8}$
1	$\frac{3}{8}$
2	$\frac{3}{8}$
3	$\frac{1}{8}$

EXAMPLE 5-2

A corporation economist developed a subjective distribution for the change in gross national product (GNP) that would take place in the following year. She established the following five categories of change and let the indicated numbers correspond to each category.

CHANGE IN GNP	NUMBER ASSIGNED
Down more than 5%	−2
Down 5% or less	−1
No change	0
Up 5% or less	1
Up more than 5%	2

On the basis of all information available to her, the economist assigned probabilities to each of these possible events as indicated in Table 5-4.

TABLE 5-4
Subjective Probability Distribution
of Change in Gross National Product

x	P(x)
−2	0.1
−1	0.1
0	0.2
1	0.4
2	0.2

As we can see from Examples 5-1 and 5-2, probability distributions may be constructed on the basis of theoretical considerations (Example 5-1) or subjective judgments (Example 5-2). They may also be established on the basis of experience and observation, as for example, in the use of death rates as probabilities. Note also that the variable denoted x must be stated as numerical values. Then a probability distribution is formed by assigning probabilities to these numerical values.

Types of probability distributions

Probability distributions are classified as either *discrete* or *continuous* depending upon the nature of the variable under consideration. The variable is usually referred to as a "random variable." We may think of a random variable roughly as a variable that takes on different values resulting from a random experiment.[1] A discrete random variable is one that can only take on distinct values. The preceding examples illustrated probability distributions of discrete random variables. For example, the random variable "number of heads" in Example 5-1 could only take on the values 0, 1, or 2. It could not take on any other values such as $1\frac{1}{2}$ or $2\frac{1}{4}$.

A random variable is said to be *continuous* in a given range if it can assume any value in that range. The term "continuous random variable" implies that variation takes place along a *continuum.* Continuous variables can be *measured* to some degree of accuracy. (With discrete variables, since only a distinct number of values are assumed, *counts* can be made at each of these values.) Examples of continuous variables include weight, length, velocity, the length of life of a product, and the duration of a rainstorm.

Characteristics of probability distributions

The characteristics of probability distributions may be formally summarized as follows:

1. $P(x) \geq 0$
2. $\sum_{x} P(x) = 1$

Property (1) simply states that probabilities are numbers equal to or greater than zero. The second property states that the sum of the probabilities in a probability distribution is equal to one. These are the same properties specified in Chapter 4 for probabilities defined on a sample space.

EXERCISES

1. State whether the following random variables are discrete or continuous.
   (a) Strength of a steel beam in pounds per square inch.
   (b) The weight of a supposedly 16-ounce box of breakfast cereal.
   (c) *X* equals 0 if the weight of a supposedly 16-ounce box of cereal is less than 16 ounces, and 1 if 16 ounces or more.
   (d) The number of defective batteries in a lot of 1000.

2. A corporation economist, in building a model to predict a company's sales, developed the following categories of change:
   Sales down more than 3%
   Sales down 3% or less
   Sales unchanged

---

[1] From a mathematical point of view, a random variable is a function that consists of the elements of a sample space and the numbers assigned to these sample points. Ordinarily, a shortcut method of referring to the concept of a random variable is used. Hence, as in the coin tossing example, we simply refer to the random variable "number of heads obtained in two tosses of a coin."

Sales up 3% or less
Sales up more than 3%

Let X be a random variable associated with sales change and assign subjective probabilities to form a probability function. Then graph the probability distribution.

3. Let $X =$ the number of minutes it takes to drain a soda filler of a particular flavor in order to change over to a new flavor. The probability distribution for X is

$$P(x) = x/15 \qquad x = 1, 2, \ldots, 5$$

   (a) Prove that $P(x)$ is a probability function.
   (b) What is the probability that it will take exactly three minutes to change over?
   (c) What is the probability that it will take at least two minutes but not more than four minutes?
   (d) What is the probability that it will at most take three minutes?
   (e) What is the probability that it will take more than two minutes?

MATHEMATICAL EXPECTATION   5.2

Mathematical expectation or expected value is a basic concept in the formulation of summary measures for probability distributions. It is a fundamental notion employed for hundreds of years in areas such as the insurance industry and currently used in fields such as operations research, management science, systems analysis, and managerial economics. Let us consider a very simple problem to obtain an understanding of this concept.

Suppose the following game of chance were proposed to you. A fair coin is tossed. If it lands heads, you win $10; if it lands tails, you lose $5. What would you be willing to pay for the privilege of playing this game?

On any particular toss, you will either win $10 or lose $5. However, let us think in terms of a repeated experiment, in which we toss the coin and play the game many times. Since the probability assigned to the event "head" is $\frac{1}{2}$ and to the event "tail" $\frac{1}{2}$, in the long run, you would win $10 on one-half of the tosses and lose $5 on one-half. Therefore, the average winnings per toss would be obtained by weighting the outcome $10 by $\frac{1}{2}$ and $-$5 by $\frac{1}{2}$ to yield a weighted mean of $2.50 per toss. In terms of equation (3.10), this weighted mean is

$$\overline{X}_w = \frac{\Sigma wX}{\Sigma w} = \frac{(\$10)(\frac{1}{2}) + (-\$5)(\frac{1}{2})}{\frac{1}{2} + \frac{1}{2}} = \$2.50 \text{ per toss}$$

Of course, when the weights are probabilities in a probability distribution, as is true in this example, the sum of the weights is equal to one. Therefore, in such cases, the formula could be written without showing the division by the sum of the weights.

The average of $2.50 per toss is referred to as the "expected value" or "mathematical expectation" of the winnings. Note that on a single toss, only two outcomes are possible, namely, win $10 or lose $5. If these two possible winnings are viewed as the values of a random variable, which occur with probabilities of $\frac{1}{2}$ each, then the expected value of the random variable can be seen to be the mean of its probability distribution. More formally, if X is a discrete random variable that takes on the value x with probability $P(x)$, then the expected value of X, denoted $E(X)$, is

(5.3) $$E(X) = \Sigma x P(x)$$

That is, to obtain the expected value of a discrete random variable, each value that the variable can assume is multiplied by the probability of occurrence of that value, and then all of these products are totaled.

Applying this formula for the expected value to the problem of tossing a coin, we have

$$E(X) = \$10(\tfrac{1}{2}) + (-\$5)(\tfrac{1}{2}) = \$2.50 \text{ per toss}$$

We see that this is the same calculation performed earlier using the weighted mean formula, except that the sum of the weights, which is equal to one, is not explicitly shown as a divisor.

A brief comment on the meaning of the "expected value of an act" is pertinent at this point. We have seen that the expected value of a random variable is a weighted average of the values that the variable can assume, where the weights are probabilities. Analogously in decision making under uncertainty, if a certain action is taken and a specific event occurs, there is a specified payoff. The expected value of an act is the weighted average of these payoffs, where the weights are the probabilities of occurrence of the various events.

Now, let us return to the question, "What would you be willing to pay for the privilege of playing this game?" It would appear reasonable that you should be willing to pay up to $2.50 on each toss. For the game to represent a perfectly "fair bet" from the probability standpoint, you should pay *exactly* $2.50 per toss. Then, in the long run, you would come out even. However, different people might be willing to pay various sums to play the game. The fact that different decisions would be made is not necessarily an irrational situation. It simply reflects the different attitudes of these people toward financial risks. A figure such as the $2.50 mathematical expectation in the above problem is often referred to as an "expected monetary value." It is not always appropriate to use expected monetary value as the basis for decision making. For the moment, let us expand on the unsuitability of expected *monetary* value as a criterion for action in certain situations. We will then return to the conditions under which expected monetary value is an appropriate criterion.

Suppose we restate the coin tossing problem in terms of choice among alternative courses of action. You are asked to choose between two options, A_1 and A_2. A_1 again involves the flipping of a fair coin only one time. If it lands heads, you win $10; if tails, you lose $5. A_2 involves a certainty of neither a gain nor a loss. Which do you prefer, A_1 or A_2? Let us assume you choose A_1. We have already seen that the expected monetary value of option A_2 is $0. The choice of A_1 certainly seems reasonable. We can also use the terminology that the expected value of alternative A_2 is $0, since the expected value of a constant is the constant itself.

Now, let us change the problem by shifting the decimal place in the gains and losses of alternative A_1. You now are asked to choose between the following two options A_1 or A_2. A_1 again involves the flipping of a fair coin. If it lands heads, you win $100,000; if tails, you lose $50,000. A_2 is the certainty of neither a gain nor a loss. Which do you now prefer? The expected monetary value of A_1 is $25,000. That is, $E(A_1) = \$100,000(\frac{1}{2}) + (-\$50,000)(\frac{1}{2}) = \$25,000$. The corresponding figure for A_2 is $0. Despite the fact that alternative A_1 has the higher expected monetary value, it would not be at all unreasonable to prefer A_2.

A loss of $50,000 might be so catastrophic to you as an individual that you simply would not be willing to take the risk. You might be willing to use expected monetary value as a criterion for decision in the first pair of alternatives, where the amounts of money involved were quite small. But you might not be willing to use this criterion for very large sums of money. In terms of business decision making, businessmen would be apt to reject expected *monetary* value as a criterion of choice when the possible extreme outcomes of an alternative are either too disastrous or too beneficial.

A basic reason expected monetary value cannot be used as a decision criterion for the entire scale of values, running from very small to very large amounts, is that it assumes a proportional relationship between money amounts and the *utility* of these amounts to the decision maker. If $10 million is not considered ten times as useful or valuable as $1 million, and twice as valuable as $5 million, then the expected *monetary* value criterion is not valid. Stating it positively, we can say that expected monetary value *is* a valid decision criterion in situations in which utility is proportional to monetary amounts. A second assumption for the appropriate use of the expected monetary value criterion is that the decision maker is neutral to risk. That is, the act of risk taking does not, in and of itself, add or subtract any utility from a given risk situation. If this assumption is not valid, and if the decision maker actively seeks out a gamble merely because he enjoys gambling or carefully avoids a gamble because he does not like gambling, then clearly he will not act on the basis of expected monetary value.

Through the use of "utility functions," a theoretical framework can be constructed for making consistent decisions regardless of whether the above two assumptions are met. In this chapter, we will assume we are dealing with situations in which expected monetary value is a valid guide, that is, the two assumptions hold. Thus, decision makers will act in such a way as to *maximize expected monetary value*. This means they will choose alternatives that maximize expected monetary gains.

Risk and uncertainty

The concept of expected value is extremely useful in the analysis of decision making under conditions of *uncertainty*. It is helpful at this point to take a closer look at what is meant by uncertainty. Many writers make a sharp distinction between "risk" and "uncertainty." They reserve the term "risk" for situations in which *objective probabilities* are available, either on the basis of a priori reasoning or relevant past frequency of occurrence experience. Hence, a priori reasoning would provide the basis for the assignment of probabilities for tosses of coins, rolls of dice, etc., where a knowledge of the physical structure of the coins or dice suffices for the establishment of the probability values. Past relative frequencies of occurrence would provide these assignments in actuarial situations, as for example, in assessing probabilities of death, accidents, sickness, or fires.

On the other hand, these writers use the term "uncertainty" to apply to situations in which only *subjective probabilities* can be applied to the events in question, and there is insufficient basis for a priori or past frequency of occurrence assignments.

Clearly most decisions in the world of business or government fall under the heading of uncertainty. However, modern usage tends to treat the distinction between risk and uncertainty as merely one of degree. At one extreme, there is the situation in which probabilities of events are treated as known; at the other extreme there is virtual or complete ignorance concerning events. The entire spectrum is referred to as one of uncertainty whether probabilities are assigned on an objective or subjective basis.

EXAMPLE 5-3

Suppose an insurance company offers a 45-year-old man a $1000 one-year term insurance policy for an annual premium of $12. Assume that the number of deaths per 1000 is five for persons of this age group. What is the expected gain for the insurance company on a policy of this type?

SOLUTION

We may think of this problem as representing a chance situation with two possible outcomes: the policy purchaser (1) lives or (2) dies during the year. Let X be a random variable denoting the dollar gain to the insurance company for these two outcomes. The probability that the man will live through the year is 0.995. In this case, the insurance company collects the premium of $12. The probability that the policy purchaser will die during the year is 0.005. In this case, the company has collected a premium of $12, but must pay the claim of $1000, for a net gain of −$988. Thus, X takes on the values $12 and −$988 with probabilities 0.995 and 0.005, respectively. The calculation of expected gain for the insurance company is displayed in Table 5-5.

TABLE 5-5
Calculation of Expected Gain for an Insurance
Company on a One-Year Term Policy

OUTCOME	x	P(x)	xP(x)
Policy Holder Lives	$12	0.995	$11.94
Policy Holder Dies	−988	0.005	−4.94
		1.000	$7.00

$$E(X) = \Sigma xP(x) = \$7.00$$

It may be noted that in setting a premium for this policy, the insurance company would have to consider usual business expenses as well as the expected gain calculation.

EXAMPLE 5-4

A financial manager for a corporation is considering two competing investment proposals. For each of these proposals, he has computed various net profit figures and has assigned subjective probabilities to the realization of these returns. For proposal A, his analysis shows net profits of $20,000, $30,000, or $50,000 with probabilities 0.2, 0.4, and 0.4, respectively. For proposal B, he concludes there is a 50-50 chance of a successful investment, estimated as producing net profits of $100,000 or an unsuccessful investment, estimated as a break-even situation involving $0 net profit. Assuming each proposal requires the same dollar investment, which is preferable from the standpoint of expected monetary return?

SOLUTION

Denoting the expected net profit on these proposals as $E(A)$ and $E(B)$, we have

$$E(A) = (\$20,000)(0.2) + (\$30,000)(0.4) + (\$50,000)(0.4)$$
$$= \$36,000$$
$$E(B) = (\$0)(0.5) + (\$100,000)(0.5) = \$50,000$$

The financial manager would maximize expected net profit by accepting proposal B.

EXERCISES

1. Let $X = -\$.10$ if a customer does not purchase from a telephone sales promotion and $X = +\$1.89$ if a customer does purchase from a telephone sales promotion. If the probability that a customer will purchase is 5%, what is the expected value of X? Interpret the result.

2. Let $X = 0$ if an item is defective and $X = 1$ if an item is not defective. If 10% of all the items are defective, what is the expected value of X? Interpret the result.

3. The number of freighters entering Hyannis Harbor between 8:00 A.M. and 2:00 P.M. on Fridays has the following probability distribution:

x	$P(x)$
0	0.10
1	0.20
2	0.40
3	0.15
4	0.10
5	0.05

What is the expected number of freighters entering this harbor during the specified time period on a given Friday?

4. The probability distribution for the number of customers entering a certain restaurant between 2:00 P.M. and 3:00 P.M. on a weekday is as follows:

x	P(x)
10	0.01
11	0.05
12	0.13
13	0.18
14	0.26
15	0.18
16	0.13
17	0.05
18	0.01

What is the expected number of customers entering the restaurant during the given time period?

5. Suppose an insurance company offers a particular home owner a $15,000 one-year term fire insurance policy on his home. Assume that the number of fires per 1000 dwellings of this particular home safety classification is two. For simplicity, assume that any fire will cause damages of at least $15,000. If the insurance company charges a premium of $36 a year, what is its expected gain from this policy? Does this mean that the company will earn that many dollars on *this particular policy*?

6. In the game of roulette there are 36 numbers, 1 through 36, and the numbers 0 and 00. On a spin of the roulette wheel it is equally probable that a ball will rest on any of the 38 numbers. A person places a $1 bet on the numbers 1–12. That is, if the ball rests on any one of these numbers, the person wins $2, whereas if it rests on any other number, he loses his $1. The person places the bet 1000 consecutive times. What is his expected profit (loss) on each roll? For the 1000 rolls?

In many situations, it is useful to represent the probability distribution of a random variable by a general algebraic expression. Probability calculations can then be conveniently made by substituting appropriate values into the algebraic model. The mathematical expression is a compact form of summarizing the nature of the process that has generated the probability distribution. Thus, the statement that a particular probability distribution is appropriate in a given situation contains considerable information about the nature of the underlying process thus described. In the following sections, we discuss the binomial and Poisson distributions, two of the most useful discrete probability distributions.

5.3   THE BINOMIAL DISTRIBUTION

The binomial distribution is undoubtedly the most widely applied probability distribution of a discrete random variable. It has been used to describe an impressive variety of processes in business and the social sciences as well as other areas. The algebraic formula for the binomial distribution is developed from a very specific set of assumptions involving the concept of a series of experimental trials.

Let us envision a process or experiment characterized by repeated trials. The trials take place under the following set of assumptions:

1. Each trial has two mutually exclusive possible outcomes, which are referred to as "success" and "failure."

2. The probability of a success, denoted p, remains constant from trial to trial. The probability of a failure, denoted q, is equal to $1 - p$.

3. The trials are independent. That is, the outcomes of any given trial or sequence of trials do not affect the outcomes on subsequent trials. The outcome of any specific trial is determined by chance.

Our aim is to develop a formula for the probability of x successes in n trials of such a process. We shall first consider a simple specific case of tossing a fair coin. We calculate the probability of obtaining exactly two heads in five tosses and then generalize the resulting expression.

Assume the appearance of a head on each toss to be a success. Of course, in these problems, the classification of one of the two possible outcomes as a "success" is completely arbitrary and involves no implication of desirability or goodness. For example, we may choose to refer to the appearance of a defective item in a production process as a success and a non-defective item as a failure. Or, if a process of births is treated as a series of experimental trials, the appearance of a female (male) may be classified as a success, and a male (female) a failure.

Suppose that the sequence of outcomes of five tosses of a fair coin is

$$HTHTT$$

As usual, H and T denote head and tail.

Since the probability of a success and a failure on a given trial are p and q, respectively, the probability of this particular sequence of outcomes is, by the multiplication rule for independent events,

$$P(H,T,H,T,T) = pqpqq = q^3p^2$$

This is the probability of obtaining the specific sequence of successes and failures, in the order in which they occurred. However, we are interested not in any specific order of results, but rather in the probability of obtaining a given number of successes in n trials. What then is the probability of obtaining exactly two successes in five tosses? The following nine other sequences also satisfy the condition of exactly two successes in five trials:

$$HHTTT$$
$$HTTHT$$
$$HTTTH$$
$$THHTT$$
$$THTHT$$
$$THTTH$$
$$TTHHT$$
$$TTHTH$$
$$TTTHH$$

By the reasoning used earlier, each of these sequences has the same probability, q^3p^2. Since the number of distinguishable sequences in this problem is 10, we may write

$$P(\text{exactly 2 successes}) = 10\ q^3\ p^2$$

The coefficient 10 in this expression represents the total number of distinguishable arrangements that can be made of the three *T*'s and two *H*'s. In this case, we were interested in obtaining "two successes in five trials." It is useful to have a symbol for the number of possible distinct arrangements. The symbol generally employed is $\binom{n}{x}$, where *n* is the number of trials and *x* is the number of successes. Hence, the number of possible sequences of five trials which would contain exactly two successes is $\binom{5}{2} = 10$.

Thus, in the coin tossing example, we may now write

$$P(\text{exactly 2 successes}) = \binom{5}{2} q^3p^2$$

Since we assign a probability of $\frac{1}{2}$ to *q* and $\frac{1}{2}$ to *p*, we have

$$P(\text{exactly two heads}) = \binom{5}{2}\left(\frac{1}{2}\right)^3\left(\frac{1}{2}\right)^2 = \frac{10}{32} = \frac{5}{16}$$

This result may be generalized to obtain the probability of exactly *x* successes in *n* trials of a process, which meets the aforementioned three assumptions. Let us assume $n - x$ failures (denoted *F*) occurred followed by *x* successes, (denoted *S*) in that order. We may then represent this sequence as

$$
\begin{array}{cc}
FFF \cdots F & SSS \cdots S \\
n - x & x \\
\text{failures} & \text{successes}
\end{array}
$$

The probability of this particular sequence is $q^{n-x}p^x$. As noted earlier, the number of possible sequences of *n* trials that would contain exactly *x* successes is $\binom{n}{x}$. Therefore, the probability of obtaining *x* successes in *n* trials is given by

(5.4) $$P(x) = \binom{n}{x}q^{n-x}p^x$$

The factors $\binom{n}{x}$, referred to as binomial coefficients because of the way they appear in the binomial expansion, are computed by the formula

(5.5) $$\binom{n}{x} = \frac{n!}{x!\,(n-x)!}$$

The symbol *n*! is read "*n* factorial," and is calculated as follows:

$$n! = (n)(n-1)\ldots(2)(1)$$

We shall only be concerned with cases for which *n* is a non-negative integer. By definition, $0! = 1$. Some other examples of factorials are

$$2! = 2 \times 1 = 2$$
$$3! = 3 \times 2 \times 1 = 6$$
$$\vdots$$
$$10! = 10 \times 9 \times 8 \times \ldots \times 1 = 3,628,800$$

Factorials obviously increase in size very rapidly. For example, how many different arrangements can be made if a deck of 52 cards is placed in a line? The answer 52! is a number that contains 68 digits.[2]

The expression in formula (5.4) is referred to as the binomial distribution. By substituting $x = 0, 1, 2, \ldots,$ and $n,$ we obtain the probabilities of $0, 1, 2, \ldots,$ and n successes in n trials. If these probabilities are added, they sum to one. Hence, the binomial distribution is a probability distribution. The term "binomial distribution" is used because we can obtain the probabilities of zero, one, and so forth successes in n trials from the respective terms of the binomial expansion of $(q + p)^n$. In Examples 5-5 and 5-6 we show how to use the binomial distribution to calculate probabilities.

[2] Warren Weaver points out in *Lady Luck* (Garden City, N.Y.: Doubleday, 1963), p. 88, about the number of possible arrangements in 52!, that, "If every human being on earth counted a million of these arrangements per second for twenty-four hours a day for lifetimes of eighty years each, they would have made only a negligible start in the job of counting all these arrangements—not a billionth of a billionth of one percent of them!"

EXAMPLE 5-5

The example of tossing a fair coin five times was used earlier in this chapter, and the probability of obtaining two heads (successes) was calculated. Compute the probabilities of all possible numbers of heads and thus establish the binomial distribution appropriate in this case.

SOLUTION

This problem is an application of the binomial distribution for $p = \frac{1}{2}$, $q = \frac{1}{2}$, and $n = 5$. If we let X represent the random variable "number of heads," the probability distribution is as follows:

x	P(x)
0	$\binom{5}{0}\left(\frac{1}{2}\right)^5\left(\frac{1}{2}\right)^0 = \frac{1}{32}$
1	$\binom{5}{1}\left(\frac{1}{2}\right)^4\left(\frac{1}{2}\right)^1 = \frac{5}{32}$
2	$\binom{5}{2}\left(\frac{1}{2}\right)^3\left(\frac{1}{2}\right)^2 = \frac{10}{32}$
3	$\binom{5}{3}\left(\frac{1}{2}\right)^2\left(\frac{1}{2}\right)^3 = \frac{10}{32}$
4	$\binom{5}{4}\left(\frac{1}{2}\right)^1\left(\frac{1}{2}\right)^4 = \frac{5}{32}$
5	$\binom{5}{5}\left(\frac{1}{2}\right)^0\left(\frac{1}{2}\right)^5 = \frac{1}{32}$

It may be noted that the probabilities represent the respective terms of the binomial $(\frac{1}{2} + \frac{1}{2})^5$.

EXAMPLE 5-6

In Chapter 4, the probability of obtaining at least one six in two rolls of a die (or in one roll of two dice) was calculated. Solve the same problem using the binomial distribution.

SOLUTION

We view the roll of the die as independent trials. If we define the appearance of a six as a success, $p = \frac{1}{6}$, $q = \frac{5}{6}$, and $n = 2$. It is instructive to examine the entire probability distribution.

x	$P(x)$
0	$\binom{2}{0}\left(\frac{5}{6}\right)^2\left(\frac{1}{6}\right)^0 = \left(\frac{5}{6}\right)^2 = \frac{25}{36}$
1	$\binom{2}{1}\left(\frac{5}{6}\right)^1\left(\frac{1}{6}\right)^1 = 2\left(\frac{5}{6}\right)\left(\frac{1}{6}\right) = \frac{10}{36}$
2	$\binom{2}{2}\left(\frac{5}{6}\right)^0\left(\frac{1}{6}\right)^2 = \left(\frac{1}{6}\right)^2 = \frac{1}{36}$

The required probability is

$$P\binom{\text{at least}}{\text{one six}} = \frac{10}{36} + \frac{1}{36} = \frac{11}{36}$$

Each pair of values of n and p establishes a different binomial distribution. Thus, the binomial is, in fact, a family of probability distributions. Since computations become laborious for large values of n, it is advisable to use special tables. Table A-1 in Appendix A gives values of $P(x) = \binom{n}{x}q^{n-x}p^x$ for $n = 2$ to $n = 20$ and $p = 0.05$ to $p = 0.50$ in multiples of 0.05. Probabilities that x is less than or greater than a given value, or that x lies between two values, and probabilities for p values greater than 0.50 can be obtained by appropriate manipulation of these tabulated values. Example 5-7 illustrates the use of the table. More extensive tables have been published by the National Bureau of Standards and Harvard University, but even such tables do not usually go beyond $n = 100$. For large values of n, approximations are available for the binomial distribution, and the exact values generally need not be determined.

EXAMPLE 5-7

An oil exploration firm is formed with enough capital to finance ten ventures. The probability of any exploration being successful is 0.10. What are the firm's chances of

(a) exactly one successful exploration?
(b) at least one successful exploration?
(c) two or less successful explorations?
(d) three or more successful explorations?

SOLUTION

Let $p = 0.10$ stand for the probability that an exploration will be successful. Then $q = 0.90$, $n = 10$, and x represents the number of successful explorations in 10 trials.

(a) $P(X = 1) = \binom{10}{1}(0.90)^9(0.10)^1$ is the required probability. If we look up $n = 10$, $x = 1$ under $p = 0.10$ in Table A-1, we find this probability to be equal to 0.3874.

(b) The event "at least one successful exploration" is the complement of the event "zero successful explorations." Therefore,

$$P(X \geq 1) = 1 - P(X = 0) = 1 - 0.3487 = 0.6513$$

(c) To obtain the probability of "two or less successful explorations," we add the probabilities of zero, one, and two successes. Thus,

$$P(x \leq 2) = 0.3487 + 0.3874 + 0.1937 = 0.9298$$

(d) The event "three or more successes" is the complement of "two or less successes." Hence, using the result from (c), we have

$$P(X \geq 3) = 1 - P(X \leq 2) = 1 - 0.9298 = 0.0702$$

It is important to note that with the binomial distribution, as with any other mathematical model, the correspondence between the real world situation and the model must be carefully established. In many cases, the underlying assumptions are obviously not met. For example, suppose that in a production process items produced by a certain machine tool are tested to determine whether they meet specifications. If the items are tested in the order in which they are produced, the assumption of independence would doubtless be violated. That is, whether an item meets specifications would not be independent of whether the preceding item(s) did. If the machine tool had become subject to wear, it is quite likely that if it produced an item that did not meet specifications, the next item would fail to conform to specifications in a similar way. Thus, whether or not an item is defective would *depend* on the characteristics of preceding items. In the coin tossing illustration, on the other hand, we conceived of an experiment in which a head and a tail on a particular toss did not affect the outcome on the next toss.

It can be seen from the underlying assumptions that the binomial is applicable to the situations of *sampling from a finite universe with replacement* or *sampling from an infinite universe,* with or without replacement. In either of these cases, the probability of success may be viewed as remaining constant from trial to trial and the outcomes as independent among trials. If the population size is large relative to sample size, that is, if the sample constitutes only a small fraction of the population, and p is neither very close to zero nor one in value, the binomial distribution is often sufficiently accurate, even though sampling may be carried out from a finite universe without replacement. It is difficult to give universal rules of thumb on what constitutes a sufficiently large sample for this purpose. Some practitioners suggest a population size at least ten times the sample size. However, the purpose of the calculations is the most important criterion for the required degree of accuracy.

EXERCISES

1. A firm bills its accounts on a 1% discount for payment within ten days and full amount due after ten days. In the past, 30% of all invoices have been paid within ten days. During the first week of July the firm sends out eight invoices. If you assume independence, what is the probability that
   (a) no one takes the discount?
   (b) everyone takes the discount?
   (c) three take the discount?
   (d) at least three take the discount?

2. A certain portfolio consists of six stocks. The investor feels the probability that each stock will go down in price is 0.4 and that these price movements are independent. What is the probability that exactly three will decline? That three or more will decline? Does the assumption of independence here seem logical? If not, is the binomial distribution the appropriate probability distribution for this problem?

3. A certain type of plastic bag in the past has burst under a pressure of 15 pounds 20% of the time. If a prospective buyer tests five bags chosen at random, what is the probability that exactly one will burst? Assume independence.

4. A manufacturer of 60-second development film advertises that 95 out of 100 prints will develop. A person buys a roll of ten prints and finds two do not develop. If the manufacturer's claim is true, what is the probability that two or more prints will not develop? Assume independence.

5. An economist claims he can predict whether a company's gross sales over the given year will or will not increase, from the previous year's annual statement of the company. He is given ten companies' reports selected randomly and asked to predict whether each company's gross sales will increase or not. He predicts six correctly. If one were to guess randomly, what is the probability that he would make six or more correct predictions?

6. A used-car salesman claims that the odds are only 3:1 against his selling a car to any particular customer. He attempts to sell automobiles to eight customers on a given day. If you assume independence, what is the probability that he will make at least one sale?

7. A physician knows from long experience that the probability is 0.25 that a patient with a certain disease will recover. At the present times he has three patients with the disease. Find the probability that
   (a) all three will recover.
   (b) at least one will recover.

8. A small manufacturer of electrical wire submits a bid on each of four different government contracts. In the past this manufacturer's bid has been the low-bid (i.e., he was awarded the contract) 15% of the time. If you assume this relative frequency is a correct probability assignment and you assume independence, what is the probability that the firm will not obtain any of the four contracts?

9. A cashier at a checkout counter of a supermarket makes mistakes on the bills of 20%

of the customers he checks out. Of the first five people he checked out, what is the probability that he made no mistakes?

10. Graph the binomial distribution for $n = 10$, $p = 0.2$, 0.5, and 0.8. What can be said about skewness and the symmetry of the binomial distribution as the value of p departs from 0.5?

THE POISSON DISTRIBUTION   5.4

Although as indicated earlier, the binomial distribution is the most widely applied probability distribution of a discrete random variable, others, such as the multinomial, the hypergeometric, and the Poisson distributions, are useful in theoretical work and for certain types of applications. We will discuss only the Poisson distribution, which has been widely used in managerial science and operations research.

The Poisson distribution was named for the Frenchman who developed it during the first half of the nineteenth century.[3] The distribution can be used in its own right and also for approximation of binomial probabilities. The former use is by far the more important one and in this context has had many fruitful applications in a wide variety of fields.

The Poisson as a distribution in its own right

The Poisson distribution has been usefully employed to describe the probability functions of phenomena such as product demand; demands for service; numbers of telephone calls that come through a switchboard; numbers of accidents; numbers of traffic arrivals such as trucks at terminals, airplanes at airports, ships at docks, and passenger cars at toll stations; and numbers of defects observed in various types of lengths, surfaces, or objects.

All these illustrations have certain elements in common. The given occurrences can be described in terms of a discrete random variable, which takes on values 0, 1, 2, and so forth. For example, product demand can be characterized by 0, 1, 2, etc., units purchased in a specified time period; number of defects can be counted as 0, 1, 2, etc., in a specified length of electrical cable. The product demand example may be viewed in terms of a process that produces random occurrences in continuous time. The defects example pertains to random occurrences in a continuum of space. In cases such as counting numbers of defects, the continuum may not only be one of length, but also one of area or volume. Thus, there may be a count of the number of blemishes in areas of sheet metal used for aircraft or the number of a certain type of microscopic particle in a unit of volume such as a cubic centimeter of a solution. In each case, there is some rate in terms of number of occurrences per interval of time or space that characterizes the process producing the outcome.

---

[3] Siméon Denis Poisson (1781–1840), was particularly noted for his applications of mathematics to the fields of electrostatics and magnetism. He wrote treatises in probability, calculus of variations, Fourier's series, and other areas.

Using as an example the occurrences of defects in a length of electrical cable, we can indicate the general nature of the process that produces a Poisson probability distribution. The length of cable has some rate of defects per interval, say, two defects per meter. If the entire length of cable is subdivided into very small subintervals, say, of one millimeter each: (1) the probability that exactly one defect occurs in this subinterval is a very small number and is constant for each such subinterval; (2) the probability of two or more defects in a millimeter is so small it may be considered to be zero; (3) the number of defects that occurs in a millimeter does not depend on where that subinterval is located; and (4) the number of defects that occurs in a subinterval does not depend on the number of events in any other non-overlapping subinterval.

In the foregoing example, the subinterval was one of length. Analogous sets of conditions would characterize examples in which the subinterval is a unit of area, volume, or time.

The nature of the Poisson distribution

As indicated, the Poisson distribution is concerned with occurrences that can be described by a discrete random variable. This random variable, denoted *X,* can take on values $x = 0, 1, 2, \ldots$, where the three dots mean "*ad infinitum.*" That is, the domain of the Poisson probability function consists of all non-negative integers. The probability of exactly x occurrences in the Poisson distribution is

(5.6)
$$P(x) = \frac{\mu^x e^{-\mu}}{x!} \text{ for } x = 0, 1, 2, \ldots$$

where μ is the mean number of occurrences per interval and $e = 2.71828 \ldots$ (the base of the Naperian or natural logarithm system).

As can be seen from (5.6), the Poisson distribution has a single parameter symbolized by the Greek letter μ ("mu"). If we know the value of μ, we can write out the entire probability distribution. The parameter μ can be interpreted as the arithmetic mean number of occurrences per interval of time or space that characterizes the process producing the Poisson distribution. Thus, μ may represent, for example, an average of 3.0 units of demand per day, 5.3 demands for service per hour, 1.2 aircraft arrivals per five minutes, and so forth.

As an illustration of the way in which probabilities are obtained for the Poisson distribution, consider Example 5-8.

EXAMPLE 5-8

A department store has determined by its inventory control system that the demand for a certain home furnishing item was distributed according to the Poisson distribution with an arithmetic mean of four units per day.
(a) Determine the probability distribution of the daily demand for this item.
(b) If the store stocks five of these items on a particular day, what is the probability that the demand will be greater than the supply?

Since calculations for the Poisson distribution can be quite tedious, it is ordinarily advisable to use tables. Table A-3 of Appendix A lists values of the Poisson distribution, that is, values of $P(x)$ or the probability of x successes for selected values of μ. As in Table A-1 for the binomial distribution, probabilities of greater than or less than a given number of successes can be obtained by appropriate addition of the tabulated values.

(a) Let $P(x)$ represent the probability that demand is x items per day. $\mu = 4$ items per day. Using $\mu = 4$, and $x = 0$, in Table A-3, we find $P(X = 0) = 0.0183$, etc. The entire probability distribution is

x	P(x)
0	0.0183
1	0.0733
2	0.1465
3	0.1954
4	0.1954
5	0.1563
6	0.1042
7	0.0595
8	0.0298
9	0.0132
10	0.0053
11	0.0019
.	.
.	.
.	.

The sum of the probabilities for demand from zero through eleven units is 0.9991. Therefore, the sum of the probabilities for twelve or more units is only 0.0009.

(b) The probability that the demand will be greater than five units is the complement of the probability that it will be five units or less. Hence, adding the probabilities given above for zero through five units, we have

$$P(X \leq 5) = 0.7852$$

and

$$P(X > 5) = 1 - 0.7852 = 0.2148$$

The Poisson as an approximation to the binomial distribution

We turn now to a consideration of the Poisson distribution as an approximation to the binomial distribution. Since computations involving the binomial distribution become quite tedious when n is large, it is useful to have a simple method of approximation.

Assume in the expression for $P(x)$ of the binomial distribution that n is permitted to increase without bound and p approaches zero in such a way that np remains constant. Let us denote this constant value for np as μ. With these assumptions, it can be shown that the binomial expression for $P(x)$ approaches the following value:

$$P(x) = \frac{\mu^x e^{-\mu}}{x!}$$

where $\mu = np$, and e is the base of the natural logarithm system. As can be seen from expression (5.6), the value approached by the binomial distribution under the given conditions is the Poisson distribution. Hence, the Poisson distribution can be used as an approximation to the binomial probability function. In this context, the Poisson distribution gives the probability of observing x successes in n trials of an experiment, where p is the probability of success on a single trial. That is, x, n, and p are interpreted in the same way as in the binomial distribution.

Because of the assumptions underlying the derivation of the Poisson distribution from the binomial distribution, the approximations to binomial probabilities are best when n is large and p is small. A frequently used rule of thumb is that the approximation is appropriate when $p \leq 0.05$ and $n \geq 20$.

Example 5-9 illustrates the use of the Poisson approximation to the binomial distribution.

EXAMPLE 5-9

An automatic machine produces washers, three percent of which are defective according to a severe set of specifications. If a sample of 100 washers is drawn at random from the production of this machine, what are the probabilities of observing
(a) exactly three defectives?
(b) between two and four defectives, inclusive?
It may be assumed that the universe of product sampled was sufficiently large to warrant binomial probability calculations.

SOLUTION

(a) With $p = 0.03$ and $n = 100$, the probability of exactly three defectives, using more complete tables for the binomial distribution than appear in this text is

$$P(x = 3) = \binom{100}{3}(0.97)^{97}(0.03)^3 = 0.227$$

An approximation to this probability is given by the Poisson distribution with

$$\mu = np = 100 \times 0.03 = 3$$

The Poisson probability of exactly three defectives is $P(x = 3) = 0.2240$, as determined from Table A-3 of Appendix A, using $\mu = 3$ and $x = 3$. Thus, the approximation is in error by about 1%.

(b) From a sufficiently extensive table of values of the binomial distribution, it can be determined that

$$P(2 \leq x \leq 4) = 0.623$$

The corresponding probability according to the Poisson distribution with $\mu = 3$ can be determined from Table A-3 of Appendix A as

$$P(2 \leq x \leq 4) = 0.2240 + 0.2240 + 0.1680 = 0.6160$$

Again, the approximation error is about 1%.

EXERCISES

1. A firm's office contains 20 typewriters. The probability that any one typewriter will not work on a given day is 0.05. With independence assumed, the binomial probability that exactly one will not work on a given day is 0.3773; that at least two will not work on a given day is 0.2642. Use the Poisson approximation to compute these probabilities and compare your results.

2. According to the manager of a certain motel, the probability that persons inquiring about a room will actually take a room is 0.1.
   (a) According to the binomial distribution, after 10 people inquire, what is the probability that one or less will take a room?
   (b) Use the Poisson approximation to solve (a) and compare your results.

3. In reference to Exercise 2, suppose the probability that someone would want to take a room was 0.05 instead of 0.1, what would be the binomial probability and the Poisson approximation? Why do you think there is a difference between the "goodness" of the approximation between the two problems?

6 Sampling Distributions

6.1   FUNDAMENTALS OF SAMPLING

Sampling is important in most applications of quantitative methods to managerial problems and, indeed, to most problems in business, political affairs, and science. Items can be selected from statistical universes in a variety of ways; for example, random or non-random methods of selection can be used. In random or probability sampling, the probability of inclusion of every element in the universe is *known.* Nonrandom sampling methods are referred to as "judgment sampling," that is, sampling in which judgment is exercised in deciding which elements of a universe to include. Such judgment samples may be drawn by choosing "typical" elements or groups of elements to represent the population.

We will deal only with random or probability sampling methods rather than judgment sampling because of the clear-cut superiority of probability selection techniques. In judgment selection there is no objective method of measuring the precision or reliability of estimates made from the sample. On the other hand, in random sampling, measures of the precision with which estimates of population values can be made are obtainable from the sample itself. Random sampling techniques thus provide an objective basis for measuring errors due to the sampling process and for stating the degree of confidence to be placed upon estimates of population values.

Judgment samples can sometimes be usefully employed in the planning and design of probability samples. For example, where expert judgment is available, a pilot sample may be selected on a judgment basis in order to obtain information that will aid in the development of an appropriate sampling frame for a probability sample.

Simple random sampling

There are many types of probability or random samples, particularly in the area of sample surveys. Experts in survey sampling have developed a large body of theory and practice to aid in the optimal design of probability samples. This highly specialized area often has an entire course or two devoted to it in a graduate statistics program. We will concentrate upon the simplest and most fundamental probability sampling method, namely, *simple random sampling.* The major body of statistical theory is based upon this method of sampling.

We first define a simple random sample for a finite population of N elements. A simple random sample of n elements is a sample drawn in such a way that *every combination of n elements has an equal chance of being the sample selected.* Since most practical sampling situations involve sampling *without replacement,* it is useful to think of this type of sample as one in which each of the N population elements has an equal probability, $1/N$, of being the one selected on the first draw, each of the remaining $N - 1$, has an equal probability, $1/(N - 1)$, of being selected on the second draw, and so on until the nth sample item has been drawn.

For example, consider a population that consists of three elements, $A, B,$ and C. Thus, $N = 3$. Suppose, we wish to draw a simple random sample of two elements. Then, $n = 2$. The total number of possible samples is three. These three possible samples contain the following pairs of elements: (A, B), (A, C), and (B, C). The probability that any one of these three samples will be the one selected is $\frac{1}{3}$.

We have defined a simple random sample for the case of sampling a finite population without replacement. If a finite population is sampled *with replacement,* the same element could appear more than once in the sample. But this sampling procedure has virtually no practical application and, therefore, will not be discussed here.

However, simple random sampling of *infinite populations* is important. The following definition corresponds to the one for finite populations. For an infinite population, a simple random sample is one in which *on every selection, each element of the population has an equal probability of being the one drawn.* This concept is difficult to visualize in terms of actual sampling from a physical population. What it means is simply that: (1) The n successive sample observations are independent, and (2) the population composition remains constant from trial to trial. Thus, we can see that in the illustration of five tosses of a coin discussed in Chapter 5, we can now refer to the five outcomes as "a simple random sample of five observations from a binomial population (or binomial distribution)."

Note that the term "random sample," although it properly refers to a sample drawn with known probabilities, is often used to mean "simple random sample."

Methods of simple random sampling

Although it is easy to define a simple random sample, it is not always obvious how such a sample is to be drawn from an actual population.

Drawing chips from a bowl. We now consider the most straightforward situation in which the population is finite and in which the elements are easily identified and can be numbered. For example, suppose a college freshman class has 100 students and we wish to draw a simple random sample of ten of these students without replacement. We could assign numbers from 1 to 100 to each of the students and place these numbers on physically similar disks (or balls, slips of paper, etc.), and put them in a bowl. We could shake the bowl to mix the disks thoroughly and then draw the sample. On drawing the first disk, we record the number written on it. We then shake the bowl again, draw the second disk, and record the result. We could repeat the process until we have drawn ten numbers. The students corresponding to these ten numbers would then constitute the required simple random sample.

Tables of random numbers. If the population size were very large, the foregoing procedure would become quite unwieldy and time-consuming. Furthermore, it may introduce biases if the disks are not thoroughly mixed. Therefore, in recent years, there has been a marked tendency to use tables of random digits for drawing such samples. These tables are also useful for the selection of other types of probability samples.

A table of random digits is simply a table of digits generated by a random process. Usually, the digits are combined, for example, into groups of five, for ease of use. Thus, a table of random digits could be generated by drawing chips from a bowl, in a process similar to the one just described. The digits 0, 1, 2, . . . , 9 could be written on disks, the disks placed in a bowl, and then drawn, one at a time, *with the selected disk replaced after each drawing.* Thus, on each selection, the population would consist of the ten digits. The recorded digits would constitute a particular sequence of random digits. These tables are now usually produced by a computer, which has been programmed to generate random sequences of digits.

Suppose there were 9241 undergraduates at a large university and we wished to draw a simple random sample of 300 of these students. Each of the 9241 students could be assigned a four-digit number, say, for convenience, from 0001 to 9241. This list of names and numbers would constitute the sampling frame. We now turn to a table of random digits, such as Table 6-1, to select a simple random sample of 300 such four-digit numbers. We may begin on any page in the table and proceed in any systematic manner to draw the sample. Assume we decided to use the first four columns of each group of five, beginning at the upper left and reading downward. The first group of five digits on the left-hand side in Table 6-1 designates the line number, so we ignore it. Starting with the second group of digits, we find the sequence 98389. Since we are using the first four digits, we have the number 9838, which exceeds the largest number in our population, 9241. Therefore, we ignore this number and read down to pick up the next four-digit number, 1724. This, then, is the number of the first student in the sample. Reading down consecutively, we find 0128, 9818 (which we ignore), 5926, and so forth until 300 four-digit numbers between 0001 and 9241 have been specified. If any previously selected number is repeated, we simply ignore the repeated appearance, and continue. In this illustration, we read downward on the page. We could have read laterally, diagonally, or in any other systematic fashion. What is important is that each four-digit number has an equal probability of selection, regardless of the systematic method of drawing used and regardless of the preceding numbers.

Methods are available for drawing samples other than simple random ones, and even for situations where the elements have not been prelisted. Many tables include instructions for their use, so we will not pursue the subject further.

In this section, we have concentrated on simple random sampling. Since the theoretical structure of statistical inference is largely based on this type of

---

TABLE 6-1

Random Digits

19300	98389	95130	36323	33381	98930	60278	33338	45778	86643	78214
19301	17245	58145	89635	19473	61690	33549	70476	35153	41736	96170
19302	01289	68740	70432	43824	98577	50959	36855	79112	01047	33005
19303	98182	43535	79938	72575	13602	44115	11316	55879	78224	96740
19304	59266	39490	21582	09389	93679	26320	51754	42930	93809	06815
19305	42162	43375	78976	89654	71446	77779	95460	41250	01551	42552
19306	50357	15046	27813	34984	32297	57063	65418	79579	23870	00982
19307	11326	67204	56708	28022	80243	51848	06119	59285	86325	02877
19308	55636	06783	60962	12436	75218	38374	43797	65961	52366	83357
19309	31149	06588	27838	17511	02935	69747	88322	70380	77368	04222
19310	25055	23402	60275	81173	21950	63463	09389	83095	90744	44178
19311	35150	34706	08126	35809	57489	51799	01665	13834	97714	55167
19312	61486	33467	28352	58951	70174	21360	99318	69504	65556	02724
19313	44444	86623	28371	23287	36548	30503	76550	24593	27517	63304
19314	14825	81523	62729	36417	67047	16506	76410	42372	55040	27431
19315	59079	46755	72348	69595	53408	92708	67110	68260	79820	91123
19316	48391	76486	60421	69414	37271	89276	07577	43880	08133	09898
19317	67072	33693	81976	68018	89363	39340	93294	82290	95922	96329
19318	86050	07331	89994	36265	62934	47361	25352	61467	51683	43833
19319	84426	40439	57595	37715	16639	06343	00144	98294	64512	19201
19320	41048	26126	02664	23909	50517	65201	07369	79308	79981	40286
19321	30335	84930	99485	68202	79272	91220	76515	23902	29430	42049
19322	33524	27659	20526	52412	86213	60767	70235	36975	28660	90993
19323	26764	20591	20308	75604	49285	46100	13120	18694	63017	85112
19324	85741	22843	16202	48470	97412	65416	36996	52391	81122	95157

SOURCE: RAND Corporation, *A Million Random Digits with One Hundred Thousand Normal Deviates* (Glencoe, Ill.: Free Press, 1955), excerpt from page 387. Used by permission.

sampling, most of the remaining chapters of this book assume this sampling procedure. However, in sample survey work, many other types of sample designs are used to accomplish increased sampling precision for fixed cost or to minimize cost for a fixed level of sampling precision. Even these designs, though, build on the foundation principles of simple random sampling.

6.2 THE BINOMIAL AS A SAMPLING DISTRIBUTION

In Chapter 3, we examined the methods by which statistics such as the arithmetic mean and standard deviation are computed from the data contained in a sample. In Section 6.1, we discussed how simple random samples can be drawn from finite and infinite populations. We now consider how such statistics differ from sample to sample if repeated simple random samples of the same size are drawn from statistical populations. For example, a given statistic, such as a proportion or a mean, will vary from sample to sample. The probability distribution of such a statistic is referred to as a *sampling distribution.* Thus, we may have a sampling distribution of a proportion, a sampling distribution of a mean, etc. These sampling distributions are a fundamental concept of modern statistical methods. We will first discuss the sampling distributions of numbers of occurrences and proportions of occurrences.

Let us assume that of the articles produced by a certain manufacturing process 10% are defective and 90% are not. We conceive of the production process as an infinite population. Thus, we may view the successive drawings of articles from the process as a series of trials for which the binomial is the appropriate probability distribution. That is, the three requirements of such a process may be interpreted in terms of this problem as follows: (1) each draw has two possible outcomes, defective or non-defective; (2) the probability of a defective (non-defective) remains constant from draw to draw; and (3) the draws are independent. We also note that the three conditions would hold equally well if we assumed we were sampling a finite shipment of articles, but replaced each article prior to drawing the next one. In summary, then, we see that our discussion pertains to the sampling of an *infinite universe* or a *finite universe with replacement.*

Now let us return to the production process. Suppose we were to draw a simple random sample of five articles and note the number of defectives in the sample. This number is a random variable, which can take on the values 0, 1, 2, 3, 4, or 5. The probabilities of obtaining these numbers of defectives may be computed by means of a binomial distribution in which $p = 0.10$, $q = 0.90$, and $n = 5$. Therefore, the respective probabilities are given by the expansion of the binomial $(0.9 + 0.1)^5$. This probability distribution is shown in Table 6-2 with the notation of Chapter 5.

We now interpret the probability distribution given in Table 6-2 as a sampling distribution. The number of defectives observed in a sample of five articles is a sample statistic. Thus, Table 6-2 displays the probability distribution of this sample statistic. Let us take a long-run relative frequency of occurrence point of view and interpret the probabilities as follows. If we took repeated simple random samples of five articles each from a manufacturing process that produces 10% defective articles, in 59% of these samples we would observe zero defectives, in 33% we would find one defective, and so forth. In summary, the probability distribution may now be called a "sampling distribution of number of occurrences."

TABLE 6-2

Probability Distribution of the Number of Defectives
in a Simple Random Sample of Five Articles from
a Process Producing 10% Defectives

NUMBER OF DEFECTIVES x	PROBABILITY $P(x)$
0	$\binom{5}{0}(0.9)^5(0.1)^0 = 0.59$
1	$\binom{5}{1}(0.9)^4(0.1)^1 = 0.33$
2	$\binom{5}{2}(0.9)^3(0.1)^2 = 0.07$
3	$\binom{5}{3}(0.9)^2(0.1)^3 = 0.01$
4	$\binom{5}{4}(0.9)^1(0.1)^4 = 0.00$
5	$\binom{5}{5}(0.9)^0(0.1)^5 = 0.00$

The mean and standard deviation of the binomial

We turn now to the properties of the binomial distribution. The mean and standard deviation of that distribution are of particular interest in statistical inference. Using the binomial distribution in Table 6-2 as an illustration, we calculate the mean by the expected value formula (5.3) as follows (the symbol μ is used here rather than $E(X)$):

$$\mu = \Sigma x P(x) = 0(0.59) + (1)(0.33) + (2)(0.07) + (3)(0.01)$$
$$+ (4)(0.00) + (5)(0.00) = 0.50 \text{ defectives}$$

Before we turn to an analogous calculation of the standard deviation, let us clarify the formula to be used. Recall that the mean of a frequency distribution of observed data as calculated by formula (3.3) is $\overline{X} = \Sigma fX/n$. We can think of the f/n values as the relative frequencies of occurrence for each class of the frequency distribution. Let us now compare this formula with the one we have just used for the mean of a probability distribution, $\mu = \Sigma x P(x)$. We see these two formulas are really the same, with the probabilities $P(x)$ playing the same role for probability distributions as the relative frequencies f/n in frequency distributions. In an analogous way, we can write the formula for a standard deviation of a probability distribution. Using notation appropriate for probability distributions, we have the following formula for the standard deviation of a probability distribution:

(6.1) $$\sigma = \sqrt{\Sigma(x - \mu)^2 P(x)}$$

Applying formula (6.1) to the binomial distribution given in Table 6-2 yields the following:

$$\sigma = \sqrt{(0 - 0.5)^2(0.59) + (1 - 0.5)^2(0.33) + \cdots + (5 - 0.5)^2(0.00)}$$
$$= 0.67 \text{ defectives}$$

As may be seen from the above, computations of the mean and standard deviation for the binomial distribution may become quite tedious from these definitional formulas, particularly for large values of n. In actual applications, such calculations are virtually never made. They are given here only to aid in your understanding of the meaning of sampling distributions. General formulas for these measures can be derived by substituting $\binom{n}{x} q^{n-x} p^x$ for $P(x)$ in the definitional formulas and performing appropriate manipulations. The results of these derivations are as follows:

MEAN AND STANDARD DEVIATION OF THE BINOMIAL DISTRIBUTION

(6.2) $$\mu = np$$

(6.3) $$\sigma = \sqrt{npq}$$

Let us illustrate the use of formulas (6.2) and (6.3) for the binomial distribution given in Table 6-2. Substituting $p = 0.10$, $q = 0.90$, and $n = 5$, we obtain

$$\mu = np = 0.50 \text{ defectives}$$
$$\sigma = \sqrt{npq} = \sqrt{5(0.10)(0.90)} = 0.67 \text{ defectives}$$

Of course, these results are the same as those obtained in the longer calculations shown earlier, and one would normally use the simpler formulas. Let us interpret these results in terms of the appropriate sampling distributions. The mean of 0.50 defectives for the binomial distribution in this example means that if repeated simple random samples of five articles each are drawn from a process containing 10% defectives, there will be, on the average, one-half a defective per sample. The standard deviation of 0.67 defectives is a measure of the variation in number of defectives that is attributable to the chance effects of random sampling. The use of such means and standard deviations of sampling distributions is described later in this chapter.

The binomial distribution in this example is skewed as shown in Figure 6-1. If p is less than 0.50, as in this case, the distribution tails off to the right as in Figure 6-1. On the other hand, if p exceeds 0.5, the skewness is to the left. If p is held fixed, and the sample size n becomes larger, the sampling distribution of the number of occurrences becomes more and more symmetrical. This is an important property of the binomial distribution from the standpoint of sampling theory and practice, which we examine further in the next section.

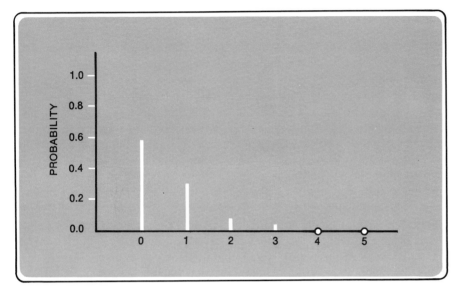

Figure 6-1
Graph of binomial distribution for $p = 0.1$, $q = 0.9$, and $n = 5$.

EXERCISES

1. An inspector of a bottling company's assembly line draws ten filled bottles at random for inspection. If a bottle contains within a half of an ounce of the proper amount it is classified as good, otherwise it is classified as defective. List the possible numbers of defective bottles he could obtain in a sample. Assuming the manufacturing process produces 10% defectives, assign probabilities to each of the possible sample outcomes. Calculate the mean and variance for the number of defectives using formulas (6.2) and (6.3).

2. Assume that 40% of the population favors a certain type of gun law legislation. If ten people selected at random are questioned in regard to the gun legislation, what would be the mean and standard deviation of the sampling distribution of the number favoring the gun legislation? If 20 people were questioned? If 40 people were questioned?

3. A political poll interviews 500 people at random and determines that X of them approve of candidate A.
   (a) What is the form of the distribution of X? Assume independence.
   (b) If 40% of the over-all population from whom the 500 were chosen favor A, what is the variance of X?

6.3 CONTINUOUS DISTRIBUTIONS

Thus far, we have dealt solely with probability distributions of *discrete* random variables. Probability distributions of continuous random variables are also of considerable importance in statistics. We turn therefore to an examination of such distributions, with particular emphasis on the meaning of their graphs. (It might be helpful to review the definitions of discrete and continuous variables given in Section 5.1.)

The binomial distribution we have been discussing in this chapter is an example of a probability distribution of a *discrete* random variable. We have graphed such distributions by erecting ordinates at distinct values along the horizontal axis. To gain better insight into the meaning of a graph of the probability distribution of a continuous random variable, let us begin by graphing a binomial distribution as a *histogram*. We assume a situation in which a true coin is tossed two times and the random variable of interest is the number of heads obtained. We have previously seen that the probabilities of zero, one, and two heads are, respectively, $\frac{1}{4}$, $\frac{1}{2}$, and $\frac{1}{4}$. This is an illustration of a binomial distribution in which $p = \frac{1}{2}$ and $n = 2$. If the coin were tossed four times, the probabilities of 0, 1, 2, 3, and 4 heads are, respectively, $\frac{1}{16}$, $\frac{4}{16}$, $\frac{6}{16}$, $\frac{4}{16}$, and $\frac{1}{16}$. This is a binomial distribution in which $p = \frac{1}{2}$ and $n = 4$. Figure 6-2 shows graphs of these distributions in the form of histograms. In these histograms, let us now interpret 0, 1, 2, 3, and 4 heads not as discrete values, but rather as midpoints of classes whose respective limits are $-\frac{1}{2}$ to $\frac{1}{2}$, $\frac{1}{2}$ to $1\frac{1}{2}$, $1\frac{1}{2}$ to $2\frac{1}{2}$, and so forth. The probabilities or relative frequencies associated with these classes are represented on the graph by the areas of the rectangles or bars. Thus, in the graph for $n = 4$, since the rectangle for the class interval

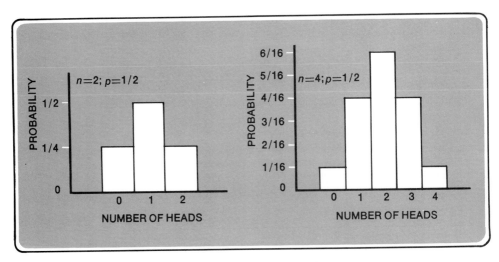

Figure 6-2
Histograms of the binomial distribution for $n = 2$, $p = \frac{1}{2}$, and $n = 4$, $p = \frac{1}{2}$.

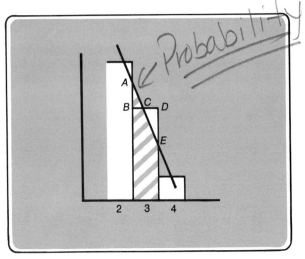

Figure 6-3
Approximation of a histogram by a continuous curve.

$2\frac{1}{2}$ to $3\frac{1}{2}$ has four times the area of that from $3\frac{1}{2}$ to $4\frac{1}{2}$, it represents four times the probability. If we were to represent the histogram for the case where $n = 4$ by means of a smooth continuous curve, the curve would pass through the rectangle for three heads as shown in Figure 6-3. It is clear that the shaded area under the curve for the class interval $2\frac{1}{2}$ to $3\frac{1}{2}$ is approximately equal to the area of the rectangle representing the probability of three heads, because the included area *ABC* is about equal to the excluded area *CDE*. In summary, in the approximation of a histogram by a smooth curve, the area under the curve bounded by the class limits for any given class represents the probability of occurrence of that class. In the foregoing illustration, if we had greatly increased n, say to 50 or 100 and decreased the width of the rectangles, we would have a visual demonstration that the histogram appeared to approach more and more closely a continuous curve. Since the total area of the rectangles in a histogram representing a probability distribution of a discrete random variable is equal to one, the total area under a continuous curve representing the probability distribution of a continuous random variable is correspondingly equal to one. *The area under the curve lying between the two vertical lines erected at points a and b on the x axis represents the probability that the random variable X takes on values in the interval a to b.* This situation is depicted in Figure 6-4.

Using mathematics beyond the scope of this book, it can be shown that, in the binomial distribution, if p is held fixed while n is increased without limit, the binomial approaches a particular continuous distribution, referred to as the normal distribution, normal curve, or Gaussian distribution, after the mathematician and astronomer Karl Gauss, and shown in Figure 6-5. Although our illustration has been in terms of $p = \frac{1}{2}$, this is not a necessary condition for the proof. Even if the binomial is not symmetrical, that is, $p \neq \frac{1}{2}$, the binomial

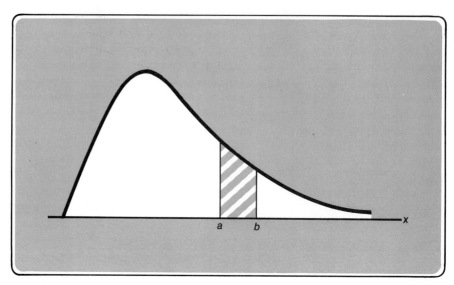

Figure 6-4
Graph of a continuous distribution: probability that the random variable *X* lies between *a* and *b*.

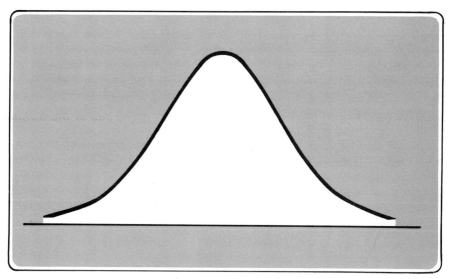

Figure 6-5
Graph of the normal distribution.

still approaches the normal distribution as *n* increases. Thus, if *p* is not close to zero or one, the binomial distribution, which is the appropriate sampling distribution for *X,* the number of successes in a sample of *n* trials, can be closely approximated by a normal curve with the same mean and standard deviation as the binomial, that is, $\mu = np$ and $\sigma = \sqrt{npq}$.

A brief comment on "standard units" is useful at this point because such units are discussed in the next section, but more generally because they are so widely employed, particularly in sampling theory and statistical inference. Standard units are merely an example of the previously mentioned "standard score" (see Section 3.14). The standard score is the deviation of a value from the mean of a frequency or probability distribution stated in units of the standard deviation. In general, it takes the form $(x - \mu)/\sigma$, where x denotes the value of the item, and μ and σ are the mean and standard deviation of the distribution, respectively. Other terms used to refer to standard scores or standard units include "standardized unit," "standardized form," and "standard form."

THE NORMAL DISTRIBUTION 6.4

The normal distribution plays a central role in statistical theory and practice, particularly in the area of statistical inference. Because of the relationship we observed in Section 6.3, the normal distribution is very useful as an approximation to the binomial in many instances where the latter distribution is the theoretically correct one. As we will see, calculations involving the normal curve are generally much easier than those involving the binomial distribution because of the simple compact form of tables of areas under the normal curve.

The normal distribution is also important in its own right in sampling applications. Before we consider such applications, let us examine the basic properties of the distribution.

Properties of the normal curve

Probability distributions of continuous random variables can be described by the same types of measures, such as means, medians, and standard deviations, as those used for discrete random variables. One of the important characteristics of the normal curve is that we need only know the mean and standard deviation to be able to compute the entire distribution.

The height of the normal curve, y, at any value x is given by the equation

(6.4)
$$y = \frac{1}{\sqrt{2\pi}\,\sigma}\, e^{-\frac{1}{2}[(x-\mu)/\sigma]^2}$$

In this equation, the mean and standard deviation, which determine the location and spread of the distribution, are denoted by μ and σ, respectively, which are said to be the two parameters of the normal distribution. This situation is analogous to that of the binomial distribution in which the parameters are n and p. π and e are simply constants that arise in the mathematical derivation and are approximately equal to 3.1416 and 2.7183, respectively.

Thus, for given values of μ and σ, if we substitute a value into equation (6.4) for x, we can compute the corresponding value for y. According to the

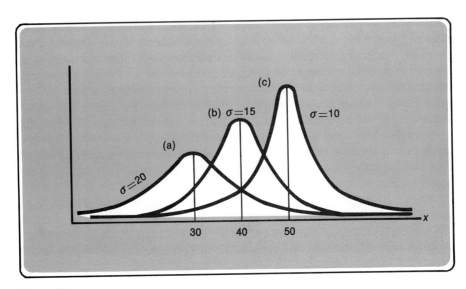

Figure 6-6
Three normal probability distributions.

usual convention, the values *x* of the variable of interest are plotted along the horizontal axis, and the corresponding ordinates *y* along the vertical.

Figure 6-6 shows three normal probability distributions, which differ in locations and spread. Thus, the mean of distribution (c), assumed equal to 50, is the largest, and distribution (a) has the smallest mean, 30. The standard deviation 10 of (c) is least, whereas 20 that of (a) is greatest. Thus, the normal distribution defined by equation (6.4) represents a family of distributions with the specific member of that family being determined by the values of the parameters μ and σ. Graphically, the normal curve is bell-shaped and symmetrical around the ordinate erected at the mean, which lies at the center of the distribution. By recalling our previous interpretation of the graph of a continuous probability distribution and since one-half of the probability lies to the left (right) of the mean, we see that the probability is 0.5 that a value of *x* will fall below (above) the mean. The values of *x* range from minus infinity to plus infinity. As we move farther away from the mean, either to the right or left, the height of the curve, *y*, decreases. Thus, moving in either direction from the mean, the curve approaches the horizontal axis, but never reaches it. However, for practical purposes, we rarely consider *x* values beyond three or four standard deviations from the mean, since virtually the entire area is included within this range. Stated differently, virtually no area in the tails of a normal distribution lies beyond 3 or 4 standard deviations from the mean.

Areas under the normal curve

We now turn to the use of the normal curve in terms of the areas under it. Although it was important to define the distribution in equation (6.4) in order to observe the relationship between *x* and *y* values, in most applications in statistical inference we are not interested in the heights of the curve. Rather, since the normal curve is a continuous distribution and emphasis is upon its use as a probability distribution, we are interested in the areas under the curve.

For convenience we shall use the term "normally distributed" for variables with normal probability distributions. Of course, the term "normal" has no implication of quality, in the sense that other distributions are in some respect "abnormal." The term merely refers to probability distributions describable by equation (6.4). Normally distributed variables occur in a variety of units, such as dollars, pounds, inches, and hours. For convenience, it is useful to transform a normally distributed variable into a form such that a single table of areas under the normal curve is applicable, regardless of the units of the original data. The transformation used for this purpose is that of the "standard unit." As we noted earlier, to express an observation of a variable in standard units, we obtain the deviation of this observation from the mean of the distribution and then state this deviation in multiples of the standard deviation.

For example, suppose a variable is normally distributed with mean 100 pounds and standard deviation ten pounds. One observation of this variable is the value 120 pounds. What is this number in standardized units? The deviation of 120 pounds from 100 pounds is +20 pounds in units of the original data. Dividing +20 pounds by 10 pounds, we obtain +2. Thus a deviation of +20 pounds from the mean lies two standard deviations above the mean if one standard deviation equals 10 pounds.

Let us state this notion in general form. The number of standard units *z* for an observation *x* from a probability distribution is defined by

(6.5)
$$z = \frac{x - \mu}{\sigma}$$

where x = the value of the observation
μ = the mean of the distribution
σ = the standard deviation of the distribution

Thus, in the illustration, $z = (120 - 100)/10 = 20/10 = +2$. The +2 indicates a value two standard deviations above the mean. If the observation had been 80, then $z = (80 - 100)/10 = -20/10 = -2$. The −2 denotes a value two standard deviations below the mean.

We now turn to another example to illustrate the use of a table of areas under the normal curve. Assume a manufacturing process that produces a certain electrical part, whose lifetime is normally distributed with an arithmetic mean of 1000 hours and a standard deviation of 200 hours. Before solving a number of probability problems concerning this distribution, we refer to Figure 6-7 to examine the relationship between values of the original variable (*x* values) and values in standard units (*z* values).

Suppose we wish to determine the proportion of parts produced by this process with lifetimes between 1000 and 1400 hours. This proportion or probability is indicated by the shaded area in Figure 6-7. We can obtain this value from Table A-5 of Appendix A, which gives areas under the normal curve lying between vertical lines erected at the mean and at specified points above the mean stated in multiples of standard deviations (z values). The left column of the table gives z values to one decimal place. The column headings give the second decimal place of the z value. The entries in the body of the table represent the area included between the vertical line at the mean and the line at the specified z value. Thus, in our example, the z value for 1400 hours is $z = (1400 - 1000)/200 = +2$. In Table A-5, we find the value 0.4772; hence, 47.72% of the area in a normal distribution lies between the mean and a value two standard deviations above the mean. In summary, 0.4772 is the proportion of parts produced by this process with lifetimes between 1000 and 1400 hours. Examples 6-1 through 6-4 further illustrate this concept.

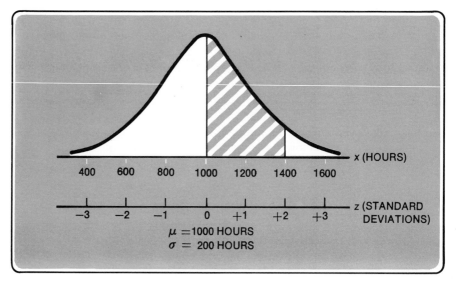

Figure 6-7
Relationship between x values and z values.

We now note a general point about the distribution of z values. Comparing the x scales and z scales in Figure 6-7, we see that for a value at the mean in the distribution of x, z is equal to zero. If an x value is at $\mu + \sigma$, that is, one standard deviation above the mean, $z = +1$, and so forth. Therefore, the probability distribution of z values, referred to as the "standard normal distribution," is simply a normal distribution with a mean of zero and a standard deviation equal to one.[1]

---

[1] For the reader with a knowledge of calculus, we note that Table A-5 gives values of the integral $\int_{0}^{z} f(z)\ dz$, where $z = (x - \mu)/\sigma$ and $f(z) = (1/\sqrt{2\pi})e^{-z^2/2}$.

EXAMPLE 6-1

What is the proportion of parts produced by the previous process with lifetimes between 600 and 1400 hours?

SOLUTION

First, we tranform to deviations from the mean in units of the standard deviation.

$$\text{If } x = 1400 \qquad z = \frac{1400 - 1000}{200} = +2$$

$$\text{If } x = 600 \qquad z = \frac{600 - 1000}{200} = -2$$

Thus, we want to determine the area in a normal distribution that lies within two standard deviations of the mean. Table A-5 gives entries only for positive z values. However, since the normal distribution is symmetrical, the area between the mean and a value two standard deviations below the mean is the same as the area between the mean and a value two standard deviations above the mean. Hence, we double the area previously determined to obtain $2(0.4772) = 0.9544$ as the required area. In summary, about 95.5% of the parts produced by this process have lifetimes between 600 and 1400 hours. We also note the generalization that about 95.5% of the area in a normal distribution lies within two standard deviations of the mean. The required area is shown in Figure 6-8(A).

EXAMPLE 6-2

What is the proportion of parts produced by the previous process with lifetimes between 1100 and 1350 hours?

SOLUTION

Both 1100 and 1350 lie above the mean of 1000 hours. We can determine the required probability by obtaining (1) the area between the mean and 1350 and (2) the area between the mean and 1100, and then subtracting (2) from (1).

$$\text{If } x = 1350 \qquad z = \frac{1350 - 1000}{200} = \frac{350}{200} = 1.75$$

$$\text{If } x = 1100 \qquad z = \frac{1100 - 1000}{200} = \frac{100}{200} = 0.50$$

Table A-5 gives 0.4599 as the area corresponding to a z value of 1.75 and 0.1915 for 0.50. Subtracting 0.1915 from 0.4599 yields 0.2684 or 26.84% as the result. This area is shown in Figure 6-8(B).

EXAMPLE 6-3

What is the proportion of parts produced by the previous process with lifetimes less than 840 hours?

SOLUTION

The observation 840 hours lies below the mean. We solve this problem by determining the area between the mean and 840 and subtracting this value from 0.5000, which is the entire area to the left of the mean.

$$\text{If } x = 840 \qquad z = \frac{840 - 1000}{200} = -0.80$$

Since only positive z values are shown in Table A-5, we look up a z value of 0.80 and find 0.2881. This is also the area between the mean and a z value of -0.80. Subtracting 0.2881 from 0.5000 gives the desired result, 0.2119. The area corresponding to this probability is given in Figure 6-8(C).

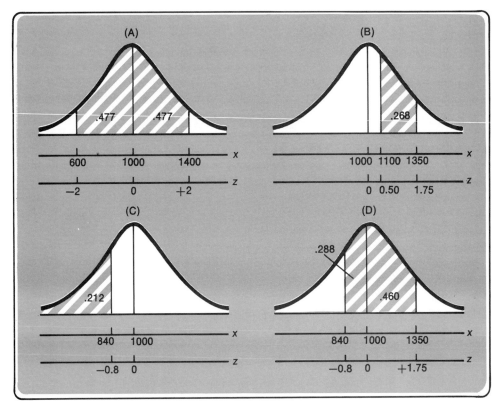

Figure 6-8
Areas corresponding to Examples 6-1–6-4.

EXAMPLE 6-4

What is the proportion of parts produced by this process with lifetimes between 840 and 1350 hours?

SOLUTION

Since 840 lies below the mean and 1350 lies above the mean, we determine (1) the area lying between 840 and the mean and (2) the area lying between 1350 and the mean, and add (1) to (2). The z values for 840 and 1350 were previously determined as −0.80 and +1.75, respectively, with corresponding areas of 0.2881 and 0.4599. Adding these two figures, we obtain 0.7480 as the proportion of parts with lifetimes between 840 and 1350 hours. The corresponding area is shown in Figure 6-8(D).

It was stated earlier that in the normal distribution, the range of the *X* variable extends from minus infinity to plus infinity. Yet, in the examples just considered (6-1 through 6-4), negative lifetimes were impossible. This illustrates the point that a variable may be said to be normally distributed provided that the normal curve constitutes a good fit to its frequency distribution within a range of about three standard deviations from the mean. Since virtually all the area is included in this range, the situation in the tails of the distribution is considered negligible.

It is useful to note the percentages of area that lie within integral numbers of standard deviations from the mean of a normal distribution. These values have been tabulated in Table 6-3. Hence, as was observed in Example 6-1, about 95.5% of the area in a normal distribution lies within plus or minus two standard deviations from the mean. Try and verify the other figures from Table A-5. Let us restate these probability figures in terms of rough statements of odds. Since about two-thirds of the area lies within one standard deviation, the odds are about two-to-one that in a normal distribution an observation will fall within that range. Correspondingly, the odds are about 95 to 5 or 19 to 1 for the two standard deviation range and 997 to 3 or about 332 to 1 for three standard deviations.

TABLE 6-3
Percentages of Area that Lie within
Specified Intervals Around the Mean
in a Normal Distribution.

INTERVAL	AREA, %
$\mu \pm \sigma$	68.3
$\mu \pm 2\sigma$	95.5
$\mu \pm 3\sigma$	99.7

EXERCISES

1. Two companies, *A* and *B*, produce a certain type of specialized steel. Consider the thickness of this steel to be normally distributed. Roughly sketch the probability distributions for the two companies on the same graph for each of the following cases:
(a) *A* has a mean of 3″ and a standard deviation of $\frac{1}{2}$″,
  B has a mean of 3″ and a standard deviation of 1″.

(b) *A* has a mean of 3″ and a standard deviation of 1″,
   B has a mean of 4″ and a standard deviation of 1″.

2. The weight of a pound can of cherry toffee produced by Kandy Corporation is normally distributed with a mean of one pound and a standard deviation of 0.04 pound. For each of the following probability questions, using graphs with both *X* and *Z* axes, indicate the corresponding area under the normal curve. If a pound can of cherry toffee is bought at random from a store, what is the probability that it will weigh
   (a) under a pound?
   (b) at most 0.95 pound?
   (c) more than 1.1 pounds?
   (d) between 0.90 and 1.1 pounds?
   (e) at least 0.90 pound?
   (f) either more than 1.1 pounds or less than 0.90 pound?

3. Let *X* represent the strength of a certain type of hemlock beam produced by Outland Lumber Company. Assume *X* is normally distributed with a mean of 2000 psi and a standard deviation of 100 psi. A beam is drawn at random, and its strength is tested. Change each of the following probability statements made in terms of *X* into statements about the standardized variable *Z*. What is the probability that the tested strength will be
   (a) at least 2150 psi?
   (b) less than 1825 psi?
   (c) between 1875 psi and 2115 psi?
   (d) between 1795 psi and 1905 psi?
   (e) either more than 2250 psi or less than 1800 psi?
   (f) at most 2100 psi?

4. The kilowatt demand at any given time on the Amgar Power Plant is normally distributed with a mean of 120,000 and a standard deviation of 10,000. If the plant can generate at most 150,000 kilowatts, what is the probability that at any given time there will be an overload?

5. The scores on an achievement test given to 231,126 high school seniors are normally distributed about a mean of 500. The distribution has a standard deviation of 90.
   (a) What is the probability that an achievement score is less than 500?
   (b) What is the probability that an achievement score is between 320 and 680?
   (c) The probability is 0.85 that a score is more than what value?

6. The weight of a box of sugar toasted rice cereal is normally distributed with $\mu = 12$ ounces and $\sigma = 0.5$ ounce.
   If we pick a box at random, what is the probability that it weighs
   (a) less than 12 ounces?
   (b) less than 12.75 ounces?
   (c) between 11.5 and 12.75 ounces?

6.5   SAMPLING DISTRIBUTION OF THE MEAN

In Section 6.2 we discussed the binomial distribution as a sampling distribution for numbers of occurrences. We now turn to another important probability distribution, namely, the sampling distribution of the arithmetic mean. For brevity, we shall use the term "the sampling distribution of the mean," or simply, "the sampling distribution of $\bar{x}$." To illustrate the nature of this distribution, let us return to the manufacturing process that produces electrical parts, whose lifetime is normally distributed with an arithmetic mean of 1000 hours and a standard deviation of 200 hours. We now interpret this process distribution as an infinite population from which simple random samples can be drawn. It is possible for us to draw a large number of such samples of a given size, for example, $n = 10$. We can compute the arithmetic mean lifetime of the ten parts in each sample. In accordance with our usual terminology, each such sample mean may be referred to as a "statistic." Since these statistics will usually differ from one another, we can construct a frequency distribution of these sample means. The universe mean of 1000 hours is the "parameter" around which these sample statistics will be distributed, with some sample means lying below 1000 and some lying above it. If we take any finite number of samples, the sampling distribution is referred to as an "empirical sampling distribution." On the other hand, if we conceive of the situation of drawing all possible samples of the given size, the sampling distribution is a "theoretical sampling distribution." It is such theoretical distributions upon which statistical inference is based, and on which we shall concentrate. These distributions are nothing more than probability distributions of the relevant statistics. In most practical situations, only one sample is drawn from a statistical population in order to test a hypothesis or to estimate the value of a parameter. The work implied in generating a sampling distribution by drawing repeated samples of the same size is virtually never carried out, except perhaps as an academic experiment. However, it is important to realize that the underlying theoretical structure for decisions based on single samples is the sampling distribution.

Sampling from normal populations

What are the salient characteristics of the sampling distribution of the mean, if samples of the same size are drawn from a population in which values are normally distributed? To obtain an answer to this question, let us begin by assuming that the sample size is five. Interpreting this in terms of our problem, let us assume that a random sample of five electrical parts is drawn from the above-mentioned population, and the mean lifetime of these five parts, denoted $\bar{x}_1$, is determined. Then, another sample of five parts is drawn, and the mean $\bar{x}_2$ is determined. Let us assume that the first mean was equal to 990 hours. Thus, it falls below the population mean. Assume the second mean was equal to 1022 hours. Hence it lies above the population mean. The theoretical frequency distribution of $\bar{x}$ values of all such simple random samples of five articles each would constitute the sampling distribution of the mean for samples of size five. Intuitively, we can see what some of the characteristics of such a distribution might be. A sample mean would be just as likely to lie above the population mean of 1000 hours as below it. Small deviations from 1000 hours would occur more frequently than large deviations. Furthermore, because of the effect of averaging, we would expect less dispersion or spread among these sample means than among the values of the individual items in the original population. That is, the standard deviation of the sampling distribution of the mean would be less than the standard deviation of the values of individual items in the population.

Several other characteristics of sampling distributions of the mean might be noted. If samples of size 50 rather than five had been drawn, another sampling distribution of the mean would be generated. Again we would expect the means of these samples to cluster around the population mean of 1000 hours. However, we would expect to find even less dispersion among these sample means than in the case of samples of size five. Thus, the standard deviation of the sampling distribution, which measures chance error inherent in the sampling process, would decrease with increasing sample size. The graphical presentation given in Figure 6-9 displays the relationships we have just discussed for the case of a normal population. For the population distribution, the horizontal axis represents values of individual items (x values). For the sampling distributions, the horizontal axis represents the means of samples of size five and 50. Since all three of the distributions are probability distributions of continuous random variables, probabilities are represented by areas under the respective curves.

Another characteristic of the sampling distributions depicted in Figure 6-9,

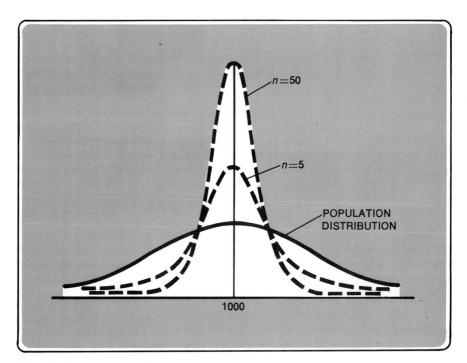

Figure 6-9
Relationship between a normal population distribution and normal sampling distributions of the mean for $n = 5$ and $n = 50$.

which is not at all intuitively obvious but can be proved mathematically, is that if the original population is normal, sampling distributions of the mean will also be normal. We state this in the form of a theorem:

THEOREM 6.1

If a population distribution is normal, the sampling distribution of the mean ($\bar{x}$) is also normal for samples of all sizes.

Theorem 6.1 has some very interesting and important implications. To discuss them, it will be useful to employ the following symbolism, which we will then use throughout the remainder of the book. The mean and standard deviation of the population are referred to as μ and σ, respectively. The mean and standard deviation of the sampling distribution of $\bar{x}$ are referred to as $\mu_{\bar{x}}$ and $\sigma_{\bar{x}}$, respectively. (A similar convention will be used for other sampling distributions where the relevant statistic is also indicated as a subscript.) Now, we can think of any sample mean, $\bar{x}$, as an estimate of the population mean, μ. The difference between the statistic $\bar{x}$ and the parameter μ, is referred to as a "sampling error." Thus, if we draw a simple random sample and observe a sample mean $\bar{x}$ exactly equal to the population mean μ, there would be no sampling error. Therefore, $\sigma_{\bar{x}}$, which is a measure of the spread of $\bar{x}$ values around μ, is a measure of *average sampling error*.

We now summarize some important properties of the sampling distribution of the mean. If the population is normally distributed with mean μ and a standard deviation σ, the sampling distribution of $\bar{x}$ has these properties:

1. It has a mean equal to the population mean. That is, $\mu_{\bar{x}} = \mu$.

2. It has a standard deviation equal to the population standard deviation divided by the square root of sample size. That is,

(6.6)
$$\sigma_{\bar{x}} = \frac{\sigma}{\sqrt{n}}$$

3. It is normally distributed.

Property (2) has some very interesting meanings. As we noted above, $\sigma_{\bar{x}}$ is a measure of average sampling error. Another interpretation is that $\sigma_{\bar{x}}$ is a measure of the *precision* with which the sample mean, $\bar{x}$, can be used to estimate the true population mean, μ. Referring to equation (6.6), we see that $\sigma_{\bar{x}}$ varies directly with the dispersion in the original population, σ, and inversely with the square root of the sample size, n. Thus, as might be expected, the greater the dispersion among the items in the original population, the greater is the expected sampling error in using $\bar{x}$ as an estimate of μ.

The fact that $\sigma_{\bar{x}}$ varies inversely with $\sqrt{n}$ means that there is a certain type of diminishing return in sampling effort. A quadrupling of sample size only halves the standard error of the mean, $\sigma_{\bar{x}}$; multiplying sample size by nine cuts the standard error to one-third its previous value.

Sampling from non-normal populations

We concluded in the foregoing discussion that if a population is normally distributed, the sampling distribution of $\bar{x}$ is also normal. However, many population distributions, particularly of business and economic data, are not normally distributed. What then is the nature of the sampling distribution of $\bar{x}$? It is a remarkable fact that for almost all types of population distributions, the sampling distribution of $\bar{x}$ is approximately normal for sufficiently large samples. This relationship between the shapes of the population distribution and the sampling distribution of the mean has been summarized in what is often referred to as the most important theorem of statistical inference, namely the *central limit theorem:*

THEOREM 6.2 (CENTRAL LIMIT THEOREM)

If a population distribution is non-normal, the sampling distribution of $\bar{x}$ may be considered to be approximately normal for large samples.

The central limit theorem assures us that no matter what the shape of the population distribution is, the sampling distribution of $\bar{x}$ approaches normality as the sample size increases. The important point is that for a wide variety of population types, samples do not have to be very large for the sampling distribution of $\bar{x}$ to be approximately normal. For most types of populations, the approach to normality is quite rapid as n increases. In fact, experiments have shown that for various types of skewed populations, the distribution of $\bar{x}$ is practically normal for sample sizes of about 10 or 20.

Some of the ideas in this section are illustrated in Examples 6-5 and 6-6.

EXAMPLE 6-5

Suppose the dollar values of the accounts receivable of a certain corporation are normally distributed with an arithmetic mean of $\$10,000$ and a standard deviation of $\$2000$.

(a) If one account is randomly selected, what is the probability that it will be between $\$9000$ and $\$12,000$ in size?

(b) If a random sample of 400 accounts is selected, what is the probability that the sample mean will be between $\$10,100$ and $\$10,200$ in size?

SOLUTION

(a) This question refers to the original population distribution, which is normal with $\mu = \$10,000$ and $\sigma = \$2000$. First, we transform to z values, or deviations from the mean in units of the standard deviation.

$$\text{For } x = \$9000; \quad z = \frac{x - \mu}{\sigma} = \frac{\$9000 - \$10,000}{\$2000} = -0.50$$

$$\text{For } x = \$12,000; \quad z = \frac{x - \mu}{\sigma} = \frac{\$12,000 - \$10,000}{\$2000} = 1.00$$

Thus, we want to determine the area in a normal distribution that lies between a value one-half of a standard deviation below the mean to a value one standard deviation above the mean. Table A-5 gives 0.3413 as the area corresponding to a

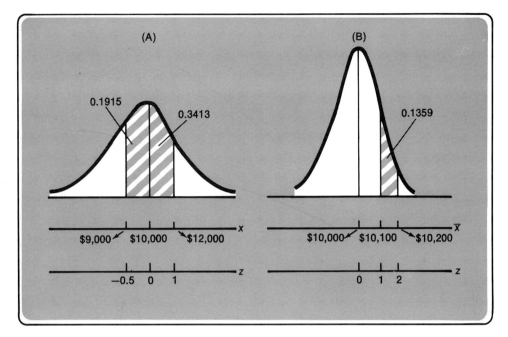

Figure 6-10
Areas corresponding to Examples 6-5 (a) and (b).

z value of 1.00. Since only positive values are shown in Table A-5, we look up a z value of 0.50 for $z = -0.50$ and find 0.1915. Adding these two areas gives 0.5328. This area is shown in Figure 6-10(A).

(b) It is important to recognize that this is a question about the sampling distribution of the mean. First we calculate the standard error of the mean.

$$\sigma_{\bar{x}} = \frac{\sigma}{\sqrt{n}} = \frac{\$2000}{\sqrt{400}} = \$100$$

For $\bar{x} = \$10,100$; $\quad z = \frac{\bar{x} - \mu}{\sigma_{\bar{x}}} = \frac{\$10,100 - \$10,000}{\$100} = 1.00$

For $\bar{x} = \$10,200$; $\quad z = \frac{\bar{x} - \mu}{\sigma_{\bar{x}}} = \frac{\$10,200 - \$10,000}{\$100} = 2.00$

Table A-5 gives 0.4772 as the area corresponding to a z value of 2.00 and 0.3413 for 1.00. Since both z values lie above the mean, we subtract 0.3413 from 0.4772 to obtain 0.1359 as the result. This area is shown in Figure 6-10(B).

EXAMPLE 6-6

The family income distribution in a certain large city is characterized by skewness to the right. A census reveals that the mean family income is $9000 and the standard deviation is $2000. If a simple random sample of 100 families is drawn, what is the probability that the sample mean family income will differ from the city mean income of $9000 by more than $200?

SOLUTION

Although the population income distribution is skewed, by the central limit theorem we may assume that the distribution of sample means for $n = 100$ is normally distributed. The standard error of the mean is

$$\sigma_{\bar{x}} = \frac{\sigma}{\sqrt{n}} = \frac{\$2000}{\sqrt{100}} = \$200$$

Hence, the question refers to the probability of observing deviations from the mean of the sampling distribution in excess of one standard error. From Table A-5, we find that the area corresponding to $z = 1.00$ is 0.3413. We subtract this figure from 0.5000 to obtain $0.5000 - 0.3413 = 0.1587$ as the probability of obtaining a deviation in excess of one standard error *above* the mean of \$9000. There is an equal probability of observing a deviation in excess of one standard error *below* the mean. Hence, $2(0.1587) = 0.3174$ is the required probability. This area is shown in Figure 6-11.

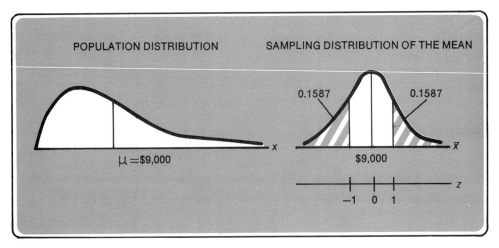

Figure 6-11
Areas corresponding to Example 6-6.

EXERCISES

1. The mean income in Orange County, North Carolina, is \$4271 with a standard deviation of \$1300.
   (a) Would it be correct to say approximately 99.7% of all people in Orange County earn between \$371 and \$8171? Why or why not?
   (b) Would it be correct to say that if simple random samples of 100 inhabitants each were repeatedly drawn from Orange County, approximately 99.7% of the time the average income of the group would be between \$3881 and \$4661? Why or why not?
   (c) Would it be correct to say that if simple random samples of 10,000 inhabitants each were repeatedly drawn from Orange County, approximately 99.7% of the time the average income of the group would be between \$3310 and \$5232? Why or why not?

(d) As an approximation, would (b) or (c) be more likely to be correct?

2. The average number of Xerox copies made per working day in a certain office is 356 with a standard deviation of 55. It costs the firm $.03 a copy. During a working period of 121 days, what is the probability that the average cost per day is more than $11.10?

3. What is the probability of drawing a simple random sample with a mean of 30 or more from a population with a mean of 28? The sample size is 100, the population variance is 81. Do we have to assume the population is normal?

4. The life of a Rollmore tire is normally distributed with a mean of 22,000 miles and a standard deviation of 1500 miles.
   (a) What is the probability that a tire will last at least 20,000 miles?
   (b) What is the probability that a tire will last more than 25,000 miles?
   (c) If a person buys four tires, what is the probability that the average life of the four tires exceeds 20,000 miles?

5. The sales on Monday of a particular small retail item are normally distributed with a mean of $1852.75 and a standard deviation of $285.15. Calculate the range in which 95.5% of the Monday sales figures will fall. If sales are averaged every month (arithmetic mean for four Mondays), calculate the range in which 95.5% of the average Monday sales would fall. If sales are averaged for the year (arithmetic mean for 52 Mondays), calculate the range in which 95.5% of the yearly average Monday sales would fall.

6. A clothing manufacturer has sales offices in Boston, New York, Washington, and Atlanta. Each office has 25 salesmen. The weekly sales for any salesman are normally distributed with a mean of $1200 and a standard deviation of $200. Within what range about the mean is the probability 0.997 that
   (a) a given salesman's weekly sales will fall?
   (b) the average weekly sales per salesman of the Atlanta sales office will fall?

7. A specification calls for a drug to have a therapeutic effectiveness for a mean period of 50 hours. The standard deviation of the distribution of the period of effectiveness is known to be 16 hours. Shipments of the drug are to be accepted if the mean period lies between 48 and 52 hours in a sample of 64 items drawn at random.
   Suppose the actual mean period of effectiveness of the drug in a given shipment is 44 hours. What is the probability that the shipment will be accepted when, in fact, it should not be?

Finite population multiplier

In our discussion of sampling distributions we have dealt with infinite populations. However, many of the populations in practical problems are finite, as for example, the employees in a given industry, the households in a city, and the counties in the United States. It turns out that as a practical matter formulas obtained for infinite populations also can be applied in most cases to finite populations. In those cases where the results for infinite populations are not directly applicable, a simple correction factor applied to the formula for the standard deviation of the relevant sampling distribution is all that is required.

In simple random sampling from an *infinite population* we have seen that the sampling distribution of $\bar{x}$ has a mean $\mu_{\bar{x}}$, which is equal to the population mean, μ, and a standard deviation, $\sigma_{\bar{x}}$, which is equal to $\sigma/\sqrt{n}$. The analogous situation in sampling from a *finite population* is that the mean $\mu_{\bar{x}}$ of the sampling distribution of $\bar{x}$ again is equal to the population mean, μ, but the standard deviation (standard error of the mean) is approximately equal to the following formula:

STANDARD ERROR OF THE MEAN FOR FINITE POPULATIONS

(6.7)
$$\sigma_{\bar{x}} = \sqrt{1 - \frac{n}{N}} \frac{\sigma}{\sqrt{n}}$$

where, as usual, n and N represent the numbers of elements in the sample and population, respectively. The ratio $f = n/N$ is referred to as the "sampling fraction," since it measures the fraction of the population contained in the sample. The quantity $\sqrt{1-f}$ is usually referred to as the "finite population correction." When the population size N is large relative to the sample size n, the fraction f is close to zero, and the finite population correction is approximately equal to one. Hence, in such cases the standard error of the mean, $\sigma_{\bar{x}}$, for sampling finite populations is for practical purposes equal to $\sigma/\sqrt{n}$ as in the case of an infinite population. A rule of thumb is that the formula $\sigma_{\bar{x}} = \sigma/\sqrt{n}$, which is strictly applicable only for infinite populations, may be used whenever the size of the population is at least 20 times that of the sample.

A very striking implication of equation (6.7) is that so long as the population is large relative to the sample, sampling precision becomes a function of sample size alone and does not depend on the relative proportion of the population sampled. For example, let us assume a situation in which we draw a simple random sample of $n = 100$ from each of two populations. Each population has a standard deviation equal to 200 units ($\sigma = 200$). In order to observe the effect of increasing the number of elements in the population, we further assume the populations are of different sizes, namely, $N = 10,000$ and $N = 1,000,000$. The standard error of the mean for the population of 10,000 elements is by equation (6.7):

$$\sigma_{\bar{x}} = \sqrt{1 - \frac{100}{10,000}} \frac{200}{\sqrt{100}} \approx \sqrt{1} \frac{200}{10} \approx 20$$

If the population is increased to 1,000,000, we have

$$\sigma_{\bar{x}} = \sqrt{1 - \frac{100}{1,000,000}} \frac{200}{\sqrt{100}} \approx \sqrt{1} \frac{200}{10} \approx 20$$

Thus, increasing the population from 10,000 to 1,000,000 has virtually no effect on the standard error of the mean, since the finite population correction is approximately equal to one in both instances. Indeed, if the population were increased to infinity, the same result would again be obtained for the standard error.

The finding that it is the absolute size of the sample and not the relative proportion of the population sampled that basically determines sampling precision is a point that many people find difficult to accept intuitively. In fact, prior to the introduction of statistical quality control procedures in American industry, arbitrary methods such as sampling 10% of the items of incoming shipments, regardless of shipment size, were quite common. There tended to be a vague feeling in these cases that approximately the same sampling precision was obtained by maintaining a constant sampling fraction. However, it is clear that widely different standard errors were the result because of large variations in the absolute sizes of the samples. The interesting principle that emerges from this discussion, for cases in which the populations are large relative to the samples, is that it is the absolute amount of work done (sample size), not the amount of work that might conceivably have been done (population size), that is important in determining sampling precision. The extent to which this finding can be applied to other areas of human activity we leave to your judgment.

In our subsequent discussion of statistical inference we will be concerned with measures of sampling error for proportions as well as for means. Therefore, we note at this point that the finite population correction can be applied to the formula for the standard error of a proportion in exactly the same way that it was applied above to the standard error of the mean.

Other sampling distributions

In this chapter, we have discussed sampling distributions of numbers of occurrences, and sampling distributions of the mean. Just as we were able to use the binomial distribution as a sampling distribution of numbers of occurrences under the appropriate conditions, other distributions may similarly be used under other sets of conditions. However, it is frequently far simpler to use normal curve methods, based on the operation of the central limit theorem. Two other continuous sampling distributions, which we have not yet examined, are important in elementary statistical methods: the student *t*-distribution and the chi-square distribution. They will be discussed at the appropriate places in connection with statistical inference.

OTHER PROBABILITY SAMPLE DESIGNS 6.6

The discussion to this point has been based on *simple random samples*. However, in many practical situations, other sample designs may be preferable to simple random sampling because they may achieve greater precision of estimation at the same cost, or the same precision at lower cost. The subject of sample survey theory and methods is very specialized, and matters such as the selection of the optimal type of sample design and estimation method require a high level of expertise. The obtaining of the advice or active involvement of a knowledgeable sampling specialist in the planning and implementation of such projects is usually a wise procedure.

Two frequently used alternatives to simple random sampling are *stratified sampling* and *cluster sampling*. In the simplest form of stratified sampling, the population is classified into mutually exclusive subgroups or *strata,* and simple random samples are drawn independently from each of these strata. Sample statistics from these strata can be combined to yield an overall estimate of a population parameter. For example, assume that we had a list of the incomes of every household in a city and we were interested in estimating the arithmetic mean household income by sampling. We could classify these households, say, into five income classes, ranging from the households with highest income in stratum one to the households with lowest income in stratum five. Suppose we draw a simple random sample of households from each stratum and calculate the sample mean income for each of these strata. Then, we weight these $\bar{x}$ values (the weights are the population numbers of households in each stratum) to obtain an estimate of the population mean. It can be shown that this estimate tends, on the average, to be closer to the actual population mean, than that which would be obtained under simple random sampling of the entire population. In other words, stratification can be used to obtain greater precision in estimation. Furthermore, sample means may be compared with one another to reveal between strata differences which may be of interest.

In our example, the property used to stratify the population, namely household income, was also the characteristic we were interested in estimating. In more realistic cases, for example, in studying consumer expenditures, we might stratify consumer units by characteristics such as disposable income, size of consumer unit, and geographic location. In estimating unemployment of the labor force, we might stratify by race, sex, education, and age. Stratification results in greater precision than simple random sampling particularly when there are extreme items that can be grouped into separate strata.

Another widely used technique is *cluster sampling* in which the population is subdivided into groups or "clusters" and then a probability sample of these clusters is drawn and studied. For example, suppose we wished to conduct a survey of expenditures on durable goods for all households in the United States. We might draw a simple random sample of counties in the United States, and then draw a simple random sample of households within the sampled counties. The counties are referred to as clusters, since for purposes of analysis, households are conceived of as being clustered into county units. Several successive stages of sampling might include clusters such as counties, political subdivisions within counties, blocks, and households. Generally, expenditures on such things as the listing of population elements, travel, interviewing and supervision are smaller when cluster sampling is used than when simple random sampling is used. In terms of cost required to obtain a fixed level of precision, a well-designed cluster sample is generally far superior to a simple random sample.

In the chapters that follow, simple random sampling is assumed. In more complex types of sampling, although the estimation formulas and methods are more complicated, the basic principles and formulas of simple random sampling represent important components of the overall methodology.

Estimation 7

The remainder of this book deals with methods by which rational decisions can be made when only *incomplete information* is available and the outcomes critical to the success of these decisions are uncertain. In our brief introduction to probability theory, we have begun to see how probability concepts can be used to deal with problems of uncertainty. The field known as *statistical inference* uses this theory as a basis for making reasonable decisions from incomplete data. Statistical inference treats two different classes of problems: (1) estimation, which is discussed in this chapter and (2) hypothesis testing, which is examined in Chapters 8 and 9. In both cases inferences are made about population characteristics from information contained in samples.

POINT AND INTERVAL ESTIMATION 7.1

The need to estimate population parameters from sample data stems from the fact that it is ordinarily too expensive or simply infeasible to enumerate complete populations to obtain the required information. The cost of complete censuses of finite populations may be prohibitive; complete enumerations of infinite populations are impossible. Statistical estimation procedures provide us with the means of obtaining estimates of population parameters with desired degrees of precision. Numerous business and economic examples can be given of the need to obtain estimates of pertinent population parameters. A marketing organization may be interested in estimates of average income and other socio-economic characteristics of the consumers in a metropolitan area, a retail chain may want an estimate of the average number of pedestrians per day who pass a certain corner, a production department may desire an estimate of the percentage of defective articles produced by a new production process, or a finance department may want an estimate of average interest rates on mortgages in a certain section of the country. Undoubtedly, in all of these cases, exact accuracy is not required, and estimates derived from sample data would probably provide appropriate information to meet the demands of the practical situation.

Two different types of estimates of population parameters are of interest: *point estimates* and *interval estimates.* A point estimate is a single number used as an estimate of the unknown population parameter. For example, the arithmetic mean income of a sample of families in a metropolitan area may constitute a point estimate of the corresponding population mean for all families in that metropolitan area. The percentage of defectives observed in a sample may be used as an estimate of the corresponding unknown percentage of defectives in a shipment from which the sample was drawn.

A distinction can be made between an *estimate* and an *estimator.* Let us consider the illustration of estimating the population figure for arithmetic mean income of all families in a metropolitan area from the corresponding sample mean. The numerical value of the sample mean is said to be an *estimate* of the population mean figure. On the other hand, the statistical measure used, that is, the method of estimation is referred to as an *estimator.* For example, the sample mean, $\bar{x}$, is an estimator of the population mean. When a specific number is calculated for the sample mean, say $8000, that number is an *estimate* of the population mean figure.

For most practical purposes, it is not sufficient to have merely a single point estimate of a population parameter. Any single point estimate will be either right or wrong. It would certainly seem useful, and perhaps even necessary, to have in addition to a point estimate, some notion of the degree of error that might be involved in using this estimate. *Interval estimation* is useful in this connection. Roughly speaking, an interval estimate of a population parameter is a statement of two values between which it is estimated that the parameter lies. Thus, an interval estimate in the example of the population arithmetic mean income of families in a metropolitan area might be $7100 to $8900. An interval estimate for the percentage of defectives in a shipment might be 3% to 5%. We may have a great deal of confidence or very little confidence that the population parameter is included in the range of the interval estimate. Therefore, it would seem necessary to attach some sort of probabilistic statement to the interval.

The procedure used to handle this problem is "confidence interval estimation." The confidence interval is an interval estimate of the population parameter. A confidence coefficient, for example, 90 or 95% is attached to this interval to indicate the degree of confidence to be placed upon the estimated interval.

We will explain the rationale of confidence interval estimation in terms of an example in which a population mean is the parameter to be estimated.

Interval estimation of a mean — rationale

Suppose a manufacturer has a very large production run of a certain brand of tires and he is interested in obtaining an estimate of their arithmetic mean lifetime by drawing a simple random sample of 100 tires and subjecting them to a forced life test. Let us assume that from long experience in manufacturing this brand of tires, he knows that the population standard deviation for a production run is $\sigma = 3000$ miles. Of course, ordinarily the standard deviation of a population is not known exactly and must be estimated from a sample, just as are the mean and other parameters. However, let us assume in this case that the population standard deviation is indeed known. The sample of 100 tires is drawn and a mean lifetime of 22,500 miles is observed. Thus, we denote $\bar{x} = 22,500$ miles. Now let us consider the reasoning involved in estimating an interval we are very confident includes the true but unknown mean lifetime of all tires in the production run.

The procedure in confidence interval estimation is based on the concept of the sampling distribution. In this example, since we are dealing with the estimation of a mean, the appropriate distribution is the sampling distribution of the mean. We will review some fundamentals of this distribution to lay the foundation for confidence interval estimation. Figure 7-1 shows the sampling distribution of the mean for simple random samples of size $n = 100$ from a population with an unknown mean, denoted μ, and a standard deviation, $\sigma = 3000$ miles. We assume that the sample is large enough so that by the central limit theorem the sampling distribution may be assumed to be normal, even if the population is non-normal. The standard error of the mean, which is the standard deviation of this sampling distribution, equals $\sigma_{\bar{x}} = \sigma/\sqrt{n}$. Strictly speaking, the finite population correction should be shown in this formula, but we will assume the population is so large relative to sample size that for practical purposes the correction factor is equal to one. The mean of the sampling distribution, $\mu_{\bar{x}}$, is equal to the population mean, μ.

In our work with the normal sampling distribution of the mean, we have learned how to make probability statements about sample means, given the value of the population mean. Thus, for example, in terms of the data of this problem, we can state that in drawing a simple random sample of 100 tires from the production run, the probability is 95% that the sample mean, $\bar{x}$, will lie within 1.96 standard error units of the mean of the sampling distribution, or from $\mu - 1.96\sigma_{\bar{x}}$ to $\mu + 1.96\sigma_{\bar{x}}$. This range is indicated on the horizontal axis of the sampling distribution in Figure 7-1. For emphasis, vertical lines have been shown at the end points of this range. As usual, we determine the 1.96 figure from Table A-5 of Appendix A, where we find that 47.5% of the area

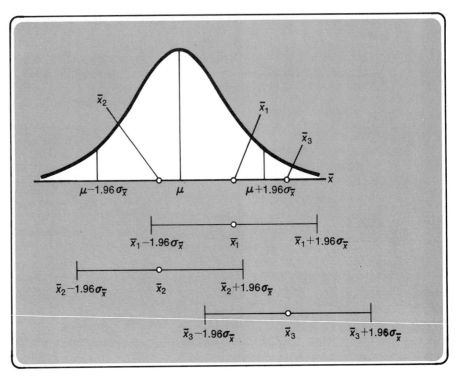

Figure 7-1
Sampling distribution of the mean and confidence interval estimates for three illustrative samples.

in a normal distribution is included between the mean of a normal distribution and a value 1.96 standard deviations to the right of the mean, and by symmetry 95% of the area is included in a range of plus or minus 1.96 standard deviations from the mean. According to a relative frequency interpretation, 95% of the $\bar{x}$ values of samples of size 100 would lie in this range if repeated samples were drawn from the given population.

How then might we construct the desired interval estimate for the population parameter? To obtain an answer to this question, let us consider again the repeated simple random samples of size 100 from the population represented by the production run of tires. Let us assume our first sample yields a mean that exceeds μ but falls between μ and $\mu + 1.96\sigma_{\bar{x}}$. The position of this sample mean, denoted $\bar{x}_1$, is shown on the horizontal axis of Figure 7-1. Suppose now that we set up an interval from $\bar{x}_1 - 1.96\sigma_{\bar{x}}$ to $\bar{x}_1 + 1.96\sigma_{\bar{x}}$. This interval is shown immediately below the graph in Figure 7-1. As can be seen in this figure, this interval, which may be written as $\bar{x}_1 \pm 1.96\sigma_{\bar{x}}$, includes the population parameter, μ. Of course, this simply follows from the fact that $\bar{x}_1$ fell within $1.96\sigma_{\bar{x}}$ from the mean of the sampling distribution, μ.

Now, let us assume our second sample from the same population yields the mean, $\bar{x}_2$, which lies on the horizontal axis to the left of μ, but again at a distance less than $1.96\sigma_{\bar{x}}$ away from μ. Again we set up an interval of the sample mean plus or minus $1.96\sigma_{\bar{x}}$, or from $\bar{x}_2 - 1.96\sigma_{\bar{x}}$ to $\bar{x}_2 + 1.96\sigma_{\bar{x}}$. This interval, shown below the graph in Figure 7-1, includes the population mean, μ.

Finally, suppose a third sample is drawn from the same population, with the mean $\bar{x}_3$, shown on the horizontal axis of Figure 7-1. This sample mean lies to the right of μ, but at a distance *greater than* $1.96\sigma_{\bar{x}}$ above μ. Now, when we set up the range $\bar{x}_3 - 1.96\sigma_{\bar{x}}$ to $\bar{x}_3 + 1.96\sigma_{\bar{x}}$, this interval does *not* include μ.

We can imagine a continuation of this sampling procedure, and we can assert that 95% of the intervals of the type $\bar{x} \pm 1.96\sigma_{\bar{x}}$ include the population parameter, μ. Now, we can get to the crux of confidence interval estimation. In the problem originally posed, as in most practical applications, only one sample was drawn from the population, not repeated samples. On the basis of the single sample, we were required to estimate the population parameter. The procedure is simply to establish the interval $\bar{x} \pm 1.96\sigma_{\bar{x}}$ and attach a suitable statement to it. The interval itself is referred to as a "confidence interval." Thus, for example, in our original problem the required confidence interval is

$$\bar{x} \pm 1.96\sigma_{\bar{x}} = \bar{x} \pm 1.96 \frac{\sigma}{\sqrt{n}} = 22,500 \pm 1.96 \frac{3000}{\sqrt{100}}$$
$$= 22,500 \pm 588 = 21,912 \text{ to } 23,088 \text{ miles}$$

We must be very careful how we interpret this confidence interval. It is incorrect to make a probability statement about this *specific* interval. For example, it is incorrect to state that the probability is 95% that the mean lifetime, μ, of all tires, falls in this interval. The population mean is not a random variable. Hence probability statements cannot be made about it. The unknown population mean, μ, either lies in the interval or it does not. We must return to the line of argument used in explaining the method and indicate that the intervals of the form $\bar{x} \pm 1.96\sigma_{\bar{x}}$ constitute the values of the random variable, not μ. Thus, the interpretation is that if repeated simple random samples of the same size were drawn from this population, and the interval $\bar{x} \pm 1.96\sigma_{\bar{x}}$ were constructed from each of them, then 95% of the statements that the interval contains the population mean, μ, would be correct. Another way of putting it is that in 95 samples out of 100 the mean, μ, would lie within intervals constructed by this procedure. The 95% figure is referred to as a "confidence coefficient" to distinguish it from the type of probability calculated when deductive statements are made about sample values from known population parameters.

Interval estimation — interpretation and use

Despite the above interpretation of the meaning of a confidence interval, where the probability pertains to the estimation procedure rather than to the specific interval constructed from a single sample, the fact remains that the investigator ordinarily must make an inference on the basis of the single sample he has drawn. He will not draw the repeated samples implied by the interpretational statement. For example, in the tire illustration, an inference is required about the production run based on the particular sample of 100 tires in hand. If the confidence coefficient attached to the interval estimate is high, then the investigator will behave as though the interval estimate is correct. In the tire example, an interval estimate of a mean lifetime of 21,912 to 23,088 miles was obtained. This interval may or may not encompass the actual value of the population parameter, μ. However, since 95% of intervals so constructed would include the value of the mean lifetime, μ, of all tires in the production run, we will behave as though this particular interval does include the actual value. It is desirable to obtain a relatively narrow interval with a high confidence coefficient associated with it. One without the other is not particularly useful. Thus, for example, in estimation of a proportion, say, the proportion of persons in the labor force who are unemployed, we can assert even without sample data that the percentage lies somewhere between 0 and 100% with a confidence coefficient of 100%. Obviously, this statement is neither very profound nor very useful because the interval is too wide. On the other hand, even if the interval is very narrow, but has a low associated confidence coefficient, say 10%, the statement would again have little practical utility.

Confidence coefficients such as 0.90, 0.95, and 0.99 and two- or three-sigma limits, such as 0.955 or 0.997 are conventionally used. Two- or three-sigma limits are those obtained by making an estimate of a population parameter and adding and subtracting two or three standard errors to establish confidence intervals. For a fixed confidence coefficient and population standard deviation, the only way to narrow a confidence interval and thus increase the precision of the statement is to increase the sample size. This is readily apparent from the way the confidence interval was constructed in the tire example. We computed $\bar{x} \pm 1.96\sigma_{\bar{x}}$, where $\sigma_{\bar{x}} = \sigma/\sqrt{n}$. If the 1.96 figure and σ remain constant, we can decrease the width of the interval only by increasing the sample size, n, since $\sigma_{\bar{x}}$ is inversely related to $\sqrt{n}$. Thus the marginal benefit of increased precision must be measured against the increased cost of sampling. In Section 7.3, we discuss a method of determining the sample size required for a specified degree of precision.

One final point may be made before turning to confidence interval estimation of different types of population parameters. Ordinarily, as was indicated in the tire example, the standard deviation of the population, σ, is unknown. Therefore, it is not possible to calculate $\sigma_{\bar{x}}$, the standard error of the mean. However, we can estimate the standard deviation of the population from a sample and use this figure to calculate an estimated standard error of the mean. We shall use this estimation technique in the examples that follow.

Interval estimation of a mean — large samples

We will use examples to discuss confidence interval estimation and will concentrate first on situations where the sample size is large. Our discussion will then focus, in turn, on interval estimation of a mean and a proportion. Finally, we will briefly treat corresponding estimation procedures for small samples.

As our first illustration of interval estimation of a mean from a large sample, let us look at Example 7-1.

EXAMPLE 7-1

A group of students working on a summer project with a social agency took a simple random sample of 120 families in a well-defined "poverty area" of a large city in order to determine the mean annual family income of this area. The sample results were $\bar{x} = \$2810$, $s = \$780$, and $n = 120$. What would be the 99% confidence interval for the mean income of all families in this poverty area?

SOLUTION

The only way this problem differs from the illustration of the mean lifetime of tires is that here the population standard deviation is unknown. The usual procedure for large samples ($n > 30$) is simply to use the sample standard deviation as an estimate of the corresponding population standard deviation. Using s as an estimator of σ, we can compute an estimated standard error of the mean $s_{\bar{x}}$. We have

$$s_{\bar{x}} = \frac{s}{\sqrt{n}} = \frac{\$780}{\sqrt{120}} = \$71.23$$

Hence, we may use $s_{\bar{x}}$ as an estimator of $\sigma_{\bar{x}}$, and because n is large we invoke the central limit theorem to argue that the sampling distribution of $\bar{x}$ is approximately normal. Again, we have assumed the finite population correction to be equal to one. The confidence interval, in general, is given by

(7.1) $\bar{x} \pm z s_{\bar{x}}$

where z is some multiple of standard errors and $s_{\bar{x}}$ now replaces $\sigma_{\bar{x}}$, which was used when the population standard deviation was known. For a 99% confidence coefficient, $z = 2.58$. Therefore, the required interval is

$$\$2810 \pm 2.58(\$71.23) = \$2810 \pm \$183.77$$

or the population mean is roughly between $2626 and $2994 with a 99% confidence coefficient. The same sort of interpretation given earlier for confidence intervals again applies here.

EXERCISES

1. A simple random sample of 400 firms within a particular industry yielded an arithmetic mean number of employees of 232 and a standard deviation of 40.
   (a) Establish a 95.5% confidence interval for the population mean.
   (b) Precisely what is the meaning of a 95.5% confidence interval?

2. Explain the following statement in detail. The standard deviation of $\bar{x}$ ($\sigma_{\bar{x}}$) is both a measure of statistical error in estimation and a measure of dispersion of a distribution.

3. A random sample of 217 of the 14,714 families in West Falls was taken in order to determine the mean family income in this depressed area. A 95% confidence interval ($3812 to $4116) was established on the basis of the sample results.
   With only the above information, which of the following statements are valid?

(a) Of all possible samples of size 217 drawn from this population, 95% of the sample means will fall in the interval.

(b) Of all possible samples of size 217 drawn from this population, 95% of the universe means will fall in the interval.

(c) Of all possible samples of size 217 drawn from this population, 95% of the confidence intervals established by the above method will contain the universe mean.

(d) 95% of families in West Falls have means between $3812 and $4116.

(e) By the above method, exactly 95% of the intervals so established will contain the sample mean $\bar{x}$.

(f) We do not know whether the universe mean is in the interval $3812 to $4116.

4. A simple random sample of 100 students at a university yielded an arithmetic mean income of $550, a modal income of $600, and a standard deviation of $120. Suppose you desire to estimate the mean income of the students of the university with 97% confidence. What would your interval estimate be?

Interval estimation of a proportion—large samples

In many situations, it is important to estimate a proportion of occurrences in a population from sample observations. For example, it may be of interest to estimate the proportion unemployed in a certain city, the proportion of eligible voters who intend to vote for a particular political candidate, or the proportion of students at a university who are in favor of changing the grading system. In all these cases, the corresponding proportions observed in simple random samples may be used to estimate the population values. Before turning to a description of how this estimation is accomplished, we will briefly establish the conceptual underpinnings of the procedure.

In Chapter 6, we saw that under certain conditions, the binomial distribution was the appropriate sampling distribution for the number of successes, x, in a simple random sample of size n. Furthermore, we noted in Section 6.3 that if p, the proportion to be estimated, is not too close to zero or one, the binomial distribution can be closely approximated by a normal curve with the same mean and standard deviation as the binomial, that is, $\mu = np$ and $\sigma = \sqrt{npq}$.

It is a simple matter to convert from a sampling distribution of *number of successes* to the corresponding distribution of *proportion of successes*. If x is the number of successes in a sample of n observations, then the proportion of successes in the sample denoted by $\bar{p}$ (read p-bar) is $\bar{p} = x/n$. Hence, dividing the formulas in the preceding paragraph by n, we find that the mean and standard deviation become

(7.2) $$\mu_{\bar{p}} = p$$

and

(7.3) $$\sigma_{\bar{p}} = \sqrt{\frac{pq}{n}}$$

Note that the subscript $\bar{p}$ has been used on the left-hand side of Equations (7.2) and (7.3) in keeping with the symbology used for the corresponding values for the sampling distribution of the mean, namely, $\mu_{\bar{x}}$ and $\sigma_{\bar{x}}$. Therefore, in summary, if p is not too close to zero or one, the sampling distribution of $\bar{p}$ can be closely approximated by a normal curve with the mean and standard deviation given in (7.2) and (7.3). It is important to observe in this connection that the central limit theorem holds for sample proportions as well as for sample means.

A couple of other points should be noted. First, the above discussion pertains to cases where the population size is large compared to the sample size. Otherwise, Formula (7.3) should be multiplied by the finite population correction $\sqrt{1 - n/N}$, as in Formula (6.7) for the standard error of the mean.

Second, in keeping with the usual terminology, $\sigma_{\bar{p}}$ is referred to as the "standard error of a proportion."

To illustrate confidence interval estimation for a proportion, we shall make assumptions similar to those in the preceding example. In Example 7-2 we assume a large simple random sample drawn from a population that is very large compared to the sample size.

EXAMPLE 7-2

In the town of Smallsville, a simple random sample of 800 automobile owners revealed that 480 would like to see the size of automobiles reduced. What are the 95.5% confidence limits for the proportion of all automobile owners in Smallsville who would like to see car size reduced?

SOLUTION

In this problem we want a confidence interval estimate for p, a population proportion. We have obtained the sample statistic $\bar{p} = 480/800 = 0.60$, which is the sample proportion who wish to see car size reduced. As noted earlier, for large sample sizes and for p values not too close to 0 or 1.00, the sampling distribution of $\bar{p}$ may be approximated by a normal distribution with mean $\mu_{\bar{p}} = p$ and $\sigma_{\bar{p}} = \sqrt{pq/n}$. Here we encounter the same type of problem as in interval estimation of the mean. The formula for the exact standard error of a proportion, $\sigma_{\bar{p}} = \sqrt{pq/n}$, requires the values of the unknown population parameters p and q. Hence, we use an estimation procedure similar to that used in the case of the mean. Just as we used s as an approximation of σ, if we substitute the corresponding sample statistics $\bar{p}$ and $\bar{q}$ for the parameters p and q in the formula for $\sigma_{\bar{p}}$, we can calculate an estimated standard error of a proportion $s_{\bar{p}} = \sqrt{\bar{p}\bar{q}/n}$. Using the same type of reasoning as that for interval estimation of the mean, we can state a two-sided confidence interval estimate for a population proportion as

(7.4) $$\bar{p} \pm z s_{\bar{p}}$$

In this problem, $z = 2$, since the confidence coefficient is 95.5% for a two-sided interval. Hence, substituting into Equation (7.4), we have as our interval estimate of the proportion of all automobile owners in Smallsville who would like to see car size reduced

$$0.60 \pm 2\sqrt{\frac{0.60 \times 0.40}{800}} = 0.60 \pm 0.0346$$

or the population proportion is estimated to be included in the interval 0.5654 to 0.6346, or roughly between 56.5% and 63.5% with a 95.5% confidence coefficient.

EXERCISES

1. In a simple random sample of 1000 stockholders of Atlas Credit Corporation, 600 were in favor of a new issue of bonds (with attached warrants for purchase of common stock) and 400 were against it. Construct a 95.5% confidence interval for the actual proportion of all stockholders who are in favor of the new issue.

2. An efficiency expert made 100 random observations of a typist. In 56 of these observations the typist was "idle." Construct a 98% confidence interval for the proportion of the time that the typist is idle.

3. A 95% confidence interval for the percentage of defective items in a lot of 100 is 10 to 12%. Is it correct therefore to say that the probability is 95% that between 10 and 12% of the items in the lot are defective?

7.3 CONFIDENCE INTERVAL ESTIMATION (SMALL SAMPLES)

The estimation methods we have discussed thus far are appropriate when the sample size is large. The distinction between large and small sample sizes is important when the population standard deviation is *unknown* and therefore must be estimated from sample observations. The main point is as follows. We have seen that the ratio $z = (\bar{x} - \mu)/\sigma_{\bar{x}}$ (where $\sigma_{\bar{x}} = \sigma/\sqrt{n}$) is normally distributed for all sample sizes if the population is normal and approximately normally distributed for large samples if the population is not normally distributed. In words, this ratio is $z =$ (sample mean $-$ population mean)/known standard error. Furthermore, in Section 7.2, we observed that for *large samples*, even if an *estimated* standard error is used in the denominator of this ratio, the sampling distribution may be assumed to be a standard normal distribution for practical purposes. However, the ratio (sample mean $-$ population mean)/estimated standard error is not approximately normally distributed for *all* sample sizes. Since this ratio is not approximately normally distributed for small samples, the theoretically correct distribution known as the *t* distribution must be used instead. Although the underlying mathematics involved in the derivation of the *t* distribution is complex and beyond the scope of our book, we can get an intuitive understanding of the nature of that distribution and its relationship to the normal curve.

The ratio $\dfrac{(\bar{x} - \mu)}{s/\sqrt{n}}$ is referred to as the *t* statistic. That is,

$$(7.5) \qquad t = \frac{\bar{x} - \mu}{s/\sqrt{n}}$$

where, as defined in Equation (3.9), the sample standard deviation $s = \sqrt{\dfrac{\Sigma(x - \bar{x})^2}{n - 1}}$ is an estimator of the unknown population standard deviation.

Let us examine the *t* statistic and its relationship to the standard normal statistic, z. As noted in the preceding paragraph, the denominator of the z ratio represents a *known* standard error, because it is based on a known population standard deviation. On the other hand, the denominator of the t statistic represents an *estimated* standard error because s is an estimator of the population standard deviation. The number $n-1$ in the formula for s is referred to as the *number of degrees of freedom,* which we will call *df.* It is not feasible to give a single simple verbal explanation of this concept. From a purely mathematical point of view, the number of degrees of freedom, *df,* is simply a parameter that appears in the formula of the t distribution. However, in the present discussion, in which s is used as an estimator of the population standard deviation σ, the $n-1$ may be interpreted as the number of independent deviations of the form $x - \bar{x}$ present in the calculation of s. Since the total of the deviations $\Sigma(x - \bar{x})$ for n observations equals zero, only $n-1$ of them are independent. This means that if we were free to specify the deviations $x - \bar{x}$, we could designate only $n-1$ of them independently. The nth one would be determined by the condition that the n deviations have to add up to zero. Therefore, in the estimation of a population standard deviation or a population variance, if an $n-1$ divisor is used in the estimator, the terminology is that there are $n-1$ degrees of freedom present.

The t distribution has been derived mathematically under the assumption of a normally distributed population. Just as is true of the standard normal distribution, the t distribution is symmetrical and has a mean of zero. However, the standard deviation of the t distribution is greater than that of the normal distribution, but approaches the latter figure as the number of degrees of freedom and, therefore, the sample size, becomes large. It can be demonstrated mathematically that in the limit, that is, for an infinite number of degrees of freedom, the t distribution and normal distribution are exactly equal. The approach to this limit is quite rapid. Hence, there is a widely applied rule of thumb that samples of size $n > 30$ may be considered large and the standard normal distribution may appropriately be used as an approximation to the t distribution, where the latter is the theoretically correct functional form. Figure 7-2 shows the graphs of several t curves for different numbers of degrees of freedom. As can be seen from these graphs, the t curves are lower at the mean and higher in the tails than is the standard normal distribution. As the number of degrees of freedom increases, the t distribution rises at the mean and lowers at the tails until for an infinite number of degrees of freedom, it coincides with the normal distribution. The use of tables of areas for the t distribution is explained in Example 7-3.

Further remarks. Early work on the t distribution was carried out by W. S. Gossett in the early 1900s. Gossett was an employee of Guinness Brewery in Dublin. Since the brewery did not permit publication of research findings by its employees under their own names, Gossett adopted "Student" as a pen name. Consequently, in addition to the term "t distribution" used here, the distribution has come to be known as "Student's distribution" or "Student's t distribution" and is so referred to in many books and journals.

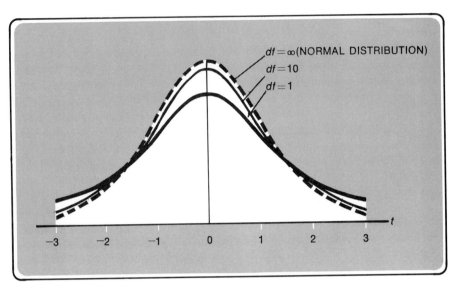

Figure 7-2
The t-distributions for $df = 1$ and $df = 10$ compared to the normal distribution ($df = \infty$).

EXAMPLE 7-3

Assume that a simple random sample of nine automobile tires was drawn from a large production run of a certain brand of tires. The mean lifetime of the tires in the sample was $\bar{x} = 22{,}010$ miles. This sample mean is the best single estimate of the corresponding population mean. The population standard deviation is unknown. Hence, an estimate of the population standard deviation was calculated by the formula $s = \sqrt{\Sigma(x - \bar{x})^2 / n - 1}$. The result was $s = 2{,}520$ miles. What are the 95% confidence limits for the mean?

SOLUTION

By the same type of reasoning we used in the case of the normal sampling distribution for means, we find that confidence limits for the population mean, using the t distribution, are given by

(7.6)
$$\bar{x} \pm t \frac{s}{\sqrt{n}}$$

where t is determined for $n - 1$ degrees of freedom. The number of degrees of freedom is one less than the sample size, that is, $df = n - 1 = 9 - 1 = 8$.

Just as the z values in Examples 7-1 and 7-2 represented multiples of standard errors, the t value in Equation (7.6) represents a multiple of estimated standard errors. We find the t value in Table A-6 in Appendix A. A brief explanation of this table is required. In the table of areas under the normal curve, areas lying between the mean and specified z values were given. However, in the case of the t distribution, since there is a different t curve for each sample size, no single table of areas can be given for all these distributions. Therefore, for compactness, a t table shows the relationship between areas and t values for only a few "percentage points" in different t distributions. Specifically, the entries in the body of the table are t values for areas in the two tails of the distribution combined. In this problem, the number of degrees of freedom is eight. Referring to Table A-6 of Appendix A under column 0.05 for eight degrees of freedom, we find $t = 2.306$. This means that, as shown in Figure 7-3, for eight degrees of freedom, a total of 0.05 of the area in the t distribution lies below a t value of -2.306 and above 2.306. Correspondingly, the probability is 0.95 that for eight degrees of freedom, the t value lies between -2.306 and 2.306.

In this problem, substituting $t = 2.306$ into Equation (7.6), we obtain the following 95% confidence limits:

$$22,010 \pm 2.306(840) = 22,010 \pm 1937.04 \text{ miles}$$

Hence to the nearest mile, the confidence limits for the estimate of the mean lifetime of all tires in the production run are 20,073 and 23,947 miles.

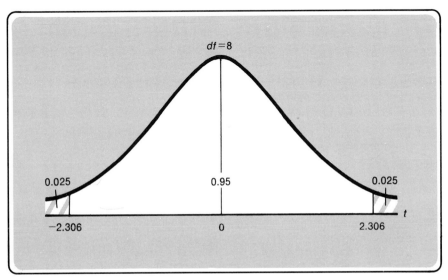

Figure 7-3
The relationship between t values and areas in the t distribution for eight degrees of freedom.

The interpretation of this interval and the associated confidence coefficient is the same as in the large-sample, normal distribution case. Comparing this procedure to the corresponding large sample method for 95% confidence limits, we note that the t value of 2.306 replaces the 1.96 figure, which is appropriate for the normal curve. Thus, we see that in the small sample case, we have a wider confidence interval leading to a vaguer result.

It may be noted that since Table A-6 shows areas in the combined tails of the t distribution, we had to look under the column headed 0.05 for the t value corresponding to a 95% confidence interval. Correspondingly, we would find t values for 90%, 98%, and 99% confidence intervals under the columns headed 0.10, 0.02, and 0.01, respectively.

7.4 DETERMINATION OF SAMPLE SIZE

In all of the examples thus far, the sample size n was given. However, an important question is how large should a sample be in a specific situation? If a larger than necessary sample is used, resources are wasted; if the sample is smaller than required, the objectives of the analysis may not be achieved.

Sample size for estimation of a proportion

Statistical inference provides the following type of answer to the size question. Let us assume an investigator desires to estimate a certain population parameter and he wants to know how large a simple random sample is required. We assume that the population is very large relative to the prospective sample size. The investigator must answer two questions in order to specify the required sample size. (1) "What degree of precision is desired?" and (2) "How probable do you want it to be that the desired precision will be obtained?" Clearly, the greater the degree of desired precision, the larger will be the necessary sample size. Also, the greater the probability specified for obtaining the desired precision, the larger will be the required sample size. We will use examples to indicate the technique of sample size determination for estimation of a population proportion and a population mean, respectively.

EXAMPLE 7-4

Let us assume a situation in which we would like to conduct a poll among eligible voters in a city in order to determine the percentage who intend to vote for the Democratic candidate in an ensuing election. We specify that we want the probability to be 95.5% that we will estimate the percentage that will vote Democratic within ±1 percentage point. What is the required sample size?

SOLUTION

An answer will be given to the question by first indicating the rationale of the procedure. Then this rationale will be condensed into a simple summary formula. The statement of the question gives a relationship between sampling error that we are willing to tolerate and the probability of obtaining this level of precision. In this problem, we have required that $2\sigma_{\bar{p}}$ be equal to 0.01. As can be seen in Figure 7-4, this means that we are willing to have a probability of 95.5% that our sample percentage $\bar{p}$ will fall within 0.01 of the true but unknown population proportion p. We may now write

$$2\sigma_{\bar{p}} = 0.01$$

or

$$2\sqrt{\frac{pq}{n}} = 0.01$$

and

$$\sqrt{\frac{pq}{n}} = 0.005$$

In all our previous problems, the sample size n was known, but here n is the unknown for which we must solve. However, it appears as though there are too many unknowns: the population parameters p and q as well as n. What we must do is estimate or guess values for p and q and then we can solve for n. Suppose we wanted to make a very conservative estimate for n. What should we guess as a value for $\bar{p}$? In this context, by a conservative estimate we mean an estimate made in such a way as to ensure that the sample size will be large enough to deliver the precision desired. In this problem, the "most conservative" estimate for n is given by assuming $p = 0.50$ and $q = 0.50$. This follows from the fact that the product pq is larger for $p = 0.50$ and $q = 0.50$ than for any other two possible p and q values, where $p + q = 1$. Thus, the largest or "most conservative" value of n is determined by substituting $p = q = 0.50$ as follows:

$$0.005 = \sqrt{\frac{0.50 \times 0.50}{n}}$$

Squaring both sides gives

$$0.000025 = \frac{0.50 \times 0.50}{n}$$

and

$$n = \frac{0.50 \times 0.50}{0.000025} = 10,000$$

Hence, to achieve the desired degree of precision, a simple random sample of 10,000 eligible voters would be required. Of course, the large size of this sample is attributable to the high degree of precision specified. If $\sigma_{\bar{p}}$ were doubled from 0.005 to 0.01, the required sample size would be cut down to one-fourth of 10,000 or 2500. This stems from the fact that the standard error varies inversely with the square root of sample size.

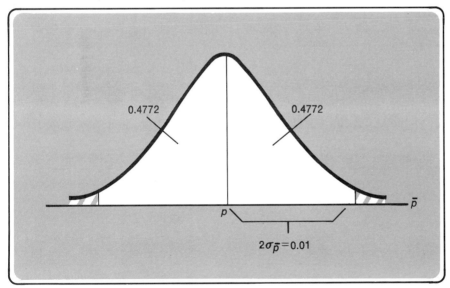

Figure 7-4
Sampling distribution of a proportion showing the relationship between error and probability of obtaining this degree of precision.

The arithmetic is simplified in these types of problems if we work in whole numbers of percentage points rather than decimals. For example, in whole numbers of percentage points, the preceding calculation becomes

$$\tfrac{1}{2} = \sqrt{\frac{50 \times 50}{n}}$$

$$(\tfrac{1}{2})^2 = \tfrac{1}{4} = \frac{50 \times 50}{n}$$

$$n = \frac{50 \times 50}{\tfrac{1}{4}} = 10{,}000$$

In this election problem, we assumed $p = q = 0.50$, although less conservative estimates are possible if it is believed that $p \neq 0.50$. In problems involving proportions, we would use whatever past knowledge we have to estimate p. For example, suppose we wanted to determine the sample size to estimate an unemployment rate, and we knew from past experience that for the community of interest the proportion of the labor force that was unemployed was somewhere between 0.05 and 0.10. We then would assume $p = 0.10$, since this would give us a more conservative estimate (larger sample size) than assuming $p = 0.05$ or any value between 0.05 and 0.10. The point is that if we assumed, say, $p = 0.07$, and in fact the true value of p was 0.09, the sample size we determined from a calculation involving $p = 0.07$ would not be large enough to give us the specified precision. On the other hand, assuming $p = 0.10$ assures us of obtaining the desired degree of precision regardless of what the true value of p is, in the range 0.05 to 0.10.

We can summarize this calculation for sample size by noting that we started with the statement that

$$e = z\sigma_{\bar{p}}$$

where e is the specified sampling error and z is the multiple of standard errors corresponding to the specified probability of obtaining this precision. Then, for infinite populations, or large populations relative to sample size, we have

(7.7)
$$n = \frac{z^2(p)(1-p)}{e^2}$$

Hence, in the preceding voting problem, applying Equation (7.7) yields

$$n = \frac{(2)^2(0.50)(0.50)}{(0.01)^2} = 10{,}000$$

EXERCISES

1. A random sample is to be selected to estimate the proportion of smokers at a university. The range of the estimate is to be kept within 5% with a confidence level of 95.5%. How large a simple random sample is required?

2. A manufacturer wishes to estimate the percentage defectives produced by one of his machines on the basis of a random sample of its output. He wishes to be 95% confident that his estimate lies within two percentage points of the true percentage defective. He feels quite certain that the machine does not produce more than 10% defectives. How large a simple random sample should the manufacturer take?

3. An estimate of p, the proportion of dwelling places that are vacant in a large city, is desired. How large a sample should be drawn if it is desired to estimate p within 0.02 with 0.98 confidence? It may be assumed that $0.01 \leq p \leq 0.10$.

Sample size for estimation of a mean

By an analogous calculation the required sample size for estimation of a mean can be determined. Suppose we wanted to estimate the arithmetic mean hourly wage rate for a group of skilled workers in a certain industry. Let us further assume that from prior studies we estimate that the population standard deviation of the hourly wage rates of these workers is about $0.15. How large a sample size would be required to yield a probability of 99.7% that we will estimate the mean wage rate of these workers within $\pm$0.03?

Since the 99.7% probability corresponds to a 3-standard error level, we can write

$$3\sigma_{\bar{x}} = \$0.03$$

or

$$\frac{3\sigma}{\sqrt{n}} = \$0.03$$

and

$$\frac{\sigma}{\sqrt{n}} = \$0.01$$

The population standard deviation, σ, is known from past experience. Hence, substituting $\sigma = \$0.15$ gives

$$\frac{\$0.15}{\sqrt{n}} = \$0.01$$

and

$$\sqrt{n} = \frac{\$0.15}{\$0.01} = 15$$

Squaring both sides yields the solution

$$n = (15)^2 = 225$$

Therefore, a simple random sample of 225 of these workers would be required. In summary, if we calculate $\bar{x}$ for the hourly wage rates of a simple random sample of 225 of these workers, we can estimate the mean wage rate of all skilled workers in this industry within $0.03 with a probability of 99.7%.

In the problem discussed above, it was assumed that an estimate of the population standard deviation was available from prior studies. This type of situation may exist in governmental agencies which conduct repeated surveys of wage rates, population, and the like. If the population standard deviations (or estimated population standard deviations) in these past studies were not erratic or excessively unstable, they would provide useful bases for estimating σ values in the above sample size computational procedures.

Of course, an estimate of the population standard deviation may not be available from past experience. It may be possible, however, to get a rough estimate of σ if there is at least some knowledge of the total range of the basic random variable in the population. For example, suppose we know that the difference between the highest and lowest paid workers is about $1.20. In a normal distribution a range of three standard deviations either side of the mean includes virtually the entire distribution. Thus, a range of 6σ includes almost all the frequencies, and we may state

$$6\sigma \approx \$1.20$$

or

$$\sigma \approx \$0.20$$

Of course, the population distribution is probably non-normal and $1.20 may not be exact, either. Consequently, the estimate of σ may be quite rough. Nevertheless, we may be able to obtain a reasonably good approximate estimate of the required sample size in a situation where in the absence of this "guestimating" procedure, we may be at a loss for a notion of a suitable number of elements for the sample.

Expressing the technique of sample size estimation in the form of an equation yields

(7.8)
$$n = \frac{z^2 \sigma^2}{e^2}$$

For the problem given earlier, this gives

$$n = \frac{(3)^2(\$0.15)^2}{(\$0.03)^2} = 225$$

In Equations (7.7) and (7.8), it is assumed that the n value determined is sufficiently large for the assumption of a normal sampling distribution to be appropriate and that the populations are large relative to this sample size.

EXERCISES

1. The Minerva Plastic Company wishes to select a random sample of plastic bars from its production process in order to estimate the mean length of bars produced by the process. On the basis of past experience, it is estimated that the process standard deviation is about 0.5 foot. How large a sample is needed to estimate the process mean within ± 0.1 foot, with a 95.5% confidence coefficient?

2. A statistician wishes to determine the average hourly earnings for employees in a given occupation in a particular state. He runs a pilot study and finds that the point estimate for the mean is $3.40 and the point estimate for the standard deviation is $0.25. He then specifies that when he takes his random sample he wants to be 95.5% confident that the maximum error of his estimate will not exceed $0.02.

(a) Discuss the sense in which he will have 95.5% confidence in his estimate.

(b) What size sample should he take?

3. In evaluating various pension plans and funding methods, the Davis Corp. must determine the mean age of its work force. Since the company has several thousand employees, a sample must be taken. Estimate how large the sample should be in order for the appropriate interval estimate of the mean age to be no more than two years wide with 90% confidence, if the standard deviation of ages is taken to be ten years.

> *"Appearances to the mind are of four kinds. Things either are what they appear to be; or they neither are, nor appear to be; or they are and do not appear to be; or they are not, and yet appear to be. Rightly to aim in all these cases is the wise man's task."*

EPICTETUS

8 *Hypothesis Testing*

Turning from the subject of estimation discussed in the preceding chapter, we now focus on hypothesis testing, the second basic subdivision of statistical inference. Hypothesis testing is a particularly central topic in modern statistical methods because it addresses itself to the very important question of how to choose among alternative propositions or among alternative courses of action. Usually these choices must be made with the realization that incorrect decisions are possible. Hypothesis testing concerns itself with the decision-making rules for choosing among alternatives while controlling and minimizing the risks of wrong decisions. At this point we will briefly and informally summarize the rationale involved in testing hypotheses and then proceed by explaining the details of these testing procedures by means of examples.

8.1 THE RATIONALE OF HYPOTHESIS TESTING

To gain some insight into the reasoning involved in statistical hypothesis testing, let us consider a nonstatistical hypothesis testing procedure with which we are all familiar. As it turns out, the basic process of inference involved is strikingly similar to that employed in statistical methodology.

Consider the process by which an accused individual is judged in a court of law under our legal system. Under Anglo-Saxon law, the man before the bar is assumed innocent. The burden of proof of his guilt rests upon the prosecution. Using the language of hypothesis testing, let us say that we want to test a hypothesis, which we denote H_0, that the man before the bar is innocent. This means that an alternative hypothesis exists, H_1, that the defendant is guilty. The jury examines the evidence to determine whether the prosecution has demonstrated that this evidence is inconsistent with the basic hypothesis, H_0, of innocence. If the jurors decide the evidence is inconsistent with H_0, they reject that hypothesis, and therefore accept its alternative, H_1, that the defendant is guilty.

If we analyze the situation that results when the jury makes its decision, we find that four possibilities exist in terms of the basic hypothesis, H_0.

1. The defendant is innocent (H_0 is true), and the jury finds that he is innocent (accepts H_0); hence the correct decision has been made.

2. The defendant is innocent (H_0 is true), but the jury finds him guilty (rejects H_0); hence an error has been made.

3. The defendant is guilty (H_0 is false), and the jury finds that he is guilty (rejects H_0); hence the correct decision has been made.

4. The defendant is guilty (H_0 is false), but the jury finds him innocent (accepts H_0); hence an error has been made.

In possibilities (1) and (3), we observe that the jury reaches the correct decision; in possibilities (2) and (4) it makes an error. Let us consider these errors in terms of conventional statistical terminology. In possibility (2), hypothesis H_0 is erroneously rejected. The basic hypothesis, H_0, tested for possible rejection is generally referred to as the "null hypothesis." Hypothesis H_1 is designated the "alternative hypothesis." To reject the null hypothesis when in fact it is true is referred to as a "Type I error." In possibility (4), hypothesis H_0 is accepted in error. To accept the null hypothesis when it is false is termed a "Type II error." It may be noted that under our legal system the commission of a Type I error has been considered far more serious than a Type II error. Thus, we feel that it is a more grievous mistake to convict an innocent man than to let a guilty man go free. Had we made H_0 the hypothesis that the defendant is guilty, the meaning of Type I and Type II errors would have been the reverse of the first formulation. What had previously been a Type I error would now become a Type II error, and Type II would now be Type I. In the statistical formulation of hypotheses, how we choose to exercise control over the two types of errors is a basic guide in stating the hypotheses to be treated. We will see in this chapter how this error control is carried out in hypothesis testing. The aforementioned situations are summarized in Table 8-1,

TABLE 8-1
The Relationship Between Actions Concerning
a Null Hypothesis and the Truth or Falsity
of Hypothesis

ACTION CONCERNING HYPOTHESIS H_0	STATE OF NATURE	
	H_0 IS TRUE	H_0 IS FALSE
Accept H_0	Correct decision	Type II error
Reject H_0	Type I error	Correct decision

where the headings are stated in modern decision theory terminology and require a brief explanation. When hypothesis testing is viewed as a problem in decision making, two alternative actions can be taken: "accept H_0," or "reject H_0." The two alternatives, truth or falsity of hypothesis H_0, are viewed as "states of nature" or "states of the world" that affect the consequences or "payoff" of the decision. The payoffs are listed in the table, and in the schematic presentation they are stated in terms of the correctness of the decision or the type of error incurred. We can see from the framework of the hypothesis testing problem that what we need is some criterion for the decision either to accept or reject the null hypothesis, H_0. Classical hypothesis testing attacks this problem by establishing decision rules based on data derived from simple random samples. The sample data are analogous to the evidence investigated by the jury. The decision procedure attempts to assess the risks of making incorrect decisions and in a sense, which we shall examine, to minimize them.

The hypothesis testing procedure

Two basic types of decision problems can be attacked by hypothesis testing procedures. The first problem is one in which we want to know whether a population parameter has changed from or differs from a particular value. Here we are interested in detecting whether the population parameter is *either* larger than or smaller than a particular value. For example, suppose that the mean family income in a certain city was determined from a census to be $9500 for a particular year and two years later we want to discover whether the mean income has *changed.* Further, let us assume that it is not feasible to take another census. Therefore, we may draw a simple random sample of families and try to reach a conclusion based on this sample. As in the court of law illustration, we can set up two competing hypotheses and choose between them. The null hypothesis H_0 would simply be an assertion that the mean family income was unchanged from the $9500 figure. In statistical language, we write this hypothesis as H_0: $\mu = \$9500$, where μ denotes the mean family income in the city. The alternative hypothesis is then that the mean family income *has* changed, or in statistical terminology, H_1: $\mu \neq \$9500$. In this example, we would observe the mean family income in the simple random sample of, say, 1000 families. We denote the sample mean as $\bar{x}$. If the value of the sample mean $\bar{x}$ differs from the population mean $\mu = \$9500$ by more than we would be willing to attribute to chance sampling error, we reject the null hypothesis H_0 and accept its alternative H_1. On the other hand, if the difference between the sample mean and population mean assumed under H_0 is small enough to be attributed to chance sampling error, we accept H_0. How do we know for what values of the sample statistic to reject H_0 and for what values to accept H_0? The answer to this question is the essence of hypothesis testing.

The hypothesis testing procedure is simply a decision rule that specifies for every possible value of a statistic observable in a simple random sample of size n whether the null hypothesis H_0 should be accepted or rejected. The set of possible values of the sample statistic is referred to as the *sample space.*

Therefore, the test procedure divides the sample space into mutually exclusive parts called the acceptance region and the rejection (or critical) region.

The nature of the division of the sample space for the example we have been discussing is illustrated in Figure 8-1. From sampling theory developed in Chapter 6, we know that given a population with a mean of $9500 and a known standard deviation, there would be sampling variation among the means of samples of the same size drawn from that population. According to the central limit theorem, regardless of the shape of the population distribution, the sampling distribution of the mean for large sample sizes may be assumed to be normal. In order to decide whether family mean income has changed from $9500, we determine two values, denoted c_1 and c_2 in Figure 8-1, which set limits on the amount of sampling variation we feel is consistent with the null hypothesis being true. Hence, as depicted in the figure, the decision rule in this case would be: If the mean income of the sample of 1000 families lies below c_1 or above c_2, reject the null hypothesis and conclude that the mean family income of the city has changed from $9500. On the other hand, if the sample mean lies between c_1 and c_2, we cannot reject H_0. Hence, we cannot conclude that the city's mean family income has changed. The type of test in which we want to determine whether a population parameter has *changed* is referred to as a two-tailed test, since the null hypothesis can be rejected by observing a statistic that falls in either of the two tails of the appropriate sampling distribution, as shown in Figure 8-1.

The second type of hypothesis test is one in which we wish to find out (1) whether the sample came from a population that has a parameter *less* than a hypothesized value; or (2) from a population that has a parameter *more* than a hypothesized value. These situations give rise to one-tailed tests. We

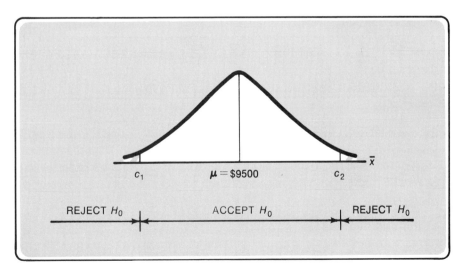

Figure 8-1
Two-tailed test: sampling distribution of the mean with acceptance and rejection regions for a null hypothesis.

will illustrate such a test with a quality control example. Suppose a construction company specified in a purchasing order that acceptable rivets must have a mean tensile strength of at least 40,000 pounds. Suppose the company decided to test the quality of an incoming shipment by taking a simple random sample of 100 rivets. Assume further that the null hypothesis H_0 to be tested is "the true mean tensile strength of all rivets in the shipment is at least 40,000 pounds." The alternative hypothesis H_1 is "the true mean tensile strength of all rivets in the shipment is less than 40,000 pounds." These two hypotheses may be expressed mathematically as follows:

$$H_0: \mu \geq 40,000 \text{ pounds}$$
$$H_1: \mu < 40,000 \text{ pounds}$$

where μ denotes the mean tensile strength of the shipment.

In this case, the decision rule for accepting or rejecting H_0 would be as follows: If the mean of the sample, $\bar{x}$, is less than some appropriate number c, reject H_0, and therefore reject the shipment. Otherwise, accept H_0 and the shipment. This decision rule is represented diagrammatically in Figure 8-2. Again, the number c represents a limit on the amount of sampling variation that we feel is consistent with the null hypothesis being true.

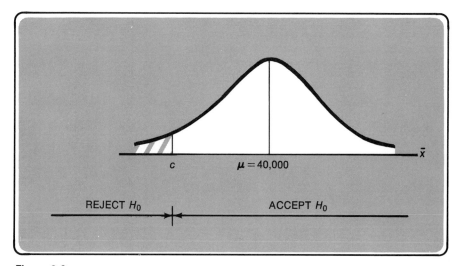

Figure 8-2
One-tailed test: sampling distribution of the mean with acceptance and rejection regions for a null hypothesis.

Concept of the null hypothesis

As can be noted from the preceding illustrations, a null hypothesis is a statement about a population parameter, as in the first example, $H_0: \mu = \$9500$, and in the second, $H_0: \mu \geq 40,000$ pounds. Note that the equality sign is present in both hypotheses. It is standard procedure to include the equal sign in the null hypothesis. By having the null hypothesis assert that the population parameter is equal to some specific value, we are then able to decide for which values of the observed sample statistic we will reject and accept the null hypothesis. For example, in the two illustrations, when we hypothesized that $\mu = \$9500$ and $\mu = 40,000$ pounds, we were then able to establish sampling distributions of the sample mean, $\bar{x}$, and to decide which values of $\bar{x}$ would lead us to reject the null hypothesis and which values would cause us to accept it. Furthermore, we are enabled to specify the risk of making Type I errors that we are willing to tolerate. For example, if we hypothesize that $\mu = 40,000$ pounds, we can compute the probability of making a Type I error, that is, the probability of erroneously rejecting that hypothesis. The application of this idea will be seen later in this chapter. The above points explain why null hypotheses are set up in the form stated earlier, rather than, say, in the form of $H_0: \mu \neq \$9500$ or $H_0: \mu < 40,000$ pounds.

We now turn to the application of some of these ideas. First, we will consider so-called "one-sample tests," which are tests of hypotheses based on data contained in a single sample. Tests involving means and proportions will be studied in that order.

ONE-SAMPLE TESTS (LARGE SAMPLES)   8.2

It is assumed in the one-sample tests discussed in this section that (1) the sample size is large ($n > 30$), (2) the sample size is decided upon before the test is conducted, and (3) the population standard deviation is known.

The significance of the assumption of large sample sizes ($n > 30$) is that we will be able to assume that the sampling distributions used in the testing procedure are normal. This is a useful simplification as compared to the small sample case, as we shall see at the end of this chapter.

As a more detailed illustration of how the hypothesis testing procedure we have discussed can be used, we consider first an example of an *acceptance sampling procedure*. The Morgan Company, a manufacturer of space vehicle components, regularly purchased a part required to withstand high temperatures. Specifications of mean heat resistance of at least 2250 degrees Fahrenheit (2250°F) were applied by the Morgan Company to shipments sent by the supplier. In the past, all shipments had met specifications. From long experience, it was found that the supplier's parts had a standard deviation of 300°F. That is, we will assume in this problem that it is *known* from experience that the standard deviation of heat resistance of parts in a shipment is 300°. In a simple random sample of 100 parts drawn from a particular shipment, a mean heat resistance of 2110°F was observed. Should the shipment be accepted, if it is desired that the risk of erroneously rejecting a shipment which meets specifications of a mean heat resistance of at least 2250°F be no more than 0.05? The shipment may be assumed to be very large relative to the sample of 100.

We proceed to convert this verbal problem to a hypothesis testing framework. As we shall see, this case is an example of a so-called "one-tailed test" or "one-sided alternative."

Tests concerning a mean: one-tailed test

We begin by stating in statistical terms the null hypothesis to be tested. Returning to our definition of a hypothesis as a quantitative statement about a population, we see that in this case the statement is about the mean of the population (shipment) sampled. It may be noted that an inference is to be drawn about the shipment, and the sample was randomly drawn from the shipment. Thus, the shipment is the population in this problem. The production process of the supplier may be viewed as a super population that gives rise to the shipment populations. Letting μ denote the mean heat resistance of the parts in this particular shipment population, we state the null and alternative hypothesis as follows:

$$H_0: \mu \geq 2250°F$$
$$H_1: \mu < 2250°F$$

where the standard deviation of the population is assumed to be 300°F. In symbols, $\sigma = 300°F$. The null hypothesis, H_0, states that the population mean heat resistance is equal to or greater than 2250°F (is at least 2250°F). If our decision procedure leads us to accept this hypothesis, our action will be to accept the shipment. On the other hand, if we reject H_0, we will reject the shipment. This means we accept the alternative hypothesis H_1, and therefore conclude that the shipment mean heat resistance is less than 2250°F. Our decision will be based on the data observed in the sample of 100 parts. Therefore, the question is simply this, "Are the sample data so inconsistent with the null hypothesis that we shall be forced to reject that hypothesis?" Before proceeding to examine this question more closely, we must discuss an important technical point. The null hypothesis must be stated in such a way that the probability of a Type I error can be calculated. This was the reason for including the equal sign in the statement of the null hypothesis, so that a particular value of μ was specified. The shipment mean heat resistance can exceed 2250°F in an infinite number of ways, one for each possible value of μ. Therefore, we cannot refer unambiguously to the probability of a Type I error in this situation as *the* probability of rejection of the null hypothesis when it is true. However, if we concentrate our attention on the single value, $\mu = 2250°F$, we can refer to a Type I error for that particular value, and we will be able to compute the probability of such an error once we have settled on a decision procedure. Hence, for the moment, we focus attention on the particular value $\mu = 2250°F$, for which the null hypothesis is true.

We now return to our question regarding the consistency of the sample data with the null hypothesis. In terms of this particular problem, we can word the question as follows, "Is a deviation as large as we have observed between a mean of 2110°F observed in a simple random sample of size 100 and a hypothesis population mean of 2250°F so great that we would be unwilling to attribute such a difference to chance errors of sampling a population with $\mu = 2250°F$?" If we conclude that the difference is so large it is improbable under the given hypothesis, then a "significant difference" is said to have been observed. An observed "significant difference" between a statistic and a parameter rejects the null hypothesis. Where the dividing line is set up for a "significant difference" versus a "non-significant difference" between the observed statistic and the parameter being tested depends on the risk we are willing to run of making a Type I error, that is, the risk of rejecting the null hypothesis when it is true. If we set up the dividing line at a point such that the probability is, say, 0.05 of erroneously rejecting the hypothesis that μ is equal to 2250°F, the test is said to have been conducted at the "5% significance level."

Conventional significance levels such as 0.05 and 0.01 are very frequently used in classical hypothesis testing, because of the desire to maintain a low probability of rejecting the null hypothesis H_0 when it is in fact true. These levels of significance are denoted by the Greek letter α. As we shall see, in one-tailed test situations α represents the *maximum* probability of a Type I error. In two-tailed test situations in which the null hypothesis consists of only one value of a population parameter, α represents *the* probability of a Type I error.

Table 8-2 summarizes the alternative hypotheses tested in the present acceptance sampling problem and the possible actions concerning these hypotheses. After a brief discussion of the meaning of Type I and Type II errors in this problem, we will examine how the hypothesis testing procedure is actually carried out quantitatively.

As we observe from Table 8-2, a Type I error in this problem, the incorrect rejection of the null hypothesis, takes the form of the rejection of a "good" shipment. The terms "good" and "bad" in this context will denote shipments which meet specifications or do not meet specifications, respectively. A

TABLE 8-2

Acceptance Sampling Problem: Relationship Between Possible Actions and Hypotheses Concerning the Quality of a Shipment

ACTION CONCERNING HYPOTHESIS H_0	STATE OF NATURE	
	H_0: $\mu \geq 2250°F$ (SHIPMENT MEETS SPECIFICATIONS)	H_1: $\mu < 2250°F$ (SHIPMENT DOES NOT MEET SPECIFICATIONS)
Accept H_0 (Accept Shipment)	No error	Type II Error Acceptance of a Shipment Which Does Not Meet Specifications
Reject H_0 (Reject Shipment)	Type I Error Rejection of a Shipment Which Meets Specifications	No Error

Type I error has been termed in quality control work the "producer's risk." Thus, the producer runs the risk of having a good shipment rejected because the data in a sample drawn from the shipment were misleading due to chance sampling error. A Type II error in this problem takes the form of the acceptance of a bad shipment and is referred to as the "consumer's risk." That is, the consumer runs the risk of accepting a bad shipment. A realistic industrial sampling plan would require an equitable balancing of these types of risks, taking into account the costs involved to the producer and consumer. However, let us proceed with the present simplified problem, in order to study the classical hypothesis testing approach.

Decision rules

We now turn to the question of how to establish a *decision rule* on which to base our acceptance or rejection of the shipment. As indicated earlier, an hypothesis testing decision rule is simply a procedure that specifies the action to be taken for each possible sample outcome. Thus, we are interested in partitioning the sample space into a region in which we will reject the null hypothesis and a region in which we accept it. In every hypothesis testing problem, the partitioning of the appropriate sample space is accomplished by considering the appropriate sampling distribution assuming the null hypothesis is true. This follows from the fact that specifying the probability of making Type I errors determines how the sample space will be partitioned. In this particular problem, since the question concerns *mean* heat resistance, the sampling distribution of means is the relevant distribution. We must use this distribution in our present problem to perform a test at the 5% significance level. Since our sample is large ($n = 100$), we use the central limit theorem and assume that the sampling distribution of means is normal. This distribution for samples of size 100 from a population, where $\mu = 2250°F$ (just in conformity with specifications) and $\sigma = 300°F$, is shown in Figure 8-3. As indicated in this graph, the striped region represents 5% of the area under the normal curve. Referring to Table A-5, we see that 5% of the area in a normal distribution lies to the right of $z = +1.65$, and correspondingly the same percentage of area lies to the left of $z = -1.65$. Thus, in our problem the point below which we would reject a shipment is a sample mean whose value is less than $\mu - 1.65\sigma_{\bar{x}}$. The standard error of the mean, $\sigma_{\bar{x}}$, is equal to

$$\sigma_{\bar{x}} = \frac{\sigma}{\sqrt{n}} = \frac{300°F}{\sqrt{100}} = 30°F$$

Thus, the critical value below which we would reject H_0 is

$$\mu - 1.65\sigma_{\bar{x}} = 2250°F - 1.65(30°F) \approx 2200°F$$

We can now see why this type of hypothesis testing situation is referred to as a one-tailed test or a one-sided alternative. Rejection of the null hypothesis takes place in only one tail of the sampling distribution.

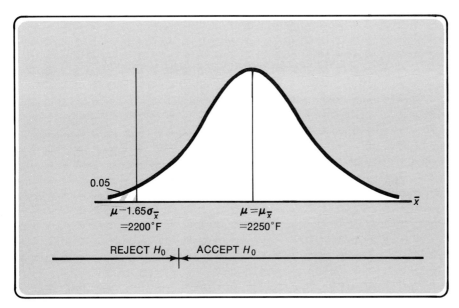

Figure 8-3
Sampling distribution of the mean showing regions of acceptance and rejection of H_0. Population parameters $\mu = 2250°F$, $\sigma = 300°F$, and sample size $n = 100$.

In summary, the Morgan Company should proceed as follows in this problem. Upon drawing a single random sample of 100 parts from the incoming shipment and observing $\bar{x}$, the sample mean heat resistance, the Morgan Company should apply this decision rule

DECISION RULE

1. If $\bar{x} < 2200°F$, reject H_0 (reject the shipment)

2. If $\bar{x} \geq 2200°F$, accept H_0 (accept the shipment)[1]

We can now answer the original question. Since the sample yielded a mean heat resistance of 2110°F, the null hypothesis H_0 should be rejected, and therefore the shipment should be rejected.

It is instructive to examine an alternative method of stating the decision rule. Instead of doing our work in terms of the original units, that is, in degrees, we could have calculated the z value in a standard normal distribution corresponding to an $\bar{x}$ value of 2110°F. If the z value lies to the left of the critical value of -1.65, H_0 is rejected; to the right of -1.65, H_0 is accepted. Thus the decision rule can be rephrased as follows:

[1] There is some ambiguity whether the equal sign should appear in the rejection or acceptance part of the decision rule. From the theoretical point of view, it is inconsequential. This follows from the fact that the normal curve is a continuous probability distribution. Thus, the probability of observing exactly $\bar{x} = 2200°F$ is zero.

DECISION RULE

1. If $z < -1.65$, reject H_0 (reject the shipment)

2. If $z \geq -1.65$, accept H_0 (accept the shipment)

The arithmetic for the $\bar{x}$ value of 2110°F is

$$z = \frac{\bar{x} - \mu}{\sigma_{\bar{x}}} = \frac{2110°F - 2250°F}{30°F} = -\frac{140°F}{30°F} = -4.67$$

Therefore, since a mean of 2110°F falls 4.67 standard error units below the mean of the sampling distribution and the dividing line is at 1.65 standard error units, the null hypothesis H_0 is rejected. This situation is displayed in Figure 8-4.

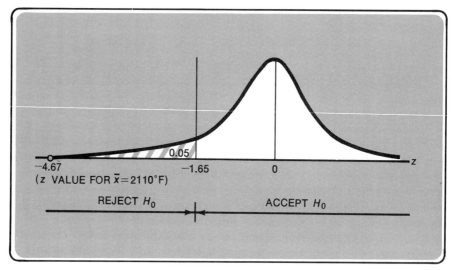

Figure 8-4
Standard normal curve for the acceptance sampling problem.

The form of the decision rule we choose is inconsequential. However, as a practical matter, if the Morgan Company repeatedly applied the same sampling procedure to incoming shipments, the first form would be simpler. Once the rule is established for a particular sample size, no further arithmetic is required. Nevertheless it is instructive in considering the rationale of the test to observe the implications of the computed z value. For example, in this case an $\bar{x}$ value of 2110°F corresponded to a z value of -4.67. Table A-5 does not give areas for z values above 4.0. Less than 0.0001 of the area in a normal distribution lies to the right of a z value of 4.0 or to the left of a z value of -4.0. Thus, interpreting the z value of -4.67 in terms of the acceptance sampling problem, if a sample of 100 parts was drawn at random from a shipment whose parts had a mean heat resistance of 2250°F and a standard deviation of 300°F, the probability of observing a sample mean of 2110°F or lower was less than 0.0001. Since this sample result is so unlikely under the hypothesis of a shipment mean of 2250°F, we reject that hypothesis.

For values of μ greater than 2250°F, the probability of a Type I error is less than 0.05. We can see this from Figure 8-3. If $\mu > 2250$°F, that is, if the sampling distribution shifts to the right, then less than 5% of the area will lie in the rejection region below 2200°F. The larger the value of μ, the lower is the probability of a Type I error. This makes sense in terms of the acceptance sampling problem. The greater the mean heat resistance in the incoming shipment, the lower is the probability that the shipment will be erroneously rejected. For $\mu = 2250$°F, the probability of a Type I error is 0.05. Now we can see the meaning of $\alpha = 0.05$, the significance level in this problem. It is the maximum probability of committing a Type I error. This sort of interpretation is typical for one-tailed tests.

Summary of procedure. In the discussion to this point, we have considered the situation in which the acceptance and rejection of the null hypothesis result in only two possible actions. Furthermore, we have concentrated on the determination of decision rules stemming from control of Type I errors without reference to the corresponding implications for Type II errors. While we shall deal with these matters subsequently, it is advisable not to clutter the present discussion with too many details. It is useful to summarize the hypothesis testing procedure discussed thus far as follows:

1. A null hypothesis and its alternatives are drawn up. The null hypothesis is framed in such a way that we can compute the probability of a Type I error.

2. A level of significance, α, is decided upon. This controls the risk of committing a Type I error.

3. A decision rule is established by partitioning the relevant sample space into regions of acceptance and rejection of the null hypothesis. This partition is accomplished by a consideration of the relevant sampling distribution. The nature of the null hypothesis and the choice of α determine the partition.

4. The decision rule is applied to the sample of size n. The null hypothesis is accepted or rejected. Rejection of the null hypothesis implies acceptance of the alternative.

Further remarks. A number of points can be made concerning the statistical theory involved in the acceptance sampling problem of the foregoing discussion. First, the normal curve was used as the appropriate sampling distribution of the mean. If the population distribution of mean heat resistances is normal, the normal curve is the theoretically correct sampling distribution. It may be noted that no statement at all was made in the acceptance sampling problem about the population distribution. The normal curve was used for the sampling distribution of $\bar{x}$ under the central limit theorem argument that no matter what the shape of the population, the sampling distribution of $\bar{x}$ would be approximately normal for a sample as large as $n = 100$.

Second, no finite population correction was used in the calculation of the standard error of the mean despite the fact that the sample of 100 parts was drawn without replacement from a finite population. However, it was stated in the problem that the population size could be assumed to be very large relative to the sample size. Therefore, the finite correction factor may be assumed to be approximately equal to one in this case.

Third, the population standard deviation, σ, was assumed to be known. If the population standard deviation is unknown, and the sample size is large, say $n > 30$, then the sample standard deviation s may be substituted for σ. Hence, instead of calculating $\sigma_{\bar{x}} = \sigma/\sqrt{n}$, an estimated standard error of the mean is computed as $s/\sqrt{n}$. In all other respects the decision procedure remains the same. In Section 8.4 we discuss how to deal with the situation in which the sample size is small and the population standard deviation is unknown.

Fourth, the nature of the z value computed in the problem is worth noting. In Chapter 6, when the idea of a standard score was discussed, it was in the context of a normally distributed *population*. In that case, $z = (x - \mu)/\sigma$ represented a deviation of the value of an individual item from the mean of the population expressed as a multiple of the population standard deviation. In our hypothesis testing problem, the z values were of the form $z = (\bar{x} - \mu)/\sigma_{\bar{x}}$. Such a z value represents a deviation of a sample mean from the mean of the sampling distribution of $\bar{x}$, stated in multiples of the standard deviation of that distribution, $\sigma_{\bar{x}}$. As we noted previously, the mean of the sampling distribution of $\bar{x}$, $\mu_{\bar{x}}$, is equal to the population mean, μ. Thus, we use μ and $\mu_{\bar{x}}$ interchangeably. As a generalization, in hypothesis testing problems, z values take the form

$$z = \frac{\text{statistic} - \text{parameter}}{\text{standard error}}$$

For example, in the hypothesis testing problem just discussed, the sample $\bar{x}$ value is the statistic; μ, the population mean, is the parameter; and $\sigma_{\bar{x}}$, the standard error of the mean, is the appropriate standard error.

Fifth, we note that the size of the sample, $n = 100$, was predetermined in our illustration. Thus, the case we discussed was one in which the sample was large, it was predetermined, and the construction of the decision rule was based on the control of only one type of incorrect decision, namely, Type I errors. The next section dealing with the power curve discusses the measurement of Type II errors for such a test.

Finally, it is important to realize that we did not *prove* that the null hypothesis was false, nor could sample evidence have proved that a null hypothesis is true. All that we can do is to discredit a null hypothesis or fail to discredit it on the basis of sample data. Actually, a single sample statistic such as $\bar{x}$ is consistent with an infinite number of hypotheses concerning μ.[2] From the standpoint of decision making and subsequent behavior, if sample data do not discredit a null hypothesis, we will act as though that hypothesis is true.

---

[2] Doubtless we have all had the disconcerting experience of observing a number of experts in disagreement after observing ostensibly the same basic set of data. Perhaps you share my experience of finding it easiest to accept and reject hypotheses when there are no data available at all.

The power curve. The hypothesis testing procedure outlined thus far has concentrated upon the control of Type I errors. The question of how well this test controls Type II errors naturally arises. That is, when the null hypothesis is false, how frequently does the decision rule lead us to accept the null hypothesis erroneously? This question is answered by means of the *power curve* also called the *power function,* which can be computed from the information of the problem and the decision rule. The Greek letter β is used to denote the probability of a Type II error; thus, β represents the probability of accepting the null hypothesis when it is false. In the acceptance sampling problem, the null hypothesis H_0 is false for each value of μ in the alternative hypothesis H_1: $\mu < 2250°F$. Therefore, for each particular value of μ less than 2250°F, we can determine a β value. Actually, by convention, the power curve gives the complementary probability to β, that is, $1 - \beta$, for each value of the alternative hypothesis. Thus, it shows the probability of rejecting the null hypothesis for each value for which the null hypothesis is false, which, of course, represents in each case the probability of selecting the correct course of action. $1 - \beta$ is referred to as the "power of the test" for each particular value of the alternative hypothesis. For completeness, in a power curve, the probabilities of rejection are also shown for each value for which the null hypothesis is true. In summary, a power curve is a function which gives the probabilities of rejecting the null hypothesis H_0 for all possible values of the parameter tested. Therefore, it shows the ability of the decision rule to discriminate between true and false hypotheses. It is useful in assessing the risks of making both Type I and Type II errors when using a decision rule.

The power curve for the acceptance sampling problem is shown in Figure 8-5. It has the typical reverse-S shape of a power curve in a one-tailed test with the rejection region in the left-hand tail. For a one-tailed test, with the rejection region in the right-hand tail, the curve would be S-shaped, or dropping from upper right to lower left on the graph. Rejection probabilities for the null hypothesis are shown on the vertical axis and possible values of the population parameter μ on the horizontal axis. Specifically, the figures plotted on the vertical axis are conditional probabilities of the form $P(\text{rejection of } H_0|\mu) = P(\bar{x} < 2200°F|\mu)$.

The nature of the power curve in Figure 8-5 can be obtained by considering a couple of the plotted values. The value of $\alpha = 0.05$ is shown for $\mu = 2250°F$, indicating the significance level of the test. We can see that this is the maximum probability of erroneously rejecting the null hypothesis H_0, because the ordinates of the curve drop off to the right as μ increases in the region where H_0 is true. The heights of the ordinates of the power curve to the right of $\mu = 2250°F$ represent the probabilities of making Type I errors. The heights of the ordinates to the left of $\mu = 2250°F$ give the values of $1 - \beta$, or the probabilities of rejecting the null hypothesis when it is false. Therefore, the complementary distance from the curve to 1.0 are values of β, or probabilities of making Type II errors. One such value is displayed in the graph for $\mu = 2200°F$. We recall that our decision rule required the rejection of H_0 if the sample mean

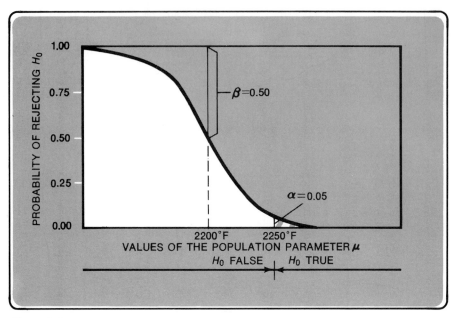

Figure 8-5
Power curve for the acceptance sampling problem.

$\bar{x}$ was less than 2200°F. If the shipment or population mean is 2200°F, obviously the probability of observing $\bar{x}$ values less than 2200°F and, therefore, of rejecting H_0 is 0.50. This situation is displayed in Figure 8-6.

Some computation is required to obtain the β figures for other μ values,

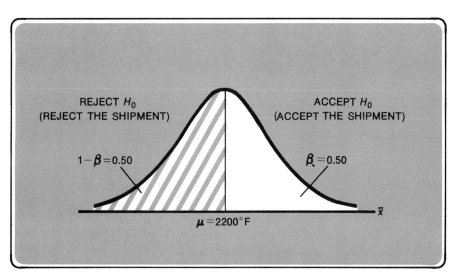

Figure 8-6
Graph illustrating Type II error probabilities for $\mu = 2200°F$.

and we will not show any such calculations here. In this type of one-tailed test, the ideal power curve would be ⌐L shaped with the vertical line occurring at $\mu = 2250°F$. Thus, the probability of rejecting H_0 would always be equal to 1.0 when H_0 is false and 0.0 when H_0 is true. However, clearly this ideal curve is unattainable when sample data are used to test hypotheses, since sampling error will always be present. We may note that there is a trade-off relationship between Type I and Type II errors for a sample of fixed size, and thus under hypothesis tests such as the one discussed here, the level of significance should be decided by a consideration of the relative seriousness of the two types of errors.

Operating characteristic curves. It is usual practice in industrial quality control to use operating characteristic curves, succinctly referred to as "O-C curves," rather than power curves to evaluate the discriminating power of a test. The O-C curve is simply the complement of the power curve. That is, the probability of acceptance rather than the probability of rejection of the null hypothesis is plotted on the vertical axis.

EXERCISES

1. Distinguish between a parameter and a statistic.

2. Support or criticize the following statement: "A statistician will make very few errors if he sets a significance level (α) very low, say 0.001; hence he should always do so."

3. Distinguish briefly between
(a) One-sided alternative and two-sided alternative.
(b) Type I and Type II errors.

4. What is meant by "the power of a test"? How is the power related to the Type II error?

5. Agree with or criticize the following statements.
(a) It is more important to control the Type I error than the Type II error, hence we should design our tests on the basis of controlling the Type I error.
(b) It is impossible to control both the Type I and Type II errors, since to decrease one increases the other.
(c) A Type I error occurs when we reject the null hypothesis incorrectly, and a Type II error occurs when we accept the null hypothesis incorrectly.
(d) Once a decision has been made, we must then consider the possibility that Type I and Type II errors may both have occurred.

6. Agree with or criticize the following statements.
(a) The power curve has the probability of rejecting the null hypothesis on one axis and the possible values of the parameter on the other, while the operating characteristic curve has the possible values of the statistics on one axis and the probability of rejecting the null hypothesis on the other.
(b) Beta represents the probability that one will reject the null hypothesis incorrectly.
(c) If a person has a choice of applying two different tests to decide between a certain null hypothesis and alternative hypothesis, both tests having the same α level, then he should use the one with the smaller β error.

7. A manufacturer claims that the average life of a certain type of transistor is at least 150 hours. It is known that the standard deviation of this type of transistor is 20 hours.

A consumer wishes to test the manufacturer's claim, and accordingly tests 100 transistors. State the null hypothesis, the alternative hypothesis, and the decision rule. Use an α level of 5%, i.e., set up a rule which would reject the manufacturer's claim only 5% of the time if it is true.

8. The specifications for a component of a spacecraft provide that the mean length should not be less than 101 millimeters with a standard deviation of 9 millimeters. A random sample of 100 components drawn from a very large shipment yields a mean length of 99 millimeters and a standard deviation of 9 millimeters. If you were the manufacturer of spacecraft who had received this shipment would you accept it? Justify the position you take.

9. A specification calls for a drug to have a therapeutic effectiveness for a mean period of at least 40 hours. The standard deviation of the period of effectiveness is known to be 15 hours. A shipment of the drug is to be accepted or rejected on the basis of a simple random sample of 100 items drawn from the shipment.
   (a) What decision rule should be used if the maximum probability of erroneously rejecting the incoming shipment is to be 0.05? State clearly the null and alternative hypotheses. Indicate the decision to be reached if a simple random sample of 100 items shows a mean period of effectiveness of 38.5 hours. Show your reasoning.
   (b) Sketch the power curve for the test you designed in (a), identifying and labeling two points on the curve.

Test concerning a proportion: two-tailed test

The preceding discussion dealt with a test of a hypothesis concerning a mean. We now turn to hypothesis testing for a proportion. As an illustration, let us consider the case of an advertising agency that developed a general theme for the commercials on a certain TV show based on the assumption that 50% of the show's viewers were over 30 years of age. The agency wished to change the general theme if the percentage had changed in either an upward or downward direction. If we use the symbol p to denote the proportion of all viewers of the show, we can state the null and alternative hypothesis as follows:

$$H_0:p = 0.50 \text{ viewers over 30 years of age}$$
$$H_1:p \neq 0.50 \text{ viewers over 30 years of age}$$

Let us assume that the agency wished to run a risk of 5% of erroneously rejecting the null hypothesis of "no change," i.e., $H_0:p = 0.50$. That is, the agency decided to test the null hypothesis at the 5% significance level. Symbolically, this may be stated as $\alpha = 0.05$.

In order to test the hypothesis, the agency conducted a survey of a simple random sample of 400 viewers of the TV show. Of the 400 viewers, 210 were over 30 years of age and 190 were 30 years or less. What conclusion should be reached?

In this problem, as contrasted with the previous one, the null hypothesis concerns a single value of p, which is a hypothetical population parameter of 0.50. The alternative hypothesis includes all other possible values of p. The reason for setting up the hypotheses this way can be seen by reflecting on how the test will be conducted. The hypothesized parameter under the null hypothesis is $p = 0.50$. We have observed in a sample a certain proportion, denoted $\bar{p}$, who were over 30 years of age. The testing procedure involves a comparison of $\bar{p}$ with the hypothesized value of p to determine whether a significant difference exists between them. If $\bar{p}$ does not differ significantly from p, and we accept the null hypothesis that $p = 0.50$, what we really mean is that the sample is consistent with a hypothesis that half the viewers of the TV show are over 30. On the other hand, if $\bar{p}$ is greater than 0.50 and a significant difference between $\bar{p}$ and p is observed, we will conclude that more than half the viewers are over 30. If the observed $\bar{p}$ were less than 0.50, and a significant difference from $p = 0.50$ were observed, we would conclude that fewer than half the viewers are over 30.

It is important to note that in hypothesis testing procedures, the alternate hypotheses and the significance level of the test must be selected before the data are examined. We can easily see the difficulty with a procedure that would permit the investigator to select α after examination of the sample data. It would always be possible to accept a null hypothesis simply by choosing a sufficiently small significance level, thereby setting up a large enough region of acceptance. Thus the first step in the present problem is the usual one of setting up the competing hypotheses, stating the null hypothesis in such a way that Type I error can be calculated. We have accomplished this by a single-valued null hypothesis, $H_0\!:\!p = 0.50$. Our next step is to set the significance level, which we have taken as $\alpha = 0.05$.

We proceed with the test. The simple random sample of size 400 is drawn, the statistic $\bar{p}$ is observed, and we can now establish the appropriate decision rule. Since the sample size is large, applying the theory developed in Chapter 6, we can use the normal curve as an appropriate approximation for the sampling distribution of the percentage, $\bar{p}$. As in the preceding problem, for illustrative purposes, we will establish the decision rule in two different forms, first in terms of $\bar{p}$ values, then in terms of the corresponding z values in a standard normal distribution. The sampling distribution of $\bar{p}$ under the assumption that the null hypothesis is true has a mean of p and a standard deviation $\sigma_{\bar{p}} = \sqrt{pq/n}$. Again, we ignore the finite population correction, because the population is so large relative to the sample size. The sampling distribution of $\bar{p}$ for the present problem in which $p = 0.50$ is shown in Figure 8-7. Also shown is the horizontal axis of the corresponding standard normal distribution in terms of z values.

Since the null hypothesis will be rejected by an observation of a $\bar{p}$ value that lies significantly below or significantly above $p = 0.50$, we clearly are dealing with a two-tailed test. The critical regions (rejection regions) are displayed in Figure 8-7. The arithmetic involved in establishing regions of acceptance and rejection of H_0 is as follows. The standard error of $\bar{p}$ is given by

$$\sigma_{\bar{p}} = \sqrt{\frac{(0.50)(0.50)}{400}} = 0.025$$

Referring to Table A-5, in Appendix A we find that 2.5% of the area in a normal distribution lies to the right of $z = +1.96$, and therefore 2.5% also lies to the left of $z = -1.96$. Thus, we establish a significance level of 5% by marking off

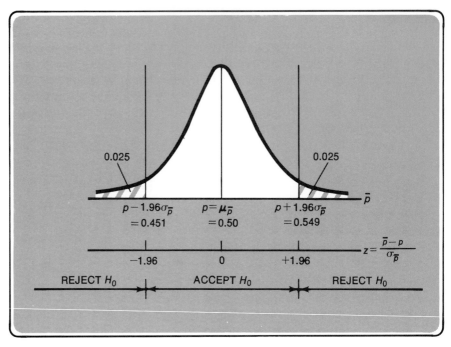

Figure 8-7
Sampling distribution of a proportion; $p = 0.50$, $n = 400$. Two-tailed test; $\alpha = 0.05$.

an acceptance range for H_0 of $p \pm 1.96\ \sigma_{\bar{p}}$. The calculation follows:

$$p + 1.96\ \sigma_{\bar{p}} = 0.50 + (1.96)(0.025) = 0.50 + 0.049 = 0.549$$
$$p - 1.96\ \sigma_{\bar{p}} = 0.50 - (1.96)(0.025) = 0.50 - 0.049 = 0.451$$

We can now state the decision rule. The agency draws a simple random sample of 400 viewers of the TV show, observes $\bar{p}$, the proportion in the sample who are over 30 years of age, and then applies the following decision rule:

DECISION RULE

1. If $\bar{p} < 0.451$ or $\bar{p} > 0.549$, reject H_0

2. If $0.451 \leq \bar{p} \leq 0.549$, accept H_0

Again, for illustrative purposes, let us restate the decision rule in terms of z values.

DECISION RULE

1. If $z < -1.96$ or $z > +1.96$, reject H_0

2. If $-1.96 \leq z \leq +1.96$, accept H_0

Applying this decision rule to the present problem, the observed sample $\bar{p}$ was

$$\overline{p} = \frac{210}{400} = 0.525$$

and

$$z = \frac{\overline{p} - p}{\sigma_{\overline{p}}} = \frac{0.525 - 0.500}{0.025} = +1.0$$

Therefore, the null hypothesis H_0 is accepted. This leads us to a rather negative conclusion. Had $\overline{p}$ fallen in the rejection region in the right-hand tail of the sampling distribution in Figure 8-7, that is, if $\overline{p}$ were greater than 0.549, we would conclude that more than half of the viewers of the TV show are over 30 years of age. If $\overline{p}$ had fallen in the left-hand tail of the rejection region, we would conclude that less than half are over 30. However, if as in this case, $\overline{p}$ lies in the acceptance region, we *cannot* conclude that more than half of the viewers are over 30 nor that less than half are over 30. The sample evidence is consistent with the hypothesis of a 50-50 split. Thus, acceptance of the null hypothesis means that on the basis of the available evidence we simply are not in a position to conclude that more than half nor less than half of the viewers are over 30 years of age. In some instances, the best course is to delay a terminal decision. Sequential decision procedures are often an appropriate technique to use where one of the alternatives is to delay the decision. Such procedures are discussed in most mathematical statistics books.

Further remarks. In this problem, the arithmetic was carried out in terms of proportion of successes. The calculations could have been made in terms of numbers of successes or percentage of successes rather than proportion. Of course, the conclusions are the same regardless of the method used. However, the arithmetic is often simpler if percentages are used. Hence, in this problem, if we work in percentage of successes, the standard error would be $\sqrt{(50)(50)/400} = \sqrt{6.25} = 2.5\%$. The fact that the standard error in this problem is 2.5% and the area in each of the rejection regions is also 2.5% is purely coincidental.

The hypothesized proportion in the null hypothesis in this problem was equal to 0.50. This stemmed from the fact that the agency was interested in whether or not more than 50% of the viewers were over 30 years of age. On the other hand, if we wanted to test the assertion that the population proportion was 0.55, 0.60, or some other number, we would have used these figures as the respective hypothesized parameters. Also in this problem, the null hypothesis was single valued ($p = 0.50$), whereas the alternative hypothesis was many valued ($p \neq 0.50$). We saw that this resulted in a two-tailed test. Whether a test of a hypothesis is one-tailed or two-tailed depends on the question to be answered. For example, suppose the problem had been framed as follows. An assertion had been made that at least 50% of the viewers of the TV show were over 30. Further assume that we wish to run a maximum risk of 5% of erroneously rejecting this assertion. The alternative hypotheses in this instance would be

$$H_0: p \geq 0.50$$
$$H_1: p < 0.50$$

This would involve a one-tailed test with a rejection region lying in the left-hand tail and containing 5% of the area under the normal curve. The critical region would be in the left tail, since only a significant difference for a $\bar{p}$ value lying below 0.50 could result in the rejection of the stated null hypothesis. The decision rule in terms of z values would be

DECISION RULE

1. if $z < -1.65$, reject H_0

2. If $z \geq -1.65$, accept H_0

On the other hand, if the assertion had been that 50% or fewer of the viewers were over 30 and if we wanted to run a maximum risk of 5% of erroneously rejecting this assertion, the rejection region would be the 5% area in the right tail. The corresponding alternative hypothesis and decision rule would be

$$H_0: p \leq 0.50$$
$$H_0: p > 0.50$$

DECISION RULE

1. If $z \geq +1.65$, reject H_0

2. If $z < +1.65$, accept H_0

Standard normal distributions with the decision rules for these one-tailed tests are depicted in Figure 8-8. When tests are conducted for means or other statistical measures, they may also be either one- or two-tailed, depending upon the context of the problem.

EXERCISES

1. Suppose you are responsible for the quality control of a certain part bought from a supplier. Inspection tests destroy the part, so you must use sampling. A 5% defective rate is tolerable but in your sample of 100 from a lot of 10,000, eight parts are defective. Is this sufficient evidence that the lot of parts has too many defectives?

2. The Constitution of the United States requires a two-thirds majority of both House and Senate to override a presidential veto. In a given situation the necessary majority was secured in the Senate, but the action of the House was uncertain. Prior to the actual vote in the House, newspaper reporters took a straw vote of 90 representatives, and 57 of them indicated their intention to vote to override the veto. If you wished to be wrong no more than five times in 100, assuming that the sample was taken at random, would you conclude that the veto would not be overridden? Justify your conclusion statistically. You may ignore the possible requirement of a finite population correction.

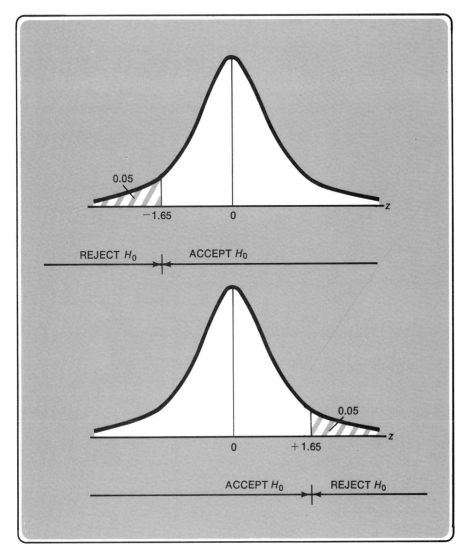

Figure 8-8
Standard normal distribution with decision rules in terms of *z* values: one-tailed tests
($\alpha = 0.05$).

3. The following test is set up: A sample of 100 people is selected at random and each
is asked if he likes a certain new product. The company conducting the survey feels
that it is necessary that at least 10% of all consumers like the product in order for the
firm to continue marketing it. The firm therefore decides that if four or fewer people
respond favorably, it will stop marketing the product. State in words the nature of the
Type I error involved here. How large is it?

4. A housing survey is to be undertaken to determine whether the proportion of substandard dwelling units in a certain large city has changed. At the time of the last
Census of Housing, 10% of the dwelling units were classified as substandard. It is

desired to maintain a 0.01 risk of erroneously concluding that a change has occurred in the proportion of substandard dwelling units. A simple random sample of 900 dwelling units is contemplated. Set up the decision rule for this test.

5. The major oil companies report that last year 5% of all credit charges for gasoline, car repairs and parts were never collected and must be written off as bad debts. Recently the oil companies have installed a central computer credit check system. That is, for any credit purchase of over $10, the local gas station must call the central computer center where after a computer search of the customer's payment record, the purchase is given a credit acceptance number or refused credit. Any sales of over $10 not given a credit acceptance number will not be honored as a credit sale by the oil company. To see if the system is effective, 1000 credit charges which were accepted were selected at random and it was found that 36 charges were uncollected and written off as bad debts. Do you think the system is effective? Use a 5% significance level.

8.3   TWO-SAMPLE TESTS (LARGE SAMPLES)

The discussion thus far has involved testing of hypotheses using data from a single random sample. Another important class of problems involves the question of whether statistics observed in two simple random samples differ significantly. Recalling that all statistical hypotheses are statements concerning population parameters, we see that this question implies a corresponding question about the underlying parameters in the populations from which the samples were drawn. For example, if the statistics observed in the two samples are arithmetic means, say $\bar{x}_1$ and $\bar{x}_2$, respectively, the question refers to whether we are willing to attribute the difference between these two sample means to chance errors of sampling. If on the basis of a test, we find that the difference is too large to attribute to chance errors, we will conclude that the populations from which the samples were drawn have *unequal means.* We illustrate these tests, first for differences between means and then for differences between proportions. In both cases, we assume as we did in Section 8.2 that we are dealing with large samples.

Test for difference between means: two-tailed test

A consulting firm conducting research for a client was asked to test whether the wage levels of unskilled workers in a certain industry was the same in two different geographical areas, referred to as Area A and Area B. The firm took simple random samples of the unskilled workers in the two areas and obtained the following sample data for weekly wages:

AREA	MEAN	STANDARD DEVIATION	SIZE OF SAMPLE
A	$\bar{x}_1 = \$90.01$	$s_1 = \$4.00$	$n_1 = 100$
B	$\bar{x}_2 = 85.21$	$s_2 = 4.50$	$n_2 = 200$

If the client wished to run a risk of 0.02 of incorrectly rejecting the hypothesis that the population means in these two areas were the same, what conclusion should be reached? Different sample sizes have been assumed in this problem in order to keep the example completely general. That is, the samples need not be of the same size.

Let us refer to the means and standard deviations of *all* unskilled workers in this industry in Areas A and B, respectively, as μ_1 and μ_2, and σ_1 and σ_2. These are the population parameters corresponding to the sample statistics $\bar{x}_1$ and $\bar{x}_2$, s_1 and s_2. The hypotheses to be tested are

$$H_0: \mu_1 - \mu_2 = 0$$
$$H_1: \mu_1 - \mu_2 \neq 0$$

That is, the null hypothesis asserts that the population parameters μ_1 and μ_2 are equal. As usual, the sample data must be compared to the hypothesis. We form the statistic $\bar{x}_1 - \bar{x}_2$, the difference between the sample means. If $\bar{x}_1 - \bar{x}_2$ differs significantly from zero, the hypothesized value for $\mu_1 - \mu_2$, we will reject the null hypothesis and conclude that the population parameters μ_1 and μ_2 are indeed different.

Since the risk of a Type I error has been set, we turn now to the determination of the decision rule that is based on the appropriate random sampling distribution. Let us examine some of the important characteristics of this distribution. The two random samples are independent, that is, the probabilities of selection of the elements in one sample are not affected by the selection of the other sample. Hence, $\bar{x}_1$ and $\bar{x}_2$ are independent random variables. It can be shown that the mean and standard deviation of the sampling distribution of $\bar{x}_1 - \bar{x}_2$ are, respectively,

(8.1)
$$\mu_{\bar{x}_1 - \bar{x}_2} = 0$$

and

(8.2)
$$\sigma_{\bar{x}_1 - \bar{x}_2} = \sqrt{\frac{\sigma_1^2}{n_1} + \frac{\sigma_2^2}{n_2}}$$

For large, independent samples, the sampling distribution of $\bar{x}_1 - \bar{x}_2$ is approximately normal by the central limit theorem argument. *In summary, therefore, if $\bar{x}_1$, and $\bar{x}_2$ are the means of two large, independent samples from populations with means μ_1 and μ_2 and standard deviations σ_1 and σ_2, respectively, and if we hypothesize that μ_1 and μ_2 are equal, then the sampling distribution of $\bar{x}_1 - \bar{x}_2$ may be approximated by a normal curve with mean $\mu_{\bar{x}_1 - \bar{x}_2} = 0$ and standard deviation $\sigma_{\bar{x}_1 - \bar{x}_2} = \sqrt{\sigma_1^2/n_1 + \sigma_2^2/n_2}$.* It is helpful to think of this sampling distribution as the frequency distribution that would be obtained by grouping the $\bar{x}_1 - \bar{x}_2$ values observed in repeated pairs of samples drawn independently from two populations with the same means.

The standard deviation $\sigma_{\bar{x}_1 - \bar{x}_2}$ is referred to as the *standard error of the difference between two means.* We see from Equation (8.2) that we must know the population standard deviations in order to calculate this standard error. However, for *large samples,* we adopt the approximate procedure of using the sample standard deviations s_1 in place of σ_1 and s_2 for σ_2. The resulting estimated

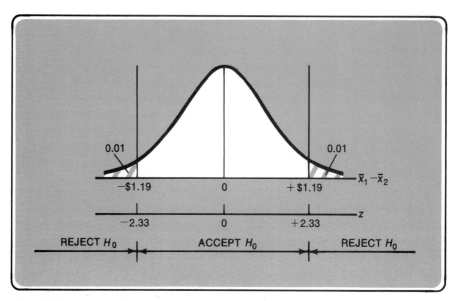

Figure 8-9
Sampling distribution of the difference between two means; two-sided test; $\alpha = 0.02$.

or approximate standard error is symbolized $s_{\bar{x}_1 - \bar{x}_2}$ and may be written

(8.3)
$$s_{\bar{x}_1 - \bar{x}_2} = \sqrt{\frac{s_1^2}{n_1} + \frac{s_2^2}{n_2}}$$

We can now proceed to establish the decision rule for the problem. The test is clearly two-tailed because the hypothesis of equal population means would be rejected if $\bar{x}_1 - \bar{x}_2$ differed significantly from zero either by lying sufficiently far above or below it. The sampling distribution of $\bar{x}_1 - \bar{x}_2$ is shown in Figure 8-9. On the horizontal scale of the distribution is shown the difference between the sample means $\bar{x}_1 - \bar{x}_2$. As indicated, the mean of the distribution is equal to zero, or in other words, under the null hypothesis, the expected value of $\bar{x}_1 - \bar{x}_2$ is equal to zero. Another way of interpreting the zero is that under the null hypothesis $H_0: \mu_1 - \mu_2 = 0$, we have assumed that the mean wages of the populations of unskilled workers are the same in Area A and Area B for the industry in question. Since the significance level is 0.02, 1% of the area under the normal curve is shown in each tail. From Table A-5 of Appendix A we find that in a normal distribution 1% of the area lies to the right of $z = +2.33$ and by symmetry 1% lies to the left of $z = -2.33$. Thus, we would reject the null hypothesis if the sample difference $\bar{x}_1 - \bar{x}_2$ fell more than 2.33 standard errors from the expected value of zero. The estimated standard error of the difference between means, $s_{\bar{x}_1 - \bar{x}_2}$ is by Equation (8.3) equal to

$$s_{\bar{x}_1 - \bar{x}_2} = \sqrt{\frac{(\$4.00)^2}{100} + \frac{(\$4.50)^2}{200}} = \$0.51$$

and

$$2.33 s_{\bar{x}_1 - \bar{x}_2} = (2.33)(\$0.51) = \$1.19$$

Thus, the decision rule may be stated as follows:

DECISION RULE

1. If $\bar{x}_1 - \bar{x}_2 < -\$1.19$ or $\bar{x}_1 - \bar{x}_2 > \$1.19$, reject H_0

2. If $-\$1.19 \leq \bar{x}_1 + \bar{x}_2 \leq \$1.19$, accept H_0

In terms of z values, we have

DECISION RULE

1. If $z < -2.33$ or $z > +2.33$, reject H_0

2. If $-2.33 \leq z \leq +2.33$, accept H_0

where

$$z = \frac{\bar{x}_1 - \bar{x}_2}{s_{\bar{x}_1 - \bar{x}_2}}$$

Note that this z value is in the usual form of the ratio (statistic − parameter)/ standard error. The difference $\bar{x}_1 - \bar{x}_2$ is the statistic. The parameter under test is $\mu_1 - \mu_2 = 0$, and thus need not be shown in the numerator of the ratio. As previously indicated, we have substituted an approximate standard error for the true standard error in the denominator.[3]

Using the decision rule in the present problem, we have

$$\bar{x}_1 - \bar{x}_2 = \$90.01 - \$85.21 = \$4.80$$

and

$$z = \frac{\bar{x}_1 - \bar{x}_2}{s_{\bar{x}_1 - \bar{x}_2}} = \frac{\$4.80}{\$0.51} = 9.4$$

Since $4.80 far exceeds $1.19 and correspondingly 9.4 far exceeds 2.33, the null hypothesis is rejected. Hence, it is extremely unlikely that these two samples were drawn from populations having the same mean. In terms of the problem, we conclude that the sample mean wages of unskilled workers in this industry *differed significantly* between Areas A and B. Since this discredits the hypothesis that the population mean wages are equal, we conclude that the population means *differ* between Areas A and B. Note that it is incorrect to use the term "significant difference" when referring to the relationship between two population parameters, which in this case, are population means. Also, it is useful to observe that in this as in all other hypothesis testing situations, we are assuming random sampling. Obviously, if the samples were not randomly drawn from the two populations, the foregoing procedure and conclusion are invalid.

[3] Some authors reserve the symbol z for the case in which the population standard deviation is known and therefore the standard error in the denominator of the z ratio is also known. However, we use that symbol also in the case of large samples, where the population standard deviation is unknown and therefore an approximate standard error is substituted for a true standard error.

EXERCISES

1. A bank is considering opening a new branch in one of two neighborhoods. One of the factors considered by the bank was whether the average family incomes in the two neighborhoods differed. From census records, the bank drew two simple random samples of 100 families each. The following information was obtained:

$$\bar{x}_1 = \$10,100 \qquad\qquad \bar{x}_2 = \$10,300$$
$$s_1 = \$300 \qquad\qquad s_2 = \$400$$
$$n_1 = 100 \qquad\qquad n_2 = 100$$

The bank wishes to test the null hypothesis that the two neighborhoods have the same mean income. What should the bank conclude? Use $\alpha = 0.05$.

2. A prospective MBA student conducted some research on two different universities that he was considering attending. He decided that the average starting salaries of the MBA graduates of the two universities were relevant. An independent study of simple random samples of the most recent MBA graduates of both schools revealed the following statistics:

	SCHOOL A	SCHOOL B
Arithmetic Mean (Monthly Starting Salary)	$1282	$1208
Standard Deviation	$80	$94
Sample Size	50	50

The student wishes to run a risk of no more than 5% saying incorrectly that the relevant universes differ with respect to mean starting salary. What should he conclude?

In answering the above question, you had to locate a sample statistic (arithmetic mean) on a random sampling distribution. Draw a rough sketch of this distribution showing
(a) The value and location of the hypothesized parameter.
(b) The value and location of the statistic.
(c) The horizontal scale description.
(d) The portion of the distribution corresponding to the probability represented by $\alpha = 0.05$.

3. Suppose that you took a *census* of the incomes of all attorneys in two towns and found:

TOWN	MEAN INCOME
A	$20,000
B	$25,000

Can you conclude statistically that the average income of attorneys in Town B exceeded that of Town A attorneys? Would you test an hypothesis? If yes, what hypothesis? If no, why not?

Test for difference between proportions: two-tailed test

Another important two-sample hypothesis testing case is one in which the observed statistics are proportions. The decision procedure is conceptually the same as when the sample statistics are means; only the computational details differ. In order to illustrate the technique, let us consider the following example. Workers in the Stanley Morgan Company and Rock Hayden Company, two firms in the same industry, were asked whether they preferred to receive a specified package of increased fringe benefits or a specified increase in base pay. For brevity in this problem, we will refer to the companies as the S. M. Company and the R. H. Company and the proposed increases as "increased fringe benefits" and "increased base pay." In a simple random sample of 150 workers in the S. M. Company, 75 indicated that they preferred increased base pay. In the R. H. Company, 103 out of a simple random sample of 200 preferred increased base pay. In each company, the sample was less than 5% of the total number of workers. It was desirable to have a very low probability of erroneously rejecting the hypothesis of equal proportions of workers in the two companies who preferred increased base pay. Therefore, a 1% level of significance was used for the test. Can it be concluded at the 1% level of significance that these two companies differed in the proportion of workers who preferred increased base pay?

Using the subscripts 1 and 2 to refer to the S. M. Company and R. H. Company, respectively, we can organize the sample data as follows:

S. M. COMPANY	R. H. COMPANY
$\bar{p}_1 = \dfrac{75}{150} = 0.50$	$\bar{p}_2 = \dfrac{103}{200} = 0.515$
$\bar{q}_1 = \dfrac{75}{150} = 0.50$	$\bar{q}_2 = \dfrac{97}{200} = 0.485$
$n_1 = 150$	$n_2 = 200$

where $\bar{p}_1$ and $\bar{q}_1$ refer to the sample proportions in the S. M. Company in favor of and opposed to increased base pay, respectively. The sample size in the S. M. Company is denoted n_1. Corresponding notation is used for the R. H. Company. If we designate the population proportions in favor of increased pay in the two companies as p_1 and p_2, then in a manner analogous to that of the preceding problem, we set up the two alternative hypotheses

$$H_0: p_1 - p_2 = 0$$
$$H_1: p_1 - p_2 \neq 0$$

The underlying theory for the test is similar to that in the two-sample test for the difference between two means. If $\bar{p}_1$ and $\bar{p}_2$ are the observed sample proportions in large simple random samples drawn from populations with parameters p_1 and p_2, respectively, then the sampling distribution of the statistic $\bar{p}_1 - \bar{p}_2$ has a mean

(8.4) $$\mu_{\bar{p}_1 - \bar{p}_2} = p_1 - p_2$$

and a standard deviation

(8.5) $$\sigma_{\bar{p}_1 - \bar{p}_2} = \sqrt{\sigma_{\bar{p}_1}^2 + \sigma_{\bar{p}_2}^2}$$

where $\sigma_{\bar{p}_1}^2$ and $\sigma_{\bar{p}_2}^2$ are the variances of the sampling distributions of $\bar{p}_1$ and $\bar{p}_2$. Under assumptions of a binomial distribution, $\sigma_{\bar{p}_1}^2 = p_1 q_1 / n$ and $\sigma_{\bar{p}_2}^2 = p_2 q_2 / n_2$. Although the sampling was conducted without replacement, since each of the samples constituted only a small percentage of the corresponding population (less than 5%), the binomial distribution assumption appears reasonable. Thus, Equation (8.5) becomes

(8.6) $$\sigma_{\bar{p}_1 - \bar{p}_2} = \sqrt{\frac{p_1 q_1}{n} + \frac{p_2 q_2}{n_2}}$$

If we hypothesize that $p_1 = p_2$, as in the null hypothesis in this problem, and if we refer to the common value of p_1 and p_2 as p, Equations (8.4) and (8.6) become

(8.7) $$\mu_{\bar{p}_1 - \bar{p}_2} = p - p = 0$$

and

(8.8) $$\sigma_{\bar{p}_1 - \bar{p}_2} = \sqrt{\frac{pq}{n_1} + \frac{pq}{n_2}} = \sqrt{pq\left(\frac{1}{n_1} + \frac{1}{n_2}\right)}$$

Since the common hypothesized proportion p under the null hypothesis is unknown, we estimate it for the hypothesis test by taking a weighted mean of the observed sample percentages. Referring to this "pooled estimator" as $\bar{p}$, we have

(8.9) $$\bar{p} = \frac{n_1 \bar{p}_1 + n_2 \bar{p}_2}{n_1 + n_2}$$

The numerator of Equation (8.9) is simply the total number of "successes" in the two samples combined, and the denominator is the total number of observations in the two samples. The standard deviation in Equation (8.8), $\sigma_{\bar{p}_1 - \bar{p}_2}$, is referred to as the *standard error of the difference between two proportions*. Substituting the "pooled estimator," $\bar{p}$ for p in (8.8), we have the following formula for the *estimated* or *approximate* standard error, $s_{\bar{p}_1 - \bar{p}_2}$.

(8.10) $$s_{\bar{p}_1 - \bar{p}_2} = \sqrt{\bar{p}\bar{q}\left(\frac{1}{n_1} + \frac{1}{n_2}\right)}$$

We can now summarize these results. Let $\bar{p}_1$ and $\bar{p}_2$ be proportions of successes observed in two large, independent samples from populations with parameters p_1 and p_2. If we hypothesize that $p_1 = p_2 = p$, we obtain a pooled estimator $\bar{p}$ for p, where $\bar{p} = (n_1 \bar{p}_1 + n_2 \bar{p}_2)/(n_1 + n_2)$. *Then, the sampling distribution of $\bar{p}_1 - \bar{p}_2$ may be approximated by a normal curve with mean $\mu_{\bar{p}_1 - \bar{p}_2} = 0$ and estimated standard deviation,*

$$s_{\bar{p}_1 - \bar{p}_2} = \sqrt{\bar{p}\bar{q}\left(\frac{1}{n_1} + \frac{1}{n_2}\right)}$$

We may think of this sampling distribution as the frequency distribution of

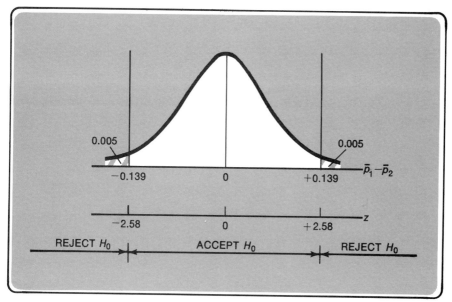

Figure 8-10
Sampling distribution of the difference between two proportions; two-sided test; $\alpha = 0.01$.

$\bar{p}_1 - \bar{p}_2$ values observed in repeated pairs of samples drawn independently from two populations having the same proportions.

Proceeding to the decision rule, we see again that the test is two-tailed because the hypothesis of equal population proportions would be rejected for $\bar{p}_1 - \bar{p}_2$ values that fall significantly above or below zero. The sampling distribution of $\bar{p}_1 - \bar{p}_2$ for the present problem is shown in Figure 8-10. Since the significance level $\alpha = 0.01$, $\alpha/2$ or $\frac{1}{2}$ of 1% of the area under the normal distribution is shown in each tail. Referring to Table A-5, we find that $\frac{1}{2}$ of 1% of the area in a normal curve lies above a z value of $+2.58$, and thus a similar percentage lies below $z = -2.58$. Hence, rejection of the null hypothesis $H_0: p_1 - p_2 = 0$ occurs if the sample difference $\bar{p}_1 - \bar{p}_2$ falls more than 2.58 standard error units from zero. By Equation (8.10), the standard error of the difference between proportions is

$$s_{\bar{p}_1 - \bar{p}_2} = \sqrt{\bar{p}\,\bar{q}\left(\frac{1}{n_1} + \frac{1}{n_2}\right)} = \sqrt{(0.51)(0.49)\left(\frac{1}{150} + \frac{1}{200}\right)}$$

$$= 0.054$$

where

$$\bar{p} = \frac{n_2\bar{p}_1 + n_2\bar{p}_2}{n_1 + n_2} = \frac{(150)(0.50) + 200(0.515)}{150 + 200} = \frac{75 + 103}{150 + 200} = 0.51$$

Hence,

$$2.58s_{\bar{p}_1 - \bar{p}_2} = (2.58)(0.054) = 0.139$$

Therefore, the decision rule is

DECISION RULE

1. If $\bar{p}_1 - \bar{p}_2 < -0.139$ or $\bar{p}_1 - \bar{p}_2 > 0.139$, reject H_0
2. If $-0.139 \leq \bar{p}_1 - \bar{p}_2 \leq 0.139$, accept H_0

In terms of z values, the rule is

DECISION RULE

1. If $z < -2.58$ or $z > +2.58$, reject H_0
2. If $-2.58 \leq z \leq +2.58$, accept H_0

where

$$z = \frac{\bar{p}_1 - \bar{p}_2}{s_{\bar{p}_1 - \bar{p}_2}}$$

Applying this decision rule yields

$$\bar{p}_1 - \bar{p}_2 = 0.500 - 0.515 = -0.015$$

and

$$z = \frac{\bar{p}_1 - \bar{p}_2}{s_{\bar{p}_1 - \bar{p}_2}} = \frac{-0.015}{0.054} = -0.28$$

Thus, the null hypothesis is accepted. In summary, the sample proportions $\bar{p}_1$ and $\bar{p}_2$ did not differ significantly, and therefore we cannot conclude that the two companies differed with respect to the proportion of workers who preferred increased base pay. Our reasoning is based on the finding that assuming the population proportions were equal, a difference between the sample proportions of the size observed could not at all be considered unusual.

Test for differences between proportions: one-tailed test

The two preceding examples illustrated two-tailed tests for cases where data are available for samples from two populations. Just as in the one-sample case, the question we wish to answer may give rise to a one-tailed test situation. In order to illustrate this point, let us examine the following problem.

Two competing drugs are available for treating a certain physical ailment. There are no apparent side-effects from administration of the first drug, whereas there are some definite side-effects of nausea and mild headaches from use of the second. A group of medical researchers has decided that it would nevertheless be willing to recommend use of the second drug in preference to the first if the proportion of cures effected by the second were higher than those by the first drug. The group felt that the potential benefits of achieving increased cures of the ailment would far outweigh the disadvantages of the possible accompanying side-effects. On the other hand, if the proportion of cures effected by the second drug was equal to or less than that of the first drug, the group would recommend use of the first. In terms of hypothesis testing, we can state the alternatives and consequent actions as

$$H_0:p_2 \leq p_1 \text{ (use the first drug)}$$
$$H_1:p_2 > p_1 \text{ (use the second drug)}$$

where p_1 and p_2 denote the population proportions of cures effected by the first and second drugs, respectively. Another way we may write these alternatives is

$$H_0:p_2 - p_1 \leq 0 \text{ (use the first drug)}$$
$$H_1:p_2 - p_1 > 0 \text{ (use the second drug)}$$

For purposes of comparison, we may note that the alternative hypotheses in the preceding problem, which was a two-tailed testing situation, were

$$H_0:p_1 = p_2$$
$$H_1:p_1 \neq p_2$$

or in the alternative form in terms of differences

$$H_0:p_1 - p_2 = 0$$
$$H_1:p_1 - p_2 \neq 0$$

Clearly, the present problem involves a one-tailed test, in which we would reject the null hypothesis only if the sample difference, $\bar{p}_2 - \bar{p}_1$ differed significantly from zero and was a positive number.

The medical researchers used the drugs experimentally on two random samples of persons suffering from the ailment, administering the first drug to a group of 80 patients and the second drug to a group of 90 patients. By the end of the experimental period, 52 of those treated with the first drug were classified as "cured," whereas 63 of those treated with the second drug were so classified. The sample results may be summarized as follows:

FIRST DRUG	SECOND DRUG
$\bar{p}_1 = \dfrac{52}{80} = 0.65$ cured	$\bar{p}_2 = \dfrac{63}{90} = 0.70$ cured
$\bar{q}_1 = \dfrac{28}{80} = 0.35$ not cured	$\bar{q}_2 = \dfrac{27}{90} = 0.30$ not cured
$n_1 = 80$	$n_2 = 90$

The pooled sample proportion cured is

$$\bar{p} = \frac{52 + 63}{80 + 90} = \frac{115}{170} = 0.676$$

and the estimated standard error of the difference between proportions is

$$s_{\bar{p}_2 - \bar{p}_1} = \sqrt{(0.676)(0.324)\left(\frac{1}{80} + \frac{1}{90}\right)} = 0.0719$$

Since the medical group wished to maintain a low probability of erroneously adopting the second drug, it selected a 1% significance level for the test. One percent of the area under the normal curve lies to the right of $z = +2.33$. Therefore, the null hypothesis would be rejected if $\bar{p}_2 - \bar{p}_1$ falls at least 2.33 standard error units above zero. In terms of proportions,

$$2.33s_{\bar{p}_2 - \bar{p}_1} = 2.33(0.0719) = 0.168$$

Hence the decision rule is

DECISION RULE

1. If $\bar{p}_2 - \bar{p}_1 > 0.168$, reject H_0
2. If $\bar{p}_2 - \bar{p}_1 \leq 0.168$, accept H_0

In terms of z values, the rule is

DECISION RULE

1. If $z > +2.33$, reject H_0
2. If $z \leq +2.33$, accept H_0

where

$$z = \frac{\bar{p}_2 - \bar{p}_1}{s_{\bar{p}_2 - \bar{p}_1}}$$

In the present problem,

$$\bar{p}_2 - \bar{p}_1 = 0.70 - 0.65 = 0.05$$

and

$$z = \frac{0.70 - 0.65}{0.0719} = 0.70$$

Thus, the null hypothesis is accepted. On the basis of the sample data, we cannot conclude that the second drug accomplishes a greater proportion of cures than the first. The sampling distribution of $\bar{p}_2 - \bar{p}_1$ is given in Figure 8-11. Note that it is immaterial whether the difference between proportions is stated as $\bar{p}_1 - \bar{p}_2$ or $\bar{p}_2 - \bar{p}_1$, but that care must be exercised concerning the correspondence between the way the hypothesis is stated, the sign of the difference between the sample proportions, and the tail of the sampling distribution in which rejection of the null hypothesis takes place.

EXERCISES

1. (a) In a simple random sample of 400 students in collegiate schools of business, 176 favored the addition of more required mathematics courses to be taken in the freshman and sophomore years. In a simple random sample of 400 students

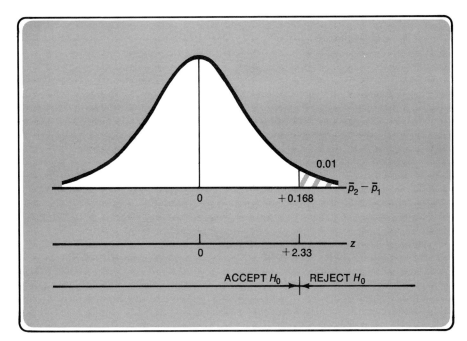

Figure 8-11
Sampling distribution of the difference between two proportions; one-sided test; $\alpha = 0.01$.

in liberal arts colleges, 144 favored the addition of more required mathematics courses. Do you believe there is a real difference in the attitude of the two groups? Justify your answer statistically and indicate the level of significance that you use.

(b) Explain specifically the meaning of a Type I error in terms of this particular problem.

2. Suppose that in a simple random sample of 400 people from one city, 188 preferred a particular brand of soap to all others, and in a similar sample of 500 people from another city, 210 preferred the same product. Is there reason to doubt the hypothesis that equal proportions of persons in the two cities preferred this brand of soap at the 5% level of significance?

3. Workers in two different industries were asked what they considered to be the most important labor-management problem in their industry. In Industry A with a total of 20,000 workers, 200 out of a random sample of 400 workers felt that a fair adjustment of grievances was the most important problem. In Industry B with a total of 30,000 workers, 60 out of a random sample of 100 workers felt that this was the most important problem.

(a) Would you conclude that these two industries differed with respect to the proportion of workers who believed that a fair adjustment of grievances was the most important problem? Demonstrate your answer statistically, and give a brief statement of your reasoning in conclusion.

(b) Draw a rough sketch (labeling the horizontal scale) of the random sampling distribution upon which your answer to (a) depends. Show on the sketch the approximate positions of the values pertinent to the solution to the above problem.

8.4 THE *t* DISTRIBUTION (SMALL SAMPLES) POPULATION STANDARD DEVIATION(S) UNKNOWN

The hypothesis testing methods discussed in the preceding sections are appropriate for large samples. In this section, we concern ourselves with the case where the *sample size* is small. The underlying theory is exactly the same as that given in Section 7.3, where confidence interval estimation for small samples was discussed. In hypothesis testing as in confidence interval estimation, the distinction between large and small sample tests becomes important when the population standard deviation is *unknown* and therefore must be *estimated* from the sample observations. The main principles will be reviewed here. The statistic $(\bar{x} - \mu)/s_{\bar{x}}$ is not approximately normally distributed for all sample sizes where $s_{\bar{x}}$ denotes an estimated standard error. As we have noted earlier, $s_{\bar{x}}$ is computed by the formula $s_{\bar{x}} = s/\sqrt{n}$ where s represents an estimate of the true population standard deviation. For large samples the ratio $(\bar{x} - \mu)/s_{\bar{x}}$ is approximately normally distributed, and we may use the methods discussed in Sections 8.2 and 8.3. However, since this statistic is not approximately normally distributed for small samples, the *t* distribution should be used instead. The use of the *t* distribution for testing a hypothesis concerning a population mean is given in Example 8.1. A small sample ($n \le 30$) is assumed and the population standard deviation is unknown.

EXAMPLE 8-1

One Sample Test of a Hypothesis about the Mean: Two-Sided Test.

The personnel department of a company developed an aptitude test for a certain type of semi-skilled worker. The individual test scores were assumed to be normally distributed. The developers of the test asserted a tentative hypothesis that the arithmetic mean grade obtained by this type of semi-skilled worker would be 100. It was agreed that this hypothesis would be subjected to a two-tailed test at the 5% level of significance. The aptitude test was given to a simple random sample of 16 of the semi-skilled workers with the following results:

$$\bar{x} = 94$$
$$s = 5$$
$$n = 16$$

The competing hypotheses are

$$H_0: \mu = 100$$
$$H_1: \mu \neq 100$$

SOLUTION

To carry out the test, the following were calculated:

$$s_{\bar{x}} = \frac{s}{\sqrt{n}} = \frac{5}{\sqrt{16}} = 1.25$$

and

$$t = \frac{\bar{x} - \mu}{s_{\bar{x}}} = \frac{94 - 100}{1.25} = -4.80$$

The significance of this *t* value is judged from Table A-6 in Appendix A. The meaning of the table was discussed in Section 7.3. We will explain the use of the table for this hypothesis testing problem. Let us set up the areas of acceptance and rejection of the hypothesis. Since the sample size is 16, the number of degrees of freedom, $df = n - 1$, is $df = 16 - 1 = 15$. Looking along the row of Table A-6 for 15 under the column 0.05, we find the *t* value, 2.131. This means that in a *t* distribution for 15 degrees of freedom, the probability is 5% that *t* is greater than 2.131 or is less than -2.131. Thus, in the present problem, at the 5% level of significance, the null hypothesis, $H_0:\mu = 100$ is rejected if a *t* value is observed that exceeds 2.131 or is less than -2.131. Since the computed *t* value in this problem is -4.80, we reject the null hypothesis. In other words, we are unwilling to attribute the difference between our sample mean of 94 and the hypothesized population mean of 100 merely to chance errors of sampling. The *t* distribution for this problem is shown in Figure 8-12.

A few remarks can be made about this problem. Since the computed *t* value of -4.80 was such a large negative number, the null hypothesis would have been rejected even at the 2% or 1% levels of significance (see Appendix Table A-6). Had the test been one-tailed at the 5% level of significance, we would have had to obtain the critical *t* value by looking under 0.10 in the caption of Table A-6, since the 0.10 figure is the combined area in both tails. Thus for a one-tailed test at the 5% level of significance and a lower tail rejection region, in the present problem, the critical *t* value would have been -1.753.

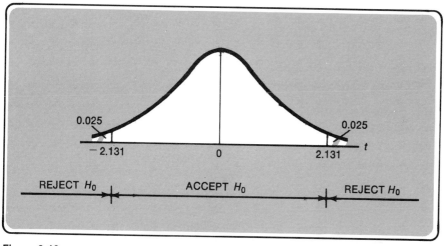

Figure 8-12
The *t* distribution for 15 degrees of freedom.

It is interesting to compare these critical *t* values with analogous critical *z* values for the normal curve. From Table A-5 of Appendix A, we find that the critical *z* values at the 5% level of significance are -1.96 and 1.96 for a two-tailed test and -1.65 for a one-tailed test with a lower tail rejection region. As we have just seen, the corresponding figures for the critical *t* values in a test involving 15 degrees of freedom are -2.131 and 2.131 for a two-tailed test and -1.753 for a one-tailed test with a lower tail rejection region.

THE *t* DISTRIBUTION (SMALL SAMPLES) POPULATION STANDARD DEVIATION(S) UNKNOWN

201

An underlying assumption in applying the t test in this problem is that the population is closely approximated by a normal distribution. Since the population standard deviation, σ, is unknown, the t distribution is the theoretically correct sampling distribution. However, if the sample size had been large, even with an unknown population standard deviation, the normal curve could have been used as an approximation to the t distribution. As we saw in this problem, for 15 degrees of freedom, a combined total of 5% of the area in the t distribution falls to the right of $t = +2.131$ and to the left of $t = -2.131$. The corresponding z values in the normal distribution are $+1.96$ and -1.96. As can be seen from Table A-6, the t value entry for 30 degrees of freedom is 2.042. The closeness of this figure to $+1.96$ gives rise to the usual rule of thumb of $n > 30$ as the arbitrary dividing line between large sample and small sample methods. We have used this convenient rule in Chapters 7 and 8. However, what constitutes a suitable approximation really depends on the context of the particular problem. Furthermore, if the population is highly skewed, a sample size as large as 100 may be required for the assumption of a normal sampling distribution of $\bar{x}$ to be appropriate.

EXAMPLE 8-2

Two-Sample Test for Means: Two-Sided Test. Small simple random samples of the freshmen and senior classes were drawn at a large university and the amounts of money (excluding checks) that these individuals had on their persons were determined. The following statistics were calculated:

FRESHMEN	SENIORS
$\bar{x}_1 = \$1.28$	$\bar{x}_2 = \$2.02$
$s_1 = \$0.51$	$s_2 = \$0.43$
$n_1 = 10$	$n_2 = 12$

SOLUTION

As in Example 7-3, the sample standard deviations were computed by Equation (3.9). At the 2% level of significance should we conclude that a significant difference was observed between the sample means?

The alternative hypotheses are

$$H_0 : \mu_1 - \mu_2 = 0$$
$$H_1 : \mu_1 - \mu_2 \neq 0$$

To test the null hypothesis, we use the t statistic

$$t = \frac{(\bar{x}_1 - \bar{x}_2) - 0}{s_{\bar{x}_1 - \bar{x}_2}} = \frac{\bar{x}_1 - \bar{x}_2}{s_{\bar{x}_1 - \bar{x}_2}}$$

where $s_{\bar{x}_1 - \bar{x}_2}$ is the estimated standard error of the difference between two means.

Unlike the case of the large sample approach it is necessary here to assume equal population variances. An estimate of this common variance is obtained by pooling the two sample variances into a weighted average, using the numbers of degrees of freedom, $n_1 - 1$ and $n_2 - 1$ as weights. This pooled estimate of the common variance, which we will denote as s^2, is given by

(8.11)
$$s^2 = \frac{(n_1 - 1)s_1^2 + (n_2 - 1)s_2^2}{n_1 + n_2 - 2}$$

The estimated standard error of the difference between two means is then

(8.12)
$$s_{\bar{x}_1 - \bar{x}_2} = \sqrt{\frac{s^2}{n_1} + \frac{s^2}{n_2}} = s\sqrt{\frac{1}{n_1} + \frac{1}{n_2}}$$

A number of alternative mathematical expressions are possible for Equation (8.12), but because of its similarity in form to previously used standard error formulas, we shall use it in this form.

We now proceed to work out the present problem. Substitution into (8.11) gives

$$s^2 = \frac{9(0.51)^2 + 11(0.43)^2}{10 + 12 - 2} = 0.2187$$

and

$$s = \sqrt{0.2187} = \$0.47$$

The estimated standard error is

$$s_{\bar{x}_1 - \bar{x}_2} = (\$0.47)\sqrt{\frac{1}{10} + \frac{1}{12}} = \$0.20$$

Thus, the t value is

$$t = \frac{\$2.02 - \$1.28}{\$0.20} = \frac{\$0.74}{\$0.20} = 3.70$$

The number of degrees of freedom in this problem is $n_1 + n_2 - 2$, that is, $10 + 12 - 2 = 20$. One way of viewing the number of degrees of freedom in this case is as follows. In the one-sample case, where the sample standard deviation was used as an estimate of the population standard deviation, there was a loss of one degree of freedom; hence, the number of degrees of freedom was $n - 1$. In the two-sample case, each of the sample variances was used in the pooled estimate of population variance; hence, two degrees of freedom were lost, and the number of degrees of freedom is $n_1 + n_2 - 2$.

The critical t value at the 2% significance level for 20 degrees of freedom is 2.528 (See Appendix Table A-6). Since the observed t value, 3.70, exceeds this critical t value, the null hypothesis is rejected and we conclude on the basis of the sample data that the population means are indeed different. In terms of the problem, we are unwilling to attribute the difference in average amount of pocket money between freshmen and seniors at this university to chance errors of sampling. On the basis of this test we tentatively conclude that seniors at the university carry more pocket change than do freshmen.

EXERCISES

1. A simple random sample of 10 delinquent charge accounts at a certain department store showed a mean of $79.00 and a standard deviation of $21.00.
   (a) Are these data consistent with a claim that the average size of delinquent charge accounts in this department store is $70? Use a two-tailed test with a 0.05 level of significance.

(b) Now instead of the above sample size, assume a sample of size 100. Would you reach the same conclusion as in (a)? Comment on the results of tests you applied in (a) and (b).

2. In a study of usage of instant coffee, by a simple random sample of 10 rural families, the consumption of such coffee was found to have an arithmetic mean of 30 ounces per family per month with a standard deviation of 10 ounces. In another similar study for 17 urban families, consumption was found to average 28 ounces with a standard deviation of 8 ounces.

(a) At the 10% level of significance, would you conclude that there was a statistically significant difference in the sample averages of consumption of instant coffee?

(b) For this problem state (1) the null and alternative hypotheses, and (2) the decision rule employed.

8.5 SUMMARY AND LOOKING AHEAD

In this chapter, we have considered some classical hypothesis testing techniques. These tests represent only a few of the simplest methods. All the cases discussed thus far have involved only one or two samples. Methods are available for testing hypotheses concerning three or more samples. The cases we have dealt with thus far have tested only one parameter of a probability distribution. Techniques are available for testing whether an entire frequency distribution is in conformity with a theoretical model, such as a specified probability distribution. The tests we have considered involved a terminal decision on the basis of the sample evidence. That is, a decision concerning the acceptance or rejection of hypotheses was reached on the basis of the evidence contained in one or two samples. Sequential decision procedures are available that permit postponement of decision pending further sample evidence. Some of these broader procedures are discussed in subsequent chapters.

Although hypothesis testing techniques of the type discussed in this chapter have been widely applied in a great many fields, it is important to note that they should not be employed in a mechanistic way. Thus, for example, in hypothesis testing, the establishing of significance levels such as 0.05 and 0.01 should not be an arbitrary procedure, but should result from a careful consideration of the relative seriousness of Type I and Type II errors. Although costs of Type I and Type II errors should theoretically be considered in hypothesis testing, as a matter of actual practice, they are rarely included explicitly in the analysis. In Bayesian decision theory, which is briefly introduced in Chapter 12, the costs of Type I and Type II errors, as well as the payoffs of correct decisions, become an explicit part of the formal analysis.

> "It is a capital mistake to theorize before one has data."

SIR ARTHUR CONAN DOYLE

Chi-Square Tests and Analysis of Variance

9

In Chapter 8, procedures were discussed for testing hypotheses using data obtained from a single simple random sample or from two such samples. For example, tests as to whether two population proportions or two population means were equal were considered. Obvious generalizations of such techniques are tests for the equality of more than two proportions or more than two means. The two topics discussed in this chapter supply these generalizations. *Chi-square tests* provide the basis for testing whether more than two population proportions may be considered equal; the *analysis of variance* tests whether more than two population means may be considered equal.

Chi-square tests of independence will be discussed first. Tests of independence constitute a method for deciding whether the hypothesis of independence between variables in different classifications is tenable. This procedure provides a test for the equality of more than two population proportions. Chi-square tests furnish a conclusion on whether a set of observed frequencies differs so greatly from a set of theoretical frequencies that the hypothesis under which the theoretical frequencies were derived should be rejected.

TESTS OF INDEPENDENCE   9.1

It is often important to determine whether a relationship exists between two bases of classification, or whether the two bases of classification may be considered independent. A very widely used test for this purpose is the so-called "chi-square" test. The general nature of the test is best explained in terms of a specific example.

In connection with an investigation of the socioeconomic characteristics of the families in a certain city, a market research firm wished to determine whether the number of telephones owned was independent of the number of automobiles owned. The firm obtained this ownership information from a simple random sample of 10,000 families who lived in the city. The results are shown in Table 9-1. This type of table, which has one basis of classification across the columns, in this case, number of automobiles owned, and another across the rows, in this case, number of telephones owned, is known as a *contingency table*. If the table has three rows and three columns, as in Table 9-1, it is referred to as a *three-by-three* (often written 3 × 3) *contingency table*. In general, in an r × c contingency table, where r denotes the number of rows and c the number of columns, there are r × c cells. For example, in the 3 × 3 table under discussion, there are 3 × 3 = 9 cells for which there are observed frequencies. In a 3 × 2 table, there are 3 × 2 = 6 cells, etc. The chi-square test consists in calculating expected frequencies under the hypothesis of independence and comparing the observed and expected frequencies.

The competing hypotheses under test in this problem may be stated as follows:

H_0: The number of automobiles owned is independent of the number of telephones owned.

H_1: The number of automobiles owned is not independent of the number of telephones owned.

Calculation of theoretical or expected frequencies

Since we are interested in determining whether the hypothesis of independence is tenable, we proceed to calculate the theoretical or expected frequencies by assuming that the null hypothesis is true. We observe from the marginal totals on the right-hand side of Table 9-1 that 2000/10,000, or 20% of the families do not own telephones. If the null hypothesis, H_0, is true, that is, if the ownership of automobiles is independent of ownership of telephones, then 2000/10,000 of the 3000 families owning no automobiles, 2000/10,000 of the 6000 families owning one automobile, and 2000/10,000 of the 1000 families owning two automobiles would be expected to have no telephones.

Thus, the expected number of "no car" families who do not own telephones is

$$\frac{2000}{10,000} \times 3000 = 600$$

This is the expected frequency that corresponds to 1000, the observed number of "no car" families who do not own telephones.

Similarly, the expected number of "one car" families who do not own telephones is

$$\frac{2000}{10,000} \times 6000 = 1200$$

TABLE 9-1
A Simple Random Sample of 10,000 Families Classified
by Number of Automobiles and Telephones Owned.

| NUMBER OF TELEPHONES | | NUMBER OF AUTOMOBILES OWNED | | | |
		(A_1) ZERO	(A_2) ONE	(A_3) TWO	TOTAL
(B_1)	Zero	1000	900	100	2000
(B_2)	One	1500	2600	500	4600
(B_3)	Two or More	500	2500	400	3400
	TOTAL	3000	6000	1000	10,000

This figure corresponds to the 900 shown in the first row.

In general, the theoretical or expected frequency for a cell in the ith row and jth column is calculated as follows:

(9.1)
$$\left(f_t\right)_{ij} = \frac{(\Sigma \text{ Row } i)(\Sigma \text{ Column } j)}{\text{Grand Total}}$$

where $\left(f_t\right)_{ij}$ = the theoretical (expected) frequency for a cell in the ith row and jth column

Σ Row i = the total of the frequencies in the ith row
Σ Column j = the total of the frequencies in the jth column
Grand Total = the total of all of the frequencies in the table

For example, the theoretical frequency in the first row and first column of Table 9-1, whose rationale of calculation was just explained, is computed by Equation (9.1) as

$$\left(f_t\right)_{11} = \frac{(2000)(3000)}{10,000} = 600$$

In order to keep the notation uncluttered, we will drop the subscripts denoting rows and columns for f_t values in the subsequent discussion.

The expected frequencies for the present problem are shown in Table 9-2. Because of the method of calculating the expected frequencies, the totals in

TABLE 9-2
Expected Frequencies for the Problem on the Relationship
Between Telephone and Automobile Ownership.

| NUMBER OF TELEPHONES | NUMBER OF AUTOMOBILES OWNED | | | |
	ZERO	ONE	TWO	TOTAL
Zero	600	1200	200	2000
One	1380	2760	460	4600
Two or More	1020	2040	340	3400
TOTAL	3000	6000	1000	10,000

the margins of the table are the same as the totals in the margins of the table of observed frequencies (Table 9-1). It is important to note that the method of computing the expected frequencies under the null hypothesis of independence is simply an application of the multiplication rule for independent events given in Equation (4.5). For example, in Table 9-1, the "zero car" and "zero telephone" categories have been denoted A_1 and B_1, respectively. Under independence $P(A_1 \text{ and } B_1) = P(A_1)P(B_1)$. The marginal probabilities $P(A_1)$ and $P(B_1)$ are given by

$$P(A_1) = \frac{3000}{10,000} = 0.30$$

$$P(B_1) = \frac{2000}{10,000} = 0.20$$

$$P(A_1 \text{ and } B_1) = P(A_1)P(B_1) = (0.30)(0.20) = 0.06$$

Multiplying this joint probability by the total frequency 10,000, we obtain the expected frequency previously derived for the upper left-hand cell

$$0.06 \times 10,000 = 600$$

The chi-square test

How great a departure from the theoretical frequencies under the assumption of independence can be tolerated before we reject the hypothesis of independence? The purpose of the chi-square test is to provide an answer to this question by comparing *observed frequencies* with the *theoretical* or *expected frequencies* derived under the hypothesis of independence. The test statistic used to make this comparison is known as chi-square, denoted χ^2. The computed value of χ^2 is

(9.2)
$$\chi^2 = \Sigma \frac{(f_o - f_t)^2}{f_t}$$

where f_o = an observed frequency
f_t = a theoretical frequency

As we can see from Equation (9.2), if every observed frequency is exactly equal to the corresponding theoretical frequency, the computed value of χ^2 is zero. This is the smallest value χ^2 can have. The larger the discrepancies between the observed and theoretical frequencies, the larger is χ^2.

The computed value of χ^2 is a variable that takes on different values from sample to sample. That is, χ^2 has a sampling distribution just as do the other test statistics discussed in Chapter 8. We wish to answer the question "Is the computed value of χ^2 so large that we are required to reject the null hypothesis?" In other words, "Are the aggregate discrepancies between the observed frequencies, f_o, and theoretical frequencies, f_t, so large that we are unwilling to attribute them to chance, and therefore have to reject the null hypothesis?"

Number of degrees of freedom

The number of degrees of freedom in the contingency table must be determined in order to apply the χ^2 test. The number of degrees of freedom in a 3×3 contingency table is equal to 4. This can be rationalized intuitively as follows. When only four of the expected frequencies have been determined, all of the others are fixed by the marginal totals. *In general, in a contingency table containing r rows and c columns, there are $(r - 1) \times (c - 1)$ degrees of freedom.* Thus, in a 2×2 table, *df* is $(2 - 1) \times (2 - 1) = 1$; in a 3×2 table, *df* is $(3 - 1)(2 - 1) = 2$, in a 3×3 table, *df* is $(3 - 1) \times (3 - 1) = 4$, etc.

The chi-square distribution

Before we apply the chi-square test, we must digress for a discussion of the appropriate sampling distribution. It can be shown that for large sample sizes the sampling distribution of χ^2 can be closely approximated by a continuous curve known as the chi-square distribution. There is a different chi-square distribution for each number of degrees of freedom. This is similar to the case of the *t* distribution. Hence, there is a family of chi-square distributions, one for each value of *df*. χ^2 is a continuous variable equal to or greater than zero. For a small value of *df*, the distribution is skewed to the right. As *df* increases, the distribution rapidly becomes symmetrical. In fact, for large values of *df*, the chi-square distribution is closely approximated by the normal curve. Figure 9-1 depicts the chi-square distribution for 1, 5, and 10 degrees of freedom.

Since the chi-square distribution is a probability distribution, the area under the curve for each value of *df* equals one. Because there is a separate distribution for each value of *df*, it is not practical to construct a detailed table of areas. Therefore, for compactness, what is generally shown in a chi-square table is the relationship between areas and χ^2 values for only a few "percentage points" in different chi-square distributions.

Table A-7 of Appendix A shows χ^2 values that correspond to selected areas in the right-hand tail of the chi-square distribution. These tabulations are shown separately for the number of degrees of freedom listed in the left-hand column. The χ^2 values are shown in the body of the table and the corresponding areas are shown in the column headings.

As an illustration of the use of the chi-square table, let us assume a chi-square distribution with 8 degrees of freedom. In Table A-7 we find a χ^2 value of 15.507 corresponding to an area of 0.05 in the right-hand tail. The relationships described in this problem are shown in Figure 9-2. Hence, in a chi-square distribution with 8 degrees of freedom, the probability that χ^2 is greater than 15.507 is 0.05. Let us give the corresponding interpretation in a hypothesis testing context. If the null hypothesis being tested is true, the probability of observing a χ^2 figure greater than 15.507 because of chance variation is equal to 0.05. Therefore, for example, if the null hypothesis were tested at the 0.05 level of significance, and we calculated $\chi^2 = 16$, we would reject the null hypothesis,

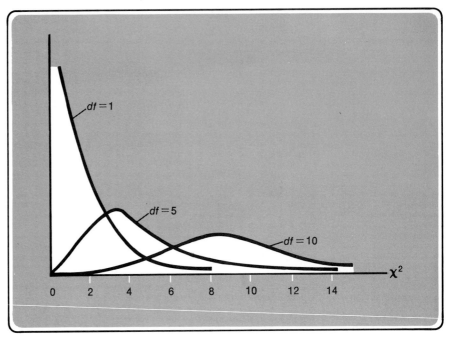

Figure 9-1
Chi-square distribution for 1, 5, and 10 degrees of freedom.

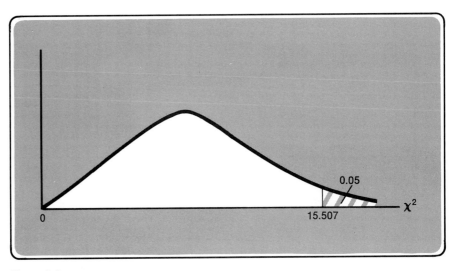

Figure 9-2
Chi-square distribution for eight degrees of freedom.

because so large a χ^2 value would occur less than five times in 100 if the null hypothesis were true.

The χ^2 test

We now return to our example to perform the chi-square test of independence. Again denoting the observed frequencies as f_o and the expected frequencies as f_t, we have shown the calculation of the χ^2 statistic in Table 9-3. No cell designations are indicated, but of course every f_o value is compared to the corresponding f_t figure. As shown at the bottom of the table, the computed value of χ^2 is equal to 794.3. The number of degrees of freedom is $(r - 1)$ $(c - 1)$ or $(3 - 1)(3 - 1) = 4$. In Table A-7 of Appendix A, we find a critical value at the 0.01 level of significance of $\chi^2_{0.01} = 13.277$. This means that if the null hypothesis is true, the probability of observing a χ^2 value greater than 13.277 is 0.01. Specifically in terms of the problem, this means that if ownership of telephones was independent of ownership of automobiles, an aggregate discrepancy between the observed and theoretical frequencies larger than a χ^2 value of 13.277 would occur only one time in 100. We can state the decision rule for this problem as follows:

1. If $\chi^2 > 13.277$, reject H_0
2. If $\chi^2 \leq 13.277$, accept H_0

Since the computed χ^2 value of 794.3 so greatly exceeds this critical value, the null hypothesis of independence between telephone and automobile ownership is emphatically rejected.

TABLE 9-3
Calculation of the χ^2 Statistic for the
Telephone and Automobile Ownership Problem.

OBSERVED NUMBER OF FAMILIES f_o	EXPECTED NUMBER OF FAMILIES f_t	$f_o - f_t$	$(f_o - f_t)^2$	$(f_o - f_t)^2/f_t$
1000	600	400	160,000	266.7
1500	1380	120	14,400	10.4
500	1020	−520	270,400	265.1
900	1200	−300	90,000	75.0
2600	2760	−160	25,600	9.3
2500	2040	460	211,600	103.7
100	200	−100	10,000	50.0
500	460	40	1600	3.5
400	340	60	3600	10.6
TOTAL 10,000	10,000	0		$\chi^2 = 794.3$

Further comments

We have seen how the chi-square test for independence in contingency tables is a means of determining whether or not a relationship exists between two bases of classification, or, in other words, whether a relationship exists between two variables. Although this type of tabulation provides a basis for testing whether there is a dependence between the two classificatory variables, it does not yield a method for estimating the values of one variable from known values or assumed values of the other. In the next chapter, where *regression* and *correlation analysis* are discussed, methods for providing such estimates are indicated. For example, regression analysis provides a method for obtaining estimates or predictions of the number of telephones owned by a family that owns a specific number of automobiles. Regression analysis, in particular, provides a very powerful tool for stating in explicit mathematical form the nature of the relationship that exists between two or more variables.

However, at least some indication may be obtained of the nature of the relationship between the two variables in a contingency table. Equivalently to the null hypothesis of independence rejected in our example, we have rejected the null hypothesis, $H_0 : p_1 = p_2 = p_3$, where p_1, p_2, and p_3 denote the population proportions of zero-, one-, and two-car families, respectively, who do not have telephones. Reference to Table 9-1 makes it obvious why the null hypothesis was rejected. Of the 3000 families who did not own automobiles, $1000/3000 = 0.33$ did not own a telephone. Let $\bar{p}_1 = 0.33$. The corresponding proportions of one- and two-car owning families who did not own telephones were $\bar{p}_2 = 900/6000 = 0.15$ and $\bar{p}_3 = 100/1000 = 0.10$. Hence, we have concluded that it is highly unlikely that these three statistics represent samples drawn from populations that have the same proportions ($p_1 = p_2 = p_3$). Clearly, the proportion of non-telephone-owning families declines as automobile ownership increases. The data suggest a strong relationship between the ownership of telephones and automobiles for the families studied.

A powerful generalization develops from the preceding discussion. It can be shown that a chi-square test applied to a 2×2 contingency table is algebraically identical to the two-sample test for difference between proportions by the methods of Section 8.3 using Equation (8.10) to calculate the estimated standard error of the difference. This means that the test of the hypothesis of independence carried out in a chi-square test for a 2×2 contingency table is identical to the testing of the following hypotheses:

$$H_0 : p_1 = p_2$$
$$H_1 : p_1 \neq p_2$$

As we have seen, in our illustrative problem involving a 3×3 contingency table, we tested the null hypothesis

$$H_0 : p_1 = p_2 = p_3$$

against the alternative that the p values were not all equal. The analogous test can be applied in general to c categories, where c is two or more.

A couple of other points may be made concerning the chi-square test for independence. The sampling distribution of the χ^2 statistic, $\chi^2 = \Sigma (f_o - f_t)^2 / f_t$ is only an approximation of the correct theoretical distribution. Hence, the sample size must be large for a good approximation to be obtained. A recommended procedure is that cells with frequencies of less than 5 should be combined.

As we have seen, in tests of independence, the null hypothesis is rejected when large enough values of χ^2 are observed. Some investigators have raised the question as to whether the null hypothesis should also be rejected when the computed value of χ^2 is too low, that is, too close to zero. This is a situation in which the observed frequencies, f_o, all appear to *agree too well* with the theoretical frequencies, f_t. The recommended course of action is to examine the data very closely to see whether errors have been made in the recording of the data. Perhaps the data rather than the null hypothesis should be rejected. An experience of one researcher is relevant to this point. He was analyzing some data on oral temperatures and found that a disturbingly large number of the recorded temperatures were equal to the "normal" figure of $98.6°F$. He suspected these data as being "too good to be true." Upon investigation, he found that the temperatures were recorded by relatively untrained nurses aides. Several of them had misread temperatures by recording the number to which the arrow on the thermometer pointed, namely, $98.6°$! Clearly, this was a case where the data rather than an investigator's null hypothesis should be rejected.

EXERCISES

1. A student looking for an easy statistics teacher was told by the departmental office that all three statistics teachers passed the same proportions of students. The student did some research and came up with the following results:

STUDENT PERFORMANCE	PROFESSOR A	PROFESSOR B	PROFESSOR C	TOTAL
Number passed	42	43	38	123
Number failed	8	5	14	27
TOTAL	50	48	52	150

Should the student believe what the office told him? Use a 0.05 level of significance.

2. A subscription service stated that preferences for different national magazines were independent of geographical location. A survey was taken in which 300 persons randomly chosen from three areas were given a choice among three different magazines. Each person expressed his or her favorite. The following results were obtained:

REGION	MAGAZINE X	MAGAZINE Y	MAGAZINE Z	TOTAL
New England	75	50	175	300
Northeastern	120	85	95	300
Southern	105	110	85	300
TOTAL	300	245	355	900

Would you agree with the subscription service's assertion? Use a 0.05 level of significance.

3. A professor wished to know if grades in a basic economics course were independent of the students' year in school. He obtained the grades of all the students taking the course in the first semester and set up the following table. Test the hypothesis that performance in the course is independent of the students' year in school using a 0.05 and 0.01 significance level. A grade in the course ranges from zero to 100.

| YEAR | GRADE RANGES | | | TOTAL |
	80–100	60–79	0–59	
Sophomore	20	46	14	80
Junior	15	40	5	60
Senior	10	20	10	40
TOTAL	45	106	29	180

4. The editor of *The Star* did research to determine whether social class has any effect on newspaper buying. He took a poll of 150 people from each of three social classes and ascertained whether they read his paper or its competitor, *The Press.* The results were

SOCIAL CLASS	THE STAR	THE PRESS	TOTAL
Lower	80	70	150
Middle	90	60	150
Upper	50	100	150
TOTAL	220	230	450

Test the hypothesis that choice of newspaper and social class are independent. Use a 0.05 level of significance.

5. Components are supplied to a television manufacturer by two subcontractors. Each component is tested with respect to five characteristics before it is accepted by the manufacturer. Records have been kept for one month on the numbers of different types of defects for each contractor, and from these the following table has been constructed.

| SUPPLIER | TYPE OF DEFECT | | | | | TOTAL |
	A	B	C	D	E	
1	70	10	10	30	0	120
2	10	10	20	20	20	80
TOTAL	80	20	30	50	20	200

Would you conclude that type of defect and supplier are independent? Use $\alpha = 0.01$.

6. The following table shows the location of a random sample of 200 members of a trade association by city type and geographic region.

CITY TYPE	GEOGRAPHIC REGION				
	(B_1) EASTERN	(B_2) SOUTHERN	(B_3) MIDWESTERN	(B_4) FAR WESTERN	TOTAL
(A_1) Large	35	10	25	25	95
(A_2) Small	15	10	15	15	55
(A_3) Suburb	25	5	10	10	50
TOTAL	75	25	50	50	200

Are city type and geographical region independent? Test at the 0.05 level of significance.

ANALYSIS OF VARIANCE: 9.2

TESTS FOR EQUALITY OF SEVERAL MEANS

In Section 9.1, we saw that the *chi-square test is a generalization of the two sample test for proportions* and enables us to test for the significance of the difference among $c, (c > 2)$, sample proportions. Conceptually, this represents a test of whether the c samples can be treated as having been drawn from the same population, or in other words, from populations having the same proportions. Similarly, in this section we consider a very ingenious technique known as the *analysis of variance,* which is a *generalization of the two sample test for means* and enables us to test for the significance of the difference among $c, (c > 2)$, sample means. Analogously to the case of the chi-square test, this represents a test of whether the c samples can be considered as having been drawn from the same population, or more precisely, from populations having the same means.

A central point to realize is that although the analysis of variance is literally a technique that analyzes or tests variances, by doing so, it provides us with a test for the significance of the difference among *means.* The rationale by which a test of variances is in fact a test for means will be explained shortly.

As an example, we consider a problem in which it is desired to test whether three methods of teaching a basic statistics course differ in effectiveness. It has been agreed that student grades on a final examination covering the work of the entire course will be used as the measure of effectiveness. The three methods of teaching are

Method 1: The lecturer does not work out nor assign problems.
Method 2: The lecturer works out and assigns problems.
Method 3: The lecturer works out and assigns problems. Students are also required to construct and solve their own problems.

The same professor taught three different sections of students, using one of the 3 methods in each class. All of the students were sophomores at the same university and were randomly assigned to the 3 sections. In order to explain the principles of the analysis without cumbersome computational detail, we will assume there were only 12 students in the experiment, 4 in each of the 3 different sections. Of course, in actual practice, a substantially larger number of observations would be required to furnish convincing results. The final examination was graded on the basis of 25 as the maximum score and 0 as the minimum score. The final examination grades of the 12 students in the 3 sections are given in Table 9-4. As shown in the table, the mean grades for students taught by methods 1, 2, and 3 were 17, 20, and 23, respectively, and the overall average of the 12 students, referred to as the "grand mean," was 20. It may be noted that the grand mean of 20 is the same figure as would be obtained by adding up all 12 grades and dividing by 12.

TABLE 9-4
Final Examination Grades of 12 Students
Taught by 3 Different Methods.

STUDENT	TEACHING METHOD		
	1	2	3
1	16	19	24
2	21	20	21
3	18	21	22
4	13	20	25
Total	68	80	92
Mean	17	20	23

$$\text{Grand mean} = \frac{17 + 20 + 23}{3} = 20$$

Notation

At this point we introduce some useful notation. In Table 9-4, there are 4 rows and 3 columns. As in the discussion of chi-square tests for contingency tests, let r represent the number of rows and c the numbers of columns. Hence, there is a total of $r \times c$ observations in the table, in this case $4 \times 3 = 12$. Let X_{ij} be the score of the ith student taught by the jth method, where $i = 1, 2, 3, 4$ and $j = 1, 2, 3$. Thus, for example, X_{12} denotes the score of student 1 taught by method 2 and is equal to 19; $X_{23} = 21$, etc. In this problem, the different methods of instruction are indicated in the columns of the table, and interest centers on the differences among the scores in the 3 columns. This is typical of the so-called "one factor (or one-way) analysis of variance," in which an attempt is made to assess the effect of only one factor (in this case, instructional method), on the observations. In the present problem, there are 3 columns. Hence, we denote the values in the columns as X_{i1}, X_{i2}, and X_{i3}, and the totals of these columns as $\sum_i X_{i1}$, $\sum_i X_{i2}$, and $\sum_i X_{i3}$. The subscript i under the summation signs indicates that the total of each of the columns is obtained by summing the entries over the rows. Adopting a simplified notation, we will refer to the means of the 3 columns as $\overline{X}_1$, $\overline{X}_2$, and $\overline{X}_3$, or in general, $\overline{X}_j$. Finally, we denote the grand mean as $\overline{\overline{X}}$ (pronounced "X double-bar"), where $\overline{\overline{X}}$ is the mean of all $r \times c$ observations.

Since each column in our example contains the same number of observations, $\overline{\overline{X}}$ can be obtained by taking the mean of the 3 sample means $\overline{X}_1$, $\overline{X}_2$, and $\overline{X}_3$. This notation is summarized in Table 9-5. It is suggested that you study Table 9-5 carefully and compare the notation with the corresponding entries in Table 9-4.

TABLE 9-5
Notation Corresponding to the Data of Table 9-4.

i	X_{i1}	X_{i2}	X_{i3}
1	X_{11}	X_{12}	X_{13}
2	X_{21}	X_{22}	X_{23}
3	X_{31}	X_{32}	X_{33}
4	X_{41}	X_{42}	X_{43}
Total	$\sum\limits_i X_{i1}$	$\sum\limits_i X_{i2}$	$\sum\limits_i X_{i3}$
Mean	$\overline{X}_1$	$\overline{X}_2$	$\overline{X}_3$

$$\overline{\overline{X}} = \frac{\overline{X}_1 + \overline{X}_2 + \overline{X}_3}{3}$$

The hypothesis to be tested

As indicated earlier, we want to test whether the effectiveness of the three methods of teaching a basic statistics course differ. We have calculated the following mean final examination scores of students taught by the 3 methods $\overline{X}_1 = 17$, $\overline{X}_2 = 20$, and $\overline{X}_3 = 23$. The statistical question is "Can the 3 samples represented by these 3 means be considered as having been drawn from populations having the same mean?" Denoting the population means corresponding to $\overline{X}_1$, $\overline{X}_2$, and $\overline{X}_3$ as μ_1, μ_2, and μ_3, respectively, we can state the null hypothesis as

$$H_0 : \mu_1 = \mu_2 = \mu_3$$

This hypothesis is to be tested against the alternative

$$H_1 : \text{The means } \mu_1, \mu_2, \text{ and } \mu_3 \text{ are not all equal}$$

Hence, what we wish to determine is whether the differences among the sample means $\overline{X}_1$, $\overline{X}_2$, and $\overline{X}_3$ are too great to attribute to the chance errors of drawing samples from populations having the same means. If we do decide that the sample means differ significantly, our substantive conclusion is that the teaching methods differ in effectiveness.

Although we will specify the assumptions underlying the test procedure at the end of the problem, we indicate one of them at this point, namely, the assumption that the variances of the 3 populations are all equal. Therefore, rewording the hypothesis slightly, we can state that we wish to test whether our 3 samples were drawn from populations having the same means and variances.

Decomposition of total variation

Before discussing the procedures involved in the analysis of variance, we consider the general rationale underlying the test. If the null hypothesis that the 3 population means, μ_1, μ_2, and μ_3, are equal is true, then both the variation among the sample means $\overline{X}_1$, $\overline{X}_2$, and $\overline{X}_3$ and the variation within the 3 groups reflect chance errors of the sampling process. The first of these types of variation is conventionally referred to as "variation between the c means," "between-group variation," or "between-column variation," (despite the English barbarism involved in using the word "between" rather than "among" when there are more than 2 groups present). The second type is referred to as "within-group variation" or simply the "between-row variation." We will use the latter term in this chapter. Between-column variation refers to variation of the sample means $\overline{X}_1$, $\overline{X}_2$, and $\overline{X}_3$ around the grand mean, $\overline{\overline{X}}$. On the other hand, between-row variation refers to the differences of the individual observations within each column from their respective means $\overline{X}_1$, $\overline{X}_2$, and $\overline{X}_3$.

Under the null hypothesis that the population means are equal, the between-column variation and the between-row variation would be expected not to differ significantly from one another, since they both reflect the same type of chance sampling errors. On the other hand, if the null hypothesis is false, and the population column means are indeed different, then the between-column variation should significantly exceed the between-row variation. This follows from the fact that the between-column variation would now be produced by the inherent differences among the column means as well as by chance sampling error. On the other hand, the between-row variation would still reflect chance sampling errors only. *Hence, a comparison of between-column variation and between-row variation yields information concerning differences among the column means.* This is the central insight provided by the analysis of variance technique.

The term "variation" is used in statistics in a very specific way to refer to a sum of squared deviations and is often referred to simply as a "sum of squares." When a measure of variation is divided by an appropriate number of degrees of freedom, as we have seen earlier in this text, it is referred to as a "variance." In the analysis of variance, such a variance is referred to as a "mean square." For example, the variation of a set of sample observations, denoted X, around their mean $\overline{X}$ is $\Sigma(X - \overline{X})^2$. Dividing this sum of squares by the number of degrees of freedom $n - 1$, where n is the number of observations, we obtain, $\Sigma(X - \overline{X})^2/(n - 1)$, the sample variance, which as indicated in Section 3.14 is an unbiased estimator of the population variance. This sample variance can also be referred to as a "mean square."

We now proceed with the analysis of variance by calculating the between-column variation and between-row variation for our problem.

Between-column variation

As indicated earlier, the between-column variation, or between-column sum of squares, measures the variation among the sample column means. It is calculated as follows:

(9.3) $$\text{Between-column sum of squares} = \sum_{j} r(\overline{X}_j - \overline{\overline{X}})^2$$

where $r =$ number of rows (sample size involved in the calculation of each column mean)[1]
$\overline{X}_j =$ the mean of the jth column
$\overline{\overline{X}} =$ the grand mean
$\sum_{j} =$ means that the summation is taken over all columns.

As indicated in Equation (9.3), the between-column sum of squares is calculated by the following steps:

1. Compute the deviation of each column mean from the grand mean.
2. Square the deviations obtained in Step 1.
3. Weight each deviation by the sample size involved in calculating the respective mean. In the illustrative example all sample sizes are the same and are equal simply to the number of rows, $r = 4$.
4. Sum up over all columns the products obtained in Step 3.

The calculation of the between-column sum of squares for the numerical example involving three different teaching methods is given in Table 9-6. As indicated in the table, the between-column variation is equal to 72.

TABLE 9-6
Calculation of the Between-Column
Sum of Squares for the Teaching
Methods Problem.

$$(\overline{X}_1 - \overline{\overline{X}})^2 = (17 - 20)^2 = 9$$
$$(\overline{X}_2 - \overline{\overline{X}})^2 = (20 - 20)^2 = 0$$
$$(\overline{X}_3 - \overline{\overline{X}})^2 = (23 - 20)^2 = 9$$
$$\sum_{j} r(\overline{X}_j - \overline{\overline{X}})^2 = 4(9) + 4(0) + 4(9) = 72$$

Between-row variation

The between-row sum of squares is a summary measure of the random errors of the individual observations around their column means. The formula for its computation is

(9.4) $$\text{Between-row sum of squares} = \sum_{j} \sum_{i} (X_{ij} - \overline{X}_j)^2$$

---

[1] In this chapter, equal sample sizes are assumed (equal numbers of rows). This treatment can be generalized to allow for differing sample sizes.

where X_{ij} = the value of the observation in the ith row and jth column

$\overline{X}_j$ = the mean of the jth column

$\sum_j \sum_i$ means that the squared deviations are first summed over all sample observations within a given column, then summed over all columns.

As indicated in Equation (9.4), the between-row sum of squares is calculated as follows:

1. Calculate the deviation of each observation from its column mean.
2. Square the deviations obtained in Step 1.
3. Add the squared deviations within each column.
4. Sum over all columns the figures obtained in Step 3.

The computation of the between-row variation for the teaching methods problem is given in Table 9-7.

TABLE 9-7

Calculation of the Between-Row Sum of Squares for the Teaching Methods Problem.

i	$(X_{i1} - \overline{X}_1)$	$(X_{i1} - \overline{X}_1)^2$	$(X_{i2} - \overline{X}_2)$	$(X_{i2} - \overline{X}_2)^2$	$(X_{i3} - \overline{X}_3)$	$(X_{i3} - \overline{X}_3)^2$
1	$(16 - 17) = -1$	1	$(19 - 20) = -1$	1	$(24 - 23) = 1$	1
2	$(21 - 17) = 4$	16	$(20 - 20) = 0$	0	$(21 - 23) = -2$	4
3	$(18 - 17) = 1$	1	$(21 - 20) = 1$	1	$(22 - 23) = -1$	1
4	$(13 - 17) = -4$	16	$(20 - 20) = 0$	0	$(25 - 23) = 2$	4
		34		2		10

$$\sum_j \sum_i (X_{ij} - \overline{X}_j)^2 = 34 + 2 + 10 = 46$$

Total variation

The between-column variation and between-row variation represent the two components of the total variation in the overall set of experimental data. The total variation or total sum of squares is calculated by adding the squared deviations of all of the individual observations from the grand mean $\overline{\overline{X}}$. Hence, the formula for the total sum of squares is

(9.5) $\qquad$ Total sum of squares $= \sum_j \sum_i (X_{ij} - \overline{\overline{X}})^2$

The total sum of squares is computed by the following steps:

1. Calculate the deviation of each observation from the grand mean.
2. Square the deviations obtained in Step 1.
3. Add the squared deviations over all rows and columns.

The total sum of squares or total variation of the 12 observations in the teaching methods problem is $(16 - 20)^2 + (21 - 20)^2 + \cdots + (25 - 20)^2 = 118$.

Referring to the results obtained in Table 9-6 and 9-7, we see that the total sum of squares, 118, is equal to the sum of the between-column sum of squares, 72, and the between-row sum of squares, 46. In general, the following relationship holds:

(9.6)   Total variation = Between-column variation + Between-row variation

Although, as we have indicated earlier, the test of the null hypothesis in a one-factor analysis of variance involves only the between-column variation and the between-row variation, it is useful to calculate also the total variation. This computation is helpful as a check procedure and is instructive in indicating the relationship between total variation and its components.

Short-cut computational formulas

The formulas we have given for calculating the between-column sum of squares (9.3), between-row sum of squares (9.4), and the total sum of squares (9.5) are the clearest ones for revealing the rationale of the analysis of variance procedure. However, the following short-cut computational formulas are often used to calculate these sums of squares.

(9.7) $\quad$ Between-column sum of squares $= \dfrac{\sum_j T_j^2}{r} - C$

(9.8) $\quad$ Between-row sum of squares $= \sum_j \sum_i X_{ij}^2 - \sum_j \dfrac{T_j^2}{r}$

(9.9) $\quad$ Total sum of squares $= \sum_j \sum_i X_{ij}^2 - C$

where C, the so-called "correction term" is given by

(9.10) $$C = \frac{T^2}{rc}$$

and where T_j is the total of the r observations in the jth column and T is the grand total of all rc observations, that is,

(9.11) $$T = \sum_j \sum_i X_{ij}$$

and all other terms are as previously defined.

These formulas are especially useful when the column means and grand mean are not integers. The short-cut formulas not only save time and computational labor, but are also more accurate because of avoidance of rounding problems which usually occur with the use of Equations (9.3), (9.4), and (9.5).

The short-cut computations for the present example are as follows:

$$C = \frac{(240)^2}{(3)(4)} = 4800$$

Between-column sum of squares $= \dfrac{(68)^2 + (80)^2 + (92)^2}{4} - 4800 = 72$

Between-row sum of squares $= (16)^2 + (21)^2 + \cdots + (25)^2 - \dfrac{(68)^2 + (80)^2 + (92)^2}{4}$

$= 46$

Total sum of squares $= (16)^2 + (21)^2 + \cdots + (25)^2 - 4800 = 118$

It is recommended that the short-cut formulas be used, particularly when carrying out computations by hand.

Number of degrees of freedom

Although the preceding discussion has been in terms of *variation* or *sums of squares* rather than *variance,* the actual test of the null hypothesis in the analysis of variance involves a comparison of the *between-column variance* with the *between-row variance,* or in equivalent terminology, a comparison of the *between-column mean square* with the *between-row mean square.* Hence, the next step in our procedure is to determine the number of degrees of freedom associated with each of the measures of variation. As stated earlier in this section, if a measure of variation, that is, a sum of squares, is divided by the appropriate number of degrees of freedom, the resulting measure is a variance, that is, a mean square.

The number of degrees of freedom associated with the between-column sum of squares is $c - 1$. We can see the reason for this by applying the same general principles indicated earlier for determining number of degrees of freedom in t tests and chi-square tests. Since there are c columns, or c group means, there are c sums of squares involved in measuring the variation of these column means around the grand mean. Because the sample grand mean is only an estimate of the unknown population mean, we lose one degree of freedom. An alternative view is that the between-column sums of squares is composed of c squared deviations of the form $(\overline{X}_j - \overline{\overline{X}})^2$. If $c - 1$ of these $\overline{X}_j$ values are assigned arbitrarily, then the cth value is determined to arrive at the figure for the between-column sum of squares. Hence, there are $c - 1$ degrees of freedom present.

The number of degrees of freedom in our example that has three different teaching methods, that is, three columns, is $c - 1 = 3 - 1 = 2$.

The number of degrees of freedom associated with the between-row variation is $rc - c = c(r - 1)$. This may be reasoned as follows. There are a total of rc observations. In determining the between-row variation, the squared deviations within each column were taken around the column mean. There are c column means, each of which is an estimate of the true unknown population column mean. Hence there is a loss of c degrees of freedom, and c must be subtracted from rc, the total number of observations.

Alternatively, there are r squared deviations in each column taken around the column mean and a total sum of squares for the column. $r - 1$ of the sums of squares can be assigned arbitrarily, and the last becomes fixed in order for the sum to equal the column sum. Since there are c columns, we have $c(r - 1)$ degrees of freedom.

In the problem, the number of degrees of freedom associated with the between-row sum of squares is $c(r - 1) = 3(4 - 1) = 9$.

The number of degrees of freedom associated with the total variation is equal to $rc - 1$. There are rc squared deviations taken from the sample grand mean, $\overline{\overline{X}}$. Since $\overline{\overline{X}}$ is an estimate of the true but unknown population mean, there is a loss of one degree of freedom. Alternatively, in the determination of the total sum of squares, there are rc squared deviations. $rc - 1$ of them may be arbitrarily assigned, but the last one is then constrained in order for the sum to be equal to the total sum of squares.

In the example, the number of degrees of freedom associated with the total variation is $rc - 1 = (4)(3) - 1 = 11$.

Just as the between-column and the between-row variation sum to the total variation, the numbers of degrees of freedom associated with the between-column and between-row variations add to the number associated with the total variation. In symbols,

(9.12)
$$rc - 1 = (c - 1) + (rc - c)$$

In the teaching methods problem, the numerical values corresponding to Equation (9.12) are $11 = 2 + 9$.

The analysis of variance table

An analysis of variance table for the teaching methods problem is given in Table 9-8. The calculations at the bottom of the table will be described presently. The table is in the standard form ordinarily employed to summarize the

TABLE 9-8
Analysis of Variance Table for the Teaching Methods Problem.

(1) SOURCE OF VARIATION	(2) SUM OF SQUARES	(3) DEGREES OF FREEDOM	(4) MEAN SQUARE
Between columns	72	2	36
Between rows	46	9	5.11
TOTAL	118	11	

$$F(2, 9) = \frac{36}{5.11} = 7.05$$

$$F_{.05}(2, 9) = 4.26$$

Since $7.05 > 4.26$, reject H_0.

results of an analysis of variance. In Columns (1), (2), and (3), respectively, are listed the possible sources of variation, the sum of squares for each of these sources, and the number of degrees of freedom associated with each of the sums of squares. We again note that both sums of squares and numbers of degrees of freedom are additive, that is, these figures for between-column and between-row sources of variation add to the corresponding figure for total variation. Dividing the sums of squares in Column (2) by the numbers of degrees of freedom in Column (3) yields the between-column and between-row variances shown in Column (4). As indicated earlier, another name for a sum of squares divided by the appropriate number of degrees of freedom is a "mean square," and it is conventional to use this term in an analysis of variance table. Thus, in our problem, the between-column mean square denoted MS_c, is equal to $72/2 = 36$. The between-row mean square, denoted MS_r, is equal to $46/9 = 5.11$. The test of the null hypothesis that the population column means are equal is carried out by a comparison of MS_c to MS_r.

Table 9-9 gives the general format of a one-factor analysis of variance table. The sums of squares have been denoted as follows:

$$SS_c = \text{Between-Column Sum of Squares}$$

$$SS_r = \text{Between-Row Sum of Squares}$$

$$SS_t = \text{Total Sum of Squares}$$

Other notation is as previously defined or as given in the next subsection.

TABLE 9-9
General Format of a One-Factor Analysis of Variance Table.

(1) SOURCE OF VARIATION	(2) SUM OF SQUARES	(3) DEGREES OF FREEDOM	(4) MEAN SQUARE
Between columns	SS_c	$df_1 = c - 1$	$MS_c = SS_c/(c - 1)$
Between rows	SS_r	$df_2 = c(r - 1)$	$MS_r = SS_r/c(r - 1)$
TOTAL	SS_t	$rc - 1$	

$$F(df_1, df_2) = \frac{MS_c}{MS_r}$$

The F test and F distribution

The comparison of the between-column mean square to the between-row mean square is made by computing their ratio, referred to as F. Hence, F is given by

(9.13)
$$F = \frac{MS_c}{MS_r}$$

In the F ratio, the between-column variance is always placed in the numerator and the between-row variance in the denominator. Under the null hypothesis that the population column means are equal, the F ratio would tend to be equal to 1. On the other hand, if the population column means do indeed differ, then the between-column mean square, MS_c, will tend to exceed the between-row mean square, MS_r, and the F ratio will then be greater than 1. In terms of our illustrative problem concerning different teaching methods, if F is large, we will reject the null hypothesis that the population mean examination scores are all equal, that is, we will reject $H_0{:}\mu_1 = \mu_2 = \mu_3$. On the other hand, if F is close to 1, we will accept the null hypothesis. The answer to how large the test-statistic F must be in order to reject the null hypothesis is given by reference to the probability distribution of the F random variable. This distribution is complex, and its mathematical basis will not be discussed here. Fortunately, critical values of the F ratio have been tabulated for frequently used significance levels, analogous to the case of the chi-square distribution.

The underlying assumptions are that two random samples are drawn from normally distributed populations with equal variances $\hat{\sigma}_1^2$ and $\hat{\sigma}_2^2$. Unbiased estimators σ_1^2 and σ_2^2 of the population variances are constructed from the sample, and

$$F = \frac{\hat{\sigma}_1^2}{\hat{\sigma}_2^2}$$

Similar to the distributions of t and χ^2, the F distribution is actually a family of distributions. Each pair of values of df_1 and df_2 specifies a different distribution, where df_1 and df_2 refer to the number of degrees of freedom in the numerator and denominator of F, respectively. F is a continuous random variable that ranges from zero to infinity. Since the variances in both the numerator and denominator of the F ratio are squared quantities, F cannot take on negative values. The F distribution has a single mode, and although the specific distribution depends on the value of df_1 and df_2, its shape is generally assymmetrical and skewed to the right. The distribution tends towards symmetry as df_1 and df_2 increase. We will use the notation $F(df_1, df_2)$ to denote the F ratio defined in Equation (9.13), that is, $F = MS_c/MS_r$, where the numerator and denominator are between-column mean squares and between-row mean squares with df_1 and df_2 degrees of freedom, respectively. Table A-8 of Appendix A presents the critical values of the F distribution for two selected significance levels, $\alpha = 0.05$ and $\alpha = 0.01$. In this table, df_1 values are listed across the columns and df_2 down the rows. There are two entries in the table corresponding to every pair of df_1, df_2 values. The upper figure in light-face type is an F value that corresponds to an area of 0.05 in the right-hand tail of the F distribution with df_1 and df_2 degrees of freedom. That is, it is an F value that would be exceeded only five times in 100 if the null hypothesis under test were true. The lower figure in boldface type is an F value corresponding to a 0.01 area in the right-hand tail.

We will illustrate the use of the F table in terms of the teaching methods problem. Assuming we wish to test the null hypothesis $H_0{:}\mu_1 = \mu_2 = \mu_3$ at the 0.05 level of significance, we find in Table A-8 of Appendix A that for $df_1 = 2$

and $df_2 = 9$ degrees of freedom an F value of 4.26 would be exceeded 5% of the time, if the null hypothesis were true. As indicated at the bottom of Table 9-8, we denote this critical value as $F_{.05}(2,9) = 4.26$. This relationship is depicted in Figure 9.3. Again referring to Table 9-8, since the computed value of the F-ratio of the between-column mean square to the between-row mean square is 7.05, and therefore greater than the critical value of 4.26, we reject the null hypothesis. Hence, we conclude that the column means, that is, the sample mean final examination scores in classes taught by the three teaching methods, differ significantly. The inference about the corresponding population means is that they are not all the same. Referring back to Table 9-4, we see that average grades under Method 3 exceed those under Method 2, which are higher than those under Method 1. Hence, based on these data, our inference is that the three teaching methods are not equally effective, and there is evidence that Method 3 is the most and Method 1 the least effective.

The foregoing example was used to illustrate the rationale involved in the analysis of variance, the statistical technique employed, and the nature of the conclusions that can be drawn. However, it is pertinent to repeat the caveat that the sample sizes in this illustration are doubtless too small for safe conclusions to be drawn about differences in effectiveness of different teaching methods. After all, only four observations were made under each teaching method. Although the risk of a Type I error was controlled at 0.05 in this problem, the risk of Type II errors may be intolerably high. Suffice it to say that

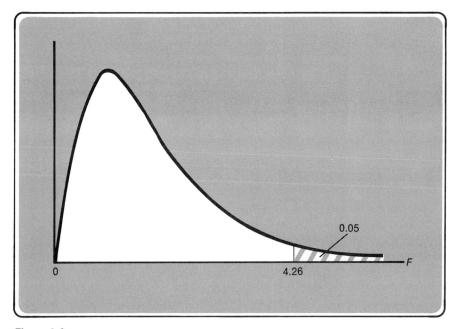

Figure 9-3
The F distribution for the teaching methods problem indicating the critical value at the 5% level of significance.

larger sample sizes are generally required, and that similar to the hypothesis testing cases for one or two sample means discussed in Chapter 8, methods are available for controlling Type II errors at specified levels in hypothesis testing for more than two sample means. The interpretation of a Type I error specifically in terms of the present problem is the erroneous rejection of the null hypothesis that all three teaching methods are equally effective. A Type II error is the acceptance of the null hypothesis when in fact the effectiveness of all three teaching methods is not the same.

Further remarks

It is useful to consider the nature of the conclusion and interpretation arrived at in the teaching methods example. We performed a one-factor analysis of variance, and because the ratio of the between-column variance to the between-row variance differed significantly from one, we rejected the hypothesis that the three population mean examination scores were the same. This conclusion was based on the reasoning that the between-row variance was a measure of random error and under the null hypothesis of equal population column means, the between-column variance also is a measure of random error. Rejection of this null hypothesis leads us to the conclusion that the variation among column means is in excess of these random or chance errors of sampling. Hence, we conclude that the population column means are indeed different. Since these mean examination scores constitute our measure of teaching effectiveness, we conclude that the three teaching methods are not equally effective.

However, let us be somewhat more critical and look at the possible interpretations of our findings more closely. We assumed that the *same teacher* taught three different sections of the basic statistics course by the three specified methods, 1, 2, and 3. Method 3 appeared most effective and Method 1 least effective. Suppose that the instructor teaches these three sections on the same days and that the class taught by Method 3 is given early in the morning. Hence, the teacher is fresh, wide awake, and enthusiastic. On the other hand, the classes taught by Methods 2 and 1 are in the middle of the day and late afternoon, respectively. Let us assume that by late afternoon, the instructor is tired, sleepy, and rather unenthusiastic. Then, it is possible that the differences in teaching effectiveness are not really attributable solely to the different teaching methods, but rather to some unknown mixture of the difference in teaching methods and the aforementioned factors associated with the time of day. We can think of experimental designs which would tend to counteract or "control" the effect of the time of day factor. Suppose the class taught by Method 1 meets one-third of the time in the early morning, one-third of the time in the middle of the day, and one-third of the time late in the afternoon; similarly for classes taught by Methods 2 and 3. Then time of class meeting would not be a factor which differentially affects teacher effectiveness. Of course, practical scheduling considerations might militate against this type of staggered class arrangement. However, the principle of explicitly controlling for the effects of the time of day factor is clear.

If the staggered scheduling system was not feasible, another approach suggests itself. If it were suspected beforehand that Method 3 is the most effective and Method 1 the least effective teaching method, the class taught by Method 3 might be scheduled late in the afternoon, and the one by Method 1 early in the morning. Then if results were obtained such as in the illustrative example, we would have even greater confidence that Method 3 was most effective and Method 1 least effective. This conclusion follows from the fact that Method 3 had to overcome the disadvantage of time of day, whereas Method 1 had a time of day advantage. However, this type of experimental design is dangerous because the effect of the factor of teaching effectiveness might be blurred by the oppositely acting effect of the time of day factor. When two factors are acting together this way and we have no way of segregating their separate effects, "confounding" is said to be present. That is, the effects of teaching method and time of day are confounded.

Other factors certainly occur to the reader that might have affected the differences among the column means, that is, the mean examination scores. For example, suppose the four students taught by Method 3 had greater aptitude for statistics than those taught by Method 2, who in turn had better aptitude than those taught by Method 1. Then, clearly again we have confounded effects. The rejection of the null hypothesis of equal population mean examination scores might not reflect differences in teaching effectiveness nearly so much as differences in student aptitudes. Of course, it was assumed in our example that the students taught by the three methods were randomly selected from a sophomore class, but nevertheless the hypothesized differences in aptitude might have been present. Again, the point is that in a more elaborate experimental design, we might attempt to control explicitly for the effect of student aptitude.

The preceding discussion applies equally well to the hypothesis testing methods considered earlier, as for example in Chapter 8, because we might have had only two teaching methods to compare rather than three. In summary, mechanistic or rote application of statistical techniques such as hypothesis testing methods must be guarded against. In this text, we consider the general principles involved in some of the simpler, basic procedures. More refined and sophisticated techniques may very well be required in particular instances. For example, we have considered only one-factor analysis of variance. More elaborate experimental designs that attempt to control and test for the effects of more factors are available, and considerable expertise is often required for their proper application.

One of the points we have attempted to convey in the preceding discussion is that statistical results are virtually always consistent with more than one interpretation. Naïve leaping to conclusions must be guarded against and careful consideration must be given to alternative interpretations and explanations. We conclude this chapter with two anonymous humorous stories relevant to the point that alternative interpretations and explanations of experimental results are often possible.

An investigator wished to determine the differential effects involved in the imbibing of various types of mixed drinks. Therefore, he had subjects drink substantial quantities of scotch and water, bourbon and water, and rye and water. All of the subjects became intoxicated. The investigator concluded that since water was the one factor common to all of these drinks, the imbibing of water makes people drunk.

The heroine of our second story is a grammar school teacher, who wished to explain the harmful effects of drinking liquor to her class of eight-year-olds. She placed two glass jars of worms on her desk. In the first jar, she poured some water. The worms continued to move about, and did not appear to have been adversely affected at all by the contact with the water. Then she poured a bottle of whiskey into the second jar. The worms became still and appeared to have been mortally stricken.

The teacher then called upon a student and asked, "Johnny, what is the lesson to be learned from this experiment?" Johnny, looking very thoughtful, replied, "I guess it proves that it is good to drink whiskey because it will kill any worms you may have in your body."

EXERCISES

1. Seven samples each of size ten are drawn from a normally distributed population. The means and variances of the seven samples are shown in the following table:

SAMPLE	MEAN	VARIANCE
A	50	5
B	47	4
C	52	5
D	51	6
E	49	5
F	52	4
G	49	6

Would you conclude that the seven samples were drawn *randomly* from the same population? Is your conclusion the same for both the 0.05 and 0.01 levels of significance?

2. A manufacturer has a choice of three subcontractors from whom to buy parts. The manufacturer, before deciding from whom he will buy, purchases five batches from each subcontractor. There are the same number in each batch. The number of defectives per batch is given in the following table:

BATCH	SUBCONTRACTOR A	SUBCONTRACTOR B	SUBCONTRACTOR C
1	35	15	25
2	25	20	40
3	30	25	40
4	35	15	35
5	20	30	30

Would you conclude that there is no real difference among these subcontractors in the average number of defectives produced per batch? Use a 0.05 level of significance.

3. The manufacturer of a new product wished to select the best advertising display for his product. Because he had a choice of five different displays he randomly selected 25 different stores and placed each type of display in five stores. The following figures are the average amount and variance per display in terms of dozens sold during the first six months.

TYPE OF DISPLAY	MEAN	VARIANCE
1	78	9
2	76	7
3	77	8
4	74	8
5	76	10

Can the manufacturer assume that it doesn't matter which display he uses? Use a 0.01 level of significance.

4. As head of a department of a consumer's research organization, you have the responsibility for testing and comparing lifetimes of light bulbs for four brands of bulbs. Suppose you test the lifetime of three bulbs of each of the four brands. Your test data are as follows, each entry representing the lifetime of a bulb, measured in hundreds of hours.

	BRAND		
A	B	C	D
20	25	24	23
19	23	20	20
21	21	22	20

Can we assume that the mean lifetimes of the four brands are equal?

MOTTO OF BIOMETRIKA

Regression and 10 Correlation Analysis

INTRODUCTION 10.1

In Chapter 9, we discussed the use of the chi-square test in contingency tables as a means of determining whether or not a relationship exists between two bases of classification, or more briefly, whether a relationship exists between two variables. However, the contingency table represents a method of classifying statistical observations by category, and, like other methods of tabulation, it does not provide a basis for estimating the values of one variable from a knowledge of the other. We now turn to a very widely used body of statistical methods, namely, *regression* and *correlation analysis,* which enables us to deal with variables stated in terms of numerical values rather than in categories. Furthermore, these methods provide the bases for estimating the values of one variable from known or assumed values of one or more other variables and for measuring the strength of the relationships among the variables.

Equations are used in mathematics to express the relationships among variables. In fields such as geometry or trigonometry, these mathematical functions or equations express the *exact relationships* present among the variables of interest. Thus, the equation $A = s^2$ describes the relationship between s, the length of the side of a square, and A, the area of the square. The equation $A = ab/2$ expresses the relationship between b, the length of any side of a triangle, a, the altitude or perpendicular distance to that side from the angle opposite it, and A, the area of the triangle. If we substitute numerical values for the variables on the right-hand sides of these equations, we can calculate the *exact values* of the quantities on the left-hand sides. In the social sciences and in fields such as business and governmental administration, exact relationships are not generally observed among variables, but rather *statistical relationships* prevail. That is, certain average relationships may be observed among variables, but these average relationships do not provide a basis for perfect predictions. For example, if we know how much money a corporation spends on television advertising, we cannot make an exact prediction of the amount of sales this promotional expenditure will generate. If we know a family's net income, we cannot make an exact forecast of the amount of money that family saves. On the other hand, we can measure statistically how sales vary, on the average, with differences in television advertising, or how family savings vary, on the average, with differences in income. Also, we can determine the amount of dispersion that exists around these average relationships. On the basis of these relationships, we may be able to estimate the values of the variables of interest closely enough for decision-making purposes. The techniques of regression and correlation analysis are important statistical tools used to accomplish this measurement and estimation process.

The term "regression analysis" refers to the methods by which estimates are made of the values of a variable from a knowledge of the values of one or more other variables, and to the measurement of the errors involved in this estimation process. The term "correlation analysis" refers to methods for measuring the strength of the association (correlation) among these variables.

We begin by discussing the case of a *two variable linear regression and correlation analysis.* The term "linear" means that an equation of a straight line of the form $Y = a + bX$, where a and b are numbers, is used to describe the average relationship that exists between the two variables and to carry out the estimation process. The factor whose values we wish to estimate is referred to as the *dependent variable* and is denoted by the symbol Y. The factor from which these estimates are made is called the *independent variable* and is denoted by X. The terms "dependent" and "independent" do not necessarily imply any cause and effect relationship between the variables. What is meant is simply that estimates of values of the dependent variable Y may be obtained for given values of the independent variable X from a mathematical function involving X and Y. In that sense, the values of Y are dependent upon the values of X. The X variable may or may not be *causing* changes in the Y variable. Hence, if we are estimating sales of a product from figures on advertising expenditures, sales is the dependent variable and advertising expenditures is the independent variable. There may or may not be a causal connection between these two factors in the sense that changes in advertising expenditures cause changes in sales. In fact, in certain situations, the cause-effect relation may be just the opposite of what appears to be the obvious one. For example, suppose a company budgets a product's advertising expenditures for the next year as a flat percentage of the sales of that product during the preceding year. Then advertising expenditures are more directly dependent on sales (with a one-year lag) than vice versa.

Let us consider some illustrative cases of variables that can reasonably be assumed related to one another, that is, correlated. If suitable data were available, we might attempt to construct an equation that would permit us to estimate the values of one variable from the values of the other. We shall assume that the first named factor in each pair is the variable to be estimated, that is, the dependent variable, and the second one is independent. Consumption expenditures might be estimated from a knowledge of income; investment in telephone equipment from expenditures on new construction; personal net savings from disposable income; commercial bank discount rates from Federal Reserve Bank discount rates; and success in college from scholastic aptitude test scores.

Of course, additional definitions are required to attach meaning to the estimation problems listed above. Thus, in the illustration of consumption expenditures and income, we would have to specify whose expenditures and whose income are involved. If we wanted to estimate family consumption expenditures from family income, the family would be said to be the "unit of association." The estimating equation would be constructed from data representing observations of these two variables for individual families. We would have to define the variables more specifically. For example, we might be interested in estimates of annual family consumption expenditures from annual family net income, where again these terms would require precise definitions.

In each of these examples, it is possible to specify other independent variables that might be included to aid in obtaining good estimates of the dependent variable. Hence, in estimating a family's consumption expenditures, we might wish to use knowledge of the size of the family in addition to information on the family's income. This would be an illustration of *multiple regression analysis,* where two independent variables, that is, (1) family income and (2) family size, are used to obtain estimates of a dependent variable. As indicated earlier in this chapter, we first consider two-variable problems involving a dependent factor and only one independent factor before turning to a brief discussion of multiple regression problems involving more than one independent variable.

SCATTER DIAGRAMS 10.2

In studying the relationship between two variables, an advisable first step is to plot the data on a graph. This allows visual examination of the extent to which the variables are related and aids in choosing the appropriate type of model for estimation. The chart used for this purpose is known as a *scatter diagram,* which is a graph on which each plotted point represents an observed pair of values of the dependent and independent variables. We will illustrate this by plotting a scatter diagram for the data given in Table 10-1. These figures represent observations for a sample of ten families of annual expenditures on consumer durables, which we shall treat as the dependent variable, Y, or the factor to be estimated, and annual net income, X, which is the independent variable or the factor from which the estimates are to be made. We shall assume that the ten families constitute a simple random sample of families with $10,000 or less of annual net income in a metropolitan area in 1973. Although only ten families are too small a sample from which to draw very useful conclusions that would apply to all such families in a metropolitan area, we shall use such a small sample in order to require only a very modest amount of arithmetic. Furthermore, we have assumed relatively low incomes in order to simplify the numerical work.

TABLE 10-1

Annual Consumer Durables Expenditures and Annual Net
Income of a Sample of Ten Families in a Metropolitan Area in 1973

FAMILY	ANNUAL CONSUMER DURABLES EXPENDITURES (HUNDREDS OF DOLLARS) Y	ANNUAL NET INCOME (THOUSANDS OF DOLLARS) X
A	23	10
B	7	2
C	15	4
D	17	6
E	23	8
F	22	7
G	10	4
H	14	6
I	20	7
J	19	6

Figure 10-1 presents the data of Table 10-1 on a scatter diagram. On the Y axis are plotted the figures on consumer durables expenditures and on the X axis annual net income. This follows the standard convention of plotting the dependent variable along the Y axis and the independent variable along the X axis. The pair of observations for each family determines one point on the scatter diagram. Thus, for family A, a point is plotted corresponding to $X = 10$ along the horizontal axis and $Y = 23$ along the vertical axis; for family B, a point is plotted corresponding to $X = 2$ and $Y = 7$, and so forth. An examination of the scatter diagram gives some useful indications of the nature and strength of the relationship between the two variables. For example, depending upon whether the Y values tend to increase as the values of X increase, or to decrease as the values of X increase, there is said to be a *direct* or *inverse* relationship, respectively, between the two variables. The configuration on Figure 10-1 indicates a general tendency for the points to run from the lower left to the upper right-hand side of the graph. Hence, as we move from low to higher income families, consumer durables expenditures tend to increase. This is an example of a *direct relationship* between the two variables. On the other hand, if the scatter of points runs from the upper left to the lower right, that is, if the Y variable tends to decrease as X increases, there is said to be an *inverse relationship* between the variables. Also, an examination of the scatter diagram gives an indication of whether a straight line appears to be an adequate descrip-

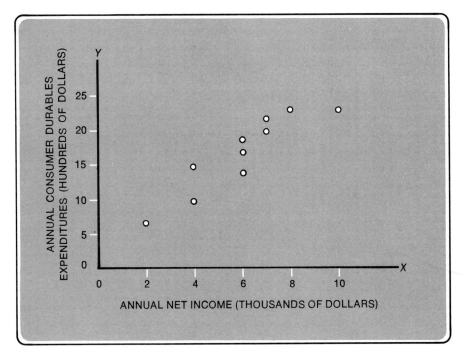

Figure 10-1
A scatter diagram of annual consumer durables expenditures and annual net income of a sample of 10 families in a metropolitan area in 1973.

tion of the average relationship between the two variables. If a straight line is used to describe the average relationship between Y and X, a *linear relationship* is said to be present. On the other hand, if the points on the scatter diagram appear to fall along a curved line rather than a straight line, a *curvilinear relationship* is said to exist. Figure 10-2 presents illustrative combinations of the foregoing types of relationships. Parts (A), (B), (C), and (D) of Figure 10-2 show, respectively, direct linear, inverse linear, direct curvilinear, and inverse curvilinear relationships. As can be seen, the points tend to follow: in (A) a straight line sloping upward, in (B) a straight line sloping downward, in (C) a curved line sloping upward, and in (D) a curved line sloping downward. Of course, the relationships are not always so obvious. In (E) the points appear to follow a horizontal straight line. Such a case depicts a situation of "no correlation" between the X and Y variables, or no evident relationship, since the horizontal line implies no change in Y, on the average, as X increases. In (F) the points follow a straight line sloping upward as in (A), but there is a much wider scatter of points around the line than in (A). In our present problem, we shall assume that a visual examination of Figure 10-1 suggests that there is a direct linear relationship between the two variables. In the next section, we shall discuss the purposes of regression and correlation analysis and the types of procedures used

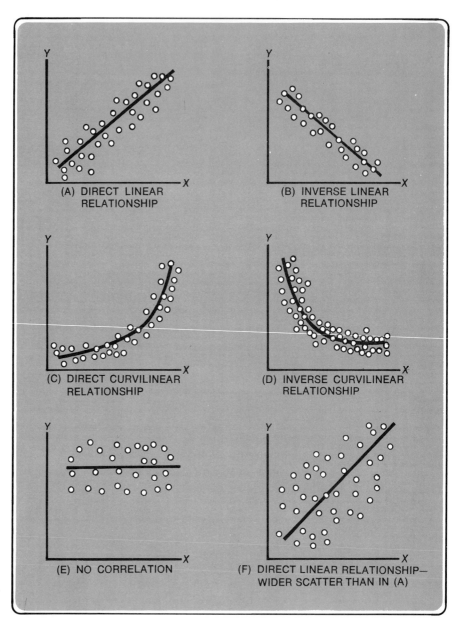

Figure 10-2
Scatter diagrams.

to accomplish these objectives. Then these procedures will be illustrated in terms of the data of Table 10-1.

What does a regression and correlation analysis attempt to accomplish in studying the relationship between two variables, such as expenditures on consumer durables and net annual income of families? We will concentrate on these basic goals, which emphasize the relationships contained in the particular sample under study. At a later point we will consider other objectives involving statistical inference, that is, inferences concerning the population from which the sample was drawn.

The first two objectives and the statistical procedures involved in their accomplishment fall under the heading of *regression analysis,* whereas the third objective and related procedures are classified as *correlation analysis.* The objectives are stated below and the statistical measures used to achieve these objectives are named. However, the mathematical definitions of these measures are postponed until the discussion of their use in the problem involving family expenditures on consumer durables and family income.

1. *The first purpose of regression analysis is to provide estimates of values of the dependent variable from values of the independent variable.* The device used to accomplish this estimation procedure is the *regression line,* which is a line fitted to the data by a method to be subsequently described. The regression line describes the average relationship existing between the X and Y variables. Somewhat more precisely, it is a line that displays mean values of Y for given values of X. The equation of this line, known as the *regression equation,* provides estimates of the dependent variable when values of the independent variable are inserted into the equation.

2. *A second goal of regression analysis is to obtain a measure of the error involved in using the regression line as a basis for estimation.* For this purpose, the "standard error of estimate" or its square, the "error variance around the regression line" are calculated. These are measures of the scatter or spread of the observed values of Y around the corresponding values estimated from the regression line. If the line fits the data closely, that is, if there is relatively little scatter of the observations around the regression line, good estimates can be made of the Y variable. On the other hand, if there is a great deal of scatter of the observations around the fitted regression line, the line will not produce accurate estimates of the dependent variable.

3. *The third objective, which we have classified as correlation analysis, is to obtain a measure of the degree of association or correlation that exists between the two variables. The coefficient of determination,* calculated for this purpose, measures the strength of the relationship that exists between the two variables. As we shall see, it assesses the proportion of variation in the dependent variable that has been accounted for by the regression equation.

10.4 ESTIMATION USING THE REGRESSION LINE

As indicated in the preceding section, to accomplish the first objective of a regression analysis, we must obtain the mathematical equation of a line that describes the average relationship between the dependent and independent variable. We can then use this line to estimate values of the dependent variable. Since the present discussion is limited to *linear* regression analysis, the line we are referring to is a straight line. Ideally, what we would like to obtain is the equation of the straight line that best fits the data. Let us defer for the moment what we mean by "best fits" and review the concept of the equation of a straight line.

The equation of a straight line is $Y = a + bX$, where a is the so-called "Y intercept," or the computed value of Y when $X = 0$, and b is the slope of the line, or the amount by which the computed value of Y changes with each one unit change in X. In regression analysis, we use the notation

(10.1) $$Y_c = a + bX$$

for the equation of the regression line. It is useful to use the different symbols Y_c, which denotes a *computed* value of the dependent variable, and Y, which denotes an *observed* value. For example, in the data given in Table 10-1, the observed value of Y for family A is 23, or $2300 annual consumer durables expenditures. The observed X value is ten, indicating that this family's net income is $10,000. When we obtain a regression equation, we may wish to estimate annual consumer durables expenditures for a family with an annual net income of $10,000. By substituting $X = 10$ into the regression equation, we can obtain the required estimate. Since the computed figure will in general be different from the observed value $Y = 23$, it is useful to have a separate symbol, such as Y_c, to denote this estimated or computed value of the dependent variable.

Let us review by means of a simple illustration the relationship between the equation $Y_c = a + bX$ and the straight line that represents the graph of the equation. Suppose the equation is

(10.2) $$Y_c = 2 + 3X$$

Thus, $a = 2$ and $b = 3$. If we substitute a value of X into this equation, we can

---

TABLE 10-2
Calculation of Pairs of X and
Y_c Values for the Line $Y_c = 2 + 3X$.

X	Y_c
0	2
1	5
2	8
3	11
4	14

---

obtain the corresponding computed value of Y_c. Each pair of X and Y_c values represents a single point. Although only two points are required to determine a straight line, several pairs of X and Y_c values are shown for the line $Y_c = 2 + 3X$ in Table 10-2. The graph corresponding to this line is shown in Figure 10-3. As can be seen on the graph, since the a value in the equation of the line is equal to two, the line intersects the Y axis at a height of two units. Also, since the b value, or slope of the line, is equal to three, we note in Table 10-2 that the Y_c values increase by three units each time X increases by one unit. This is shown

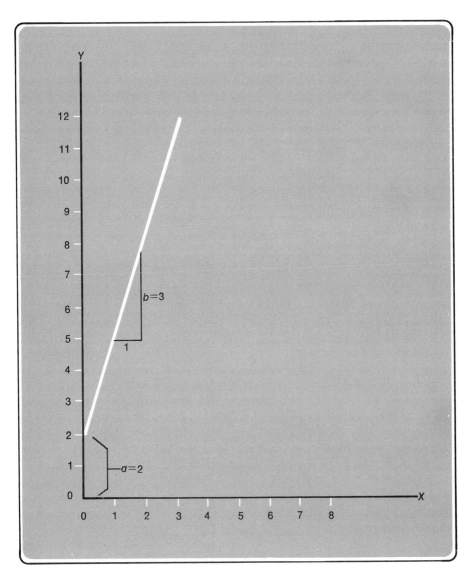

Figure 10-3
Graph of the line $Y_c = 2 + 3X$.

graphically in Figure 10-3 as a rise of three units in the line when X increases by one unit.

The terms "regression line" and "regression equation" for the estimating line and equation stem from the pioneer work in regression and correlation analysis of the British biologist, Sir Francis Galton in the nineteenth century. The lines that he fitted to scatter diagrams of data on heights of fathers and sons in this early work came to be known as "regression lines" and the equations of these lines as "regression equations," because Galton found that the heights of the sons "regressed" toward an average height. Unfortunately, the terminology has persisted. Thus, these terms for the estimating line and estimating equation are used in the wide variety of fields in which regression analysis is applied, despite the fact that the original implication of a regression toward an average is not necessarily present in terms of the phenomena under investigation.

We now turn to the question of obtaining a best fitting line to the data plotted on a scatter diagram in a two-variable linear regression problem. The fitting procedure to be discussed is the *method of least squares,* which undoubtedly is the most widely applied curve fitting technique in statistics.

The method of least squares

In order to establish a best fitting line to a set of data on a scatter diagram, we must have criteria concerning what constitutes *goodness of fit.* A number of such criteria, which might at first thought seem reasonable, turn out to be unsuitable. For example, we might entertain the idea of fitting a straight line to the data in such a way that one-half of the points fall above the line and one-half below. However, this requirement can be easily dismissed, since such a line may represent a quite poor fit to the data if, say, the points that fall above the line lie very close to it whereas the points below deviate considerably from it.

Let us now consider the most generally applied curve fitting technique in regression analysis, namely, the *method of least squares.* This method imposes the requirement that the *sum of the squares* of the deviations of the observed values of the dependent variable from the corresponding computed values on the regression line must be a minimum. Thus, if a straight line is fitted to a set of data by the method of least squares it is a "best fit" in the sense that the sum of the squared deviations, $\Sigma(Y - Y_c)^2$, is less than it would be for any other possible straight line. Another useful characteristic of the least squares straight line is that it passes through the point of means, $(\overline{X}, \overline{Y})$, and therefore, makes the total of the positive and negative deviations equal to zero. In summary, the least squares straight line possesses the following mathematical properties:

(10.3) $\qquad\qquad\qquad\qquad \Sigma(Y - Y_c)^2$ is a minimum

(10.4) $\qquad\qquad\qquad\qquad \Sigma(Y - Y_c) = 0$

By using calculus methods to apply the condition that the sum of the squared deviations from a straight line must be a minimum, two equations are derived,[1] which can be solved for the values of *a* and *b* in the regression equation $Y_c = a + bX$. The values of *a* and *b* can then be determined from the following general solution for *a* and *b*:

(10.5)
$$a = \bar{Y} - b\bar{X}$$

(10.6)
$$b = \frac{\Sigma XY - n\bar{X}\bar{Y}}{\Sigma X^2 - n\bar{X}^2}$$

where $\bar{X}$ and $\bar{Y}$ are the arithmetic means of the *X* and *Y* variables.

Let us return to the problem involving the sample of ten families and assume that after examination of the scatter diagram in Figure 10-1, we decide to fit a *straight line* to the data. From the original observations, we can determine the various quantities n, ΣY, ΣX, ΣXY, and ΣX^2 required in Equation (10.5) and (10.6), where n is the number of pairs of *X* and *Y* values, in this case, ten. For our illustration the computation of the required totals is shown in Table 10-3. Although ΣY^2 is not needed for the calculation of *a* and *b*, its computation is also shown. This figure is useful for calculating the standard error of estimate to be discussed shortly.

From Table 10-3, we compute the means of *X* and *Y* to be

$$\bar{X} = \frac{\Sigma X}{n} = \frac{60}{10} = 6 \text{ (thousand dollars)}$$

$$\bar{Y} = \frac{\Sigma Y}{n} = \frac{170}{10} = 17 \text{ (hundred dollars)}$$

---

[1] The method of deriving the equations is as follows. We denote the sum of squared deviations that must be minimized as some function of the unknown quantities *a* and *b*. Thus, let

$$F(a, b) = \Sigma (Y - Y_c)^2$$

Substituting the right-hand expression for $Y_c = a + bX$ into the above equation gives

$$F(a, b) = \Sigma (Y - a - bX)^2$$

We impose the condition of minimizing $F(a, b)$ by obtaining its partial derivatives with respect to *a* and *b* and setting them equal to zero. Thus.

$$\frac{\partial F(a, b)}{\partial a} = -2\Sigma (Y - a - bX) = 0$$

$$\frac{\partial F(a, b)}{\partial b} = -2\Sigma (Y - a - bX)(X) = 0$$

Solving these equations yields

$$\Sigma Y = na + b\Sigma X$$

$$\Sigma XY = a\Sigma X + b\Sigma X^2$$

from which Equations (10.5) and (10.6) can be derived.
A check that the second derivatives of $F(a, b)$ are positive reveals that a minimum has been found.

TABLE 10-3
Computations Required for a Regression and Correlation Analysis for the Data Shown in Table 10-1.

FAMILY	Y	X	XY	X^2	Y^2
A	23	10	230	100	529
B	7	2	14	4	49
C	15	4	60	16	225
D	17	6	102	36	289
E	23	8	184	64	529
F	22	7	154	49	484
G	10	4	40	16	100
H	14	6	84	36	196
I	20	7	140	49	400
J	19	6	114	36	361
	170	60	1122	406	3162

Substituting the additional quantities $n = 10$, $\Sigma XY = 1122$, and $\Sigma X^2 = 406$ from Table 10-3 into Equations (10.5) and (10.6), we obtain the following values for a and b

$$b = \frac{1122 - (10)(6)(17)}{406 - 10(6)^2} = \frac{102}{46} = 2.22$$

$$a = 17 - 2.22(6) = 17 - 13.32 = 3.68$$

Hence, the least squares regression line is

(10.7)
$$Y_c = 3.68 + 2.22X$$

If a family drawn from the same population had an annual net income of $8000 in 1973, its estimated annual consumer durables expenditures from Equation (10.7) would be

$$Y_c = 3.68 + 2.22(8) = 21.44 \text{ (hundred dollars)}$$

By plotting the point thus determined ($X = 8$, $Y_c = 21.44$) and one other point, or by plotting any two points derived from the regression equation, we can graph the regression line. The line is shown in Figure 10-4, along with the original data. Hence, in this case $a = 3.68$ means that the estimated annual consumer durables expenditures for a family whose income is zero dollars in 1973 is $3.68 (hundreds) or $368. Since no families in the original sample had incomes less than $2000, it would be extremely hazardous to make predictions for families with incomes less than the $2000 figure. Prediction outside the range of the original observations is discussed in a later section of this chapter.

The b value in the regression equation is often referred to as the "regression coefficient" or "slope coefficient." The figure of $b = 2.22$ indicates first

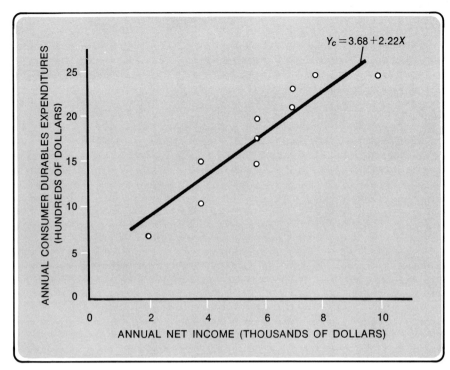

Figure 10-4
Least squares regression line for annual consumer durables expenditures and annual net income of a sample of 10 families in a metropolitan area in 1973.

of all that the slope of the regression line is positive. Thus, as income increases, estimated consumer durables expenditures increase. Taking into account the units in which the X and Y variables are stated, $b = 2.22$ means that for two families whose annual net income differs by $1000, the *estimated difference* in their annual consumer durables expenditures is $222. This is an interpretation in terms of the *regression line*. If we think of the figure $b = 2.22$ in terms of the *sample studied,* we can say that for two families whose annual net incomes differed by $1000, *on the average,* their annual consumer durables expenditures differed by $222.

EXERCISES

1. Look at the following scatter diagram and regression line showing the relationship between earnings in 1973 and price per share at end of 1973 for selected common stocks:

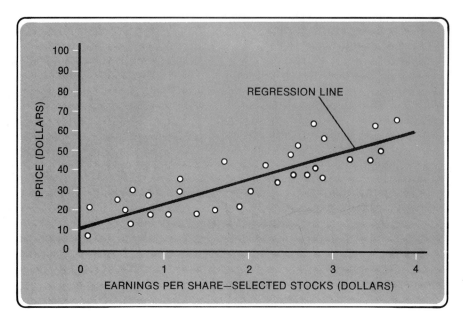

On the basis of the above chart, estimate the values of a and b in the equation of the line of regression $Y_c = a + bX$. Use specific numbers.

2. The following data represent observations on ten families randomly selected in a given poverty area:

FAMILY	CLOTHING EXPENDITURES (Y) (HUNDREDS OF DOLLARS)	INCOME (X) (HUNDREDS OF DOLLARS)	ALSO
A	.	.	$\Sigma X^2 = 10{,}000$
.	.	.	
.	.	.	
.	.	.	$\Sigma XY = 15{,}000$
J	$\overline{42}$	$\overline{200}$	

(a) Estimate the constants of a linear equation from which family clothing expenditures can be estimated, given family income.
(b) What is the meaning of the term, "method of least squares"?

It is important to note that up to now no assumptions have been made concerning the probability distributions of the X and Y variables. Thus, only relatively limited interpretations of our results can be given. The fitting process we have carried out may be interpreted as one that minimized the sum of the squared errors of estimation for the sample of ten families studied. However, since we assumed the sample was randomly drawn from families with $10,000 or less of annual net income in a metropolitan area in 1973, we may say that our computed regression equation should provide about as accurate estimates for such families not in the sample as for those which happened to be selected.

Now that we have seen that the regression equation is used for estimation, we can turn to the second objective, referred to in Section 10.3, that of obtaining a measure of the error involved in using the regression line for estimation. If there is a great deal of scatter of the observed Y values around the line, estimates of Y values based on computed values on the regression line will not be very close to the observed Y values. On the other hand, if every point falls on the regression line, insofar as the sample observations are concerned, perfect estimates of the Y values can be made from the fitted regression line. Just as the variance was used as a measure of variability of a set of observations about the mean, it would seem that an analogous measure of variability of observed Y values around the regression line would be $\Sigma(Y - Y_c)^2/n$. Similarly, the square root of this figure would be a measure of dispersion. The measure of dispersion, referred to as the *standard error of estimate,* that is conventionally used is

(10.8)
$$s_{Y \cdot X} = \sqrt{\frac{\Sigma(Y - Y_c)^2}{n - 2}}$$

where, as previously, n is the size of sample.

The standard error of estimate measures the scatter of the observed values of Y around the corresponding computed Y_c values on the regression line. The sum of squared deviations is divided by $n - 2$ because this divisor makes $s^2_{Y \cdot X}$ an unbiased estimator of the variance around the true population regression line. The latter line is the one that would have prevailed had a complete census rather than a sample of families been studied. The $n - 2$ represents the number of degrees of freedom around the fitted regression line. In general, the denominator is $n - k$ where k is the number of constants in the regression equation. In the case of a straight line, the denominator is $n - 2$ because two degrees of freedom are lost when a and b are used as estimates of the corresponding constants in the population regression line.

It is useful to consider the nature of the notation for the standard error of estimate, $s_{Y \cdot X}$. In the discussion of dispersion in Section 3.14, the symbol s was used to denote the standard deviation of a sample of observations. The use of the letter s in $s_{Y \cdot X}$ is analogous since, as explained in the preceding paragraph, $s_{Y \cdot X}$ is also a measure of dispersion computed from a sample. However, since both the variables Y and X are present in a two-variable regression and correlation analysis, subscript notation is required to distinguish among the various possible dispersion measures. Hence, as we have seen, the notation for the standard error of estimate, where Y and X are, respectively, the dependent and independent variables, is $s_{Y \cdot X}$. The letter to the left of the period in the subscript is the dependent variable, the letter to the right denotes the independent variable. Subscripts are also required to distinguish standard deviations around the means of the two variables. Thus, s_Y denotes the standard deviation of the Y values of a sample around the mean $\overline{Y}$, and s_X denotes the standard deviation of the X values around their mean, $\overline{X}$.

In a realistic problem containing large numbers of observations, the computation of the standard error of estimate using Equation (10.8) clearly involves a great deal of arithmetic. The calculation of Y_c for each X value in the sample is required, and then the arithmetic implied by the formula must be carried out. It is useful to have a short-cut method, which only involves quantities already computed. It is easy to see how such a short-cut formula might be derived. Since $Y_c = a + bX$, the right-hand expression in this equation can be substituted for Y_c in Equation (10.8) and the resulting expression may be simplified in various ways. A convenient form of such a short-cut formula is given by

(10.9)
$$s_{Y \cdot X} = \sqrt{\frac{\Sigma Y^2 - a\Sigma Y - b\Sigma XY}{n - 2}}$$

All quantities required by Equation (10.9) were calculated for our illustrative problem in Table 10-3, or were computed in obtaining the constants of the regression line. Hence, the standard error of estimate for these data is

$$s_{Y \cdot X} = \sqrt{\frac{3162 - (3.68)(170) - (2.22)(1122)}{10 - 2}} = 2.39 \text{ (hundreds of dollars)}$$

The standard error of estimate has been indicated in Figure 10-5. The scatter diagram for the illustrative problem, the regression line, and bands of one and two $s_{Y \cdot X}$ in width have been shown above and below the regression line.

To this point, no probability assumptions have been introduced for either the dependent or independent variables. However, in order to use a measure of dispersion around the regression line as an indicator of error of estimation, some assumption concerning the distribution of points around the regression line is necessary. If we assume that the observed Y values are normally distributed around the regression line, about 68% of the points will fall within the band made by parallel lines a distance of one standard error of estimate above and below the regression line, about 95.5% of the points will fall within a distance of two standard errors of estimate, and so forth. A visual interpretation of the normality assumption is given in Figure 10-5. Let us be somewhat more specific concerning this assumption. For every given X value, we can compute Y_c, our estimate of the value of the dependent variable. We are assuming that

1. The actual Y values for the given X form a normal distribution with Y_c as the arithmetic mean of the distribution and $s_{Y \cdot X}$ as the standard deviation of the distribution.
2. At every value of X, the dispersion of the Y values around the computed Y_c is the same; that is, it is assumed that the probability distributions of the Y values have the same standard deviation, $s_{Y \cdot X}$, for every X value within the observed range.

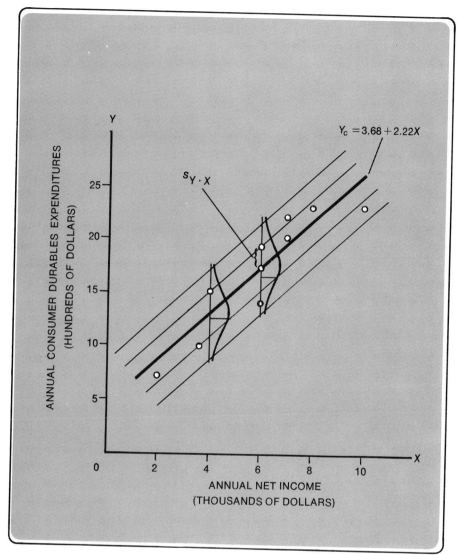

$$Y_c = 3.68 + 2.22X$$

$s_{Y \cdot X}$

ANNUAL CONSUMER DURABLES EXPENDITURES
(HUNDREDS OF DOLLARS)

ANNUAL NET INCOME
(THOUSANDS OF DOLLARS)

Figure 10-5
Least squares regression line and bands at distances of one and two standard errors of estimate.

Prediction intervals in regression analysis

The use of the standard error of estimate can be illustrated employing the assumptions of the preceding paragraph. We observed in Section 10.4 that using the regression equation derived in the illustrative problem, the estimated consumer durables expenditure for a family with $8000 of income was 21.44 hundreds of dollars ($2144). Under the assumptions given above, we can estimate that this family's consumer durables expenditures would be equal to

$$Y_c + s_{Y \cdot X} = 21.44 \pm (2.39)$$
$$= 19.05 \text{ to } 23.83 \text{ (hundreds of dollars)}$$

with a probability of about 68% (odds of about 2 to 1) of being correct, and

$$Y_c \pm 2s_{Y \cdot X} = 21.44 \pm 2(2.39)$$
$$= 16.66 \text{ to } 26.22 \text{ (hundreds of dollars)}$$

with a probability of 95.5% (odds of about 19 to 1) of being correct.

This type of interval established in regression analysis is usually called a "prediction interval." Its interpretation is analogous to that of a confidence interval in statistical inference (see page 151).

The preceding procedure provides an approximate prediction interval, and the method is a rough one appropriate only in the case of *large samples* ($n > 30$). Again, we repeat the caveat that the sample in the illustrative problem is small ($n = 10$) and is used here merely for convenience of exposition and arithmetic.

A more exact prediction interval can be obtained by using the t distribution. Since $s_{Y \cdot X}$ is an *estimated* standard deviation, the t distribution is the appropriate one for establishing the required interval. That is, the present example is a case in which the sample size is small ($n \leq 30$) and $s_{Y \cdot X}$ is an *estimate* of a population standard deviation rather than a *known* population standard deviation. For a prediction interval with an associated confidence coefficient of 95% (odds of 19 to 1), we find from Table A-6, for 8 degrees of freedom, the value of t is 2.306. Thus, the prediction interval is

$$Y_c \pm t \, s_{Y \cdot X} = 21.44 \pm 2.306 \, (2.39)$$
$$= 21.44 \pm 5.51 \text{ (hundreds of dollars)}$$
$$= \$1593 \text{ to } \$2695$$

If n is large, the prediction intervals obtained by the two methods discussed to this point would for practical purposes be about equal for the same confidence coefficient. That is, if n is large, the z values for the normal distribution are quite close to the t values in the t distribution for the same coefficient.

The usefulness of any prediction interval depends on the purposes for which it is used. For example, for long-range planning purposes, relatively wide limits may be appropriate and useful. On the other hand, for short-term operational decision making, narrower and therefore more precise intervals may be required.

10.6   CORRELATION ANALYSIS – MEASURES OF ASSOCIATION

In the preceding two sections, regression analysis was discussed, with emphasis on estimation and measures of error in the estimation process. We now turn to correlation analysis, in which the basic objective is to obtain a measure of the degree of association that exists between two variables. In this analysis, interest centers on the strength of the relationship between the variables, or, in other words, on how well the variables are correlated.

The coefficient of determination

A measure of the amount of correlation that exists between Y and X can be developed in terms of the relative variation of the Y values around the regression line and the corresponding variation around the mean of the Y variable. The term "variation," as used in statistics, conventionally refers to a sum of squared deviations.

The variation of Y values around the regression line is measured by

(10.10) $$\Sigma(Y - Y_c)^2$$

The variation of Y values around the mean of the Y variable is measured by

(10.11) $$\Sigma(Y - \overline{Y})^2$$

The first of these expressions (10.10) is the sum of the squared vertical deviations of the Y values from the regression line. The second expression is the sum of the squared vertical deviations from the horizontal line $Y = \overline{Y}$. The relationship between the variations around the regression line and mean can be summarized in a single measure to indicate the degree of association between X and Y. The measure used for this purpose is the *sample coefficient of determination,* defined as follows:

(10.12) $$r^2 = 1 - \frac{\Sigma(Y - Y_c)^2}{\Sigma(Y - \overline{Y})^2}$$

As we shall see from the subsequent discussion, r^2 may be interpreted as the percentage of variation in the dependent variable Y that has been accounted for or "explained" by the relationship between Y and X expressed in the regression line. Hence, it is a measure of the degree of association or correlation between Y and X.

In order to present the rationale of this measure of strength of the relationship between Y and X, we will consider two extreme cases, (1) zero linear correlation and (2) perfect direct linear correlation. The term "linear" indicates that a straight line has been fitted to the X and Y values and the term "direct" indicates that the line is inclined from the lower left to the upper right-hand side of a scatter diagram.

Two sets of data are presented in Table 10-4 labeled (a) and (b). They are shown in the form of scatter diagrams in Figure 10-6. As we shall see, the data in (a) and (b) illustrate the case of zero linear correlation and perfect direct linear correlation, respectively. In the discussion that follows, we will assume the collections of units of association shown in (a) and (b) represent simple random samples from their respective universes. Therefore, we employ notation pertinent to the samples. An analogous argument can be presented assuming the observations represent population data. The notation would

TABLE 10-4

Two Sets of Data Displaying (a) Zero Linear Correlation and
(b) Perfect Direct Linear Correlation.

	(a)			(b)	
UNIT OF ASSOCIATION	X	Y	UNIT OF ASSOCIATION	X	Y
A	1	4	A	1	2
B	1	6	B	2	4
C	2	4	C	3	6
D	2	6	D	4	8
E	3	4	E	5	10
F	3	6	F	6	12
G	4	4	G	7	14
H	4	6	H	8	16
		$\overline{Y} = 5$			$\overline{Y} = 9$

change correspondingly. The calculations given below the scatter diagrams in Figure 10-6 will be explained in terms of the data displayed in the charts.

Case (a) represents a situation in which $\overline{Y}$, the mean of the Y values, coincides with a least squares regression line fitted to these data. Even without doing the arithmetic, we can see why this is so. The slope of the regression line is equal to zero, because the same Y values are observed for $X = 1, 2, 3$, and 4. Thus, the regression line would coincide with the mean of the Y values, balancing deviations above and below the regression line. Another way of observing this relationship is in terms of the first of the two equations used to solve for a and b. In Equation (10.5), i.e., $a = \overline{Y} - b\overline{X}$, since $b = 0$, $a = \overline{Y}$. Hence, the regression line has a Y intercept equal to $\overline{Y}$. Since it is also a horizontal line, the regression line coincides with $\overline{Y}$. From the point of view of estimation of the Y variable, the regression line represents no improvement over the mean of the Y values. This can be shown by a comparison of $\Sigma(Y - Y_c)^2$, the variation around the regression line and $\Sigma(Y - \overline{Y})^2$, the variation around the mean of the Y values. In this case, the two variations are equal. These variations may be interpreted graphically as the sum of the squares of the distances between the points on the scatter diagram and $\overline{Y}$, shown in Figure 10-6 (A).

Now, let us consider case (b). This is a situation in which the regression line is a perfect fit to the data. The regression equation is a very simple one, which can be determined by inspection. The Y intercept is equal to zero since the line passes through the origin (0,0). The slope is equal to 2, because for every unit increase in X, Y increases by two units. Hence, the regression equation is $Y_c = 2X$, and all of the data points lie on the regression line. Insofar as the data in the sample are concerned, perfect predictions are provided by this regression line. Given a value of X, the corresponding value of Y can be correctly estimated from the regression equation indicating a perfect linear relationship between the two variables. Again, a comparison can be made of $\Sigma(Y - Y_c)^2$ and $\Sigma(Y - \overline{Y})^2$. Since all points lie on the regression line, the variation around the line, $\Sigma(Y - Y_c)^2$, is equal to zero. On the other hand, the variation around the mean, $\Sigma(Y - \bar{Y})^2$, is some positive number, in this case, 168.

As indicated in Figure 10-6 (A), when there is no linear correlation between X and Y, the sample coefficient of determination, r^2, is equal to zero. This follows from the fact that, since $\Sigma(Y - Y_c)^2$ and $\Sigma(Y - \bar{Y})^2$ are equal, the ratio $\Sigma(Y - Y_c)^2/\Sigma(Y - \bar{Y})^2$ equals one. Hence r^2 equals zero, because the computation of the coefficient of determination requires subtraction of this ratio from one.

On the other hand, as indicated in Figure 10-6 (B), when there is perfect linear correlation between X and Y, the sample coefficient of determination, r^2, is equal to one. In this case, the variation around the regression line is equal to zero, while the variation around the mean is some positive number. Thus, the ratio $\Sigma(Y - Y_c)^2/\Sigma(Y - \bar{Y})^2$ equals zero. Hence, r^2 equals one when the value of this ratio is subtracted from one.

In realistic problems, r^2 falls somewhere between the two limits, zero and one. A value close to zero suggests that there is not much linear correlation between X and Y; a value close to one connotes a strong linear relationship between X and Y.

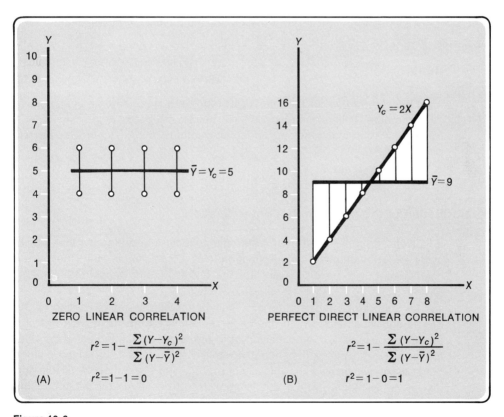

Figure 10-6
Scatter diagrams representing zero linear correlation and perfect direct linear correlation.

Population coefficient of determination

The measure r^2 has been referred to as the *sample* coefficient of determination. That is, it pertains only to the sample of n observations studied. The regression line computed from the sample may be viewed as an estimate of the true population regression line, which may be denoted as

(10.13)
$$Y' = A + BX$$

The corresponding population coefficient of determination is defined as

(10.14)
$$\rho^2 = 1 - \frac{\sigma_{Y.X}^2}{\sigma_Y^2}$$

The use of the symbol ρ^2 (rho squared) adheres to the usual convention of employing a Greek letter for a population parameter corresponding to the same letter in our alphabet that denotes a sample statistic. In the definition of ρ^2, $\sigma_{Y.X}^2$ is the variance around the population regression line $Y' = A + BX$ and σ_Y^2 is the variance around the population mean of the Ys, denoted μ_Y. Hence, both the sample and population coefficients of determination are equal to one minus a ratio of the variability around the regression line to the variability around the mean of the Y values.

A slightly different form of the *sample* coefficient of determination, which is directly parallel to Equation (10.14), is

(10.15)
$$r_c^2 = 1 - \frac{s_{Y.X}^2}{s_Y^2} = 1 - \frac{\Sigma(Y - Y_c)^2/(n-2)}{\Sigma(Y - \bar{Y})^2/(n-1)}$$

r_c^2 is referred to as the "corrected" or "adjusted" sample coefficient of determination. This terminology is used because $s_{Y.X}^2$ and s_Y^2 are estimators of $\sigma_{Y.X}^2$ and σ_Y^2, that make the appropriate corrections or adjustments for degrees of freedom.[2]

Interpretation of the coefficient of determination

It is useful to consider in more detail the specific interpretations that may be made of coefficients of determination. For convenience, only the sample coefficient r^2 will be discussed, but the corresponding meanings for ρ^2 are obvious.

An important interpretation of r^2 may be made in terms of variation in the dependent variable Y, which has been explained by the regression line. The problem of estimation is conceived of in terms of "explaining" or accounting for the variation in the dependent variable Y. Figure 10-7, on which a single point is shown, gives a graphical interpretation of the situation. In this context, if $\bar{Y}$, the mean of the Y values were used to estimate the value of Y, the

---

[2] The relationship between r_c^2 and r^2 is given by $r^2 = 1 - (1 - r_c^2)\left(\frac{n-1}{n-2}\right)$. For large sample sizes, $\left(\frac{n-1}{n-2}\right)$ is close to one and r_c^2 and r^2 are approximately equal.

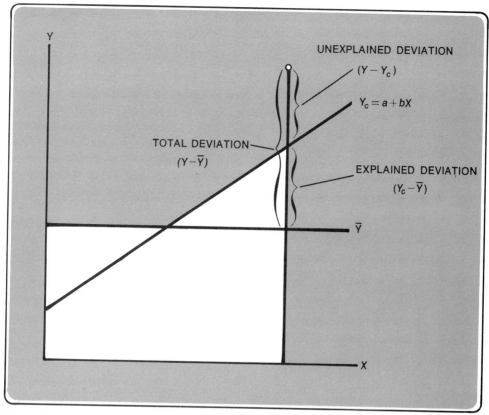

Figure 10-7
Graphical representation of total, explained, and unexplained variation.

total deviation would be $Y - \overline{Y}$. We can think of this deviation as being made up as follows:

Total deviation = Explained deviation + Unexplained deviation
$$(Y - \overline{Y}) \quad = \quad (Y_c - \overline{Y}) \quad + \quad (Y - Y_c)$$

If the regression line were used to estimate the value of Y, we would now have a closer estimate. As shown in the figure, there is still an "unexplained deviation" of $(Y - Y_c)$, but we have explained (or accounted for) $(Y_c - \overline{Y})$ of the total deviation.

In an analogous manner, we can partition the total variation of the dependent variable, (or total sum of squares), $\Sigma(Y - \overline{Y})^2$ as follows:

Total variation = Explained variation + Unexplained variation
$$\Sigma(Y - \overline{Y})^2 \quad = \quad \Sigma(Y_c - \overline{Y})^2 \quad + \quad \Sigma(Y - Y_c)^2$$

The ratio $\Sigma(Y - Y_c)^2/\Sigma(Y - \overline{Y})^2$ is the proportion of total variation that remains unexplained by the regression equation; correspondingly $1 - [\Sigma(Y - Y_c)^2/\Sigma(Y - \overline{Y})^2]$ represents the *proportion of total variation in Y* that has been explained by the regression equation. These ideas may be summarized as follows:

$$r^2 = 1 - \frac{\Sigma(Y - Y_c)^2}{\Sigma(Y - \bar{Y})^2} = 1 - \frac{\text{unexplained variation}}{\text{total variation}}$$

(10.16) $\qquad r^2 = \dfrac{\text{explained variation}}{\text{total variation}}$

A simple numerical example helps to clarify these relationships. Let $\Sigma(Y - \bar{Y})^2 = 10$ and $\Sigma(Y - Y_c)^2 = 4$. Thus, $r^2 = 1 - \frac{4}{10} = \frac{6}{10} = 60\%$. In this problem ten units of total variation in Y have to be accounted for. After fitting the regression line, the residual variation or unexplained variation amounts to four units. Hence, 60% of the total variation in the dependent variable is explained by the relationship between Y and X expressed in the regression line.

Calculation of the sample coefficient of determination

The computation of r^2 from the definitional formula, Equation (10.12), becomes quite tedious, particularly when there are a large number of observations in the sample. Just as in the case of the standard error of estimate, shorter methods of calculation are ordinarily used. These short-cut formulas are particularly helpful when computations are carried out by hand or on a calculating machine, but even when computers are used, they represent more efficient methods of computation. Such a formula, which only involves quantities already calculated, is

(10.17) $\qquad r^2 = \dfrac{a\Sigma Y + b\Sigma XY - n\bar{Y}^2}{\Sigma Y^2 - n\bar{Y}^2}$

Substituting into Equation (10.17), we obtain

$$r^2 = \frac{(3.68)(170) + (2.22)(1122) - 10(17)^2}{(3162) - 10(17)^2} = 0.832$$

Thus, for our sample of ten families, about 83% of the variation in annual consumer durables expenditures was explained by the regression equation, which related such expenditures to annual net income.

The coefficient of correlation

A widely used measure of the degree of association between two variables is the coefficient of correlation, which is nothing more than the square root of the coefficient of determination. Thus, the population and sample coefficients of correlation are, respectively,

(10.18) $\qquad\qquad\qquad\qquad \rho = \sqrt{\rho^2}$

and

(10.19) $\qquad\qquad\qquad\qquad r = \sqrt{r^2}$

Again, for convenience, our discussion will relate only to the sample value.

The algebraic sign attached to r is the same as that of the regression coefficient, b. Thus, if the slope of the regression line, b, is positive, then r is given a plus sign also; if b is negative, r is given a minus sign.[3] Hence, r ranges from a value of minus one to plus one. A figure of $r = -1$ indicates a perfect inverse linear relationship; $r = +1$ indicates a perfect direct linear relationship, and $r = 0$ indicates no linear relationship.

Illustrative scatter diagrams for the cases of $r = +1$ ($r^2 = 1$) and $r = 0$ ($r^2 = 0$) were given in Figure 10-6. A corresponding scatter diagram for $r = -1$, the case of perfect inverse linear correlation, is shown in Figure 10-8. As indicated in the graph, the slope of the regression line is negative and every point falls on the line. Thus, for example, if the slope on the regression line, b, were equal to -2, this would mean that with each increase of one unit in X, Y would decrease by two units. Since all points fall on the regression line in the case of perfect inverse correlation, $\Sigma(Y - Y_c)^2 = 0$. Therefore, substituting into Equation (10.12) to compute the sample coefficient of determination, we have

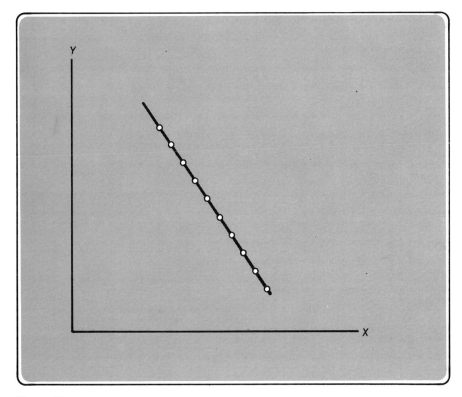

Figure 10-8
Scatter diagram representing perfect inverse linear correlation, $r = -1$.

[3] An interesting relationship between r and b is given by $b = r(\Sigma(Y - \bar{Y})^2/\Sigma(X - \bar{X})^2)$. Since $\Sigma(Y - \bar{Y})^2$ and $\Sigma(X - \bar{X})^2$ are positive numbers, b has the same sign as r. We can also note that when the regression line is horizontal, $b = 0$ and also $r = 0$.

$r^2 = 1 - 0 = 1$. Taking the square root, we obtain $r = \sqrt{1} = \pm 1$. However, since the b value is negative, that is, X and Y are inversely correlated, we assign the negative sign to r, and $r = -1$.

In our problem,

$$r = \sqrt{0.832} = 0.912$$

The sign is positive because b was positive, indicating a direct relationship between consumer durables expenditures and net income.

Despite the rather common use of the coefficient of correlation, it is preferable for interpretation purposes to use the coefficient of determination. As we have seen, r^2, the coefficient of determination, can be interpreted as a proportion or a percentage figure. When the square root of a percentage is taken, the specific meaning becomes obscure. Furthermore, since r^2 is a decimal value (unless it is equal to zero or one), its square root, or r, is a larger number. Thus, the use of r values to indicate the degree of correlation between two variables tends to give the impression of a stronger relationship than is actually present. For example, an r value of $+0.7$ or -0.7, seems to represent a reasonably high degree of association. However, since $r^2 = 0.49$, less than one-half of the total variance in Y has been explained by the regression equation.

It is useful to observe that the values of r and r^2 do not depend on the units in which X and Y are stated nor on which of these variables is selected as the dependent or independent variable. Whether a value of r or r^2 may be considered high depends somewhat upon the specific field of application. With some types of data it is relatively unusual to find r values in excess of about 0.80. On the other hand, particularly in the case of time series data, r values in excess of 0.90 are quite common. In the following section, we consider the matter of determining whether the observed degree of correlation in a sample is sufficiently large to justify a conclusion that correlation between X and Y actually exists in the population.

10.7   INFERENCE ABOUT POPULATION PARAMETERS IN REGRESSION AND CORRELATION

In the procedures discussed to this point, computation and interpretation of *sample* measures have been emphasized. However, as we know from our study of statistical inference, sample statistics ordinarily differ from corresponding population parameters because of chance errors of sampling. Therefore, it is useful to have a protective procedure against the possible error of concluding from a sample that an association exists between two variables, while actually no such relationship exists in the population from which the sample was drawn. A hypothesis testing technique, such as those discussed in Chapter 8, can be employed for this purpose.

Inference about the population correlation coefficient ρ

Let us assume a situation in which we take a simple random sample of n units from a population and make paired observations of X and Y for each unit. The sample correlation coefficient, r, as defined in Equation (10.19), is calculated. The procedure involves a test of the hypothesis that the population correlation coefficient, ρ, is equal to zero in the universe from which the sample was drawn. In keeping with the language used in Chapter 8, we wish to test the null hypothesis that $\rho = 0$ versus the alternative $\rho \neq 0$. Symbolically, we may write

$$H_0 : \rho = 0$$

$$H_1 : \rho \neq 0$$

If the computed r values in successive samples of the same size from the hypothesized population were distributed normally around $\rho = 0$, we would only have to know the standard error of r, σ_r, to perform the usual test involving the normal distribution. Although r values are not normally distributed, a similar procedure is provided by the statistic

(10.20)
$$t = \frac{r - \rho}{s_r} = \frac{r}{\sqrt{(1 - r^2)/(n - 2)}}$$

which has a t distribution for $n - 2$ degrees of freedom. $s_r = \sqrt{(1 - r^2)/(n - 2)}$ is the estimated standard error of r, and $\rho = 0$ by the null hypothesis. It may be noted that despite the previous explanation that r^2 is easier to interpret than r, the hypothesis testing procedure is in terms of r rather than r^2 values. The reason is that under the null hypothesis, $H_0 : \rho = 0$, the sampling distribution of r leads to the t statistic, which is a well-known distribution and is relatively easy to work with. On the other hand, under the same hypothesis of no correlation in the universe, r^2 values, which range from zero to one, would not even be symmetrically distributed, and the sampling distribution would be more difficult to deal with. Suppose we wish to test the hypothesis that $\rho = 0$ at the 5% level of significance for our problem involving ten families. Since $r = 0.912$ and $n = 10$, substitution into (10.20) yields

$$t = \frac{0.912}{\sqrt{\dfrac{1 - 0.832}{10 - 2}}} = 6.3$$

Referring to Table A-6, we find a critical t value of 2.306 at the 5% level of significance, for eight degrees of freedom. Therefore, the decision rule is

1. If $-2.306 \leq t \leq 2.306$, accept H_0
2. If $t < -2.306$ or $t > 2.306$, reject H_0

Since our computed t value is 6.3, far in excess of the critical value, we conclude that the sample r value differs significantly from zero. We reject the hypothesis that $\rho = 0$, and we conclude that there is a positive linear relationship between annual consumer durables expenditures and annual net income in the population from which our sample was drawn. Since the critical t value is 3.355 at the 1% level of significance (the smallest level shown in Table A-6), it is extremely unlikely that an r value as high as 0.91 would have been observed in a sample of ten items drawn from a population in which X and Y were uncorrelated.

A few comments may be made concerning this hypothesis testing procedure. First of all, this technique is valid only for a hypothesized universe value of $\rho = 0$. Other procedures must be used for assumed universe correlation coefficients other than zero.

Second, only Type I errors are controlled by this testing procedure. That is, when the significance level is set at, say, 5%, the test provides a 5% risk of incorrectly rejecting the null hypothesis of no correlation. No attempt is made to fix the risks of Type II errors (i.e., the risk of accepting $H_0: \rho = 0$ when $\rho \neq 0$) at specific levels.

Third, even though the sample r value is significant according to this test, the amount of correlation may not be considered substantively important. For example, in a large sample, a quite low r value may be found to differ significantly from zero. However, since relatively little correlation has been found between the two variables, we may be unwilling to use the relationship observed between X and Y for decision-making purposes. Furthermore, prediction intervals based on the use of the applicable standard errors of estimate may be too wide to be of practical use.

Fourth, the distributions of t values computed by Equation (10.20), as in previously discussed cases, approach the normal distribution as sample size increases. Hence, for large sample sizes, the t value is approximately equal to z in the standard normal distribution, and critical values applicable to the normal distribution may be used instead. For example, in the preceding illustration, in which the critical t value was 2.306 for eight degrees of freedom at the 5% level of significance, the corresponding critical z value would be 1.96 at the same significance level. For large sample sizes, these values would be much closer.

Inference about the population regression coefficient B

In many cases, a great deal of interest is centered on the value of b, the slope of the regression line computed from a sample. Statistical inference procedures involving either hypothesis testing or confidence interval estimation are often useful for answering questions concerning the size of the population regression coefficient, B in the population regression equation, $Y' = A + BX$.

In order to illustrate the hypothesis testing procedure for a regression coefficient, let us return to the data in our problem, in which $b = 2.22$. We recall that the interpretation of this figure was that the estimated difference in annual consumer durables expenditures for two families whose annual net income in 1973 differed by $1000 was $222. Suppose that on the basis of similar studies in the same metropolitan area, it had been concluded that in previous years a valid assumption for the true population regression coefficient was $B = 2$. Can we conclude that the population regression coefficient had changed?

To answer this question, we use a familiar hypothesis testing procedure. We establish the following null and alternative hypotheses:

$$H_0: B = 2$$
$$H_1: B \neq 2$$

Assume that we were willing to run a 5% risk of erroneously rejecting the null hypothesis that $B = 2$. The procedure involves a t test in which the estimated standard error of the regression coefficient, denoted s_b, is given by

(10.21)
$$s_b = \frac{s_{Y.X}}{\sqrt{\Sigma(X - \overline{X})^2}}$$

Hence, s_b, the estimated standard deviation of the sampling distribution of b values, is a function of the scatter of points around the regression line and the dispersion of the X values around their mean. The t statistic computed in the usual way is given by

(10.22)
$$t = \frac{b - B}{s_b}$$

We calculate s_b according to (10.21) as follows:

$$s_b = \frac{2.39}{\sqrt{46}} = 0.35$$

Substituting this value for s_b into (10.22) gives

$$t = \frac{2.22 - 2}{0.35} = 0.63$$

Since the same level of significance and the same number of degrees of freedom are involved as in the preceding test for the significance of r, the decision rule is identical. With a critical t value of 2.306 at the 5% level of significance, we cannot reject the null hypothesis that $B = 2$. Hence, we cannot conclude that the regression coefficient has changed from $B = 2$ for families in the given metropolitan area.

The corresponding confidence interval procedure involves setting up the interval

(10.23)
$$b \pm ts_b$$

In this problem the 95% confidence interval for B is $2.22 \pm (2.306)(0.35) = 2.22 \pm 0.81$.

Therefore, we can assert that the population B figure is included in the interval 1.41 to 3.03 with an associated confidence coefficient of 95%.

It may be noted that we used the t distribution in both the hypothesis testing and confidence interval procedures just discussed. As in previous examples, we may note that normal curve procedures can be used for large sample sizes. Hence, for two-tailed hypothesis testing at the 5% level of significance and for 95% confidence interval estimation, the 2.306 t value given in the preceding examples would be replaced by a normal curve z value of 1.96.

10.8 CAVEATS AND LIMITATIONS

Regression analysis and correlation analysis are very useful and widely applied techniques. However, it is important to understand the limitations of the methods and to interpret the results with care.

Cause and effect relationships

In correlation analysis, the value of the coefficient of determination, r^2, is calculated. This statistic measures the degree of association between two variables. Neither this quantity nor any other statistical technique that measures or expresses the relationship among variables can prove that one variable is the cause and one or more other variables are the *effects*. Indeed, there has been philosophical speculation and debate through the centuries as to the meaning of cause and effect, and as to whether such a relationship can ever be demonstrated by experimental methods. In any event, a measure such as r^2 does not prove the existence of a cause-effect relationship between two variables X and Y.

In a situation in which a high value of r^2 is obtained, X may be producing variations in Y, or third and fourth (etc.) variables W and Z may be producing variations in both X and Y. Numerous examples, frequently humorous in nature, have been given to demonstrate the pitfalls in attempting to draw cause-effect conclusions in such cases. For example, if the average salaries of ministers are associated with the average price of a bottle of scotch whiskey over time, that is, time represents the unit of association, a high degree of correlation between these two variables will probably be observed. Doubtless, we would be reluctant to conclude that it is fluctuations in ministers' salaries that cause the variations in the price of a bottle of scotch, or vice versa. This is a case where a third variable, which we may conveniently designate as the general level of economic activity, operates to produce variations in both of the aforementioned variables. From the economic standpoint, salaries of ministers represent the price paid for a particular type of labor; the cost of a bottle of scotch is also a price. When the general level of economic activity is high, both of these prices will tend to be high. When the general level of economic activity is low, as in periods of recession or depression, both of these prices tend to be lower than they were during more prosperous times. Thus, the high degree of correlation between the two variables of interest is produced by a third variable (and possibly others); certainly neither of the two variables is *causing* the variations in the other.

Furthermore, it is important to keep in mind the problem of sampling error. As we have seen, it is conceivable that in a particular sample a high degree of correlation, either direct or inverse, may be observed, while in fact there is no correlation (or very little correlation) between the two variables in the population.

Finally, in applying critical judgment to the evaluation of observed relationships, one must be on guard against "nonsense correlations" where no meaningful unit of association is present. For example, suppose we record in a column labeled X the distance from the ground of the skirt hemlines of the first 100 women who pass a particular street corner. In a column labeled Y, we record 100 observations of the heights of the Himalaya mountains along a certain latitude at five-mile intervals. It is possible that a high r^2 value might be obtained for these data. Clearly the result is nonsensical because there is no meaningful unit or entity through which these data are related. In the illustrative example used in this chapter, expenditures and income were observed for the same family. We have seen that the unit of association might be a time period or some other entity. There must be a reasonable link between the variables studied that is embodied in the unit of association.

Extrapolation beyond the range of observed data

In regression analysis, an estimating equation is established on the basis of a particular set of observations. A great deal of care must be exercised in making predictions of values of the dependent variable based on values of the independent variable outside the range of the observed data. Such predictions are referred to as *extrapolations*. For example, in the problem considered in this chapter, a regression line was computed for families whose annual net income ranged from $2000 to $10,000. It would be extremely unwise to make a prediction of consumer durables expenditures for a family with an annual net income of $25,000 using the computed regression line. To do so would imply that the straight-line relationship could be projected up to a value of $25,000 for the independent variable. Clearly, in the absence of other information, we simply do not know whether the same functional form of the estimating equation is valid outside the range of the observed data. In fact, in certain cases, unreasonable or even impossible values may result from such extrapolations. For example, suppose a regression with a negative slope had been computed relating the percentage of defective articles produced, Y, with the number of weeks of on-the-job training received, X, by a group of workers. An extrapolation for a large enough number of weeks of training would produce a negative value for the percentage of articles produced, which is an impossible result. Clearly in this case, although the computed estimating equation may be a good description of the relation between X and Y within the range of the observed data, an equation with different parameters or even a completely different functional form is required outside this range. Without a specific investigation, one simply does not know what the appropriate estimating device is outside the range of observed data. The maxim, "To know how many teeth a horse has, you must open his mouth and count his teeth," is relevant here.

However, sometimes the exigencies of a situation require an estimate, and obtaining additional data is either impractical or impossible. Extrapolations and alternative methods of prediction have to be engaged in, but the limitations and risks involved must be kept constantly in mind.

Other regression models

To this point, we have considered only one particular form of the regression model, namely, that of a straight-line equation relating the dependent variable Y to the independent variable X. Sometimes theoretical considerations indicate that this is the model required. On the other hand, a linear model is often used because either the theoretical form of the relationship is unknown and a linear equation appears to be adequate, or the theoretical form is known but is rather complex, and a linear equation may provide a sufficiently good approximation.

Often the straight-line model is not an adequate description of the relationship between two variables. In some applications, a curvilinear regression function may be more appropriate than a linear one. For example, in Chapter 13 the fitting of a second degree parabola to time series data is discussed. In other applications, models that involve transformations of one or both of the variables (for example, to logárithmic form) may provide better fits to the data. In all cases, the determination of the most appropriate regression model should be the result of a combination of theoretical reasoning, practical considerations, and careful screening of the available data.

Thus far, the discussion in this chapter has been limited to two-variable regression and correlation analysis. In many problems, the inclusion of more than one independent variable in a regression model may be required to provide useful estimates of the dependent variable. Suppose, for example, that in the illustration involving family consumer durables expenditures and family income, poor predictions were made based on the single independent variable income. Other factors such as family size, age of the head of the family, and number of employed persons in the family might be considered as possible additional independent variables to aid in the estimation of consumer durables expenditures. When two or more independent variables are utilized, the problem is referred to as a *multiple regression and correlation analysis*. Although a detailed discussion of this type of analysis is beyond the scope of this text, a brief description of the technique is included in the next section.

EXERCISES

Note: For simplicity in these exercises, substitute r_c for r in the t test of the null hypothesis $H_0{:}\rho = 0$. That is, use the t statistic

$$t = \frac{r_c}{\sqrt{(1 - r_c^2)/(n - 2)}}$$

1. As personnel director of a large manufacturing firm, you are given the following information collected from a simple random sample of employees:

$$Y_c = 50 + 0.4X$$

X is score on aptitude test
Y is the quality rating by company officials at the end of two years of service
Unit of association is employee
X varies between 40 to 110
$n = 100$
$s_Y^2 = 50$
$s_{Y.X}^2 = 20$

(a) Calculate r_c^2 and interpret your answer.
(b) Do you believe there is any correlation (for all employees of the firm) between their score on the aptitude test and the quality rating by company officials at the end of two years of service? Justify statistically.
(c) In terms of this problem, explain precisely the meaning of the regression coefficient, $b = +0.4$.
(d) Do you think an employee who scored 125 on the aptitude test will receive a quality rating (after two years of service) in excess of 100? Why or why not?

2. A competent research worker concluded that two variables were correlated in the universe. Another competent research worker, using the same sample information, was unwilling to conclude that the two variables were correlated in the universe. Does this seem possible to you? Why or why not?

3. A least squares linear regression and correlation analysis was conducted on the data obtained for a simple random sample of 102 sales persons of the International Conglomerate Company with the following results:

$$Y_c = -10 + X$$
X is age in years
Y is annual commissions in thousands of dollars
X ranges between 30 and 50 years
$n = 102$
$s_Y = 5$ (thousands of dollars)
$s_{Y.X} = 3$ (thousands of dollars)

(a) The sales manager objected to the results of the equation, because the value $a = -10$ did not seem reasonable. Explain briefly how you would reply to the sales manager.
(b) Interpret the regression coefficient, $b = 1$, specifically in terms of this problem.
(c) Would you conclude that there is any correlation *for all 30- to 50-year-old sales people* of this company between age and annual commissions? Carry out the necessary computations to justify statistically.
(d) Would a 40-year-old sales person who received $18,000 in annual sales commissions be considered a "poor performer" or just average? Why?

4. A sample survey of 102 families gave a linear regression equation of

$$Y_c = 8.0 - 0.2X$$

where Y was the percentage of income spent for medical care and X was the family income *in thousands of dollars*. Individual family incomes in the sample ranged from $2500 to $15,000. The following values were calculated:

$$s_Y = 0.6\%$$
$$s_{Y.X} = 0.4\%$$

(a) Compute the coefficient of determination, r_c^2, and state its meaning in terms of this problem.

(b) To draw inferences about the relationship between X and Y for all families in the universe, one could test the significance of r. What is the null hypothesis for this test? Distinguish clearly the conclusions made when the null hypotheses is (1) accepted and (2) rejected.

(c) A family with an income of $10,000 spent $1000 on medical care. Should this be considered an unusual expenditure? Justify statistically.

5. Based upon a sample of 32 of its sales people, a large corporation finds the following relationship:

$$Y_c = -4 + 20X$$
$$s_Y = 90$$
$$s_{Y.X} = 30$$

where X = number of calls on prospects by sales people
Y = sales made, in hundreds of dollars
$n = 32$

(a) Explain precisely the meaning of the -4 and the 20 in the above equation.

(b) Calculate and interpret r_c^2 in terms of this problem.

(c) Test the significance of r. Indicate clearly the hypothesis that is tested and give the meaning of your conclusion in terms of the problem.

(d) Distinguish clearly between the measures, s_Y, $s_{Y.X}$, and the standard error of r with particular reference to the distributions involved.

6. A personnel director devised a test of manual dexterity that she wished to use to screen job applicants in order to predict performance in assembly work. To determine the effectiveness of the test she selected a sample of 50 workers in the assembly department, determined their outputs per hour and their scores on the test. The results of a regression analysis of the data were as follows:

$$Y_c = 19.7 + 0.6X$$
X, the test scores, ranged from 25 to 74
Y ranged from 31 to 59 units of output per hour
$$s_Y = 10.00$$
$$s_{Y.X} = 6.0$$

(a) Would you be willing to conclude that a relationship exists between test scores and units of output per hour? Justify your answer statistically, using a 0.05 risk of a Type I error.

(b) Assuming a relationship exists between the two variables, estimate with 95% confidence the output per hour of a job applicant who scored 50 on the dexterity test.

7. The following results were obtained by correlating dollar volume of sales, Y, with number of employees, X, for a random sample of 66 marketing establishments of a given type. These firms had between 10 and 100 employees.

$$Y_c \text{ (thousands of dollars)} = 40 + 12X$$
$$s_{Y.X} = 4 \text{ (thousands of dollars)}$$
$$r_c = +0.9$$

(a) Interpret specifically in terms of the problem the value of the regression coefficient, $b = 12$.

(b) Would you be willing to estimate the dollar volume of sales for an establishment with 200 employees? Discuss briefly.

(c) Is it reasonable to conclude that the degree of correlation between volume of sales and number of employees is attributable to chance variation? Demonstrate statistically.

8. An insurance company wished to examine the relationship between income and amount of life insurance held by heads of families. The company drew a simple random sample of ten family heads and obtained the following results:

FAMILY	AMOUNT OF LIFE INSURANCE ($000 OMITTED)	INCOME ($000 OMITTED)
A	9	4
B	20	8
C	22	9
D	15	8
E	17	8
F	30	12
G	18	6
H	25	10
I	10	6
J	20	9

(a) Determine the linear regression equation using the method of least squares with income as the independent variable.
(b) What is the meaning of the regression coefficient b in this case?
(c) Test the hypothesis that the population regression coefficient is equal to zero. State your conclusion.
(d) Compute the standard error of estimate $s_{Y.X}$.
(e) What is your estimate of the amount of life insurance carried by a family head from the same population whose income is $10,000? Give a 95.5% prediction interval around this estimate. For simplicity, assume the sample is large, although n is only equal to ten. Use the standard error calculated in (d).

9. Assume that from a random sample of 102 new products brought to the market, the following least squares regression equation was determined:

$$Y_c = 1.0 + 4.0X$$

Y is "demand" measured by first year sales in millions of dollars
X is "awareness" measured by the proportion of consumers who had heard of the product by the third month after introduction of the product; $0 \leq X \leq 1$

$$s_Y^2 = 1$$
$$s_{Y.X}^2 = 0.2$$

(a) Would you be willing to conclude that a relationship exists between demand and awareness? Justify your conclusion statistically using a 0.05 risk of Type I error.
(b) Explain in your own words the meaning of a Type I error specifically in terms of this problem.
(c) Assuming a relationship between demand and awareness, estimate with 98% confidence the demand for a new product for which 30% of consumers had heard of the product by the third month after introduction.

10.9   MULTIPLE REGRESSION AND CORRELATION ANALYSIS

Multiple regression analysis represents a logical extension of two-variable regression analysis. Instead of a single independent variable, two or more independent variables are used to estimate the values of a dependent variable. However, the fundamental concepts in the analysis remain the same. Thus, just as in the analysis involving the dependent and only one independent variable, there are the same three general purposes of multiple regression and correlation analysis.

The first purpose is accomplished by deriving an appropriate regression equation by the method of least squares. The second purpose is achieved through the calculation of a standard error of estimate, which just as in two-variable analysis is simply the standard deviation of the observed values of the dependent variable around the estimated values computed from the regression equation. The third purpose is accomplished by computing the multiple coefficient of determination, which is analogous to the coefficient of determination in the two-variable case and measures the proportion of variance in the dependent variable explained by the independent variables.

As an example, let us return to our problem in which family consumer durables expenditures were estimated from family net income, both variables being stated on an annual basis. As indicated in the preceding section, the use of additional variables to income might be considered to obtain improvement in the prediction of the dependent variable. Let us assume that family size is selected as a second independent variable. Estimates of consumer durable expenditures may now be made from the following multiple regression equation:

(10.24)
$$Y_c = a + b_1 X_1 + b_2 X_2$$

where Y_c = family consumer durables expenditures (estimated)
   X_1 = family net income
   X_2 = family size

and a, b_1, and b_2 are numerical constants, which must be determined from the data in a manner analogous to that of the two-variable case. For simplicity, we have assumed a linear regression function.

As an example, we will discuss a multiple regression and correlation analysis, fitting the linear regression equation (10.24) to data for the indicated variables. The basic data for family consumer durables expenditures, family income, and family size are shown in the first three columns of Table 10-5. The data for the first two of these variables are the same as those given in Table 10-1 for the two-variable problem previously solved. The data on family size represent the total number of persons in each of the families in the sample.

We begin the analysis by using the method of least squares to obtain the best fitting three-variable linear regression equation of the form given in (10.24). In the two-variable regression problem, the method of least squares was used to obtain the best fitting straight line. In the present problem, the analogous geometric interpretation is that the method of least squares is used to obtain the best fitting plane. In a three-variable regression problem, the points can be plotted in three dimensions, along the X_1, X_2, and Y axes analogous to the case of a two-variable problem, in which the points are plotted in two dimensions along an X and Y axis. The best fitting plane would pass through the points as shown in Figure 10-9, with some falling above and some below the plane in such a way that $\Sigma(Y - Y_c)^2$ is a minimum. Whereas in our previous illustration involving two variables, two equations resulted from the minimization procedure, now three equations must be solved to determine the values of a, b_1, and b_2:[4]

(10.25)
$$\Sigma Y = na + b_1\Sigma X_1 + b_2\Sigma X_2$$
$$\Sigma X_1 Y = a\Sigma X_1 + b_1\Sigma X_1^2 + b_2\Sigma X_1 X_2$$
$$\Sigma X_2 Y = a\Sigma X_2 + b_1\Sigma X_1 X_2 + b_2\Sigma X_2^2$$

The calculations of the required sums are shown in Table 10-5. Substituting into Equation (10.25) gives

$$170 = 10a + 60b_1 + 40b_2$$
$$1122 = 60a + 406b_1 + 267b_2$$
$$737 = 40a + 267b_1 + 182b_2$$

Solving these three equations simultaneously, we obtain the following values for a, b_1, and b_2:

$$a = 3.92$$
$$b_1 = 2.50$$
$$b_2 = -0.48$$

When large numbers of variables and observations are present, the calculations are apt to be too laborious. The utilization of electronic computing equipment in such cases may represent the only feasible alternative.

The multiple regression equation may now be written as

(10.26)
$$Y_c = 3.92 + 2.50X_1 - 0.48X_2$$

Let us illustrate the use of this equation for estimation. Suppose we want to estimate consumer durables expenditures for a family from the same population as the sample studied. The family's income is $6000 and there are four

[4] In a manner similar to that of the two-variable case, a function of the form

$$F(a, b_1, b_2) = \Sigma(Y - Y_c)^2 = \Sigma(Y - a - b_1X_1 - b_2X_2)^2$$

is set up. This function is minimized by the standard calculus method of taking its partial derivatives with respect to a, b_1, and b_2 and equating these derivatives to zero. This procedure results in the three equations of (10.25).

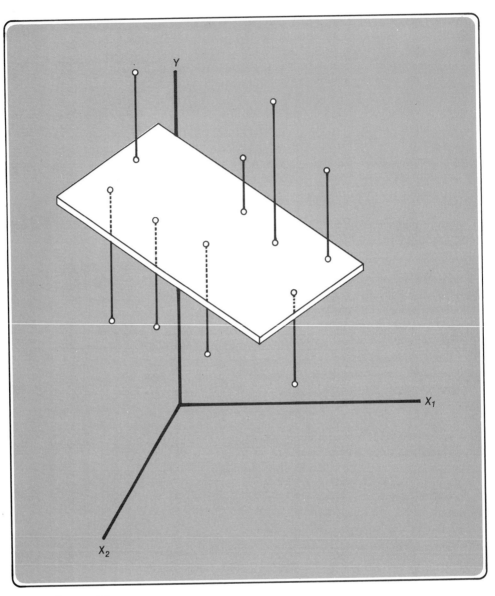

Figure 10-9
Graph of a multiple regression plane for data on the variables Y, X_1, and X_2.

persons in the family. Substituting $X_1 = 6$ and $X_2 = 4$ yields the following estimated expenditures on consumer durables:

$$Y_c = 3.92 + 2.50(6) - 0.48(4)$$
$$= 17.00 \text{ (hundreds of dollars)}$$
$$= 1700 \text{ (dollars)}$$

TABLE 10-5

Computations for Linear Multiple Regression Analysis:
Expenditures on Consumer Durables, Y, Family Income, X_1, and Family Size, X_2.

FAMILY	EXPENDITURES ON CONSUMER DURABLES (HUNDREDS OF DOLLARS) Y	INCOME (THOUSANDS OF DOLLARS) X_1	FAMILY SIZE X_2	X_1Y	X_2Y	X_1X_2	Y^2	X_1^2	X_2^2
A	23	10	7	230	161	70	529	100	49
B	7	2	3	14	21	6	49	4	9
C	15	4	2	60	30	8	225	16	4
D	17	6	4	102	68	24	289	36	16
E	23	8	6	184	138	48	529	64	36
F	22	7	5	154	110	35	484	49	25
G	10	4	3	40	30	12	100	16	9
H	14	6	3	84	42	18	196	36	9
I	20	7	4	140	80	28	400	49	16
J	19	6	3	114	57	18	361	36	9
TOTAL	170	60	40	1122	737	267	3162	406	182
MEAN	17	6	4						

In two-variable analysis, we discussed the interpretation of the constants a and b in the regression equation. Let us consider the analogous interpretation of the constants a, b_1, and b_2 in the multiple regression equation. The constant a is again the Y intercept. However, now it is interpreted as the value of Y_c when X_1 and X_2 are both equal to zero. The b values are referred to in multiple regression analysis as *net regression coefficients.* The b_1 coefficient measures the change in Y_c per unit change in X_1 when X_2 is held fixed, and b_2 measures the change in Y_c per unit change in X_2 when X_1 is held fixed.

Hence, in the present problem, the b_1 value of 2.50 indicates that if a family has an income which is $1000 greater than another's (a one unit change in X_1) and *the families are of the same size* (X_2 is held constant), then the estimated expenditures on consumer durables of the higher income family exceed those of the other by 2.5 hundreds of dollars or $250. Similarly, the b_2 value of -0.48 means that if a family has one person more than another (a one unit change in X_2) and *the families have the same income* (X_1 is held constant), then the estimated expenditures of the larger family are less than those of the smaller by $48.

Two properties of these net regression coefficients are worth noting. The b_1 value of 2.50 hundreds of dollars implies that an increment of one unit in X_1, or a $1000 increment in income, occasions an increase of $250 in Y_c, estimated expenditures of consumer durables, regardless of the size of the family (for families of the sizes studied). Hence, an increase of $1000 in income adds $250 to estimated consumer durables expenditures, regardless of whether there are two or six people in the family. An analogous interpretation holds for b_2. These interpretations follow from the fact that a *linear* multiple regression equation was used, and are embodied in the assumption of linearity.

A second property of regression coefficients is apparent from a comparison of the b value of 2.22 in the simple regression equation (10.7), $Y_c = 3.68 + 2.22X$, previously obtained when family income, X, was the only independent variable, with the b_1 value of 2.50, the net regression coefficient of income in the multiple regression equation $Y_c = 3.92 + 2.50X_1 - 0.48 X_2$, when the family size variable is included in the regression equation. The coefficient $b = 2.22$ in the simple two-variable regression equation makes no explicit allowance for family size. The net regression coefficient $b_1 = 2.50$, on the other hand, "nets out" the effect of family size. A net regression coefficient may in general be greater or less than the corresponding regression coefficient in a two-variable analysis.

In this problem, the families with larger incomes were also the ones with larger family sizes. The positive correlation between income and family size is indicated by the correlation coefficient $r = 0.85$ for these two variables. The foregoing pattern exemplifies an important characteristic of regression coefficients, regardless of the number of independent variables which have been included in the study. That is, a regression coefficient for any specific independent variable, for example, income, measures not only the effect on the dependent variable of income, but also the effect that is attributable to any other independent variables which happen to be correlated with it but have not been explicitly included in the analysis. This is true for both two-variable and multiple regression analyses.

When independent variables are highly correlated, rather odd results may be obtained in a multiple regression analysis. For example, a regression coefficient that is positive (negative) in sign in a two-variable regression equation may change to a negative (positive) sign for the same independent variable in a multiple regression equation containing other independent variables which are highly intercorrelated with the one in question. For example, in this problem, the dependent variable, consumer durables expenditures, Y, is positively correlated with family size, X_2. Hence, the regression coefficient for family size would also be positive in sign. However, as we have seen, the net regression coefficient for family size, b_2, in the three-variable regression equation is equal to -0.48 and is thus negative in sign.

In multiple regression analysis, the regression coefficients for highly intercorrelated independent variables tend to be unreliable. The importance of this is that when independent variables are highly intercorrelated, it is extremely difficult to separate out the individual influences of each variable. This can be seen by considering an extreme case. Suppose a two-variable regression and correlation analysis is carried out between a dependent variable, denoted Y, and an independent variable, denoted X_1. Further, let us assume that we introduce another independent variable X_2, which has perfect positive correlation with X_1; that is, the correlation coefficient between X_1 and X_2 is $+1$. We now conduct a three-variable regression and correlation analysis. It is clear that X_2 cannot account for or explain any additional variance in the dependent variable Y after X_1 has been taken into account. The same argument could be made if X_1 were introduced after X_2. As indicated in the ensuing discussion of statistical inference in multiple regression, the net regression coefficients, b_1 and b_2, in cases of high intercorrelation between X_1 and X_2 will tend not to differ significantly from zero. Yet, if separate two-variable analyses had been run between Y and X_1 and Y and X_2, the individual regression coefficients might have differed significantly from zero. In fields such as econometrics and applied statistics, there is a great deal of concern with this problem of intercorrelation among independent variables, often referred to as *multicollinearity*. One of the simplest solutions to the problem of two highly correlated independent variables is merely to discard one of the variables.

The illustration in this section used only two independent variables. The general form of the linear multiple regression function for $k - 1$ independent variables $X_1, X_2, \ldots, X_{k-1}$ is

(10.27) $Y_c = a + b_1X_1 + b_2X_2 + \cdots + b_{k-1}X_{k-1}$

In this formulation, there are k variables in the regression equation, one dependent and $k - 1$ independent. The linear function fitted to data for two variables is referred to as a straight line, for three variables a plane, for four or more variables, a *hyperplane*. Although we cannot visualize a hyperplane, its linear characteristics are analogous to those of the linear functions of two or three variables. With the use of computers it is possible to test and include large numbers of independent variables in a multiple regression analysis. However, good judgment and knowledge of the logical relationships involved must always be used as a guide to deciding which variables to include in the construction of a regression equation.

Measures of dispersion and correlation

As noted earlier it is beyond the scope of this book to discuss the detailed nature of the calculations in multiple regression and correlation analysis. However, a brief indication will be given here of the measures of dispersion and correlation analogous to the two-variable case.

The variance around the regression equation given in Equation (10.27) is

(10.28) $$S^2_{Y.12\,\cdots\,(k\,-\,1)} = \frac{\Sigma(Y - Y_c)^2}{n - k}$$

where n is the number of observations and k is the number of constants in the regression equation. As in the two-variable case, the standard error of estimate is given by the square root of this variance, and prediction intervals for individual estimates of Y values would be obtained in a manner similar to that of two-variable analysis by calculating

(10.29) $$Y_c \pm tS_{Y.12} \cdots {\scriptstyle (k-1)}$$

The number of degrees of freedom for the t value is $n - k$.

An analogous measure to the two-variable sample coefficient of determination is given by the *coefficient of multiple determination,* denoted $R^2_{Y.12} \cdots {\scriptstyle (k-1)}$, and defined as

(10.30) $$R^2_{Y.12 \cdots (k-1)} = 1 - \frac{S^2_{Y.12 \cdots (k-1)}}{s^2_Y}$$

where

$$s^2_Y = \frac{\Sigma (Y - \overline{Y})^2}{n - 1}$$

as in the two-variable case.

Hence, in a manner completely similar to the interpretation of r_c^2, the sample coefficient of determination adjusted for degrees of freedom, we may interpret $R^2_{Y.12 \cdots (k-1)}$ as measuring the proportion of variance in the dependent variable, which is explained by the regression equation.

Selected general considerations

In the discussion of multiple regression and correlation analysis, we have confined ourselves to the case of a linear model. Of course, frequently when a linear regression equation is used, it simply represents a convenient approximation to the unknown "true" relationship. Where linear relationships are inadequate, curvilinear regression equations may be required.

The quest to provide a good fit of the regression equation to the data leads to adding more and more independent variables. However, cost considerations, difficulties of providing data in the implementation and monitoring of the model, and the search for a reasonable simple model ("parsimony") point toward the use of as few independent variables as possible. Since no mechanistic statistical procedure exists to resolve this dilemma and many other problems of multiple regression and correlation analysis, subjective judgment inevitably plays a large role.

The use of computers in multiple regression analysis

The use of high-speed electronic computers has greatly simplified the testing and analysis of statistical relationships among variables. Nowadays, the libraries of most computer centers contain programs for various types of multivariate analysis including multiple regression analysis (for brevity, we will use that term rather than the longer "multiple regression and correlation analysis"). In the past the cost and tedious labor involved in multiple regression analyses involving more than two or three independent variables severely restricted the analyst's ability to test and experiment. With the use of modern computer programs, the analyst now has a much wider range of choice in selection of variables, in options for performing transformations, in adding and deleting variables at various stages of the analysis, and in testing curvilinear as well as linear relationships.

Stepwise regression analysis is a versatile form of multiple regression analysis for which computers are particularly useful, and for which there are a number of available computer programs. In this type of analysis, at the first stage, the computer determines which of the independent variables (as many as about 30 may be included in some programs) is most highly correlated with the dependent variable. The computer printout then displays all of the usual statistical measures for the two variable relationship. At the next stage, the program selects that independent variable which accomplishes the greatest reduction in the unexplained variance remaining from the two variable analysis. As in the previous stage, the computer printout then displays all of the usual statistical measures for the three variable relationship. The program continues in this stepwise fashion, at each stage entering the "best" independent variable in terms of ability to reduce unexplained variance. Analysis of variance tables and lists of residuals $(Y - Y_c)$ are provided at each stage. It is evident that without the use of computers and "canned" programs, the corresponding time to perform such an analysis by hand would be prohibitive.

A number of other types of multiple regression analysis programs are available in which the analyst initially includes a certain number of independent variables (for example, as many as 30 or so), and then can delete and add variables as desired. As an illustration, some of the computer results of the expenditures vs. income and family size problem discussed earlier in this section are shown (Table 10-6). The data were run on the IBM 370/165 at the University of Pennsylvania Computer Center using the UCLA BMDØ3R multiple linear regression program as modified by Wharton Computational Services. Explanations of the computer output for this problem are included in the Study Guide that accompanies the text.

One final comment is in order. Because of the remarkably increased possibilities in multiple regression analysis opened up by the use of computers, it becomes even more important that care and good judgment be exercised to avoid misuse of methods and misinterpretation of findings.

TABLE 10-6

THE FIRST   10 OBSERVATIONS ARE PRINTED OUT TO SERVE AS A DIAGONISTIC AID

OBS. NO.			
1	23.00000	10.00000	7.00000
2	7.00000	2.00000	3.00000
3	15.00000	4.00000	2.00000
4	17.00000	6.00000	4.00000
5	23.00000	8.00000	6.00000
6	22.00000	7.00000	5.00000
7	10.00000	4.00000	3.00000
8	14.00000	6.00000	3.00000
9	20.00000	7.00000	4.00000
10	19.00000	6.00000	3.00000
SUMS	170.00000	60.00000	40.00000

SUM OF SQUARES

3162.00000	406.00000	182.00000

CROSS PRODUCT SUMS

VARIABLE 1
3162.00000	1122.00000	737.00000

VARIABLE 2
1122.00000	406.00000	267.00000

VARIABLE 3
737.00000	267.00000	182.00000

CORRELATION COEFFICIENTS

VARIABLE 2
1.0000	0.8487	0.9119

VARIABLE 3
0.8487	1.0000	0.7368

VARIABLE 1
0.9119	0.7368	1.0000

SAMPLE SIZE   10
NO. OF VARIABLES 3   NO. OF VARIABLES DELETED 0   (FOR VARIABLES DELETED, SEE BELOW)
DEPENDENT VARIABLE IS NOW NO. 1

COEFFICIENT OF DETERMINATION   0.8364
ADJUSTED COEF. OF DETERMINATION 0.7663

MULTIPLE CORR. COEFFICIENT   0.9146

SUM OF SQUARES ATTRIBUTABLE TO REGRESSION   227.51237
SUM OF SQUARES OF DEVIATION FROM REGRESSION   44.48763

VARIANCE OF ESTIMATE   6.35538
STD. ERROR OF ESTIMATE   2.52099

INTERCEPT (A VALUE)   3.91873

```
              ANALYSIS OF VARIANCE FOR THE MULTIPLE
                       LINEAR  REGRESSION
    SOURCE OF VARIATION          D.F.    SUM OF        MEAN        F
                                         SQUARES       SQUARES     VALUE
DUE TO REGRESSION.............     2    227.51237     113.75618    17.8992
DEVIATION ABOUT REGRESSION...      7     44.48753       6.35538
                 TOTAL...          9    272.00000
```

VARIABLE NO.	MEAN	STD. DEVIATION	REG. COEFF.	STD.ERROR OF REG.COE.	COMPUTED T VALUE	PARTIAL CORR. COE.	SUM OF SQ. ADDED	PROP. OF TOT VAR.
2	6.00000	2.26078	2.49117	0.70289	3.54416	0.80134	226.17391	0.83152
3	4.00000	1.56347	-0.46643	1.01638	-0.45891	-0.17090	1.33845	0.00492
1	17.00000	5.49747						

```
COMP. CHECK ON FINAL COEFF.      -0.46643
```

```
              TABLE OF RESIDUALS
```

OBSERVATION	Y VALUE	Y ESTIMATE	RESIDUAL
1	23.00000	25.56537	-2.56537
2	7.00000	7.50177	-0.50177
3	15.00000	12.95053	2.04947
4	17.00000	17.00000	0.00000
5	23.00000	21.04947	1.95053
6	22.00000	19.02473	2.97527
7	10.00000	12.48410	-2.48410
8	14.00000	17.46643	-3.46643
9	20.00000	19.49117	0.50883
10	19.00000	17.46643	1.53357

11 *Nonparametric Statistics*

11.1   INTRODUCTION

Most of the methods discussed thus far have involved assumptions about the distributions of the populations sampled. For example, as we have seen, when certain hypothesis testing techniques are used it is assumed that the observations are drawn from normally distributed populations. Recently a number of very useful techniques that do not make these restrictive assumptions have been developed. Such procedures are referred to as *nonparametric* or *distribution-free* tests. Many writers prefer the latter term, because it emphasizes the fact that the techniques are free of assumptions concerning the underlying population distribution. However, the two terms are generally used synonymously.

In addition to making less restrictive assumptions that the corresponding so-called "parametric" methods, nonparametric procedures are generally easy to carry out and understand. Furthermore, as is implied by the lack of underlying assumptions, they are applicable under a very wide range of conditions. Many nonparametric tests are in terms of the *ranks* or *order* rather than the numerical values of the observations. Sometimes, even ordering is not required. Hence, nonparametric procedures may be employed at levels of measurement where parametric procedures cannot.

However, when distribution-free procedures are applied where parametric techniques are possible, the nonparametric methods have some disadvantages. Since these nonparametric procedures may use ordering or ranking as opposed to the actual numerical values of the observations, they are in effect, guilty of ignoring a certain amount of information. As a result, nonparametric tests are somewhat less efficient than the corresponding standard tests. This means that in testing at a given level of significance, say $\alpha = 0.05$, the probability of a Type

II error, β, would be greater for the nonparametric than for the parametric test. Advocates of nonparametric tests argue, however, that despite the lessened efficiency, the analyst can have more confidence in these tests than in the standard ones, because of the restrictive and somewhat unrealistic assumptions often required in the latter procedures.

In this chapter, we will consider a few of the simple and widely applied nonparametric techniques.

THE SIGN TEST 11.2

As we have seen in Chapter 8, the solution to many problems in public and business administration and in social science research centers on a comparison between two different samples. In some of the hypothesis testing techniques previously discussed, very restrictive assumptions about the populations sampled were necessary. For example, the t test for the difference between two sample means assumes that the populations are normally distributed and have equal variances. Sometimes, one or both of these assumptions may be unwarranted. Furthermore, situations often exist in which quantitative measurements are impossible. In such cases, it may be possible to assign ranks or scores to the observations in each sample. In these situations, the *sign test* can be used. The name of the test indicates that the signs of observed differences, that is, positive or negative signs, are used rather than quantitative magnitudes.

We will illustrate the sign test in terms of data obtained from a panel of 60 beer-drinking consumers. Let us assume a blindfold test in which the tasters were asked to rate a glass of each of two brands of beer, Wudbeiser and Diller, on a scale from 1 to 5 with 1 representing the best taste (excellent) 5 the worst taste (poor), and the other scores denoting the appropriate intermediates. Table 11-1 shows a partial listing of the scores assigned by the panel members in this taste test.

TABLE 11-1

Ranking Scores Assigned to Taste of Two Brands of Beer by a Panel.

PANEL MEMBER (1)	SCORE FOR WUDBEISER (2)	SCORE FOR DILLER (3)	SIGN OF DIFFERENCE (4)
A	3	2	+
B	4	1	+
C	2	4	−
D	3	3	0
E	1	2	−
⋮	⋮	⋮	⋮

NOTE: 1 denotes best score; 5 denotes worst. Hence, a plus sign means Diller is preferred; a minus sign means Wudbeiser is preferred.

Column (4) shows the signs of the differences between the scores assigned by each participant in columns (2) and (3). As indicated, a plus sign means a higher numerical score was assigned to Wudbeiser than to Diller beer, a minus sign means the reverse, and a zero denotes a tie score. Let us assume the following results were obtained:

+ scores	35
− scores	15
ties	10
Total	60

Method

By means of the sign test, we can test the null hypothesis of no difference in rankings of the two brands of beer. That is, more specifically, we can test the hypothesis that plus and minus signs are equally likely for the differences in rankings. If this null hypothesis were true, we would expect about equal numbers of plus and minus signs. We would reject the null hypothesis if too many of one type occurred. If we use p to denote the probability of obtaining a plus sign, we can indicate the hypotheses as

$$H_0 : p = 0.50$$
$$H_1 : p \neq 0.50$$

Since tied cases are excluded in the sign test, the data used for the test consist of 35 pluses and 15 minuses. The problem is conceptually the same as one in which a coin has been tossed 50 times, yielding 35 heads and 15 tails, and we wish to test the hypothesis that the coin is fair. The binomial is the theoretically correct distribution. However, we can use the large sample method of Section 8.2 consisting of the normal curve approximation to the binomial. In terms of proportions, the mean and standard deviation of the sampling distribution are

$$\mu_{\bar{p}} = p = 0.50$$

$$\sigma_{\bar{p}} = \sqrt{\frac{pq}{n}} = \sqrt{\frac{(0.50)(0.50)}{50}} = 0.071$$

Assuming that the test is performed at the 5% level of significance ($\alpha = 0.05$), we would reject the null hypothesis if z is less than -1.96 or is greater than 1.96.

Since, in this problem, the observed proportion of pluses is $\bar{p} = 35/50 = 0.70$, then

$$z = \frac{\bar{p} - p}{\sigma_{\bar{p}}} = \frac{0.70 - 0.50}{0.071} = 2.82$$

Hence, we reject the null hypothesis that plus and minus signs are equally likely. Since the plus signs exceeded minus signs in the observed data, our

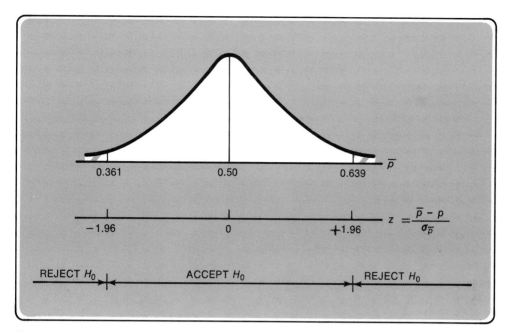

Figure 11-1
Sampling distribution of a proportion for beer tasting problem: $p = 0.50$, $n = 50$. Two-tailed test; $\alpha = 0.05$.

interpretation of the experimental data is that Diller beer is preferred to Wudbeiser, according to the rank scores given by the consumer panel.

We note that the arithmetic could have been carried out in terms of critical limits for $\bar{p}$, rather than for z values. The critical limits for $\bar{p}$ are

$$p + 1.96\sigma_{\bar{p}} = 0.50 + (1.96)(0.071) = 0.639$$
$$p - 1.96\sigma_{\bar{p}} = 0.50 - (1.96)(0.071) = 0.361$$

Since the observed $\bar{p}$ of 0.70 exceeds 0.639, we reach the same conclusion of rejecting H_0. The testing procedure is shown in the usual way in Figure 11-1.

Two points may be made concerning the techniques used in this illustration of the sign test. First, although a two-tailed test was appropriate for the problem, the sign test can also be used in one-tailed test situations. Second, a normal curve approximation to the binomial was used. For small samples, binomial probability calculations and tables should be used.

General comments

As we have seen, the sign test is very simple to apply. In the example, it was not feasible to obtain quantitative data for beer tasting, so rank scores were used. Because of its simplicity, the sign test is sometimes used instead of a standard test even when quantitative data are available. For ease of reference, let us refer to the beer taste rankings for any particular individual as observations from a "matched pair." We might have matched pair observations where the data are, say, weights before and after a diet, grades on a preliminary scholastic aptitude test and on the regular aptitude test, etc. In applying the sign test, only the signs of the differences in the matched pair observations would be used, rather than the actual magnitudes of the differences. Of course, as noted earlier, there would be some loss in the efficiency of the test as a result.

In addition to being simple to apply, the sign test is applicable in a wide variety of situations. As in the example given above, the two samples do not even have to be independent. Indeed, when matched pair observations are used, the elements in the first sample are usually matched as closely as possible with the corresponding elements in the second sample. Furthermore, the sign test may be used in cases of qualitative classifications where it may even be difficult to use a ranking scheme such as that used above. For example, an experimenter may classify subjects after a treatment has been applied as improved (+), worse (−), or the same (0), and then use the sign test.

EXERCISES

1. A panel of dermatologists participated in an experimental test of a new method of treatment for acne. They reported the following results for the new treatment as compared to a standard previous method of treatment:

	NUMBER OF PATIENTS
New treatment better	62
Previous treatment better	28
No difference	10
	100

Use the method of Section 11.2 to test the null hypothesis of no difference in effectiveness of the two methods of treatment. Use a two-tailed test at the 0.01 level of significance.

2. In a certain production process, the specified standard weight for a critical component was 14.5 pounds. In a production run of 200 of these components, 110 weighed in excess of 14.5 pounds, and 90 weighed less. Replacing sample values in excess of specifications by a plus sign and those less than specifications by a minus sign, test the null hypothesis that p, the proportion of pluses, equals 0.50. Use a two-tailed test at the 0.05 level of significance.

3. A marketing executive wanted to determine his salesmen's opinions on a proposed promotional program for a certain product as compared to a current program. He asked a randomly selected sample of 35 salesmen to assign ratings from zero to 10 for the effectiveness of each program, with 10 being the highest possible score. The salesmen's ratings are given below. For example, the first salesman assigned a 9 to the new program and a 7 to the old, etc.

New program: 9, 7, 10, 6, 9, 8, 5, 9, 7, 8, 6, 4, 10, 8, 9, 7
8, 6, 9, 8, 9, 7, 6, 8, 9, 7, 6, 5, 8, 7, 8, 6, 4
8, 8

Old program: 7, 8, 8, 7, 9, 7, 4, 5, 8, 6, 3, 7, 8, 5, 7, 4, 8
7, 6, 9, 6, 7, 8, 7, 8, 8, 5, 6, 7, 9, 6, 5, 6, 7
9

Match the scores, assigning a plus sign when the rating for the new program exceeds that of the old, etc. Since the executive wished to place the burden of proof on the new program, he decided to use a one-tailed test of the following type:

$$H_0:p \leq 0.50$$
$$H_1:p > 0.50$$

where p is equal to the probability of obtaining a plus sign. Carry out the appropriate sign test using a 0.05 level of significance.

4. A research agency was interested in determining whether television viewers felt that there had been a decrease in the amount of violence depicted on television programs during a one-year period as compared to the preceding year. In a simple random sample of 400 viewers, 178 felt there had been a decrease, 120 felt there had been an increase, and 102 thought that the amount of depicted violence was about the same. Using a plus sign to represent a perceived decrease, a minus sign for a perceived increase, and a zero for "the same," apply a one-tailed sign test at the 0.01 level of significance. Use the null hypothesis $H_0:p \leq 0.50$, where p is equal to the probability of a perceived decrease in violence.

MANN-WHITNEY U TEST (RANK SUM TEST)   11.3

Another very useful nonparametric technique involving a comparison of data from two samples is the Mann-Whitney U test, often referred to as the rank sum test. This procedure is used to test whether two independent samples have been drawn from the same population, or equivalently from two different populations having the same mean. Hence, the rank sum test may be viewed as a substitute for the parametric t test or the corresponding large sample normal curve test for the difference between two means. Since the rank sum test explicitly takes into account the rankings of measurements in each sample, in that sense it uses more information than does the sign test.

As an illustration of the use of the rank sum test, we will consider the data shown in Table 11-2. These data represent the grades obtained on an aptitude test given by a large corporation to management training program applicants. The samples consist of graduates of two different universities, referred to as H and W.

Method

The first step in the rank sum test is to merge the two samples, arraying the individual scores in rank order as shown in Table 11-3. The test is then carried out in terms of the sum of the ranks of the observations in either of the two samples. The following symbolism is used:

n_1 = number of observations in sample number one
n_2 = number of observations in sample number two
R_1 = sum of the ranks of the items in sample number one
R_2 = sum of the ranks of the items in sample number two

Treating the data for H University as sample number one, R_1 is equal to the sum of ranks 1, 2, 3, 4, 5, 6, 7, 8, 10, 11, 12, 15, 23, 24, and 27 or 158. Correspondingly R_2 is equal to 307.

TABLE 11-2
Grades on an Aptitude Test Obtained
by Graduates of Two Universities.

H UNIVERSITY	W UNIVERSITY
50	70
51	76
53	77
56	80
57•	81
63	82
64	83
65	86
71	87
73	88
74	92
78	93
89	96
90	98
95	99

If the null hypothesis that the two samples were drawn from the same population were true, we would expect the totals of the ranks (or equivalently, the mean ranks) of the two samples to be about the same. In order to carry out the test, a new statistic, U, is calculated. This test statistic, which depends only on the number of items in the samples and the total of the ranks in one of the samples, is defined as follows:

(11.1)
$$U = n_1 n_2 + \frac{n_1(n_1 + 1)}{2} - R_1$$

The statistic U provides a measurement of the difference between the ranked observations of the two samples and yields evidence as to the difference between the two population distributions. Very large or very small values of U constitute evidence of the separation of the ordered observations of the two samples. Under the above stated null hypothesis, it can be shown that the sampling distribution of U has a mean equal to

(11.2)
$$\mu_U = \frac{n_1 n_2}{2}$$

and a standard deviation of

TABLE 11-3

Array of Grades on an Aptitude Test Obtained by Graduates of Two Universities.

RANK	GRADE	UNIVERSITY
1	50	H
2	51	H
3	53	H
4	56	H
5	57	H
6	63	H
7	64	H
8	65	H
9	70	W
10	71	H
11	73	H
12	74	H
13	76	W
14	77	W
15	78	H
16	80	W
17	81	W
18	82	W
19	83	W
20	86	W
21	87	W
22	88	W
23	89	H
24	90	H
25	92	W
26	93	W
27	95	H
28	96	W
29	98	W
30	99	W

(11.3)
$$\sigma_U = \sqrt{\frac{n_1 n_2 (n_1 + n_2 + 1)}{12}}$$

Furthermore, it can be shown that the sampling distribution approaches normality very rapidly and may be considered approximately normal when both n_1 and n_2 are in excess of about ten items.

In terms of the data in the present problem, substituting into equations (11.1), (11.2), and (11.3), we have

$$U = (15)(15) + \frac{(15)(15 + 1)}{2} - 158 = 187$$

$$\mu_U = \frac{(15)(15)}{2} = 112.5$$

$$\sigma_U = \sqrt{\frac{(15)(15)(15 + 15 + 1)}{12}} = 24.1$$

MANN-WHITNEY U TEST (RANK SUM TEST)   **283**

Hence, proceeding in the usual manner, we calculate the standardized normal variate

$$z = \frac{U - \mu_U}{\sigma_U} = \frac{187 - 112.5}{24.1} = 3.09$$

Thus, if the test were carried out at, say, the 5% or 1% level of significance with critical absolute values for z of 1.96 and 2.58, respectively, we would reject the null hypothesis that the samples were drawn from the same populations. Referring to the original data in Tables 11-2 and 11-3, we can observe that H University has more lower grades and hence smaller ranks (1, 2, etc.), whereas W University has a heavier concentration of higher grades and larger ranks. Therefore, in terms of the original data on aptitude test scores, we conclude that the population of applicants from W University has a higher average aptitude test score than does the corresponding population from H University.

General comments

The above test was carried out in terms of the sum of the ranks for sample number one. That is, the U statistic was defined in terms of R_1. It could similarly have been defined in terms of R_2 as

(11.4) $$U = n_1 n_2 + \frac{n_2(n_2 + 1)}{2} - R_2$$

The subsequent test would have yielded the same z value as the one previously calculated, except that the sign would change. Of course, the conclusion is exactly the same.

There were no ties in rankings in the example given. However, if there are such ties, the average rank value is assigned to the tied items. For example, suppose the sixth and seventh grades in the array were identical. Then, a rank of $(6 + 7)/2 = 6.5$ would be assigned to each item. The analogous procedure is used if more than two items are tied. A correction is available for the calculation of σ_U when ties occur, but for large samples, the effect is generally negligible.

As mentioned earlier, the rank sum test may be viewed as a substitute for the t test for the difference between two means. The rank sum test may be particularly useful in this connection when the underlying assumptions of the t test are not met. For example, in the example given, there is much more variability in the H University data than in the W University figures. Furthermore, the t-test assumption of population normality may not be valid in this case.

In the case of parametric tests, the analysis of variance represents a generalization to k samples of the two-sample test for the difference between means. Similarly, there are nonparametric tests analogous to the analysis of variance in the sense that they test the null hypothesis that k independent random samples have been drawn from identical populations. Such tests are described in the texts on nonparametric statistics referred to in the bibliography at the back of the book.

EXERCISES

1. A college track star was interested in comparing his times for running the 100-yard dash during his senior and junior years. His times were recorded under essentially similar noncompetitive situations during the two years. When the times were merged for the two years and ranked, the following results were observed:

   Senior year: 1, 3, 4, 6, 8, 9, 12, 14, 16, 17, 18, 20, 23
   Junior year: 2, 5, 7, 10, 11, 13, 15, 19, 21, 22, 24, 25, 26, 27, 28

   Use the rank sum test for the null hypothesis that there is no difference between the true average times during the senior and junior years. Use $\alpha = 0.05$.

2. A market research director drew simple random samples of 15 salesmen from each of two sales regions of his company in order to compare sales figures. When last year's dollar values of sales made by these salesmen were arrayed for the two regions combined, the following rankings emerged:

   Region A: 1, 2, 4, 7, 8, 10, 12, 13, 14, 17, 21, 24, 26, 27, 28
   Region B: 3, 5, 6, 9, 11, 15, 16, 18, 19, 20, 22, 23, 25, 29, 30

   Use the rank sum test at the 0.01 level of significance to determine whether there is a significant difference in the average level of sales in the two samples.

3. The following data represent the weight losses of twenty-eight different people, fourteen of whom had used one form of diet during a one-week period and fourteen had used the other.

   Diet 1: 10.4, 9.7, 9.6, 9.3, 8.9, 8.7, 8.2, 7.7, 7.5, 6.9, 6.2, 5.8, 5.5, 5.1
   Diet 2: 9.8, 9.5, 8.8, 8.6, 8.4, 8.3, 7.9, 7.8, 7.6, 7.2, 7.1, 6.8, 5.4, 5.3,

   Use the rank sum procedure to test the hypothesis that the two samples were drawn from populations having the same average weight loss. Use $\alpha = 0.01$.

ONE-SAMPLE RUNS TESTS   11.4

As we have seen in Chapters 7 and 8, estimation procedures and parametric tests of hypotheses are predicated on the fact that the observed data have been obtained from random samples. Indeed, in many instances, evidence of nonrandomness can represent an important phenomenon. As an example, in the frontier days of the old Wild West, rather serious consequences were predictable if a card player suspected the randomness of the hands of cards dealt by another player. In many less exotic contexts also, the randomness of selection of sampled items is of considerable import.

Let us consider a rather oversimplified situation. Suppose that in a certain city, there were about 50% whites and 50% blacks on the rolls of persons eligible for jury duty. Further, let us assume that the following sequence represents the order in which the first 48 persons were drawn from the rolls (B = Black, W = White):

BBBBBBBBBBBB WWWWWWWWWWWW BBBBBBBBBBBB
                                    WWWWWWWWWWWW

Would you suspect, on an intuitive basis, that there was a real question concerning the randomness of selection of these persons? Undoubtedly your answer is in the affirmative, but why? Note that there are 24 Bs and 24 Ws. Hence, the observed proportion of blacks (or whites) is 50 percent, which does not differ from the known population proportion. Your suspicions concerning nonrandomness doubtlessly stem from the *order* of the items listed, rather than from their frequency of occurrence. Similarly we would find a perfectly alternating sequence of BWBWBWBW . . . suspect with respect to randomness of order of occurrence. The *theory of runs* has been developed in order to test samples of data for randomness, with emphasis on the order in which these events occur.

A *run* is defined as a sequence of identical occurrences (symbols) that are followed and preceded by different occurrences (symbols) or by none at all. Hence, in the listing of 48 symbols, there are four runs, the first run consisting of the first 12 Bs, the second run of 12Ws, etc. Our intuitive feeling is that this represents too few runs. Analogously, in the perfectly alternating series, BWBWBWBW . . . , we would feel that there are too many runs to have occurred on the basis of chance alone.

Method

We will illustrate the analytical procedure for the runs test in terms of a somewhat less extreme illustration than that considered above. Let us assume that the following 42 symbols represent the successive occurrences of births of males (M) and females (F) in a certain hospital.

MM F M FFF MM FF M F MMM FF M FFF MM FF MM FF

MMM FF MM FF MMM

Using the symbol r to denote the number of runs, we have $r = 21$. The runs, which, of course, may be of differing lengths, have been indicated by separation of sequences and by underlining. As we have noted, if there are too few or too many runs, we have reason to doubt that their occurrences are random. The runs test is based on the idea that if there are n_1 symbols of one type and n_2 symbols of a second type, and r denotes the total number of runs, the sampling distribution of r has a mean

(11.5)
$$\mu_r = \frac{2n_1n_2}{n_1 + n_2} + 1$$

and a standard deviation

(11.6)
$$\sigma_r = \sqrt{\frac{2n_1n_2(2n_1n_2 - n_1 - n_2)}{(n_1 + n_2)^2(n_1 + n_2 - 1)}}$$

If either n_1 or n_2 is larger than 20, the sampling distribution of r is closely approximated by the normal distribution. Hence, we can compute

$$z = \frac{r - \mu_r}{\sigma_r}$$

and proceed with the test in the usual manner.

In the present problem, where there are $n_1 = 22$ Ms, $n_2 = 20$ Fs, and $r = 21$, we have

$$\mu_r = \frac{(2)(22)(20)}{22 + 20} + 1 = 21.95$$

$$\sigma_r = \sqrt{\frac{(2)(22)(20)[(2)(22)(20) - 22 - 20]}{(22 + 20)^2(22 + 20 - 1)}} = 3.19$$

Therefore,

$$z = \frac{21 - 21.95}{3.19} = -0.30$$

Testing at a significance level of (say) 5%, where a critical absolute value of 1.96 for z is required for rejection of the null hypothesis, we find that the randomness hypothesis cannot be rejected. In other words, the number of runs in this case is neither small enough nor large enough to lead us to conclude that the sequence of male and female births is nonrandom.

General comments

The runs test has many applications, including cases where the sequential data are numerical rather than in the form of attributes such as the symbols of the above illustrations. Thus, for example, runs tests could be applied to sequences of random numbers, such as those in Table 6-1 on page 121. Such tests might be applied, say, to sequences of random numbers generated on computers. One form of the test might be in terms of runs above and below the median. For the digits 0, 1, 2, 3, 4, 5, 6, 7, 8, and 9, the median is 4.5. Hence, runs could be determined for digits that fall above and below the median. Also the test could be applied in terms of runs of odd-numbered digits versus even-numbered ones. Another alternative is to group the numbers into pairs of digits where the possible occurrences are 00, 01, . . . , 99. Here the median is 49.5, and similar runs tests could be applied to those suggested for the one digit case. It is clear that with a bit of imagination an analyst can devise many useful and easily applied versions of the runs test.

EXERCISES

1. In the table of random numbers given on page 121, consider the 50 digits on the first line, excluding the line number 19300. Label the even digits a and the odd digits b. Carry out a runs test at both the 0.05 and 0.01 levels of significance.

2. Consider the same 50 digits as in Exercise 1. Now label the digits a or b depending upon whether they are above or below the theoretical median 4.5. Carry out a runs test at both the 0.05 and 0.01 levels of significance.

3. Toss a coin 40 times and record heads and tails as H and T, respectively. Test these data for randomness at the 0.05 level of significance.

4. The following figures represent the monthly numbers of on-the-job accidents occurring in a certain factory over a 24-month period:

2,2,1,3,4,4,1,3,1,2,4,4
2,3,3,2,5,6,4,5,4,6,5,5

Determine the median of this set of 24 figures. Label the numbers above and below the median a and b, respectively. Perform a runs test at the 0.01 level of significance on the series of as and bs. This type of test of runs above and below the median is particularly useful for determining the existence of trend patterns in data. If there is a trend, as will tend to appear in the early part of the series, and bs in the later part, or vice versa.

11.5 RANK CORRELATION

Nonparametric procedures can be useful in correlation analysis where the basic data are not available in the form of numerical magnitudes but where rankings can be assigned. If two variables of interest can be ranked in separate ordered series, a *rank correlation coefficient* can be computed, which is a measure of the degree of correlation that exists between the two sets of ranks. We will illustrate the method in terms of a simple random sample of individuals for whom such rankings have been established for two variables referring to ability in two different sports activities.

Method

For illustrative purposes, we will consider two extreme cases, the first representing perfect *direct* correlation between two series, the second perfect *inverse* correlation. Table 11-4 displays data on the rankings of a simple random sample of ten individuals according to playing abilities in baseball and tennis. Clearly, this represents a case in which it would be extremely difficult, if not impossible, to obtain precise quantitative measures of these abilities, but where rankings may be feasible. In rank correlation analysis, the rankings may be assigned in order from high to low, with 1 representing the highest rating, 2 next highest, etc. or 1 may represent the lowest rank, 2 the next to lowest, etc. The computed rank correlation coefficient will be the same regardless of the rank ordering used. Let us assume in this case that 1 represents the highest or best rank, 2 second highest, etc.

The rank correlation coefficient (also referred to as the Spearman rank correlation coefficient) can be derived mathematically from one of the formulas for r, the sample correlation coefficient discussed in Chapter 10 where ranks are used for the observations of X and Y. We will use the symbol r_r to denote the rank correlation coefficient. It is computed by the following formula:

(11.7)
$$r_r = 1 - \frac{6\Sigma d^2}{n(n^2 - 1)}$$

TABLE 11-4

Rank Correlation of Baseball Playing Ability with Tennis Playing Ability (Perfect Correlation Case).

INDIVIDUAL	RANK IN BASEBALL ABILITY X	RANK IN TENNIS ABILITY Y	DIFFERENCE IN RANKS $d = X - Y$	$d^2 = (X - Y)^2$
A	1	1	0	0
B	2	2	0	0
C	3	3	0	0
D	4	4	0	0
E	5	5	0	0
F	6	6	0	0
G	7	7	0	0
H	8	8	0	0
I	9	9	0	0
J	10	10	0	0
TOTAL				0

$$r_r = 1 - \frac{6\Sigma d^2}{n(n^2 - 1)} = 1 - \frac{6(0)}{10(10^2 - 1)} = 1$$

where d = difference between the ranks for the paired observations

n = number of paired observations

The calculations of the rank correlation coefficients for the two extreme cases mentioned earlier are shown in Tables 11-4 and 11-5. In Table 11-4, there is perfect direct correlation in the rankings. That is, the individual who

TABLE 11-5

Rank Correlation of Baseball Playing Ability with Tennis Playing Ability (Zero Correlation Case).

INDIVIDUAL	RANK IN BASEBALL ABILITY X	RANK IN TENNIS ABILITY Y	DIFFERENCE IN RANKS $d = X - Y$	$d^2 = (X - Y)^2$
A	1	10	−9	81
B	2	9	−7	49
C	3	8	−5	25
D	4	7	−3	9
E	5	6	−1	1
F	6	5	1	1
G	7	4	3	9
H	8	3	5	25
I	9	2	7	49
J	10	1	9	81
TOTAL				330

$$r_r = 1 - \frac{6\Sigma d^2}{n(n^2 - 1)} = 1 - \frac{6(330)}{10(10^2 - 1)} = -1$$

ranks highest in baseball playing ability is also best in tennis, etc. On the other hand, in Table 11-5, there is perfect inverse correlation. That is, the individual who ranks highest in baseball playing ability is worst in tennis, etc. It may be noted from the calculations shown in the two tables that in the case of perfect direct correlation between the ranks, $r_r = 1$; in the case of perfect inverse correlation, $r_r = -1$. This is not surprising, because the rank correlation coefficient is derived mathematically from the sample correlation coefficient, r. Hence, the range of possible values of these coefficients is the same. An r_r value of zero would analogously indicate no correlation between the ranks. Tied ranks are handled in the calculations by averaging in the usual way.

The significance of the rank correlation may be tested in the same way as the sample correlation coefficient, r. That is, we compute the statistic

(11.8)
$$t = \frac{r_r}{\sqrt{(1 - r_r^2)/(n - 2)}}$$

which has a t distribution for $n - 2$ degrees of freedom. Hence, for example, suppose in an example such as the one above, r_r had been computed to be 0.90. Then, substituting into (11.8) yields

$$t = \frac{0.90}{\sqrt{(1 - 0.81)/(10 - 2)}} = 5.84$$

Let us assume we are using a two-tailed test of the null hypothesis of zero correlation in the ranked data of the population. Then referring to Table A-6, we find critical t values of 2.306 and 3.355 at the 5% and 1% levels of significance, respectively. Hence, we would reject the hypothesis of no rank correlation at both levels and we would conclude that there is a positive linear relationship between the rankings in baseball playing ability and tennis playing ability.

Rank correlation can be useful in cases where there are one or more extreme observations in the original numerical valued data. Consider, for example, the scatter diagram of Figure 11-2. Assume that the plotted points represent numerical values of X and Y. The situation is one in which all of the points, except for one extreme observation, exhibit a strong linear inverse relationship. However, in the case of the extreme observation, a large value of Y is associated with a large value of X. The effect of this extreme observation may be so great as to cause the regression line to have a positive slope, and, therefore, the correlation coefficient, r, to be positive. It may be noted that the regression line is not a good description of the relationship that exists among all points but one, nor would it provide a good estimation of the Y value for the extreme observation.

How should this problem be handled? If it can be demonstrated that the extreme observation really represents an observation from a conceptually different statistical universe than the other points, it may be eliminated from the analysis. However, if no such demonstration can be made, the observation must be included. In such a case, if ranks are substituted for the numerical

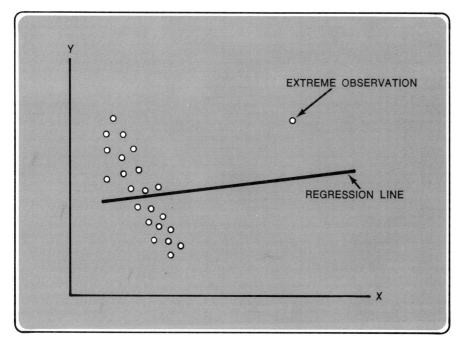

Figure 11-2
Scatter diagram with one extreme observation.

values of X and Y, the effect of the extreme observation will be considerably reduced.

EXERCISES

1. The following were the rankings of a league of 10 baseball teams in the pre-season and regular season competitions.

TEAM	PRE-SEASON RANKING	REGULAR SEASON RANKING
A	1	4
B	2	2
C	3	8
D	4	6
E	5	5
F	6	3
G	7	1
H	8	7
I	9	10
J	10	9

Calculate the rank correlation coefficient for the above set of data.

2. The data given below represent the rankings of a simple random sample of the seniors in a high school on college entrance verbal and mathematics aptitude tests:

STUDENT	VERBAL TEST	MATHEMATICS TEST
A	7	5
B	8	11
C	4	2
D	3	1
E	12	9
F	11	10
G	9	8
H	1	3
I	2	4
J	5	6
K	6	7
L	10	12

Calculate the rank correlation coefficient for the above set of data.

3. Calculate the rank correlation coefficient for the data in Exercise 8 on page 265. Assign average rank values to tied items. Test the null hypothesis of no correlation at the 0.01 level of significance.

11.6   GOODNESS OF FIT TESTS

In Chapter 9, we considered the use of chi-square tests for deciding whether the hypothesis of independence between classificatory variables is tenable. In this section, we discuss another important use of chi-square tests, that of testing whether a particular probability distribution is appropriate to use, based on a sample of observations.

One of the major problems in the application of the theory of probability, statistics, and mathematical models in general is that the real world phenomena to which they are applied usually depart somewhat from the assumptions embodied in the theory or models. For example, let us consider use of the binomial probability distribution in a particular problem. As indicated in Section 5.3, two of the assumptions involved in the derivation of the binomial distribution are

1. The probability of a success, p, remains constant from trial to trial.
2. The trials are independent.

Let us consider whether these assumptions are met in the following problem.

A firm bills its accounts on a 2% discount basis for payment within ten days and full amount due for payment after ten days. In the past, 40% of all invoices have been paid within ten days. In a particular week, the firm sends out 20 invoices. Is the binomial distribution appropriate for computing the probabilities that 0, 1, 2, . . . , 20 firms will take the discount for payment within ten days?

Considering the possible use of the binomial distribution, we can let $p = 0.40$ represent the probability that a firm will take the discount and $n = 20$ firms the number of trials. Does it seem logical to assume constancy of p, that is, the probability of taking the discount is 0.40 for each firm? Past relative frequency data on the taking of discounts for *each* firm could be brought to bear on the answer to this question. In most practical situations of this sort we would probably find that the practices of individual firms vary widely, with some firms virtually always taking discounts, some firms virtually never taking discounts, and most firms falling somewhere between these two extremes.

Does the assumption of independence seem tenable in this problem; that is, does it seem logical that whether one firm takes the discount is independent of whether another firm takes the discount? Probably not, since general monetary conditions doubtlessly affect many of the firms in a similar way. For example, when money is "tight" and it is difficult for many firms to acquire adequate amounts of working capital, the fact that one firm does not take the discount is related to rather than *independent* of whether other firms have taken it. Also, there may be traditional practices in certain industries concerning the taking or not taking of discounts and other factors which would interfere with the independence assumption.

How great a departure from the assumptions underlying a probability distribution, or more generally, from the assumptions embodied in any theory or mathematical model, can be tolerated before we should conclude the distribution, theory, or model is no longer applicable? This is a very complex question and not one that can be readily answered by any simple universally applicable rule. The purpose of chi-square "goodness of fit" tests is to provide one type of answer to the preceding question by comparing *observed frequencies* with *theoretical* or *expected frequencies* derived under specified probability distributions or hypotheses.

The sequence of steps in performing goodness of fit tests is very similar to previously discussed hypothesis testing procedures. The following are the steps in testing for goodness of fit:

1. A null and alternative hypothesis are established, and a significance level is selected for rejection of the null hypothesis.

2. A random sample of observations is drawn from a relevant statistical population.

3. A set of expected or theoretical frequencies is derived under the assumption that the null hypothesis is true. This generally takes the form of assuming that a particular probability distribution is applicable to the statistical population under consideration.

4. The observed frequencies are compared to the expected, or theoretical, frequencies.

5. If the aggregate discrepancy between the observed and theoretical frequencies is too great to attribute to chance fluctuations at the selected significance level, the null hypothesis is rejected.

We will illustrate goodness of fit tests and discuss some of the underlying theory for an example involving a uniform probability distribution. A discrete random variable is said to have a uniform distribution if equal probabilities are assigned to all of the possible values of the random variable.

Suppose a consumer research firm wished to determine whether or not there was a real preference in taste by coffee drinkers in a certain metropolitan area among five brands of coffee. The firm took a simple random sample of 1000 coffee consumers in the area and conducted the following experiment. Each consumer was given five cups of coffee, one of each brand A, B, C, D, and E without identification of the individual brands. The cups were presented to each consumer in a random order determined by sequential selection from five paper slips, each containing one of the letters A, B, C, D, and E. In Table 11-6 are shown the numbers of coffee consumers who stated that they liked the indicated brands best.

TABLE 11-6
Number of Coffee Consumers in a
Certain Metropolitan Area Who Most
Preferred the Specified Brand of Coffee.

BRAND PREFERENCE	NUMBER OF CONSUMERS
A	210
B	312
C	170
D	85
E	223
	1000

Denoting the true proportions of preference for each brand as p_A, p_B, p_C, p_D, and p_E, we can state the null and alternative hypotheses as follows:

$$H_0: p_A = p_B = p_C = p_D = p_E = 0.20$$

H_1: The p's are not all equal

That is, if in the population from which the sample was drawn there were no differences in preference among the five brands, 20% of coffee drinkers would prefer each brand. An equivalent way of stating these hypotheses is

H_0: The probability distribution is uniform

H_1: The probability distribution is not uniform

In other words, the question we are raising is, "Should we consider the sample of 1000 coffee drinkers to be a random sample from a population in which the proportions who prefer each of the five brands are equal?" Of course, this hypothesis is only one of many that could conceivably be formulated. One of the strengths of the goodness of fit test discussed in this section is that it permits a variety of different hypotheses to be raised and tested.

If the null hypothesis of no difference in preference were true, the *expected* or *theoretical* number of the 1000 coffee drinkers in the sample who would prefer each brand would be $0.20 \times 1000 = 200$. Hence, the expected frequency that corresponds to each of the observed frequencies in Table 11-6 is 200. We can now compare the set of observed frequencies with the set of theoretical frequencies derived under the assumption that the null hypothesis is true. The test statistic that is computed to make this comparison is the same chi-square statistic used in tests of independence. The chi-square statistic, (χ^2), given earlier in Equation (9.2) is shown again here for convenience.

(9.2)
$$\chi^2 = \sum \frac{(f_0 - f_t)^2}{f_t}$$

where $f_0 =$ an observed frequency

$f_t =$ a theoretical or expected frequency

We recall that if the observed frequencies are exactly the same as the corresponding theoretical frequencies, the computed value of χ^2 is zero. As the discrepancies between the observed and the theoretical frequencies increase, the computed value of χ^2 increases. We wish to determine whether the computed value of χ^2 is so large that we will be required to reject the null hypothesis. The calculation of χ^2 for the present problem is shown in Table 11-7. We now turn to a discussion of the rules for determining the number of degrees of freedom involved in a chi-square goodness of fit test.

TABLE 11-7
Calculation of Chi-Square Statistic for the Coffee Tasting Problem

BRAND PREFERENCE	(1) OBSERVED FREQUENCY f_0	(2) THEORETICAL (EXPECTED) FREQUENCY f_t	(3) $(f_0 - f_t)$	(4) $(f_0 - f_t)^2$	(5) COLUMN (4) COLUMN (2) $\frac{(f_0 - f_t)^2}{f_t}$
A	210	200	10	100	0.5
B	312	200	112	12,544	62.7
C	170	200	−30	900	4.5
D	85	200	−115	13,225	66.1
E	223	200	23	529	2.6
TOTAL	1000	1000	0		136.4

$$\chi^2 = \sum \frac{(f_0 - f_t)^2}{f_t} = 136.4$$

Number of degrees of freedom

As discussed earlier, a chi-square goodness of fit test involves a comparison of a set of observed frequencies denoted f_0, with a set of theoretical frequencies denoted f_t. Let the symbol k equal the number of classes for which these comparisons are made. For example, in the coffee tasting problem $k = 5$ because these are five classes for which we computed relative deviations of the form $(f_0 - f_t)^2/f_t$. To determine the number of degrees of freedom, we must reduce *k* by 1 for each restriction imposed. In the coffee tasting example, the number of degrees of freedom is equal to $df = k - 1 = 5 - 1 = 4$. The rationale for this computation follows.

In the calculations shown in Table 11-7, there are five classes for which f_0 and f_t values are to be compared. Hence, we start with $k = 5$ degrees of freedom. However, we have forced the total of the theoretical frequencies, Σf_t, to be equal to the total of the observed frequencies, Σf_0; that is, 1000. Therefore, we have reduced the number of degrees of freedom by 1 and there are now only 4 degrees of freedom. That is, once the total of the theoretical frequencies is fixed, only four of the f_t values may be freely assigned to the classes. When these four have been assigned, the fifth class is immediately determined because the theoretical frequencies must total 1000.

If any further restrictions are imposed in the calculation of the theoretical frequencies, the number of degrees of freedom is reduced by one for each such restriction. Hence, for example, if a sample statistic, such as the sample mean $\bar{x}$, is used as an estimate of an unknown population parameter, such as the population mean, μ, there would be a reduction of one degree of freedom. In summary, we can state the following rules for determining *df*, the number of degrees of freedom is a chi-square test in which *k* classes of observed and theoretical frequencies are compared.

1. If the only restriction is $\Sigma f_t = \Sigma f_0$, the number of degrees of freedom is $df = k - 1$.

2. If in addition to the above restriction *m* parameters are replaced by sample estimates, the number of degrees of freedom is $df = k - 1 - m$.

Decision procedure

We can now return to the coffee tasting example to perform the goodness of fit test. Let us assume we wish to test the null hypothesis at the 0.05 level of significance. Since the number of degrees of freedom is 4, we find the critical value of χ^2, which we denote as $\chi^2_{.05}$, to be 9.488 (Table A-7, Appendix A). This means that if the null hypothesis is true, the probability of observing a χ^2 value greater than 9.488 is 0.05. Specifically in terms of the problem, this means that if there were no difference in preference among brands, an aggregate discrepancy between the observed and theoretical frequencies larger than a χ^2 value of 9.488 would occur only five times in 100. We can state the decision rule for this problem in which $\chi^2_{.05} = 9.488$ as follows:

1. If $\chi^2 > 9.488$, reject H_0
2. If $\chi^2 \leq 9.488$, accept H_0

Since the computed χ^2 value in this problem is 136.4 and is thus very much larger than the critical $\chi^2_{.05}$ value of 9.488, we reject the null hypothesis. A reference to Table A-7 of Appendix A indicates that the computed χ^2 value of 136.4 is also far in excess of the critical value at the 0.01 level, namely, $\chi^2_{.01} = 13.277$. Hence, although in this problem we tested the null hypothesis H_0 at the 0.05 level of significance, we would have rejected H_0 at the 0.01 level as well.

Therefore, our conclusion is that there exist real differences in consumer preference among the brands of coffee involved in the experiment. In statistical terms, we cannot consider the 1000 coffee drinkers in the experiment to be a simple random sample from a population having equal proportions who prefer each of the five brands. In terms of goodness of fit, we reject the null hypothesis that the probability distribution is uniform. Hence we conclude that the uniform distribution is decidedly not a "good fit" to the sample data.

EXERCISES

1. A consumer research firm wished to determine whether there was a real difference in preference in taste by iced tea drinkers in a certain metropolitan area among five brands of iced tea. The firm took a simple random sample of 100 iced tea consumers in the area and constructed the following experiment. Five glasses of iced tea were given to each consumer. The glasses were marked A, B, C, D, and E and contained one of each of the individual brands. The cups were presented to each consumer in a random order determined by sequential selection from five paper slips, each containing one of the letters A, B, C, D, and E. In the table are shown the numbers of iced tea consumers who stated that they most preferred the indicated brands.

BRAND PREFERRED	NUMBER OF CONSUMERS
A	27
B	16
C	22
D	18
E	17

Using a chi-square test determines whether the null hypothesis

H_0: the probability distribution is uniform

should be rejected. Test using both a 0.05 and a 0.01 level of significance.

2. A set of five coins was tossed 1000 times. The number of times that 0, 1, 2, 3, 4, and 5 heads were obtained is shown in the table.

NUMBER OF HEADS	NUMBER OF TOSSES
0	36
1	138
2	348
3	287
4	165
5	26
TOTAL	1000

Determine whether the binomial distribution is a good fit to these data. Assume the probability of a head is 0.5 and use a 0.05 level of significance. (Hint: Determine the binomial probabilities of 0, 1, 2, 3, 4, and 5 heads for $p = 0.5$ and $n = 5$.) Multiply these probabilities by 1000 to obtain the expected frequencies. The number of degrees of freedom is $df = k - 1 = 6 - 1 = 5$ because the only restriction is $\Sigma f_t = \Sigma f_0$.

12 Decision Making Under Uncertainty

12.1 INTRODUCTION

In recent years there has been a reorientation of classical or traditional statistical inference, with the modern emphasis being placed on the problem of decision making under conditions of uncertainty. This modern formulation has come to be known as *statistical decision theory* or *Bayesian decision theory*. The latter term is often used as a means of emphasizing the role of Bayes' theorem in this type of decision analysis. The two ways of referring to modern decision analysis have come to be used interchangeably and will be so used henceforth in this book. Statistical decision theory has developed into an important model for the making of rational selections among alternative courses of action when information is incomplete and uncertain. It is a prescriptive theory rather than a descriptive one. That is, it presents the principles and methods for making the best decisions under specified conditions, but it does not purport to present a description of how actual decisions are made in the real world.

12.2 STRUCTURE OF THE DECISION-MAKING PROBLEM

Managerial decision making has increased in complexity as the economy of the United States and the business units within it have grown larger and more intricate. However, Bayesian decision theory is based on the assumption that regardless of the type of decision—whether it involves long-range or short-range consequences; whether it is in finance, production, or marketing, or some other area; whether it is at a relatively high or low level of managerial responsibility—certain common characteristics of the decision problem can be discerned. These characteristics constitute the formal description of the problem and provide the structure for a solution. The decision problem under study may be represented by a model in terms of the following elements:

1. The decision maker. He is charged with the responsibility for making the decision. The decision maker is viewed as an entity and may be a single individual, a corporation, a government agency, etc.

2. Alternative courses of action. The decision involves a selection among two or more alternative courses of action, referred to simply as "acts." The problem is to choose the best of these alternative acts. Sometimes the decision maker's problem is to choose the best of alternative "strategies," where each strategy is a decision rule indicating which act should be taken upon observation of a specific type of experimental or sample information.

3. Events. Occurrences that affect the achievement of the objectives. These are viewed as lying outside the control of the decision maker who does not know for certain which event will occur. The events constitute a mutually exclusive and complete set of outcomes. Hence, one and only one of the specified events can occur. Events are also synonymously referred to as "states of nature," "states of the world," or simply "outcomes."

4. Payoff. A measure of net benefit received by the decision maker. These payoffs are summarized in a so-called "payoff table" or "payoff matrix," which displays the consequences of the action selected and the event that occurs.

5. Uncertainty. The indefiniteness concerning which events or states of nature will occur. This uncertainty is indicated in terms of probabilities assigned to events. One of the distinguishing characteristics of Bayesian decision theory is the use of personalistic or subjective probabilities as well as other types of probabilities for these assignments.

The payoff table, expressed symbolically in general terms, is given in Table 12-1. It is assumed that there are n alternative acts, denoted $A_1, A_2, \ldots, A_n$. These different possible courses of action are listed as column headings in the table. There are m possible events or states of nature denoted $\theta_1, \theta_2, \ldots, \theta_m$. The payoffs resulting from each act and event combination are designated by the symbol u with appropriate subscripts. The letter u has been

TABLE 12-1
The Payoff Table.

EVENTS	A_1	A_2	$\cdots$	A_n
		ACTS		
θ_1	u_{11}	u_{12}	$\cdots$	u_{1n}
θ_2	u_{21}	u_{22}	$\cdots$	u_{2n}
.	.	.	$\cdots$	.
.	.	.	$\cdots$	.
.	.	.	$\cdots$	.
θ_m	u_{m1}	u_{m2}	$\cdots$	u_{mn}

used because it is the first letter of the word "utility." The net benefit or payoff of selecting an act and having a state of nature occur can be treated most generally in terms of the utility of this consequence to the decision maker. How these utilities are arrived at is a technical matter that is discussed subsequently in this chapter. In summary, the utility of selecting act A_1 and having event θ_1 occur is denoted u_{11}; the utility of selecting act A_2 and having event θ_1 occur is u_{12}, and so forth. It may be noted that the first subscript in these utilities indicates the event that prevails and the second subscript denotes the act chosen. A convenient general notation is the symbol u_{ij}, which denotes the utility of selecting act A_j if subsequently event θ_i occurs. The rows of a table (or matrix) are commonly denoted by the letter i, where i can take on values $1, 2, \ldots, m$ and the columns by j, where j can take on values $1, 2, \ldots, n$.

If the event that will occur were known with certainty beforehand by the decision maker, for example, θ_3, then he could simply look along row θ_3 in the payoff table and select that act which yields the greatest payoff. However, in the real world, the context of decision problems is such that since the states of nature lie beyond the control of the decision maker, he ordinarily does not know with certainty which specific event will occur. The choice of the best course of action in the face of this uncertainty is the crux of the decision maker's problem.

12.3   AN ILLUSTRATIVE EXAMPLE

In order to illustrate the ideas discussed in the preceding section, we will take a simplified business decision problem. (This problem will also be continued in later sections to exemplify other principles of decision analysis.) A man has invented and patented a new device. A bank is willing to lend him the money to manufacture the device himself. After some preliminary investigation, it is decided that the next five years is a suitable planning period for the comparison of payoffs from this invention. According to the inventor's analysis, if sales are strong, he anticipates profits of $800,000 over the next five years; if sales are average he expects to make $200,000; and if weak to lose $50,000. A company, Nationwide Enterprises, Inc., has offered to purchase the patent rights from him. Based on the royalty arrangement offered to him, the inventor estimates that if he sells the patent rights and if sales are strong he can anticipate a net profit of $400,000; if sales are average, $70,000; and if weak, $10,000. The payoff table for the inventor's problem is given in Table 12-2.

In this problem, the alternative acts, denoted A_1 and A_2, respectively, are for the inventor to manufacture the device himself or to sell the patent rights. The events or states of nature denoted θ_1, θ_2, and θ_3, respectively, are strong sales, average sales, and weak sales for the five-year planning period. The payoffs are in terms of the net profits which would accrue to the inventor under each act–event combination. In order to keep the numbers rather simple in this problem, the payoffs have been stated in units of $10,000; hence a net

TABLE 12-2

Payoff Table for the Inventor's Problem
(in units of $10,000 profit).

EVENTS	A_1 MANUFACTURE DEVICE HIMSELF	A_2 SELL PATENT RIGHTS
θ_1: Strong Sales	$80	$40
θ_2: Average Sales	20	7
θ_3: Weak Sales	−5	1

profit of $800,000 has been recorded as $80, a net loss of $50,000 has been entered as −$5, etc.[1]

The types of events or states of nature used in the inventor's problem are, of course, very simplified. Generally, an unlimited number of possible events could occur in the future relating to such matters as the customers, technological change, competitors, etc., which lie beyond the decision maker's control. These may all be considered to be states of nature that affect the potential payoffs of the alternative decisions to be made. However, in order to cut our way through the maze of complexities involved, and to construct a manageable framework of analysis for the problem, we can think of the variable "demand" as the resultant of all of these other underlying factors.

In the inventor's problem, three different levels of demand have been distinguished — namely, strong, average, and weak. It is helpful in this regard to think of "demand" as a variable. In the present problem, demand is a discrete random variable that can take on three possible values. Demand could have been conceived of as a discrete random variable taking on any finite or infinite number of values. For example, it could have been stated in numbers of units demanded or, say, in hundreds of thousands of units demanded. Demand can also be treated as a continuous rather than discrete variable. The conceptual framework of the solution to the decision problem remains the same, but the required mathematics differs somewhat from the case where the events are stated in the form of a discrete variable.

CRITERIA OF CHOICE   12.4

Assuming the inventor in our illustrative problem has carried out the thinking, experiments, data collection, etc., required to construct the payoff matrix (Table 12-2), how should he now compare the alternative acts? Neither act is preferable to the other under all states of nature. For example, if event θ_1

[1] It is good practice in the comparison of economic alternatives to compare the present values of discounted cash flows or, what amounts to the same thing, equivalent annual rates of return. Both of these methods take into account the time value of money; that is, the fact that a dollar received today is worth more than a dollar received in some future period. These are conceptually the types of monetary payoff values that should appear in the payoff table. This point is amply discussed in standard texts dealing with economy studies or investment analysis. To avoid a lengthy tangential discussion, we will not elaborate on the point here.

occurs, that is, if sales are strong, the inventor would be better off to manu-facture the device himself (act A_1), realizing a profit of $800,000, as compared to selling the patent rights (act A_2), which would yield a profit of only $400,000. On the other hand, if event θ_3 occurs, and sales are weak, the preferable course of action would be to sell the patent rights, thereby earning a profit of $10,000, as compared to a loss of $50,000. If the inventor knew with *certainty* which event was going to occur, his decision procedure would be very simple. He would merely have to look along the row represented by that event and select the act which yielded the highest payoff. However, it is the uncertainty with regard to which state of nature will prevail that makes the decision problem an interesting one.

Maximin criterion

A number of different criteria for selecting the best act have been suggested. One of the earliest suggestions, made by the mathematical statistician Abraham Wald,[2] is known as the *maximin* criterion. Under this method, the decision maker is supposed to assume that once he has chosen a course of action, nature might be malevolent and hence might select the state of nature that minimizes the decision maker's payoff. According to Wald, the decision maker should choose the act that maximizes his payoff under this pessimistic as-sumption concerning nature's activity. In other words, Wald suggested that a selection of the "best of the worst" is a reasonable form of protection. By applying this criterion to our illustration, if the decision maker chose act A_1, nature would cause event θ_3 to occur and the payoff would be a loss of $50,000. If the decision maker chose A_2, nature would again cause θ_3 to occur, since that would yield the worst payoff, in this case, a profit of $10,000. Comparing these worst or minimum payoffs, we have

	ACTION	
	A_1	A_2
Minimum Payoffs (in units of $10,000)	−5	1

The decision maker is now supposed to do the best he can in the face of this sort of perverse nature, and select that act which yields the greatest minimum payoff, act A_2. That is, he should sell the patent rights, for which the minimum payoff is $10,000. Thus, the proposed decision procedure is to choose the act that yields the maximum of the minimum payoffs—hence the use of the term "maximin."

Obviously, the maximin is a very pessimistic type of criterion. It is not reasonable to suppose that the usual businessman would or should make his decisions in this way. By following this decision procedure, he would always

[2] Abraham Wald, *Statistical Decision Functions* (New York: John Wiley and Sons, 1950).

be concentrating on the worst things that could happen to him. In most situations, the maximin criterion would freeze the businessman into complete inaction and would imply that he should go out of business entirely. For example, let us consider an inventory stocking problem, where the events are possible levels of demand, the acts are possible stocking levels, that is, the numbers of items to be stocked, and the payoffs are in terms of profits. If zero items are stocked, the payoffs will be zero for every level of demand. For each of the other numbers of items stocked, we can assume that for some levels of demand, losses will occur. Since the worst that can happen if no items are stocked is that no profit will be made, whereas under all other courses of action the possibility of a loss exists, the maximin criterion would require the businessman to carry no stock or, in effect, go out of business. Such a procedure is not necessarily irrational, and it might be consistent with the attitudes toward risk of certain people. However, for the businessman who is willing to take some risks in the pursuit of his objectives, such an arbitrary decision rule would be completely unacceptable. A number of other decision criteria have been suggested by various writers, but, to avoid a lengthy digression, they will not be discussed here.[3]

It seems reasonable to argue that a decision maker should take into account the probabilities of occurrence of the different possible states of nature. As an extreme example, if the state of nature that results in the minimum payoff for a given act has (say) only one chance in a million of occurring, it would seem unwise to concentrate on this possible occurrence. The decision procedures we will focus upon include the probabilities of states of nature as an important part of the problem.

Expected profit under uncertainty

In a realistic decision problem, it would be reasonable to suppose that a decision maker would have some idea of the likelihood of occurrence of the various states of nature and that this knowledge would help him choose a course of action. For example, in our illustrative problem, if the decision maker felt very confident that sales would be strong this would tend to move him toward manufacturing the device himself, since the payoff under that act would exceed that of selling the patent rights. By the same reasoning, if he were very confident that sales would be weak, he would be influenced to sell the patent rights. If there are many possible events and many possible courses of action, the problem becomes complex, and the decision maker clearly needs some orderly method of processing all the relevant information. Such a systematic procedure is provided by the computation of the *expected* monetary value of each course of action, and the selection of that act which yields the highest of these expected values. As we shall see, this procedure yields reasonable results in a wide class of decision problems. Furthermore, we will see how this method can be adjusted for the computation of expected utilities rather

[3] See, for example, Chapter 5 of D. W. Miller and M. K. Starr, *Executive Decisions and Operations Research* (Englewood Cliffs, N.J.: Prentice-Hall, 1960).

than expected monetary values in cases where the maximization of expected monetary values is not an appropriate criterion of choice.

We now return to the inventor's problem to illustrate the calculations for decision making by maximization of the expected monetary value criterion. In this case, the maximization takes the form of selecting that act which yields the largest expected profit. Let us assume that the inventor carries out the following probability assignment procedure. On the basis of extensive investigation of the experience with similar devices in the past, and on the basis of interviews with experts, the inventor comes to the conclusion that the odds are 50-50 that sales will be average; that is, the event we previously designated as θ_2 will occur. Furthermore, he concludes that it is somewhat less likely that sales will be strong (event θ_1) than that they will be weak (event θ_3). On this basis, the inventor assigns the following subjective probability distribution to the events in question:

EVENTS	PROBABILITY
θ_1: Strong sales	0.2
θ_2: Average sales	0.5
θ_3: Weak sales	0.3
	1.0

In order to determine the basis for choice between the inventor's manufacturing the device himself (act A_1) and selling the patent rights (act A_2), we compute the expected profit for each of these courses of action. These calculations are shown in Table 12-3. As indicated in that table, profit is treated as a variable, which takes on different values depending upon which event occurs. We compute its expected value in the usual way, according to Equation (5.3). As noted earlier in Chapter 5, the "expected value of an act" is the weighted average of the payoffs under that act, where the weights are the probabilities of the various events that can occur.

We see from Table 12-3 that the inventor's expected profit if he manufactures the device himself is $245,000, whereas if he sells the patent rights, his expected profit is only $118,000. If he acts on the basis of maximizing his expected profit, the inventor would select A_1; that is, he would manufacture the device himself.

It is useful to have a brief term to refer to the expected benefit of choosing the optimal act under conditions of uncertainty. We shall refer to the expected value of the monetary payoff of the best act as the *expected profit under uncertainty*. Hence, in the foregoing problem, the expected profit under uncertainty is $245,000.

We can summarize the method of calculating the *expected profit under uncertainty* as follows:

1. Calculate the expected profit for each act as the weighted average of the profits under that act, where the weights are the probabilities of the various events that can occur.

TABLE 12-3
Inventor's Expected Profits (in units of $10,000 profit).

ACT A_1: MANUFACTURE DEVICE HIMSELF

EVENTS	PROBABILITY	PROFIT	WEIGHTED PROFIT
θ_1: Strong sales	0.2	$80	$16.0
θ_2: Average sales	0.5	20	10.0
θ_3: Weak sales	0.3	−5	−1.5
	1.0		$24.5

Expected Profit = $24.5 (ten thousands of dollars)
= $245,000

ACT A_2: SELL PATENT RIGHTS

EVENTS	PROBABILITY	PROFIT	WEIGHTED PROFIT
θ_1: Strong sales	0.2	$40	$ 8.0
θ_2: Average sales	0.5	7	3.5
θ_3: Weak sales	0.3	1	0.3
	1.0		$11.8

2. The expected profit under uncertainty is the maximum of the expected profits calculated under Step 1.

We now turn to an analysis of the same problem from another important point of view, that of "opportunity loss." The relationship between the results obtained by the two alternative methods of solution, that is, comparison of expected payoffs and comparison of expected opportunity losses is important in statistical decision theory.

EXERCISES

1. If possible states of nature are: competitor will set his price
   (a) higher,
   (b) the same, or
   (c) lower,
   what is wrong with assessing prior probabilities as 0.6, 0.3, and 0.2, respectively?

2. A new appliance store finds that in its first week of business it sold five major appliances, ten home appliances, and 30 small appliances. Based solely on this past knowledge, what prior probability distribution would you formulate for the type of appliance to be sold?

3. As manager of a plant you must decide to invest in either a cost reduction program or a new advertising campaign. Assume that you know the cost reduction program will increase the profit-to-sales ratio, from the present 10% to 11%. The sales campaign, if successful, is expected to increase the present $2 million of sales by 12%. The probability that the campaign will be successful is 0.8. What would be the better course of action?

4. R.B.A., Inc. is given the opportunity to submit a closed bid to the government to build certain electronic equipment. An examination of similar proposals made in the past

revealed that the average profit per successful bid was $175,000, and that R.B.A., Inc. received the contract (i.e., had the lowest bid) on 10% of its submitted bids. The cost of preparing a bid is, on the average, $10,000. Should R.B.A., Inc. prepare a bid?

5. In Problem 4, suppose R.B.A., Inc. chose to prepare a bid. For this particular proposal, assume the company finds it can submit only the following four bids: $1,600,000, $1,700,000, $1,800,000, or $1,900,000. At $1,600,000, expected profit is $160,000. Each successive bid yields an increase in profit equal to the increase in the bid. From an examination of past accounting records, R.B.A., Inc. assesses the probabilities that the bids will be the lowest ones to be 0.4, 0.3, 0.2, 0.1, respectively. Which bid should be submitted?

6. A retailer must decide how much inventory he should carry, which is dependent upon demand. Since the stock is perishable, a loss occurs when he is overstocked. Because of space limitations, the retailer can stock at most five items. The cost per item is $1 and the selling price is $5. The profit table and probabilities are:

PROBABILITY	NUMBER DEMANDED	NUMBER OF UNITS STOCKED					
		0	1	2	3	4	5
2/20	0	$0	$-1	$-2	$-3	$-4	$-5
3/20	1	0	4	3	2	1	0
5/20	2	0	4	8	7	6	5
5/20	3	0	4	8	12	11	10
4/20	4	0	4	8	12	16	15
1/20	5	0	4	8	12	16	20

What is the optimal stocking level and what is the retailer's expected profit?

7. An operations research team is trying to decide whether to put the predictions of the ten leading investment advice newsletters into an information system it is building. The cost of including the predictions is $4850 a year. It is estimated that in 20 decisions to be made in a year, the added information would result in a new decision only once. However, the change in decision would result on the average in a saving of $75,000. Should the team include the newsletter in its information system?

8. A company has $100,000 available to invest. The company can either build a new plant or put the money in the bank at 4% interest. If business conditions remain good, the company expects to make 10% on its investment in a new plant, but if there is a recession, the investment is expected to return only 2%. What probability must management assign to the occurrence of a recession to make the two investments equally attractive? Assume that the only two possible states for business conditions are "good" and "recession."

Expected opportunity loss

A useful concept in the analysis of decisions under uncertainty is that of "opportunity loss." An opportunity loss is the loss incurred because of failure to take the best possible action. Opportunity losses are calculated separately for each event that might occur. Given the occurrence of a specific event, we can determine the best possible act. For a given event, the opportunity loss of an act is the difference between the payoff of that act and the payoff for the best act that could have been selected. Thus, for example, in the inventor's problem, if event θ_1 (strong sales) occurs, the best act is A_1 for which

TABLE 12-4
Payoff Table and Opportunity Loss Table for the Inventor's Problem
(in units of $10,000).

EVENTS	PAYOFF TABLE ACTS		OPPORTUNITY LOSS TABLE ACTS	
	A_1	A_2	A_1	A_2
θ_1: Strong sales	$80*	$40	$0	$40
θ_2: Average sales	20*	7	0	13
θ_3: Weak sales	−5	1*	6	0

the payoff is $80 (in units of ten thousand dollars). The opportunity loss of that act is $80 − $80 = $0. The payoff for Act A_2 is $40. The opportunity loss of Act A_2 is the amount by which the payoff of the best act, $80, exceeds the $40 payoff of Act A_2. Hence, the opportunity loss of Act A_2 is $80 − $40 = $40.

It is convenient to asterisk the payoff of the best act for each event in the original payoff table so as to denote that opportunity losses are measured from these figures. Both the original payoff table and the opportunity loss table are given in Table 12-4 for the inventor's problem.

We can now proceed with the calculation of expected opportunity loss in a manner completely analogous to the calculation of expected profits. Again, we use the probabilities of events as weights and determine the weighted average opportunity loss for each act. Our goal is to select that act which yields the *minimum* expected opportunity loss. The calculation of the expected opportunity losses for the two acts in the inventor's problem is given in Table 12-5. The symbol EOL will be used to represent expected opportunity loss. Hence, EOL(A_1) and EOL(A_2) denote the expected opportunity losses of acts

TABLE 12-5
Expected Opportunity Losses in the Inventor's Problem (in units of $10,000).

ACT A_1: MANUFACTURE DEVICE HIMSELF

EVENTS	PROBABILITY	OPPORTUNITY LOSS	WEIGHTED OPPORTUNITY LOSS
θ_1: Strong sales	0.2	0	0
θ_2: Average sales	0.5	0	0
θ_3: Weak sales	0.3	6	1.8
	1.0		1.8

$$EOL(A_1) = 1.8 \text{ (ten thousands of dollars)}$$
$$= \$18,000$$

ACT A_2: SELL PATENT RIGHTS

EVENTS	PROBABILITY	OPPORTUNITY LOSS	WEIGHTED OPPORTUNITY LOSS
θ_1: Strong sales	0.2	40	8.0
θ_2: Average sales	0.5	13	6.5
θ_3: Weak sales	0.3	0	0
	1.0		14.5

$$EOL(A_2) = 14.5 \text{ (ten thousands of dollars)}$$
$$= \$145,000$$

A_1 and A_2, respectively. The inventor's EOL if he manufactures the device himself is $18,000, and if he sells the patent rights, his EOL is $145,000. If he selects the act that minimizes his EOL, he will choose A_1, that is, to manufacture the device himself. This is the same act that he selected under the criterion of maximizing expected profit. It can be proved that the best act according to the criterion of maximizing expected profit is also best if the decision maker follows the criterion of minimizing expected opportunity loss. The relationship between the maximum expected profit and the minimum expected opportunity loss will be examined later. It should be noted that opportunity losses are not losses in the accountant's sense of profit and loss, because as we have seen, they even occur where only profits of different actions are compared for a given state. They represent foregone opportunities rather than incurred monetary losses.

12.5 EXPECTED VALUE OF PERFECT INFORMATION

Thus far, in our discussion we have considered situations in which the decision maker chooses among alternative courses of action on the basis of *prior information* without attempting to gather further information before he makes his decision. That is, the probabilities used in computing the expected value of each act, as shown in Table 12-3, are termed "prior probabilities" to indicate that they represent probabilities established prior to obtaining additional information through sampling. The procedure of calculating expected values of each act based on these prior probabilities and selecting the optimal act is referred to in Bayesian decision theory as *prior analysis.* In Section 12.8 we consider how courses of action may be compared after these prior probabilities are revised on the basis of sample information, experimental data, or information resulting from tests of any sort. However, the analysis carried out to this point provides a yardstick for measuring the value of perfect information concerning which events will occur. This yardstick will be referred to as the "expected value of perfect information." In order to determine this value, we must calculate the *expected profit with perfect information.* Then, if we subtract the *expected profit under uncertainty,* whose calculation we previously examined, we will have the *expected value of perfect information.* These concepts will be explained in terms of the inventor's problem. We begin with the idea of expected profit with perfect information.

The calculation of the expected profit of acting with perfect information is based on the expected payoff if the decision maker has access to a perfect predictor. It is assumed that if this perfect predictor forecasts that a particular event will occur, then indeed that event will occur. The expected payoff under these conditions for the inventor's problem is given in Table 12-6. In order to understand the meaning of this calculation, it is necessary to adopt a long-run relative frequency point of view. If the forecaster says the event "strong sales" will prevail, the decision maker can look along that row in the payoff table

TABLE 12-6

Calculation of Expected Profit with Perfect Information
for the Inventor's Problem (in units of $10,000 profit).

PREDICTED EVENT	PROFIT	PROBABILITY	WEIGHTED PROFIT
θ_1: Strong sales	$80	0.2	$16.0
θ_2: Average sales	20	0.5	10.0
θ_3: Weak sales	1	0.3	0.3
			$26.3

Expected Profit with Perfect Information = $26.3 (ten thousands of dollars)
= $263,000

and select the act that yields the highest profit. In the case of strong sales, the best act is A_1, which yields a profit of $800,000. Hence the figure $80 is entered under the profit column in Table 12-6. The same procedure is used to obtain the payoffs for each of the other possible events. The probabilities shown in the next column are the original probability assignments to the three states of nature. From a relative frequency viewpoint, these probabilities are now interpreted as the proportion of times the perfect predictor would forecast that each of the given states of nature would occur if the present situation were faced repetitively. Each time the predictor makes his forecast the decision maker selects the optimal payoff. The expected profit with perfect information is then calculated as shown in Table 12-6 by weighting these best payoffs by the probabilities and totaling the products. The expected profit with perfect information in the inventor's problem is $263,000. This figure can be interpreted as the average profit the inventor could realize from this type of device if he were faced with this decision problem repeatedly under identical conditions, and if he always took the best action after receiving the forecast of the perfect indicator. Expected profit with perfect information has sometimes been called the "expected profit under certainty," but this term is clearly somewhat misleading. The inventor is not *certain* to earn any one profit figure. The expected profit with perfect information is to be interpreted as indicated in this discussion.

The expected value of perfect information, abbreviated as EVPI, is defined as *the expected profit with perfect information* minus *the expected profit under uncertainty*. Its calculation is shown in Table 12-7.

TABLE 12-7

Calculation of Expected Value of Perfect
Information for the Inventor's Problem.

Expected Profit with Perfect Information	$263,000
Less: Expected Profit under Uncertainty	245,000
Expected Value of Perfect Information (EVPI)	$ 18,000

EVPI = EOL of the Optimal Act under Uncertainty = $18,000

The interpretation of the EVPI is clear from its calculation. In the inventor's problem, his expected payoff if he selects the optimal act under conditions of uncertainty is $245,000. (See Table 12-3.) On the other hand, if the perfect predictor were available and the inventor acted according to his predictions, the expected payoff would be $263,000. (See Table 12-6.) The difference of $18,000 represents the increase in profit attributable to the use of the perfect forecaster. Hence, the expected value of perfect information may be interpreted as the most the inventor should be willing to pay to get perfect information on the level of sales for his device.

The expected opportunity loss of selecting the optimal act under conditions of uncertainty in the inventor's problem was also shown earlier to be $18,000. That is, this figure represented the minimum value among the expected opportunity losses associated with each act. As shown in Table 12-7, this figure is equal to the expected value of perfect information. It can be mathematically proved that this equality is true in general. Another term used for the expected opportunity loss of the optimal act under uncertainty is the *cost of uncertainty*. This term highlights the "cost" attached to the making of a decision under conditions of uncertainty. Expected profit would be larger if a perfect predictor were available and this uncertainty were removed. Hence this cost of uncertainty is also equal to the expected value of perfect information. In summary, the following three quantities are equivalent:

Expected value of perfect information
Expected opportunity loss of the optimal act under uncertainty
Cost of uncertainty

12.6   DECISION DIAGRAM REPRESENTATION

It is useful to represent the structure of a decision problem under uncertainty by a "decision tree diagram," "decision diagram," or briefly "tree." The problem can be depicted in terms of a series of choices made in alternating order by the decision maker and "Chance." Forks at which the decision maker is in control of choice are referred to as *decision forks;* those at which Chance is in control as *chance forks*. Decision forks will be represented by a little square, whereas no special designation will be used for chance forks. Forks may also be referred to as branching points or junctures.

A simplified decision diagram is given for the inventor's problem in Figure 12-1. After explaining this skeletonized version, we will insert some additional information to obtain a completed diagram. As we can see from Figure 12-1, the first choice is the decision maker's at branching point 1. He can follow either branch A_1 or branch A_2; that is, he can choose either act A_1 or A_2. Assuming he follows path A_1, he comes to another juncture, which is a chance fork. Chance now determines whether the event which will occur is θ_1, θ_2, or θ_3. If Chance takes him down the θ_1 path, the terminal payoff is $800,000; the corresponding payoffs are indicated for the other paths. An analogous interpretation holds if

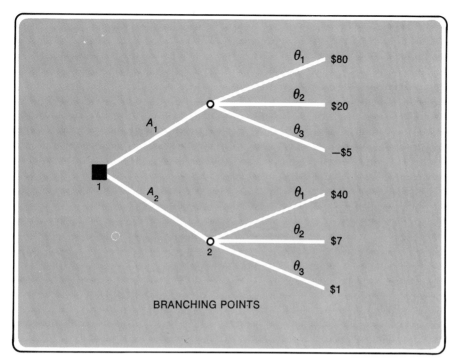

θ_1 $80

θ_2 $20

θ_3 −$5

A_1

1

A_2

θ_1 $40

θ_2 $7

2

θ_3 $1

BRANCHING POINTS

Figure 12-1
Simplified decision diagram for inventor's problem (payoffs are in units of $10,000 profit).

he chooses to follow branch A_2. Thus the decision diagram depicts the basic structure of the decision problem in schematic form. In Figure 12-2, additional information is superimposed on the diagram to represent the analysis and solution to the problem.

The decision analysis process represented by Figure 12-2 (and in other decision diagrams to be considered at later points) is known as *backward induction.* We imagine ourselves as located at the right-hand side of the tree diagram, where the monetary payoffs are. Let us consider first the upper three paths denoted θ_1, θ_2, and θ_3. To the right of these symbols we enter the respective probability assignments 0.2, 0.5, and 0.3 as given in Table 12-3. These represent the probabilities assigned by Chance to following these three paths, after the decision maker has selected act A_1. Moving back to the chance fork from which these three paths emanate, we can calculate the expected monetary value of being located at that fork. This expected monetary value is $24.5 (in units of $10,000 as are the other obvious corresponding numbers) and is calculated in the usual way, that is,

$$\$24.5 = (0.2)(\$80) + (0.5)(\$20) + (0.3)(-\$5).$$

This figure is entered at the chance fork under discussion. It represents the value of standing at that fork after choosing act A_1, as Chance is about to select one of the three paths. The analogous figure entered at the lower chance fork

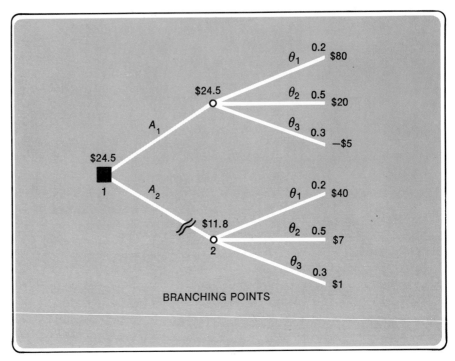

θ_1 0.2 $80
$24.5 θ_2 0.5 $20
A_1 θ_3 0.3 −$5
$24.5
1 A_2 θ_1 0.2 $40
$11.8 θ_2 0.5 $7
2 θ_3 0.3 $1

BRANCHING POINTS

Figure 12-2
Decision diagram for inventor's problem (payoffs are in units of $10,000 profit).

is $11.8. Therefore, imagining ourselves as being transferred back to branching point 1, where the little square represents a fork at which the decision maker can make a choice, we have the alternatives of selecting act A_1 or act A_2. Each of these acts leads us down a path at the end of which is a risky option whose expected profit has been indicated. Since following path A_1 yields a higher expected payoff than path A_2, we block off A_2 as a non-optimal course of action. This is indicated on the diagram by the two vertical lines. Hence, A_1 is the optimal course of action, and it has the indicated expected payoff of $24.5.

Thus, the decision tree diagram reproduces in compact schematic form the analysis given in Table 12-3. An analogous diagram could be constructed in terms of opportunity losses to reproduce the analysis of Table 12-5. However, it is much more customary to use tree diagrams to portray analyses in terms of payoffs rather than opportunity losses, and we will follow the usual practice in this text.

EXERCISES

1. Explain the meaning of expected value of perfect information.

2. Explain the difference between expected opportunity loss and expected value of perfect information

3. Given an opportunity loss table, can you compute the corresponding payoff table? Explain why or why not.

4. The following is a payoff matrix in units of $1000:

| | PRICE THE ITEM AT | | | |
COMPETITOR'S PRICE	A_1 $0.90	A_2 $0.95	A_3 $1.00	A_4 $1.05
S_1: $1.00	10	6	3	1
S_2: 0.95	5	8	4	6
S_3: 0.90	12	9	8	5
S_4: 0.85	8	10	12	14

The prior probabilities are

STATE OF NATURE	PROBABILITY
S_1	.3
S_2	.2
S_3	.4
S_4	.1

Compute the EVPI by two different methods.

5. The following is a payoff table in units of $1000:

| | ACTION | | | |
DEMAND IS	A_1	A_2	A_3	A_4
S_1: Above average	18	15	16	11
S_2: Average	8	12	12	10
S_3: Below average	2	5	3	8

where A_1 = Keep store open weekdays, evenings, and Saturday
A_2 = Keep store open weekdays plus Wednesday evening
A_3 = Keep store open weekdays and Saturday
A_4 = Keep store open only weekdays

The prior probability distribution of demand is

S_i	$P(S_i)$
S_1	0.5
S_2	0.3
S_3	0.2

(a) Find the expected profit under certainty.
(b) Find the expected profit under uncertainty.
(c) How much would you pay for information which yields the true state of nature?

6. Trivia Press, Inc. has been offered an opportunity to publish a new novel. If the novel is a success, the firm can expect to earn $8 million over the next five years; if a failure, to lose $4 million over the next five years. After reading the novel, the publisher assesses the probability of success as $\frac{1}{3}$. Should she publish the book? What is the expected value of perfect information?

7. An advertising firm submits for acceptance a campaign costing $55,000. The company's marketing manager estimates that if the campaign is received well by the public, profits will increase by $175,000; if it is received moderately well, profits will increase by $55,000; and if it is received poorly, profits will remain unchanged. Compute the appropriate opportunity loss table.

8. Assume that there are ten urns, seven of type A and three of type B. Type A urns contain five white balls and five black balls. Type B urns contain eight white balls and two black balls. One of the ten urns is to be selected at random. You are required to guess whether the urn selected is of type A or B. Assume you are willing to act on the basis of expected monetary value. You will receive payoffs and penalties according to the payoff table given below.

TRUE STATE OF NATURE	YOUR GUESS TYPE A	TYPE B
Type A	+$500	−$ 40
Type B	− 300	+ 800

Find and interpret the expected value of perfect information.

9. As personnel manager of Lemon Motors you must decide whether to hire a new salesman. Depending on his performance, the payoff to the firm is

SALES	PAYOFF
High	$10,000
Average	3000
Low	− 13,000

(a) If judging from his application, you feel the probabilities attached to his possible performances are high, 0.3; average, 0.4; and low, 0.3; should you hire him?

(b) How much would you be willing to pay a perfect predictor to tell you what the salesman's performance would be?

10. As marketing manager of a firm, you are trying to decide whether to open a new region for a product. Success of the product depends on demand in the new region. If demand is high, you expect to gain $100,000, if average $10,000, and if low to lose $80,000. From your knowledge of the region and your product you feel the chances are four out of ten that sales will be average, and equally likely that they will be high or low. Should you open the new region? How much would you be willing to pay to know the true state of nature?

11. In Problem 7, if the president of the company, after examining the proposed campaign, feels that the probabilities that it would be received "well" and "moderately well" are 0.4 and 0.2, respectively, what is the expected opportunity loss for each action? What is the optimal decision?

12. A brewer presently packages beer in old style cans. He is debating whether to change the packaging of his beer for next year. He can adopt A_1, an easy-open aluminum can; A_2, a lift-top can; A_3, a new wide-mouth screw-top bottle; or retain A_4, the same old style cans. Profits resulting from each move depend on what the brewer's competitor does for the next year. The payoff matrix and prior probabilities, measured in $10,000 units, are as follows:

PRIOR PROBABILITY	COMPETITOR USES	A_1	A_2	A_3	A_4
0.5	Old Style Bottles	15	14	13	16
0.2	Easy-Open Cans	12	11	10	8
0.1	Lift-Top Cans	6	9	8	6
0.2	Screw-Top Bottles	5	6	8	5

(a) Find the expected opportunity loss for each act.
(b) Determine EVPI.
(c) Determine the optimal decision.

INCORPORATION OF SAMPLE INFORMATION   12.7

The discussion in the preceding sections may be referred to as *prior analysis,* that is, decision making in which expected payoffs of acts are computed on the basis of prior probabilities. In this section we discuss *posterior analysis,* in which expected payoffs are calculated with the use of *posterior probabilities,* which are revisions of prior probabilities on the basis of additional sample or experimental evidence. Bayes' theorem is utilized to accomplish the revision of the prior probabilities. The terms "prior" and "posterior" in this context are relative ones. For example, subjective prior probabilities may be revised to incorporate the additional evidence of a particular sample. The revised probabilities then constitute posterior probabilities. If these probabilities are in turn revised on the basis of another sample, they represent prior probabilities relative to the new sample information, and the revised probabilities are "posteriors."

The basic purpose of attempting to incorporate more evidence through sampling is to reduce the expected cost of uncertainty. If the expected cost of uncertainty (or the expected opportunity loss of the optimal act) is high, then it will ordinarily be wise to engage in sampling. Sampling in this context is understood to include statistical sampling, experimentation, testing, and any other methods used to acquire additional information.

POSTERIOR ANALYSIS   12.8

The general method of incorporating sample evidence into the decision-making process will be illustrated in terms of the inventor problem of Section 12.3. The example is somewhat artificial, because no particular sample size is assumed. However, the problem illustrates the general method of posterior analysis in a very straightforward way.

TABLE 12-8

Computation of Posterior Probabilities in the Inventor's Problem
for the Sample Indication of an Average Level of Sales.

EVENTS θ_i	PRIOR PROBABILITY $P(\theta_i)$	CONDITIONAL PROBABILITY $P(x_2\|\theta_i)$	JOINT PROBABILITY $P(\theta_i) P(x_2\|\theta_i)$	POSTERIOR PROBABILITY $P(\theta_i\|x_2)$
θ_1: Strong Sales	0.2	0.1	0.02	0.042
θ_2: Average Sales	0.5	0.8	0.40	0.833
θ_3: Weak Sales	0.3	0.2	0.06	0.125
	1.0		0.48	1.000

Suppose the inventor decided not to rely solely on prior probabilities concerning the demand for his new device, but to have a market research organization conduct a sample survey of potential consumers to gather additional evidence for the probable level of sales for his product. Let us assume that the survey can result in three types of sample results denoted x_1, x_2, and x_3, corresponding to the three states of nature, sales levels, θ_1, θ_2, and θ_3. Specifically, the possible results may be

x_1: sample indicates strong sales

x_2: sample indicates average sales

x_3: sample indicates weak sales

The survey is conducted and the sample gives an indication of an average level of sales, that is, x_2 is observed. Assume that on the basis of previous surveys of this type, the market research organization can assess the reliability of the sample evidence in the following terms. In the past, when the actual level of sales after a new device was placed on the market was average, sample surveys properly indicated an average level of demand 80% of the time. However, when the actual level was strong sales, about 10% of the sample surveys incorrectly indicated demand as average and when the actual level was weak sales, about 20% of the sample surveys gave an indication of average sales. These relative frequencies represent conditional probabilities of the sample evidence "average sales," given the three possible underlying events concerning sales level, and can be symbolized as follows:

$$P(x_2|\theta_1) = 0.1$$

$$P(x_2|\theta_2) = 0.8$$

$$P(x_2|\theta_3) = 0.2$$

The revision by means of Bayes' theorem of the prior probabilities assigned to the three sales levels on the basis of the observed sample evidence x_2 (average sales), is given in Table 12-8. In terms of Equation (4.9) for Bayes' theorem, x_2 plays the role of B, the sample observation; and θ_i replaces A_i, the possible events, or states of nature. In the usual way, after the joint probabilities are calculated, they are divided by their total, in this case, 0.48, to yield posterior or revised probabilities for the possible events. The effect of the weighting given to the sample evidence by Bayes' theorem in the revision of the prior probabilities may be noted by comparing the posterior probabilities with the corresponding "priors" in Table 12-8. With a sample indication of average sales, the prior probability of the event "average sales," 0.5 was revised upward to 0.833. Correspondingly, the probabilities of events "strong sales" and "weak sales" declined from 0.2 to 0.042 and from 0.3 to 0.125, respectively.

Decision making after the observation of sample evidence

The revised probabilities calculated in Table 12-8 can now be used to compute the "posterior expected profits" of the inventor's alternative courses of action. In Table 12-3, expected payoffs were computed based on the subjective prior probabilities assigned to the possible events. These can now be referred to as "prior expected profits." The calculation of the posterior expected profits (using the revised or posterior probabilities as weights) is displayed in Table 12-9. It is customary to denote prior probabilities as $P_0(\theta_i)$ and posterior probabilities as $P_1(\theta_i)$. That is, the subscript zero is used to denote prior probabilities and the subscript one to signify posterior probabilities.

Since the posterior expected profit of act A_1 exceeds that of A_2, the better of the two courses of action remains that of the inventor manufacturing the device himself. However, after the sample indication of "average sales," the

---

TABLE 12-9
Calculation of Posterior Expected Profits in the Inventor's Problem
Using Revised Probabilities of Events (in units of $10,000).

ACT A_1: MANUFACTURE DEVICE HIMSELF

EVENTS	PROBABILITY $P_1(\theta_i)$	PROFIT	WEIGHTED PROFIT
θ_1: Strong Sales	0.042	$80	3.360
θ_2: Average Sales	0.833	20	16.660
θ_3: Weak Sales	0.125	−5	−.625
	1.000		19.395

Posterior Expected Profit $A_1 = \$19.395$ (ten thousands of dollars) $= \$193,950$

ACT A_2: SELL PATENT RIGHTS

EVENTS	PROBABILITY $P_1(\theta_i)$	PROFIT	WEIGHTED PROFIT
θ_1: Strong Sales	0.042	$40	1.680
θ_2: Average Sales	0.833	7	5.831
θ_3: Weak Sales	0.125	1	0.125
	1.000		7.636

Posterior Expected Profit $A_2 = 7.636$ (ten thousands of dollars) $= \$76,360$

---

expected profit of act A_1 has decreased from $245,000 based on the prior probabilities to $193,950 based on the revised probabilities. Also, the difference in the expected profits of the two acts has narrowed somewhat. The $245,000 and $193,950 figures are, respectively, the *prior expected profit under uncertainty* and the *posterior expected profit under uncertainty*. It is entirely possible for the optimal course of action under a posterior analysis to change from that of the prior analysis. In the present example, if the sample indication had been "weak sales," with appropriate conditional probabilities, it would have been possible for the posterior expected profit of A_2 to have exceeded that of A_1. (Assume some figures and demonstrate this point.)

EXERCISES

1. Given:

STATE OF NATURE	$P(\theta)$	$P(X\|\theta)$	$P(\theta)P(X\|\theta)$	$P(\theta\|X)$
θ_1: Housing starts will increase next year	0.5	0.6		
θ_2: Housing starts will remain at the same level or will decline		0.4		

Fill in the blanks and interpret the data, if X is the result of a survey of 100 construction companies.

2. A prior probability function is

S	$P_0(S)$
S_1	0.1
S_2	0.2
S_3	0.3
S_4	0.4

and a sample observation X occurs which has the following properties: $P(X\|S_1) = 0.8$, $P(X\|S_2) = 0.6$, $P(X\|S_3) = 0.5$, and $P(X\|S_4) = 0.2$. What are the revised probabilities?

3. A firm is trying to decide whether to embark on a new advertising campaign. Management assigns the following prior probability distribution:

STATE OF NATURE	PROBABILITY
S_1: Successful	0.5
S_2: Unsuccessful	0.5

A sample result is observed which has the following probability of occurring:

0.3 if S_1 is true,
0.6 if S_2 is true.

Revise the prior probability distribution in light of this new information.

4. Let p be the true percentage of customers who will purchase a new product. Assume p can take on the values given below with the respective prior probabilities.

p	$P_0(p)$
0.10	0.8
0.20	0.2

A simple random sample of ten customers is drawn. The customers are asked whether they would purchase the product. What are the revised prior probabilities if
(a) one customer would purchase?
(b) two customers would purchase?
(c) three customers would purchase?

5. There are two actions to take, A_1 and A_2, and there are two states of nature S_1 and S_2. A_1 is preferred if S_1 is true, and A_2 is preferred if S_2 is true. If the prior probabilities are $P(S_1) = 0.7$ and $P(S_2) = 0.3$ and you observe a sample observation S such that the $P(S|S_1) = 0.9$ and $P(S|S_2) = 0.2$, can you conclude that A_1 is the better act? Explain your answer.

6. Let p be the proportion of defective cigarette lighters in a lot offered to you by a jobber. Given

p	$P_0(p)$	OPPORTUNITY LOSS OF ACCEPTING LOT
0.10	0.5	0
0.20	0.3	100
0.30	0.2	200

A sample of ten is taken and two are found defective. What is the expected opportunity loss of the action "accept,"
(a) if action is taken before sampling?
(b) if action is taken after sampling?

7. Management's prior probability assessment of demand for a newly developed product is: high, 0.6; low, 0.2; and average, 0.2. A survey, taken to help determine the true demand for the product, indicates demand is average. The reliability of the survey is such that it will indicate "average" demand 70% of the time when it is really high, 95% of the time when it is really average, and 10% of the time when it is actually low. In light of this information, what would be the reassessed probabilities of the three states of nature?

DECISION MAKING BASED ON EXPECTED UTILITY 12.9

In the decision analysis discussed up to this point, the criterion of choice was the maximization of expected monetary value. This criterion can be interpreted as a test of preferredness that selects as the optimal act the one that yields the greatest long-run average profit. That is, in a decision problem such as our example involving the inventor's choices, the optimal act is the one that would result in the largest long-run average profit if the same decision had to be made repeatedly under identical environmental conditions. In general, in such decision-making situations, as the number of repetitions becomes large, the observed average payoff approaches the theoretical expected payoff. Gamblers, baseball managers, insurance companies, and others who engage in what is colloquially called "playing the percentages," may often be characterized as using the aforementioned criterion. However, many of the most important personal and business decisions are made under unique sets of conditions and in some of these occasions it may not be realistic to think in terms of many repetitions of the same decision situation. Indeed, in the business world, many of management's most important decisions are unique, high-risk, high-stake choice situations, whereas the less important, routine, repetitive decisions are ones customarily delegated to subordinates. Therefore, it is useful to have an apparatus for dealing with one-time decision making. Utility theory, which we discuss in this section, provides such an apparatus, as well as providing a logical method for repetitive decision making too.

Whether an individual, a corporation, or other entity would be willing to make decisions on the basis of the expected monetary value criterion depends upon the decision maker's attitude toward risk situations. Several simple choice situations are presented in Table 12-10 to illustrate that in choosing between two alternative acts we might select the one with the lower expected value. The reason we might make such a choice is our feeling that the incremental expected gain of the act with greater expected monetary value does not sufficiently reward us for the additional risk involved.

Table 12-10 gives three choice situations for alternative acts grouped in pairs. For each pair a decision must be made between the two alternatives.

The illustrative choices are to be made once and only once. That is, there is to be no repetition of the decision experiment. We shall assume for simplicity that all monetary payoffs are tax free. Suppose you choose acts A_2, B_1, and C_1. In the case of the choice between A_1 and A_2, you might argue as follows. "The expected value of act A_1 is $0; the expected value of A_2 is $E(A_2) = (\frac{1}{2})(\$.60) + (\frac{1}{2})(-\$.40) = \$.10$. A_2 has the higher expected value, and since if I incur the loss of $.40, I can sustain such a loss with equanimity, I am willing to accept the risk involved in the selection of this course of action."

A useful way of viewing the choice between A_1 and A_2 is to think of A_2 as

---

TABLE 12-10
Alternative Courses of Action with Different Expected Monetary Payoffs.

A_1: Receive $0 for certain. That is, you are certain to incur neither a gain nor a loss.	or	A_2: Receive $.60 with probability $\frac{1}{2}$ and lose $.40 with probability $\frac{1}{2}$.
B_1: Receive $0 for certain. That is, you are certain to incur neither a gain nor a loss.	or	B_2: Receive $60,000 with probability $\frac{1}{2}$ and lose $40,000 with probability $\frac{1}{2}$.
C_1: Receive a $1,000,000 gift for certain.	or	C_2: Receive $2,100,000 with probability $\frac{1}{2}$ and receive $0 with probability $\frac{1}{2}$.

an option in which a fair coin is tossed. If it lands "heads" you receive a payment of $.60. If it lands "tails" you must pay $.40. The choice of act A_1 means that you are unwilling to play the game involved in flipping the coin; hence, you neither lose nor gain anything.

On the other hand, you might very well choose act B_1 rather than B_2, even though the respective expected values are

$$E(B_1) = \$0$$

$$E(B_2) = (\tfrac{1}{2})(\$60,000) + (\tfrac{1}{2})(-\$40,000) = \$10,000$$

In this case, you might reason that even though act B_2 has the higher expected monetary value, a calamity of no mean proportions would occur if the coin landed tails, and you incurred a loss (say, a debt) of $40,000. Your present level of assets might cause you to view such a loss as intolerable. Hence, you would refuse to play the game. If you look at the difference between the two choices just discussed, A_1 versus A_2 and B_1 versus B_2, you will note that the only difference is that in A_2 we had the payoffs $.60 and −$.40. In B_2 the decimal point has been moved five places to the right for each of these numbers, making the monetary gains and losses much larger than in A_2. In all other respects, the wording of the choice between A_1 and A_2 and between B_1 and B_2 is the same. Nevertheless, as we shall note after the ensuing discussion of the choice between acts C_1 and C_2, it is not necessarily irrational to select act A_2 over A_1 where A_1 has the greater expected monetary value and B_1 over B_2 where B_1 has the smaller expected monetary value.

In the choice between acts C_1 and C_2, you would probably select act C_1, which has the lower expected monetary value. That is, most people would doubtless prefer a gift of $1,000,000 for certain to a 50-50 chance at $2,100,000 and $0, for which the expected payoff is

$$E(C_2) = (\tfrac{1}{2})(\$2,100,000) + (\tfrac{1}{2})(\$0) = \$1,050,000$$

In this case, you might argue that you would much prefer to have the $1,000,000 for certain, and go home to contemplate your good fortune in peace than to play a game where on the flip of a coin you might receive nothing at all. You might further feel that there are relatively few things that you could do with $2,100,000 that you could not accomplish with $1,000,000. Hence, the incremental satisfaction to be derived even from winning on the toss of the coin in C_2 might not convince you to take the risk involved as compared to the "sure thing" of $1,000,000 in the selection of act C_1.

From the above discussion, we may conclude that it is logical to depart sometimes from the criterion of maximizing expected monetary values in making choices in risk situations. We cannot specify how a person *should* choose among alternative courses of action involving monetary payoffs, given only the type of information contained in Table 12-10. His decisions will clearly depend upon his *attitude toward risk*, which in turn will depend on a combination of factors such as his level of assets, his liking or distaste for gambling, and his psycho-emotional constitution. If we single out the factor of level of assets, for example, it is evident that a large corporation with a substantial level of assets may choose to undertake certain risky ventures that a smaller corporation with smaller assets would avoid. In the case of the larger corporation an outcome of a loss of a certain number of dollars might represent an unfortunate occurrence but as a practical matter would not materially change the nature of operation of the business, whereas in the case of the smaller corporation a loss of the same magnitude might constitute a catastrophe and might require the liquidation of the business. Hence, large and small corporations do and should have different attitudes toward risk. It may be noted that reverse attitudes toward risky ventures to those just indicated might be present in comparing a dynamic management of a small company with a highly conservative management of a large company.

To recapitulate, we can summarize the problem concerning decision making in problems involving payoffs that depend upon risky outcomes. Monetary payoffs are sometimes inappropriate as a calculation device, and it appears appropriate to substitute some other set of values or "numeraire," which reflects the decision maker's attitude toward risk. A clever approach to this problem has been furnished by Von Neumann and Morganstern, who developed the so-called Von Neumann and Morganstern utility measure. In the next section, we consider how these utilities may be derived, and the procedures for using them in decision analysis.

Construction of utility functions

We have seen that in certain risk situations we might prefer one course of action to another even though the first act has a lower expected monetary value. In the language of decision theory, the reason for preferring the first act is that it possesses greater expected utility than does the second act.[4] The procedure used to establish the utility function of a decision maker requires him to respond to a series of choices in each of which he receives with certainty an amount of money denoted M (for money) as opposed to a gamble in which he would receive an amount M_1 with probability p and an amount M_2 with probability $1 - p$. The question the decision maker must answer is what probability p would he require for consequence M_1 in order to be indifferent between receiving M for certain and partaking in the gamble involving the receipt of M_1 with probability p and M_2 with probability $1 - p$. This probability assessment provides the assignment of a utility index to the monetary value M. The data obtained from the series of questions posed to the decision maker result in a set of utility-money pairs, which can be plotted on a graph, and constitutes the decision maker's utility function for money. We will illustrate the procedure for constructing an individual's utility function by returning to one of the examples given in Table 12-10.

[4] The term "utility" as used by Von Neumann and Morganstern and as used in this text differs from the economist's use of the same word. In traditional economics, utility referred to the inherent satisfaction delivered by a commodity and was measured in terms of psychic gains and losses. On the other hand, Von Neumann and Morganstern conceived of utility as a measure of value used in the assessment of situations involving risk, which provides a basis for choice making. The two concepts can give rise to widely differing numerical measures of utility.

Assume we ask the decision maker which of the following he prefers:

B_1: Receive $0 for certain.   or B_2: Receive $60,000 with probability $\frac{1}{2}$, lose $40,000 with probability $\frac{1}{2}$.

Suppose he responds that he prefers to receive $0 for certain. Our task then is to find out what probability he would require for the receipt of the $60,000 to make him just indifferent between the gamble and the certain receipt of $0. This will enable us to determine the utility he assigns to $0. The first step is the arbitrary assignment of "utilities" to the monetary consequences in the gamble as, for example,

$$U(\$60,000) = 1$$

$$U(-\$40,000) = 0$$

where the symbol U denotes "utility" and $U(\$60,000) = 1$ is read "the utility of $60,000 is equal to one." It should be emphasized that the assignment of the numbers 0 and 1 as the utilities of the lowest and highest outcomes of the gamble is entirely arbitrary. Any other numbers could have been assigned, just so the utility assigned to the higher monetary outcome is greater than that assigned to the lower outcome. Thus, the utility scale has an arbitrary zero point, just like the 0° mark in temperature, which corresponds to different conditions depending upon whether the Centigrade or Fahrenheit scale is used.

The expected utility of the indicated gamble is

$$E[U(B_2)] = \tfrac{1}{2}[U(\$60,000)] + \tfrac{1}{2}[U(-\$40,000)] = \tfrac{1}{2}(1) + \tfrac{1}{2}(0) = \tfrac{1}{2}$$

Therefore, since the decision maker has indicated that he prefers $0 for certain to this gamble, it follows that the utility he assigns to $0 is greater than $\frac{1}{2}$ or $U(\$0) > \frac{1}{2}$. In order to aid the decision maker in deciding how much greater than $\frac{1}{2}$ the utility is which he assigns to the monetary outcome of $0, we introduce the concept of a hypothetical lottery for use in calibrating his utility assessment.

Let us assume we have a box with 100 balls in it, 50 of which are black and 50 white. The balls are identical in all other respects. Further, we assume that if a ball is drawn at random from the box and its color is black, the decision maker receives a payoff of $60,000. On the other hand, if the ball is white, his payoff is −$40,000. We now have constructed a physical counterpart of the gamble denoted B_2. The question now is, "If we retain the total number of balls in the calibrating box at 100, but vary the composition in terms of the number of black and white balls, how many black balls would be required for a decision maker to say that he is indifferent between receiving $0 for certain and participating in the gamble?" With 50 black balls in the box, the decision maker prefers $0 for certain. With 100 black balls (and zero white balls), the decision maker would obviously prefer the gamble, since it would result in a payoff of $60,000 with certainty. For some number of black balls between 50 and 100 the decision maker should be indifferent, that is, he would be at the threshold beyond which he would prefer the gamble to the certainty of the $0 payoff. Suppose we begin replacing white balls by black balls, and for some time the decision maker is still unwilling to participate in the gamble. Finally, when there are 70 black balls and 30 white balls, he announces that the point of indifference has been reached. We now can calculate the utility he has assigned to $0 as follows:

$$U(\$0) = 0.70[U(\$60,000)] + 0.30[U(-\$40,000)] = 0.70(1) + 0.30(0) = 0.70$$

This utility calculation is a particular case of the general relationship

(12.1) $$U(M) = pU(M_1) + (1 - p)U(M_2)$$

where M is an amount of money received for certain and M_1 and M_2 are component prizes received in a gamble with probabilities p and $1 - p$, respectively.

We have now determined three money-utility pairs: $(-\$40,000, 0)$, $(\$60,000, 1)$, and $(\$0, 0.70)$, where the first figure in the ordered pair represents a monetary payoff in dollars and the second figure the utility index assigned to this amount. The utility figures for other monetary payoffs between $-\$40,000$ and $\$60,000$ can be assessed in exactly the same way as for $0, assuming the patience of our long-suffering decision maker holds out. Of course, as a practical matter, a relatively small number of points could be determined and the rest of the function interpolated. Suppose the utility function shown in Figure 12-3 results from the indifference probabilities assigned by the decision maker in the set of gambles proposed to him. The one point whose determination was illustrated ($0, 0.70) is depicted on the graph. This utility function can now be used to evaluate risk alternatives that might be presented to the decision maker. He can calculate the expected utility of an alternative by reading off the utility figure corresponding to each monetary outcome and then weighting these utilities by the probabilities that pertain to the outcomes. In other words, the utility figures can now be used by the decision maker in place of the original monetary values, for calculation of expected utilities, whereas for a person with his type of utility function, calculation of expected monetary values is clearly an inadequate guide for decision making.

Characteristics and types of utility functions

The utility function depicted in Figure 12-3 rises consistently from the lower left to the upper right side of the chart. That is, the utility curve has a positive slope throughout its extent. This is a general characteristic of utility functions; it simply implies that people ordinarily attach greater utility to a larger sum of money than to a smaller sum.[5] Economists have noted this psychological trait in traditional demand theory and have referred to it as a "positive marginal utility for money." The concave downward shape shown in Figure 12-3 illustrates the utility curve of an individual who has a diminishing marginal utility

[5] The almost infinite variety of types of human behavior is attested to by the fact that conduct which runs counter to a generalization of this sort is even occasionally observed. A few years ago, newspapers carried an account of an heir to a fortune of $30 million who committed suicide at the age of 23, indicating in a final letter that his great wealth prevented him from living a normal life.

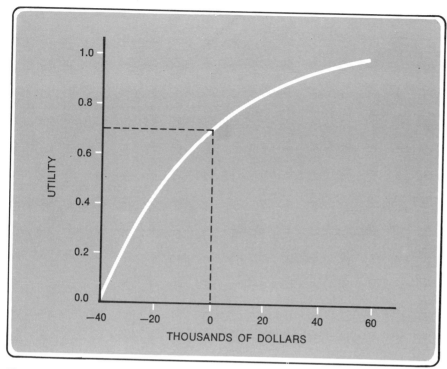

Figure 12-3
A utility function.

for money, although the marginal utility is always positive. This type of utility curve is characteristic of a "risk avoider," and is so indicated in Figure 12-4A. A person characterized by such a utility curve would prefer a small but certain monetary gain to a gamble whose expected monetary value is greater but may involve a large but unlikely gain, or a large and not unlikely loss. The linear function in Figure 12-4B depicts the behavior of a person who is "neutral" or "indifferent" to risk. For such a person every increment of, say, a thousand dollars has an associated constant increment in utility. This type of individual would use the criterion of *maximizing expected monetary value* in decision making because by so doing he would also *maximize expected utility*.

In Figure 12-4C is shown the utility curve for a "risk preferrer." This type of person willingly accepts gambles that have a smaller expected monetary value than an alternative payoff received with certainty. In the case of such an individual, the attractiveness of a possible large payoff in the gamble tends to outweigh the fact that the probability of such a payoff may indeed be very small.

Empirical research suggests that most individuals have utility functions in which for small changes in money amounts the slope does not change very much. Hence, over these ranges of money outcomes, the utility function may approximately be considered as linear and as having a constant slope. How-

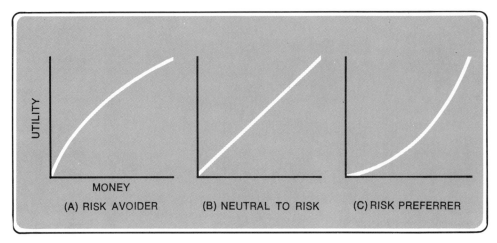

Figure 12-4
Various types of utility functions.

ever, in considering courses of action in which one of the consequences is very adverse or in which one of the payoffs is very large, individuals can be expected to depart from the maximization of expected monetary values as a guide to decision making. For many business decisions, where the monetary consequences may represent only a small fraction of the total assets of the business unit, the use of maximization of expected monetary payoff may constitute a reasonable approximation to the decision-making criterion of maximization of expected utility. In other words, in such cases, the utility function may often be treated as approximately linear over the range of monetary payoffs considered.

Assumptions underlying utility theory

The utility measure we have discussed was derived by evoking from the decision maker his preferences between sums of money obtainable with certainty and lotteries or gambles involving a set of basic alternative monetary outcomes. This procedure entails a number of assumptions.

It is assumed that an individual when faced with the types of choices discussed can determine whether an act, say A_1 is preferable to another act, A_2; whether these acts are indifferently regarded or whether A_2 is preferred to A_1. If A_1 is preferred to A_2, then the utility assigned to A_1 should exceed the utility assigned to A_2.

Another behavioral assumption is that if the individual prefers A_1 to A_2 and he also prefers A_2 to A_3, then he should prefer A_1 to A_3. This is referred to as the principle of *transitivity*. The assumption extends also to indifference relationships. Hence, if the decision maker is indifferent between A_1 and A_2 and between A_2 and A_3, he should be indifferent between A_1 and A_3.

Furthermore, it is assumed that if a payoff or consequence of an act is replaced by another, and the individual is indifferent between the former and new consequences, he should also be indifferent between the old and new acts. This is often referred to as the principle of *substitution.*

Finally, it is assumed that the utility function is bounded. This means that utility cannot increase or decrease without limit. As a practical matter, this simply means that the range of possible monetary values is limited. For example, at the lower end the range may be limited by a bankruptcy condition.

It may be argued that human beings do not always exhibit the type of consistency in their choice behavior that is implied by these assumptions. However, the point is that if in the construction of an individual's utility function it is observed that he is behaving inconsistently and these incongruities are indicated to him, and if he is "reasonable" or "rational," he should adjust his choices accordingly. If he insists on being irrational and refuses to adjust his choices that violate the underlying assumptions of utility theory, then a utility function cannot be constructed for him and he cannot use maximization of expected utility as a criterion of rationality in his choice making. It is important to keep in mind that the type of theory discussed here does not purport to describe the way people actually *do behave* in the real world, but rather specifies how they *should behave* if their decisions are to be consistent with their own expressed judgments as to preferences among consequences. Indeed, it may be argued that since human beings are fallible and do make mistakes, it is useful to have normative procedures which police their behavior and provide ways in which it can be improved.

A brief note on scales

Von Neumann-Morganstern utility scales are examples of *interval scales.* Such scales have a constant unit of measurement, but an arbitrary zero point. Differences between scale values can be expressed as multiples of one another, but individual values cannot.

EXAMPLE 12-1

The familiar scales for temperature are examples of interval scales. We cannot say that 100°C is twice as hot as 50°C. The corresponding Fahrenheit measures would not exhibit a ratio of 2 to 1. On the other hand, we can say that the intervals or differences between 100°C and 50°C and 75°C and 50°C are in a two-to-one ratio. Thus, using the relationship $F = (9/5)C + 32°$:

$$C = 100°; \quad F = (9/5)(100°) + 32° = 212°$$
$$C = 75°; \quad F = (9/5)(75°) + 32° = 167°$$
$$C = 50°; \quad F = (9/5)(50°) + 32° = 122°$$

The difference between 100° and 50° = 50°; between 75° and 50° = 25°.
The ratio of 50° to 25° is two-to-one.

The difference between 212° and 122° = 90°; between 167° and 122° = 45°.
The ratio of 90° to 45° is two-to-one.

In decision making using utility measures, if a different zero point and a different scale are selected, the same choices will be made. A constant can be added to each utility value, and each utility value can be multiplied by a constant, without changing the properties of the utility function. Thus, if a is any constant and b is a positive constant, and x is an amount of money,

$$U_2(x) = a + bU_1(x)$$

and $U_2(x)$ is as legitimate a measure of utility as $U_1(x)$.

EXERCISES

1. (a) The expected monetary return of the decision to buy life insurance is negative. Thus it is irrational to buy life insurance. Do you agree or disagree? Explain.
   (b) If the A.T. & T. Corporation does not carry automobile insurance, why do you think this is so?

2. If the following prospects have the given utilities:

PROSPECT	UTILES
A	10
B	8
C	5
D	3
E	2

would you prefer C for certain to
   (a) a chance of getting A with 0.4 probability and E with 0.6 probability?
   (b) a chance of getting B with 0.5 probability and E with 0.5 probability?
   (c) a chance of getting A with 0.3 probability and D with 0.7 probability?
   (d) a chance of getting B with 0.4 probability and E with 0.6 probability?

3. You have a choice of placing your money in the bank and receiving interest equal to ten utiles or investing in Rerox stock. With a probability of 0.4, Rerox will yield gains equal to 45 utiles and with a probability of 0.6, it will cause a loss of 15 utiles.
   (a) What is the expected utility of the prospect, "buy the stock"?
   (b) Should you buy Rerox or put the money in the bank?

4. Drillwell Oil Company is debating what it should do with an option on a parcel of land. If it takes the option, the firm can drill with 100% interest or with 50% interest (i.e., all costs and profits are split with another firm). It costs $50,000 to drill a well and $20,000 to operate a producing well until it is dry. The oil is worth $1 a barrel. Assume the well is either dry or produces 200,000 or 500,000 barrels of oil. The firm assesses the probability of each outcome as 0.8, 0.1, and 0.1, respectively.
   (a) Based on expected monetary return, what is the best action?
   (b) Suppose Drillwell's management has the following utility function:

DOLLARS	UTILES
−50,000	−30
−25,000	−10
65,000	25
130,000	60
215,000	120
430,000	200

What is the best decision based on expected utility?

5. I.R.S. has audited your last year's income tax and has sent you a bill for $225 for back taxes. You now have the choice of paying the bill or disputing the audit. If you dispute it, it will cost you $20 for an accountant's fee to prepare your case. After preliminary talks with your accountant, you feel the chances of your winning the dispute are five in one hundred.

(a) Should you dispute the case based on monetary expectations?

(b) Assume large losses of money are disastrous to you as a struggling student. This is reflected in your utility function, which indicates that $U(-\$20)$ is −4 utiles, $U(-\$225)$ is −425 utiles, and $U(-\$245)$ is −440 utiles. Based on expected utility, what is your best course of action?

6. A drug manufacturer has developed a new drug named Thalidimous. Tests have shown it to be extremely effective with almost no side effects. However, it has only been tested for three years and long-range side effects are really unknown. The research department feels the probability the drug will have any serious long-range effects is 1 in 100. The Food and Drug Administration (FDA) must first clear the drug for sale. Assume the FDA evaluates the loss to society because of serious long-range side effects as −900,000 utiles, the gain to society because of the use of the drug as 8,000 utiles and the gain attributable to the economic advantages of production of a new drug as 1000 utiles. If the FDA accepts the firm's appraisal of the probability of long-range side effects should it "accept" the drug?

> *"In the space of one hundred and seventy-six years the Lower Mississippi has shortened itself two hundred and forty-two miles. That is an average of a trifle over one mile and a third per year. Therefore, any calm person, who is not blind or idiotic, can see that in the Old Oölitic Silurian Period just a million years ago next November, the Lower Mississippi River was upward of one million three hundred thousand miles long and stuck out over the Gulf of Mexico like a fishing-rod. And by the same token any person can see that seven hundred and forty-two years from now the Lower Mississippi will be only a mile and three-quarters long, and Cairo and New Orleans will have joined their streets together, and be plodding comfortably along under a single mayor and a mutual board of aldermen. There is something fascinating about science. One gets such wholesome returns of conjecture out of such a trifling investment of fact."*

MARK TWAIN

13 Time Series

13.1   INTRODUCTION

The decisions of business management determine whether a firm will thrive and expand or whether its position will deteriorate, and, in the worst instance, pass out of existence. These vital decisions are based on perceptions of future outcomes that will affect the payoffs of possible alternative courses of action. Since these outcomes occur in the future, they must be forecast. Not only must businessmen forecast, but they must plan and think through the nature of the activities which will permit them to accomplish their objectives. However, it is clear that business planning and decision making are inseparable from forecasting.

Methods of forecasting vary considerably. For example, in customer demand forecasting, the range of prevailing methods includes informal "seat of the pants" estimating, executive panels and composite opinions, consensus of sales force opinions, combined user responses, statistical techniques, and various combinations of these methods.

Because of such factors as the stepped-up complexity of business operations, the need for greater accuracy and timeliness, the dependence of outcomes on so many different variables, and the demonstrated utility of the techniques, management is increasingly turning to formal models, such as those provided by statistical methods, for assistance in the difficult task of peering into the future. A very widely applied and extremely useful set of procedures is *time series analysis.* A time series is a set of statistical observations arranged in chronological order. Examples include a weekly series of end-of-week stock prices, a monthly series of steel production, and an annual series of national income. Such time series are essentially historical series, whose values at any point in time are the resultants of the interplay of large numbers of diverse economic, political, social, and other factors.

A first step in the prediction of any series involves an examination of past observations. Time series analysis deals with the methods for analyzing these past data and for projecting them to obtain estimates of future values. The traditional or "classical" methods of time series analysis, which we will primarily discuss, are descriptive in nature and do not provide for probability statements concerning future events. These time series models, although admittedly only approximate and not highly refined, have proven their worth when cautiously and sensibly applied. It is very important to realize that these methods cannot simply be used mechanistically but must at all times be supplemented by sound subjective judgment.

Although the preceding discussion has referred to the use of time series analysis for the purpose of forecasting and its usefulness for planning and control, these procedures also are often used for the simple purpose of historical description. Hence, for example, they may be usefully employed in an analysis in which interest centers upon the comparative differences in the nature of variations in different time series. The general nature of the classical time series model is described in the next section.

THE CLASSICAL TIME SERIES MODEL 13.2

If we wished to construct a mathematical model of an economic time series that was ideally satisfying, we might seek to define and measure the many determinants of the variations in the time series and then proceed to state the mathematical relationships between these determinants and the particular series in question. However, the determinants of change in an economic time series are multitudinous, including such factors as changes in population, consumer tastes, technology, investment or capital-goods formation, weather, customs, and numerous other variables both of an economic and non-economic nature. The enormity and impracticability of the task of measuring all of the aforementioned factors and then relating them mathematically to an economic time series whose variations we wish to account for and predict militates against the use of this direct approach to time series analysis. Hence, it is not surprising that a more indirect and practical approach, such as classical methodology, has come into use. Classical time series analysis is essentially a descriptive method, which attempts to break down an economic time series into distinct components that represent the effects of the operation of groups of explanatory factors such as those given earlier. These component variations are

1. Trend.
2. Cyclical fluctuations.
3. Seasonal variations.
4. Irregular movements.

Trend refers to the smooth upward or downward movement characteristic of a time series over a long period of time. Such movements are thought of as requiring a minimum of about 15 or 20 years to describe, and as being attributable to factors such as population change, technological progress, and large-scale shifts in consumer tastes.

Cyclical fluctuations, or business cycle movements, are recurrent up and down movements around trend levels that have a duration of anywhere from about 2 to 15 years. There is no single simple explanation of business cycle activity, and there are different types of cycles of varying length and size. Therefore, it is not surprising that no generally satisfactory mathematical model has been constructed for either describing or forecasting these cycles, and perhaps none will ever be.

Seasonal variations are a type of cycle that completes itself within the period of a calendar year and then continues in a repetition of this basic pattern. The major factors in producing these annually repetitive patterns of seasonal variations are weather and customs, where the latter term is broadly interpreted to include observance of various holidays such as Easter and Christmas. Series of monthly data and quarters of the year are ordinarily used to examine these seasonal variations. Hence, regardless of trend or cyclical levels, one can observe in the United States that each year more ice cream is sold during the summer months than during the winter, whereas more fuel oil for home heating purposes is consumed in the winter than during the summer months. Both of these cases illustrate the effect of weather or climatic factors in determining seasonal patterns. Also, department store sales generally reveal a minor peak during the months in which Easter occurs and a larger peak in December, when Christmas occurs, reflecting the shopping customs of consumers associated with these dates. The techniques of measurement of seasonal variations we will discuss are particularly well suited to the measurement of relatively stable patterns of seasonal variations, but can be adapted to cases of changing seasonal movements as well.

Irregular movements are fluctuations in time series that are short in duration, erratic in nature, and follow no regularly recurrent or other discernible pattern. These movements are sometimes referred to as *residual variations,* since, by definition, they represent what is left over in an economic time series after trend, cyclical, and seasonal elements have been accounted for. These irregular fluctuations result from sporadic, unsystematic occurrences such as earthquakes, accidents, strikes, and the like. Whereas in the classical time series model, the elements of trend, cyclical, and seasonal variations are viewed as resulting from systematic influences leading to either gradual growth, de-

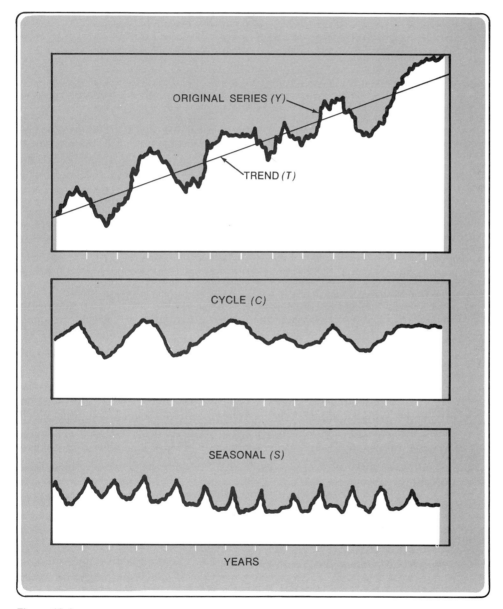

Figure 13-1
The components of a time series.

cline, or recurrent movements, irregular movements are considered to be so erratic that it would be fruitless to attempt to describe them in terms of a formal model.

Figure 13-1 presents the typical pattern of the trend, cyclical, and seasonal components of a time series.

13.3   DESCRIPTION OF TREND

As pointed out in the preceding section, the classical model involves the separate statistical treatment of the component elements of a time series. We shall begin our discussion by indicating how the description of the underlying trend is accomplished.

Before the trend of a particular time series can be determined, it is generally necessary to subject the data to some preliminary treatment. The amount of such adjustment depends somewhat on the time period for which the data are stated. For example, if the time series is in monthly form, certain reconciliations for calendar differences may be required. For example, it is often necessary to revise the monthly data to take account of the differing number of days per month. This may be accomplished by stating the data for each month on a per day basis by dividing the monthly figures by the number of days in the respective months, or by the number of working days per month.

Even when the original data are in annual form, which is often the case where primary interest is centered on the long term trend of the series, the data may require a considerable amount of preliminary treatment before a meaningful analysis can be carried out. Adjustments for changes in population size are often made by dividing the original series by population figures to state the series in per capita form. Frequently, comparisons of trends in these per capita figures are far more meaningful than corresponding comparisons in the unadjusted figures.

It is particularly important to scrutinize a time series and adjust it for differences in definitions of statistical units, the consistency and coverage of the reported data, and similar items. It is important to realize that one cannot simply proceed in a mechanical fashion to analyze a time series. Careful and critical preliminary treatment of such data is required to ensure the meaningfulness of the results.

Purposes for fitting trend lines

The trend in a time series can be measured by the free-hand drawing of a line or curve that seems to fit the data, by fitting appropriate mathematical functions, or by the use of moving average methods. Moving averages are discussed later in this chapter in connection with seasonal variations.

A free-hand curve may be fitted to a time series by visual inspection. When this type of characterization of a trend line is employed, the investigator is usually interested in a quick description of the underlying growth or decline in a series, without any careful further analysis. In many instances, this rapid graphic method may suffice. However, it clearly has certain disadvantages. Different investigators would surely obtain different results for the same time series. Indeed, even the same analyst would probably not sketch in exactly the same trend line in two different attempts on the same series. This excessive amount of subjectivity in choice of a trend line is especially disadvantageous if further quantitative analysis is planned. In the ensuing discussion, we will concentrate on mathematically fitted trend lines.

Even in the case of the mathematical measurement of long run trend, the purpose of the analysis is of considerable importance in the selection of the appropriate trend line. Several different types of purposes can be specified.

1. Trend lines may be fitted for the purpose of historical description. If so, any good fitting line will, in general, suffice. The line need not have logical implications for forecasting purposes, nor should it be evaluated primarily by characteristics that might be desirable for other purposes.

2. A second purpose is that of prediction or projection into the future. In this case, particularly if long-term projection is desired, the selected line should have logical implications when it is extended into the future. The analyst, when engaging in prediction, must always carefully weigh the implications of the models he projects into the future as regards their reasonableness for the phenomena being described and predicted.

3. A third purpose for which trend lines are fitted to economic data is to describe and eliminate trend movements from the series in order that the non-trend elements may be studied. Thus, if the analyst's primary interest is to study cyclical fluctuations, freeing the original data of trend enables him to examine cyclical movements undisturbed by the presence of the trend factor. For this purpose, any type of trend line that does a reasonably good job of bisecting the individual business cycles in the data would be appropriate.

THE FITTING OF TREND LINES BY THE METHOD OF LEAST SQUARES 13.4

For situations in which it is desirable to have a mathematical equation to describe the secular trend of a time series, the most widely used method is the fitting of some type of polynomial function to the data. In this section, we illustrate the general method by means of very simple examples, fitting in turn a straight line and a second degree parabola by the method of least squares to time series data.

The method of least squares

The method of least squares, when used to fit trend lines to time series data, is employed mainly because it is a simple, practical method which provides best fits according to a reasonable criterion. However, it should be recognized that the method of least squares does not have the same type of theoretical under-pinning when applied to fitting trend lines as when used in regression and correlation analysis, as described in Chapter 10. The major difficulty is that the usual probabilistic assumptions present in regression and correlation analysis are simply not met in the case of time series data. For example, in regression analysis, the dependent variable is assumed to be a random variable. Therefore, the model assumes probability distributions of this random variable around the computed values of the dependent variable that fall along the regression line. A number of assumptions are implicit in this type of model. Deviations from the regression line are considered to be random errors describable by a probability distribution. The successive observations of the dependent variable are assumed to be independent. For example, in the illustrative problem in Chapter 10 Family B's expenditures were assumed to be independent of Family A's and so forth.

Clearly, in the fitting of trend lines to time series data, the probabilistic assumptions of the method of least squares are not met. If a trend line is fitted, for example, to an annual time series of department store sales, time is treated as the independent variable X and department store sales is the dependent variable Y. It is not reasonable to think of the deviation of actual sales in a given year from the computed trend value as a random error. Indeed, if the original data are annual, then deviations from trend would be considered to represent the operation of cyclical and irregular factors. Seasonal factors would not be present in annual data because by definition they complete themselves within a year. Finally, the assumption of independence is not met in the case of time series data. A department store's sales in a given year surely are not independent of what they were in the preceding year. In summary, returning to the point made at the outset of this discussion, the method of least squares when used to fit trend lines is employed primarily because of its practicality, simplicity, and good fit characteristics rather than because of its justification from a theoretical viewpoint.

Fitting an arithmetic straight-line trend

As an example, we will fit a straight line by the method of least squares to an annual series on value added by manufacture in the United States from 1950 to 1970. Although we wrote the equation of a straight line in the discussion of regression analysis in Chapter 10 as $Y_c = a + bX$, in time series analysis we will use the equation

(13.1) $$Y_t = a + bx$$

The computed trend value is denoted Y_t, with the subscript t standing for trend. That is, Y_t is the computed trend figure for the time period x. In time series analysis, the computations can be simplified by transforming the X variable, which is the independent variable time, to a simpler variable with fewer digits. This is accomplished by stating the time variable in terms of deviations from the arithmetic mean time period, which is simply the middle time period.

The transformed time variable is denoted by lower case x. Hence in the illustrative example in Table 13-1, $x = 0$ in 1960, the middle year in the time series which runs from 1950 through 1970. The x values (or $X - \overline{X}$ figures) for years before and after 1960 are, respectively, $-1, -2, -3, \ldots$, and $1, 2, 3, \ldots$. For example, the x value for 1961 is equal to 1 because $X - \overline{X} = 1961 - 1960 = 1$. The constants in the trend equation are interpreted in a similar way to those in the straight line discussed in regression analysis; a is the computed trend figure for the period when $x = 0$, in this case, 1960; b is the slope of the trend line, or the amount of change in Y_t per unit change in x, or per year in the present example. Because as indicated in Section 3.11, the sum of the deviations of a set of observations from their mean is equal to zero, $\Sigma x = 0$. This property makes the computation of the constants for the trend line simpler than in the corresponding case of the straight-line regression equation. In Chapter 10, the equations for fitting a straight line were given as

(13.2)
$$a = \overline{Y} - b\overline{X}$$

(13.3)
$$b = \frac{\Sigma XY - n\overline{X}\overline{Y}}{\Sigma X^2 - n\overline{X}^2}$$

In the least squares fitting of a straight-line trend equation, x is substituted for X. Since $\Sigma x = 0$, and therefore, $\bar{x} = 0$, the equations become

(13.4)
$$a = \overline{Y} = \frac{\Sigma Y}{n}$$

(13.5)
$$b = \frac{\Sigma xY}{\Sigma x^2}$$

Hence, the constant a is simply equal to the mean of the Y values and b is calculated by a division of two numbers easily determined from the original data. The calculations for fitting a straight-line trend to the time series on value added by manufacture are given in Table 13-1. Columns (2) through (5) contain the basic computations for determining the values of a and b. As indicated in the calculation of these constants at the bottom of the table, $a = 181.2$ and $b = 10.4$. The trend equation is $Y_t = 181.2 + 10.4x$. An identification statement such as the one given below the trend equation that $x = 0$ in 1960 and Y is in billions of dollars should always accompany the equation, since it is not possible to interpret fully the meaning of the trend line without it. The trend figures are determined by substituting the appropriate values of x into the trend equation. Hence, for example, the trend figure for 1950 is

$$Y_{t,1950} = 181.2 + 10.4\,(-10) = \$77.2 \text{ billions}$$

Since the b value measures the change in Y_t per year, it can be added to each trend value to obtain the following year's figure. The trend figures are given in Column (6) of Table 13-1.

The trend line is graphed in Figure 13-2. Any two points can be plotted to determine the line. Interpreting the values of $a = 181.2$ and $b = 10.4$, we have a computed trend figure for value added by manufacture in 1960 of 181.2 billions and an increase in trend of 10.4 billions per year. As can be seen from the graph, the trend line fits the data rather well. Since the line was fitted by the method of least squares, the sum of the squared deviations of the actual data from the

TABLE 13-1

Straight-Line Trend Fitted by the Method of Least Squares to Values Added by Manufactures in the United States, 1950–1970.

YEAR (1)	x (2)	VALUE ADDED BY MANUFACTURE[a] (BILLIONS OF DOLLARS) Y (3)	xY (4)	x^2 (5)	Y_t (6)	PERCENT OF TREND $\frac{Y}{Y_t} \cdot 100$ (7)
1950	−10	89.7	−897.0	100	77.2	116.2
1951	−9	102.1	−918.9	81	87.6	116.6
1952	−8	109.2	−873.6	64	98.0	111.4
1953	−7	121.6	−851.2	49	108.4	112.2
1954	−6	117.0	−702.0	36	118.8	98.5
1955	−5	135.0	−675.0	25	129.2	104.5
1956	−4	144.9	−579.6	16	139.6	103.8
1957	−3	147.8	−443.4	9	150.0	98.5
1958	−2	141.5	−283.0	4	160.4	88.2
1959	−1	161.5	−161.5	1	170.8	94.6
1960	0	163.9	0	0	181.2	90.5
1961	1	164.3	164.3	1	191.6	85.8
1962	2	179.1	358.2	4	202.0	88.7
1963	3	192.1	576.3	9	212.4	90.4
1964	4	206.2	824.8	16	222.8	92.5
1965	5	226.9	1134.5	25	233.2	97.3
1966	6	250.9	1505.4	36	243.6	103.0
1967	7	262.0	1834.0	49	254.0	103.1
1968	8	285.0	2280.0	64	264.4	107.8
1969	9	305.9	2753.1	81	274.8	111.3
1970	10	298.3	2983.0	100	285.2	104.6
TOTALS	0	3804.9	8028.4	770		

$$a = \frac{3804.9}{21} = 181.2$$

$$b = \frac{8028.4}{770} = 10.4$$

$$Y_t = 181.2 + 10.4x$$

$$x = 0 \text{ in } 1960$$

x is in one-year intervals; Y is in billions of dollars

[a] Value added is obtained by subtracting the cost of materials, supplies, containers, fuel, purchased electrical energy, and contract work from the value of shipments for products manufactured plus receipts for services rendered.

In general, the "value added" by a business firm is the sales of that firm minus its costs of materials and costs of products purchased from other firms. Hence, this difference represents the "value added" to the national product by this particular firm.

SOURCE: *Annual Survey of Manufactures* (individual years), U.S. Department of Commerce, and *Statistical Abstract of the U.S.*, 1971.

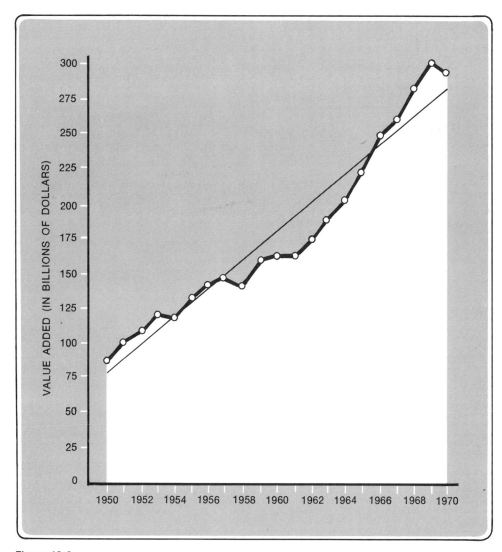

Figure 13-2
Straight-line trend fitted to value added by manufacture in the United States, 1950–1970.

trend line is less than from any other straight line that could have been fitted to the data, and the total of the deviations above the line is equal to the total below the line.

A couple of technical points concerning the fitting procedure may be noted. Since the present illustration contained an odd number of years, the time period at which $x = 0$, or the x origin, coincided with one of the years of data, and the x values were stated as 1, 2, 3, . . . for years after the x origin and -1, -2, -3, . . . for years before the origin. On the other hand, if there had been an even number of years, the mean time period at which $x = 0$ would fall midway between the two central years. For example, suppose, there had been one less year of data and the value added figures were available only for 1950–1969. Then there would be 20 annual figures and $x = 0$ at $1959\frac{1}{2}$. The two central years 1959 and 1960 deviate from this origin by $-\frac{1}{2}$ and $+\frac{1}{2}$, respectively. To avoid the use of fractions, it is usual to state the deviations in terms of one-half year intervals rather than a year. Hence, the x values for 1960, 1961, 1962, . . . , would be 1, 3, 5, . . . , and for 1959, 1958, 1957, . . . , they would be $-1, -3, -5, ...$ The computation of the constants a and b would proceed in the usual way. However, now a would be interpreted as the computed trend figure for a time point midway between the two central years and the b value would be the amount of change in trend per one-half year.

If the time intervals of the original data were not annual, the transformed time variable x would have to be appropriately interpreted. For example, if the data were stated in the form of five-year averages and there were an odd number of such figures, then x would be in five-year intervals. If there were an even number of figures, and the non-fractional method of stating x referred to above were used, then x would be in $2\frac{1}{2}$ year intervals.

Projection of the trend line

Projections of the computed trend line can be obtained by substituting the appropriate values of x into the trend equation. For example, if a projected trend figure for 1975 were desired for value added by manufacture, it would be computed by substituting $x = 15$ in the previously determined trend equation. Hence,

$$Y_{t,1975} = 181.2 + 10.4(15) = \$337.2 \text{ billions}$$

A rougher estimate of this trend figure would be obtained by extending the straight line graphically in Figure 13-2 to the year 1975. It must be remembered that these projections are estimates of only the trend level in 1975 and not of the actual figure for value added by manufacture in that year. If a prediction of the latter figure were desired, estimates of the non-trend factors would have to be combined with the trend estimate. This means that a prediction of cyclical fluctuations would have to be made and incorporated with the trend figure. Accurate forecasts of this type are difficult to make over extended time periods. However, insofar as managerial applications of trend analysis are concerned, for long-range planning purposes often all that is desired is a projection of the trend level of the economic variable of interest. For example, a good estimate of the trend of demand would be adequate for a business firm planning a plant expansion to anticipate demand many years into the future. Accompanying predictions of business cycle standings many years into the future would not be required nor, for that matter, would they be realistically feasible.

Cyclical fluctuations

As was previously indicated, when a time series consists of annual data, it contains trend, cyclical, and irregular elements. The seasonal variations are absent, since they occur within a year. Hence, deviations of the actual annual data from a computed trend line are attributable to cyclical and irregular factors. Since the cyclical element is the dominant factor, a study of these deviations from trend essentially represents an examination of business cycle fluctuations. The deviations from trend are most easily observed by dividing the original data by the corresponding trend figures for the same time period. By convention, the result of this division of an original figure by a trend value is multiplied by 100 to express the figure as a percent of trend. Hence, if the original figure is exactly equal to the trend figure, the percent of trend is 100; if the original figure exceeds the trend value, the percent of trend is above 100; if the original figure is less than the trend value, the percent of trend is below 100.

The formula for percent of trend figures is

(13.6) $$\text{Percent of trend} = \frac{Y}{Y_t} \cdot 100$$

where Y = annual time series data
Y_t = trend values

In summary, the original annual data contain trend, cyclical, and irregular factors. Since the data are annual, the seasonal component is not included. When converted to percent of trend, these numbers contain only cyclical and irregular movements, since the division by trend eliminates that factor. The rationale of this procedure is easily seen by using a so-called multiplicative model for the analysis. That is, the original annual figures are viewed as representing the combined effect of trend, cyclical, and irregular factors. In symbols, let T, C, and I represent trend, cyclical, and irregular factors, respectively, and Y and Y_t mean the same as in Equation (13.6). Then dividing the original time series by the corresponding trend values yields

(13.7) $$\frac{Y}{Y_t} = \frac{T \times C \times I}{T} = C \times I$$

The percents of trend for the series on value added by manufacture are given in Column (7) of Table 13-1 and are plotted in Figure 13-3. As may be seen from the chart, the underlying upward trend movement is no longer present. Instead, the percent of trend series fluctuates about the line labeled 100, which is the trend level. These percents of trend are sometimes referred to as cyclical relatives; that is, the original data are stated relative to the trend figure. Of course, strictly speaking, Y/Y_t is the cyclical relative, and the multiplication by 100 converts the relative to a percentage figure. Another way of depicting cyclical fluctuations is in terms of relative cyclical residuals, which are percentage deviations from trend, and are computed by the formula

(13.8) $$\text{Relative cyclical residual} = \frac{Y - Y_t}{Y_t} \cdot 100$$

Hence, for example, if we refer to the value added data in Table 13-1 for 1970, the actual figure is 298.3, the computed trend value is 285.2, and the per-

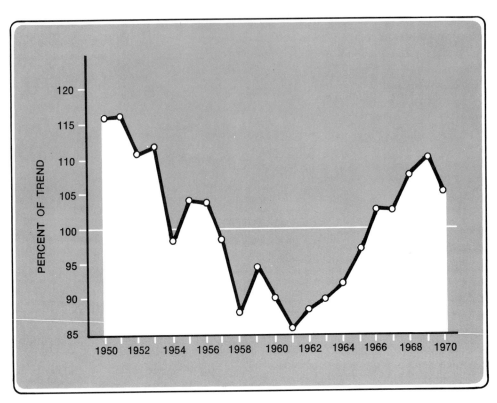

Figure 13-3
Percents of trend for value added by manufacture in the United States, 1950–1970.

cent of trend is 104.6. The relative cyclical residual in this case is +4.6%, indicating that the actual value added figure is 4.6% above the trend figure because of cyclical and irregular factors. These residuals are positive or negative depending on whether the actual time series figures fall above or below the computed trend values. The graph of relative cyclical residuals is visually identical to that of the percent of trend values except that relative cyclical residuals are shown as fluctuations around a zero base line rather than around a base line of 100%.

The familiar charts of business cycle fluctuations that often appear in publications such as the financial pages of newspapers and business periodicals are usually graphs of either percents of trend or relative cyclical residuals. These charts may be studied for timing of peaks and troughs of cyclical activity, for amplitude of fluctuations, for duration of periods of expansion and contraction, and for other relevant items of interest to the business cycle analyst.

Fitting a second degree trend line

The preceding discussion on the fitting of a straight line pertains to the case in which the trend of the time series can be characterized as increasing or decreasing by constant amounts per time period. Actually very few economic time series exhibit this type of constant change over a long period of time, say, over a period of several business cycles. Therefore, it generally is necessary to fit other types of lines or curves to the given time series. Polynomial functions are particularly convenient to fit by the method of least squares. Frequently a second degree parabola provides a good description of the trend of a time series. In this type of curve, the amounts of change in the trend figures, Y_t, may increase or decrease per time period. Hence, a second degree parabola may provide a good fit to a series whose trend is increasing by increasing amounts, increasing by decreasing amounts, etc. The procedure of fitting a parabola by the method of least squares involves the same general principles as the fitting of a straight line, but entails somewhat more arithmetic.

EXAMPLE 13-1

We illustrate the method of fitting a second degree parabola to a time series in terms of a very simple illustration. The reader is warned that the time period in this example is entirely too short to permit a valid description of trend. However, the illustration is given for expository purposes only to indicate the procedure involved. In Table 13-2 is given a time series on the number of persons employed in anthracite coal mining in a certain coal region from 1967 to 1973. This series is graphed in Figure 13-4. The trend of these data may be described as decreasing by decreasing amounts. The general form of a second degree parabola is $Y_t = a + bX + cX^2$. Analogous to the method of stating the equation for a straight-line trend, the trend line for a second degree parabola may be written

(13.9)
$$Y_t = a + bx + cx^2$$

where
$$Y_t = \text{the trend values}$$
$$a, b, c = \text{constants to be determined}$$
$$x = \text{deviations from the middle time period}$$

If the transformed variable x, representing deviations from the mean time period is used, the equations for fitting a second degree parabola are

(13.10)
$$\Sigma Y = na + c\Sigma x^2$$

(13.11)
$$\Sigma x^2 Y = a\Sigma x^2 + c\Sigma x^4$$

(13.12)
$$b = \frac{\Sigma x Y}{\Sigma x^2}$$

Hence, the constant b is determined by the same equation as in fitting the straight line. The constants a and c are found by solving simultaneously the Equations (13.10) and (13.11).

In the present problem, since there are an odd number of years, $x = 0$ in the middle year, 1970. Solving for b by substituting the appropriate totals from Table 13-2, we have

$$b = \frac{-336}{28} = -12$$

Substituting into Equations (13.10) and (13.11) gives

$$266 = 7a + 28c$$
$$1232 = 28a + 196c$$

THE FITTING OF TREND LINES BY THE METHOD OF LEAST SQUARES **343**

Dividing the second equation by 4 to equate the coefficients of a, we obtain

$$266 = 7a + 28c$$
$$308 = 7a + 49c$$

Subtracting the first equation from the second,

$$42 = 21c$$

and

$$c = \frac{42}{21} = 2$$

Substituting this value for c into the first equation

$$266 = 7a + 28(2)$$
$$a = 30$$

Therefore, the equation of the second degree parabola fitted to the employment time series is

(13.13) $$Y_t = 30 - 12x + 2x^2$$

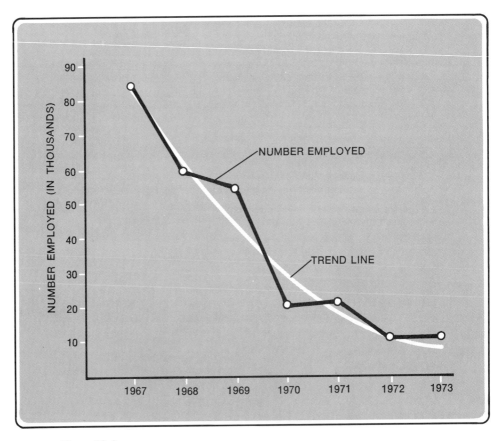

Figure 13-4
Second degree parabola fitted to the number of persons employed in anthracite coal mining in a certain coal region, 1967–1973.

where $x = 0$ in 1970

     x is in one-year intervals

     Y is in thousands of persons

The trend figures, Y_t, shown in Column (8) of Table 13-2 are obtained by substituting the appropriate values of x into Equation (13.13). The constants a, b, and c may be interpreted as follows: a is the computed trend figure at the time origin, that is when $x = 0$; b is the slope of the parabola at the time origin; and c indicates the amount of acceleration or deceleration in the curve, or the amount by which the slope changes per time period.[1]

Although the second degree parabola appears from Figure 13-4 to provide a reasonably good fit to the data in this example, the dangers of a mechanistic projection of a trend line are clearly illustrated. The parabola would begin to turn upward after 1973, and the projected trend figure for each year would be higher than the preceding year's figure. Therefore, only if an analysis of all of the underlying factors determining the trend of this series revealed reasons for a reversal of the observed decline should one be willing to entertain the notion of extending the trend line into the future for forecasts, even for relatively short periods.

TABLE 13-2

Second Degree Parabola Fitted by the Method of Least Squares to the Number of Persons Employed in Anthracite Coal Mining in a Certain Coal Region, 1967–1973.

YEAR (1)	x (2)	NUMBER EMPLOYED (IN THOUSANDS) Y (3)	xY (4)	x^2Y (5)	x^2 (6)	x^4 (7)	Y_t (8)
1967	-3	83	-249	747	9	81	84
1968	-2	60	-120	240	4	16	62
1969	-1	54	-54	54	1	1	44
1970	0	21	0	0	0	0	30
1971	1	22	22	22	1	1	20
1972	2	13	26	52	4	16	14
1973	3	13	39	117	9	81	12
TOTALS	0	266	-336	1232	28	196	

SOURCE: Hypothetical data.

The equations of trend lines embody assumptions concerning the type of change that takes place over time. Hence, the arithmetic straight line assumes a trend that increases or decreases by constant amounts, whereas the second

---

[1] In calculus terms, the derivative of the second degree parabola trend equation is

$$\frac{dY_t}{dx} = b + 2cx$$

Hence, the slope of the curve differs at each time period x. When $x = 0$, $\frac{dY_t}{dx} = b$. Therefore, the slope at the time origin

is b. The second derivative is $\frac{d^2Y_t}{dx^2} = 2c$. Thus the acceleration or rate of change in the slope is $2c$ per time period.

degree parabola assumes that the change in these amounts of change is constant per unit time. It is often useful to describe the trend of an economic time series in terms of the percentage rates of change that are taking place. Logarithmic trend lines are useful for this purpose. The fitting of such lines will not be illustrated here, but discussions of these techniques may be found in many of the statistics books listed in the bibliography.

EXERCISES

1. The following series shows the total U.S. national income in billions of dollars from 1933 to 1937:

YEAR	NATIONAL INCOME (BILLIONS OF DOLLARS)
1933	35.0
1934	40.2
1935	44.0
1936	49.9
1937	55.0

(a) For each year, compute the Y_t values for the equation

$$Y_t = a + bx$$

(b) Are these Y_t values a good description of the secular trend of national income? Why or why not?

(c) Compute the relative cyclical residual for 1935 and explain what it means.

2. Assume that the following trend equation resulted from the fitting of a least squares parabola to the size of the labor force in a Southern county.

$$Y_t = 49.17 + 4.23x - 0.19x^2$$
$x = 0$ in 1945
x is in $2\frac{1}{2}$-year intervals
Y is the size of the labor force in thousands

(a) Assume the above trend line is "a good fit." What generalizations can you make concerning the way in which the labor force of this county has grown in absolute amounts? Concerning the percentage rate at which it has grown?

(b) In (a) you were instructed to assume the trend line was "a good fit." However, the actual size of the labor force in 1970 was 95,185. This is rather striking evidence that the equation is not "a good fit." Do you agree? Discuss.

13.5   MEASUREMENT OF SEASONAL VARIATIONS

For long-range planning and decision making, in terms of time series components, executives of a business or governmental enterprise concentrate primarily on forecasts of trend movements. For intermediate planning periods, say from about two to five years, business cycle fluctuations are of critical importance, too. For shorter range planning, operational decision and control purposes, seasonal variations must also be taken into account.

Seasonal movements, as indicated in Section 13.2, are periodic patterns of variation in a time series. Strictly speaking, the terms "seasonal movements" or "seasonal variations" can be applied to any regularly repetitive movements that occur in a time series where the interval of time for completion of a cycle is one year or less. Hence, under this classification are subsumed movements such as daily cycles in utilization of electrical energy and the weekly cycles in the use of public transportation vehicles. However, seasonal movements generally refer to the annual repetitive patterns of economic activity which are associated with climatic and custom factors. As noted earlier, these movements are generally examined by using series of monthly or quarterly data.

Purpose of analyzing seasonal variations

Just as was true in the case of the study of trend movements, seasonal variations may be studied because *interest is primarily centered upon these movements,* or they may be measured merely *in order that they may be eliminated,* so that business cycle fluctuations can be more clearly revealed. For example, as an illustration of the first purpose, a company might be interested in analyzing the seasonal variations in sales of a product it produces in order to iron out variations in production, scheduling, and in personnel requirements. Another reason a company's interest may be primarily focused on seasonal variations is to budget a predicted annual sales figure by monthly or quarterly periods based on observed seasonal patterns in the past.

On the other hand, as an illustration of the second purpose, an economist may wish to eliminate the usual month-to-month variations in series such as personal income, unemployment rates, and housing starts in order to study the underlying business cycle fluctuations present in these data.

Rationale of the ratio-to-moving average method

There are a number of techniques by which seasonal variations can be measured, but only the most widely used one, the so-called "ratio-to-moving average method" will be discussed here. It is most frequently applied to monthly data, but we will illustrate its use for a series of quarterly figures, thus reducing substantially the required number of computations.

It is helpful in acquiring an understanding of the rationale of the measurement of seasonal fluctuations to begin with the final product, the seasonal indices. The object of the calculations when the raw data are for quarterly periods and a stable or regular seasonal pattern is present is to obtain four seasonal indices, each one indicating the seasonal importance of a quarter of the year. The arithmetic mean of these four indices is 100.0. Hence, if the seasonal index for, say, the first quarter is 105, this means that the first quarter averages 5% higher than the average for the year as a whole. If the original data had been monthly, there would be twelve seasonal indices which average 100.0, and each index would indicate the seasonal importance of a particular month. These indices are descriptive of the recurrent seasonal pattern in the original series.

As an example of how these seasonal indices might be used, we can refer to budgeting a predicted annual sales figure, say, by quarterly periods. Suppose that $40,000,000 of sales of particular products was budgeted for the next year, or an average of $10,000,000 per quarter. If the quarterly seasonal indices based on an observed stable seasonal pattern in the past were 97.0, 110.0, 85.0, and 108.0, respectively, for the four quarters of the year, then the amounts of sales budgeted for each quarter would be

$$\text{First quarter } 0.97 \times \$10 \text{ million} = \$ 9.7 \text{ million}$$
$$\text{Second quarter } 1.10 \times 10 \text{ million} = 11.0 \text{ million}$$
$$\text{Third quarter } 0.85 \times 10 \text{ million} = 8.5 \text{ million}$$
$$\text{Fourth quarter } 1.08 \times 10 \text{ million} = 10.8 \text{ million}$$

The essential problem in the measurement of seasonal variations is eliminating from the original data the non-seasonal elements in order to isolate the stable seasonal component. In trend analysis, when annual data were used and it was desired to arrive at cyclical fluctuations, a similar problem existed. It was solved by obtaining measures of trend and using these as base line or reference figures. Deviations from trend were then measures of cyclical (and irregular) movements. Analogously, when we have monthly or quarterly original data, which consist of all of the components of trend, cycle, seasonal, and irregular movements, ideally we would like to obtain a series of base line figures that contain all the non-seasonal elements. Then deviations from the base line would represent the pattern of seasonal variations. Unsurprisingly, this ideal method of measurement is not feasible. However, the practical method is to obtain a series of moving averages that roughly include the trend and cycle components. Dividing the original data by these moving average figures eliminates the trend and cyclical elements and yields a series of figures, which contain seasonal and irregular movements. These data are then averaged by months or by quarters to eliminate the irregular disturbances and to isolate the seasonal factor. This method of describing a pattern of stable seasonal movements is explained below.

Ratio-to-moving average method

In order to derive a set of seasonal indices from a series characterized by a stable seasonal pattern, about five to eight years of monthly or quarterly data are required. A stable seasonal pattern means that the peaks and troughs generally occur in the same months or quarters year after year.

The ratio-to-moving average method of computing seasonal indices for quarterly data may be summarized as consisting of the following steps:

1. Derive a four-quarter moving average which contains the trend and cyclical components present in the original quarterly series. A four-quarter

moving average is simply an annual average of the original quarterly data successively advanced one quarter at a time. For example, the first moving average figure contains the first four quarters. Then the first quarter is dropped, and the second through fifth quarterly figures are averaged. The computation proceeds this way until the last moving average is calculated, containing the last four quarters of the original series. In the actual calculation an adjustment is made in order to center the moving average figures so their timing corresponds to that of the original data.

The reason these moving averages include the trend and cyclical components may perhaps be most easily understood by considering what these averages do not contain. Since they are annual averages, they do not contain seasonal movements, since such fluctuations, by definition, average out over a one-year period. Also, the irregular movements that tend to raise the figures for certain months or quarters and to lower them in others tend to cancel out when averaged over the year. Thus, only the trend and cyclical elements tend to be present in the moving averages.

2. Divide the original data for each quarter by the corresponding moving average figure. These "ratio-to-moving average" numbers contain only the seasonal and irregular movements, since the trend and cyclical components were eliminated in the division by the moving average.

3. Arrange the ratio-to-moving average figures by quarters, that is, all the first quarters in one group, all the second quarters in another, and so forth. Average these ratio-to-moving average figures for each quarter in an attempt to eliminate the irregular movements, and thus to isolate the stable seasonal component. The type of average used for this procedure is referred to as a "modified mean." This is an arithmetic mean of the ratio-to-moving average figures after dropping the highest and lowest extreme values.

4. Make an adjustment to force the four modified means to total 400, and thus average out to 100.0. The resultant four figures, one for each quarter of the year constitute the seasonal indices for the series in question.

In symbols, this procedure may be summarized as follows. Let Y be the original quarterly observations; MA the moving average figures; and T, C, S, I, the trend, cyclical, seasonal, and irregular components, respectively.

Then, dividing the original data by the moving average values gives

(13.14)
$$\frac{Y}{MA} = \frac{T \times C \times S \times I}{T \times C} = S \times I$$

Averaging these ratio-to-moving average figures (Y/MA) accomplishes an elimination of the irregular movements that tend to make the Y/MA values too high in certain years and too low in others. Hence, if the eliminations of the non-seasonal elements were perfect, the final seasonal indices would reflect only the effect of seasonal variations. Of course, since the entire method is a rather rough and approximate procedure, the non-seasonal elements are generally not completely eliminated. The moving average usually contains the trend and *most* of the cyclical fluctuations. Therefore the cyclical component is usually not completely absent in the Y/MA values. Also, the modified means

TABLE 13-3

Feed Grain Index Numbers of Average Prices Received by Farmers by Quarters, 1959–1966: Computations for Seasonal Indices and Deseasonalizing of Original Data.

QUARTER (1)	FEED GRAIN PRICE INDEX NUMBERS (1957–1959 = 100) (2)	FOUR-QUARTER MOVING TOTAL (3)	TWO-OF-A-FOUR-QUARTER MOVING TOTAL (4)	MOVING AVERAGE COL(5) = COL(4) × 1/8 (5)	ORIGINAL DATA AS PERCENT OF MOVING AVERAGE [COL(2) ÷ COL(5)] × 100 (6)	SEASONAL INDEX (7)	DESEASONALIZED FEED GRAIN PRICE INDEX NUMBERS [COL(2) ÷ COL(7)] × 100 (8)
1959							
I	96					98.8	97.2
II	103					102.9	100.1
III	100	391	779	97.38	102.69	102.6	97.5
IV	92	388	771	96.38	95.46	95.7	96.1
1960							
I	93	383	762	95.25	97.64	98.8	94.1
II	98	379	752	94.00	104.26	102.9	95.2
III	96	373	744	93.00	103.23	102.6	93.6
IV	86	371	737	92.13	93.35	95.7	89.9
1961							
I	91	366	733	91.63	99.31	98.8	92.1
II	93	367	741	92.63	100.40	102.9	90.4
III	97	374	750	93.75	103.47	102.6	94.5
IV	93	376	756	94.50	98.41	95.7	97.2
1962							
I	93	380	759	94.88	98.02	98.8	94.1
II	97	379	758	94.75	102.37	102.9	94.3
III	96	379	762	95.25	100.79	102.6	93.6
IV	93	383	771	96.38	96.49	95.7	97.2

1963							
I	97	388	786	98.25	98.73	98.8	98.2
II	102	398	801	100.13	101.87	102.9	99.1
III	106	403	809	101.13	104.82	102.6	103.3
IV	98	406	814	101.75	96.31	95.7	102.4
1964							
I	100	408	813	101.63	98.40	98.8	101.2
II	104	405	813	101.63	102.33	102.9	101.1
III	103	408	823	102.88	100.12	102.6	100.4
IV	101	415	838	104.75	96.42	95.7	105.5
1965							
I	107	423	851	106.38	100.58	98.8	108.3
II	112	428	853	106.63	105.04	102.9	108.8
III	108	425	848	106.00	101.89	102.6	105.3
IV	98	423	842	105.25	93.11	95.7	102.4
1966							
I	105	419				98.8	106.3
II	108					102.9	105.0

SOURCE: Agricultural Handbook No. 325, U.S. Department of Agriculture, 1966.

do not ordinarily remove all of the erratic disturbances attributed to the irregular component. Nevertheless, in the case of series with a stable seasonal pattern, the computed seasonal indices generally isolate the underlying seasonal pattern quite well.

In Table 13-3 is given a quarterly series of feed grain price index numbers of average prices received by farmers from 1959 to 1966. The base period of the index number series is 1957–1959. As is indicated in the section on index numbers later in this chapter, this means that the average level of prices during this period is designated as 100. Index figures above and below 100 represent price levels that are higher and lower, respectively, than during the base period. Examination of this series reveals that feed grain prices tend to be highest during the second and third quarters, that is, during the spring and summer months, and lowest during the first and fourth quarters, or during the fall and winter. The calculation of quarterly seasonal indices will be illustrated in terms of this series.

The feed grain price indices have been listed in Column (2) of Table 13-3 from the first quarter of 1959 through the second quarter of 1966. The inclusion of the first two quarters of 1966 permits the computation of the moving averages for all four quarters of 1965. Our first task is the calculation of the four-quarter moving average. This moving average would simply be calculated as indicated above by averaging four quarters at a time, continually moving the average up by a quarter. However, because of a problem of centering of dates, a slightly different type of average, a so-called "two-of-a-four-quarter moving average" is calculated. The problem is as follows. An average of four quarterly figures would be centered halfway between the dating of the second and third figures and would thus not correspond to the date of either of those figures. For example, the average of the four quarters of 1959, the first figures shown in Column (2) of Table 13-3, would be centered midway between the second and third quarter dates, or at the center of the year, July 1, 1959. The original quarterly figures are centered at the middles of their respective time periods, or, for simplicity, say, February 15, May 15, August 15, and November 15. Hence, the dates of a simple four-quarter moving average would not correspond to those of the original data. This problem is easily solved by averaging the moving averages two at a time. For example, as we have seen, the first moving average obtainable from Table 13-3 is centered at July 1, 1959. The second moving average, which contains the last three quarters of 1959 and the first quarter of 1960, is centered at October 1, 1959. Averaging these two figures yields a figure centered at August 15, the same as the dating of the third quarter.

The easiest way to calculate this properly centered moving average is given in Columns (3) through (5) of Table 13-3. In Column (3) is given a four-quarter moving total. The first figure, 391, is the total of the first four quarterly figures, 96, 103, 100, and 92. This figure is listed opposite the third quarter, 1959, although actually it is centered at July 1. The next four-quarter moving total is obtained by dropping the figure for the first quarter, 1959 and including the first quarter, 1960 figure. Hence, 388 is the total of 103, 100, 92, and 93.

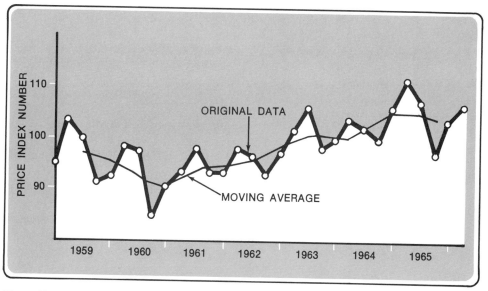

Figure 13-5

Feed grain index numbers of average prices received by farmers by quarters, 1959–1966.

The total of 391 and 388 or 779 is the first entry in Column (4). This represents the total for the eight months which would be present in the averaging of the first two simple four-quarter moving averages. Dividing this total by 8 yields the first two-of-a-four-quarter moving average figure of 97.38, properly centered at the middle of the third quarter, 1959.[2]

The moving averages given in Column (5) of Table 13-3 are shown in Figure 13-5 along with the original data. It is very useful to examine graphs in the calculation of seasonal indices because we can observe visually what is accomplished in each major step of the procedure. We have noted earlier that the original data, if stated in monthly or quarterly form, contain all of the components of trend, cycle, seasonal, and irregular movements. Although the time period is too short for trend to be revealed, we can observe in the series of feed grain price index numbers some effects of cyclical fluctuations as the data move into a trough at the end of 1960 and continue in an expansion swing thereafter. The repetitive annual rhythm of the seasonal movements is clearly discernible. Irregular movements are also present. The moving average which runs smoothly through the original data can be observed to follow the cyclical fluctuations rather closely and if the series were long enough, we would be able to see how the moving average describes trend movements as well. Another way to view this point is to note that the seasonal variations and to a large degree the irregular movements are absent from the smooth line which traces the path of the moving average. It should be noted that there

[2] If the computations are carried out on a calculating machine, it is most efficient to place the reciprocal of 8, or ⅛ in the keyboard and then multiply it by the totals in Column (4) to yield the desired moving average.

are no moving average figures corresponding to the first two and the last two quarters of original data. Correspondingly, if the original data were in monthly form and a twelve-month moving average were computed, there would be no moving averages to correspond to the first six months of data nor to the last six months of data.

The "ratio-to-moving average" figures, or original data, Column (2), divided by the moving average, Column (5), are given in Column (6) of Table 13-3. As is customary, these figures have been multiplied by 100 to express them in percentage form. They are often referred to as "percent of moving average" values, and may be represented symbolically as $(Y/MA) \times 100$. These values are graphed in Figure 13-6. As can be seen in the graph, the trend and cyclical movements are no longer present in these figures. The 100-base line represents the level of the moving average or the trend-cycle base. The fluctuations above and below this base line clearly reveal the repetitive seasonal movement of feed grain prices. As noted earlier, the irregular component is also present in these figures.

The next step in the procedure involves the attempt to remove the effect of irregular movements from the $(Y/MA) \times 100$ values. This is accomplished by averaging the percents of moving average figures for the same quarter. That is, the first quarter $(Y/MA) \times 100$ values are averaged, the second quarter values are averaged, and so forth. The average customarily used in this procedure is a modified mean, which is simply the arithmetic mean of the percents of moving average figures for each quarter over the different years, after eliminating the lowest and highest figures. It is desirable to make these deletions particularly when the highest and lowest figures tend to be atypical because

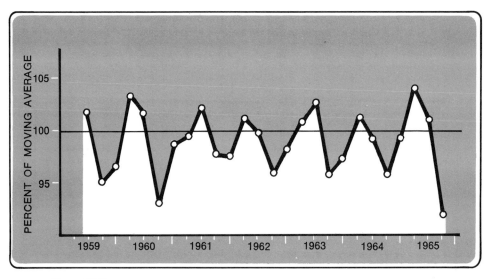

Figure 13-6
Percent of moving averages for feed grain price index numbers, 1959–1966.

TIME SERIES

TABLE 13-4

Feed Grain Price Index Numbers: Calculation
of Quarterly Seasonal Indices from Percent
of Moving Average Figures.

PERCENT OF MOVING AVERAGES
QUARTER

	I	II	III	IV
1959			102.69	95.46
1960	~~97.64~~	104.26	103.23	93.35
1961	99.31	~~100.40~~	103.47	~~98.41~~
1962	98.02	102.37	100.79	96.49
1963	98.73	101.87	~~104.82~~	96.31
1964	98.40	102.33	~~100.12~~	96.42
1965	~~100.58~~	~~105.04~~	101.89	~~93.11~~
Modified				
Means	98.6	102.7	102.4	95.6

Total of Modified Means = 399.3
Adjustment Factor   = 400/399.3 = 1.0018
          Seasonal Indices

I	II	III	IV
98.8	102.9	102.6	95.7

of erratic or irregular factors such as strikes, work stoppages, or other unusual occurrences.

The percent of moving average figures for each quarter are listed in Table 13-4. The highest and lowest figures have been designated as deleted by a line drawn through them, and the modified means of the remaining values are shown for each quarter. These means are 98.6, 102.7, 102.4, and 95.6, respectively, for the first through fourth quarters. The total of these modified means is 399.3. Since it is desirable that the four indices total 400, in order that they average 100%, each of them is multiplied by the adjustment factor of 400/399.3. This adjustment has the effect of forcing a total of 400 by raising each of the unadjusted figures by the same percentage. The final quarterly seasonal indices are shown on the bottom line of Table 13-4. In the case of monthly seasonal indices, a similar adjustment is made in order for the 12 monthly indices to total 1200; thus, the average monthly index equals 100%.

As indicated earlier, if interest centers on the pattern of seasonal variations itself, the four quarterly indices represent the final product of the analysis. On the other hand, sometimes the purpose of measuring seasonal variations is to eliminate them from the original data in order to examine, for example, the cyclical movements. The method of "deseasonalizing" the original data or adjusting these figures for seasonal movements is simply to divide them by the appropriate seasonal indices. This adjustment is shown in Table 13-3 for the feed grain price data by the division of the original figures in Column (2) by the seasonal indices in Column (7). The result is multiplied by 100, since the seasonal index is stated as a percentage rather than as a relative.

Let us illustrate the meaning of a deseasonalized figure by reference to the first line of figures in Table 13-3. The feed grain price index in the first quarter of 1959 was 96. Dividing this figure by the seasonal index for the first quarter of 98.8 and multiplying by 100 yields 97.2. This is the feed grain price index for the first quarter of 1959 adjusted for seasonal variations. *That is, it represents the level that food prices would have attained if there had not been the depressing effect of seasonality in the first quarter of the year.* All time series components, other than seasonal variations, are present in these de-seasonalized figures. This idea can be expressed symbolically as follows in terms of the aforementioned multiplicative model of the time series analysis:

(13.15) $$\frac{Y}{SI} = \frac{T \times C \times S \times I}{S} = T \times C \times I$$

The figures for the feed grain price index numbers adjusted for seasonal movements are graphed in Figure 13-7. It can be seen that the underlying cyclical movement is present in these data, irregular movements are indicated, and if a sufficiently long period had been used, say, at least a couple of business cycles, the trend would also be apparent. It may be noted that as compared to the plot of the original data in Figure 13-5, most of the repetitive seasonal movements are no longer present in the decentralized figures. However, ordinarily, as in this case too, the adjustment for seasonality is not perfect. To the extent that seasonal indices do not completely portray the effect of seasonality, division of original data by seasonal indices will not entirely remove these influences.

Seasonal indices are often used for the adjustment purpose just discussed.

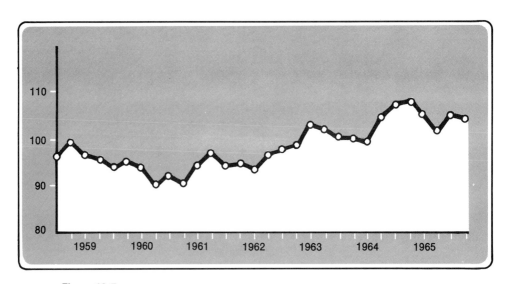

Figure 13-7
Deseasonalized figures for feed grain index numbers of average prices received by farmers by quarters, 1959–1966.

Economic time series adjusted for seasonal variations are often charted in the *Federal Reserve Bulletin,* the *Survey of Current Business,* and other publications. Also quarterly gross national product figures are often given as "seasonally adjusted at annual rates." These are simply deseasonalized quarterly figures multiplied by four to state the result in annual terms.

FORECASTING METHODS   13.6

We have seen how classical methods are used in analyzing the separate components of an economic time series. These methods involve an implicit assumption that the various components act independently of one another. For example, there were no specific procedures established for taking into account cyclical influences on seasonal variations, or long-run changes in the structure of business cycles. Special procedures can be established to gauge some of these interactions, but basically the model used in classical time series analysis assumes that there are independent sources of variation in economic time series and measures these sources separately. This decomposition or separation process, although often very useful for descriptive or analytical purposes, is nevertheless artificial. Therefore, it is not surprising that for a complex problem such as economic forecasting, it virtually never suffices simply to make mechanistic extrapolations based on classical time series analysis alone. However, time series analysis frequently is a very helpful starting point and an extremely useful supplement to other analytical and judgmental methods of forecasting.

In short-term forecasting, often a combined trend-seasonal projection provides a convenient first step. For example, as a first approximation, say, in a company's forecast of next year's sales by months, a projection of a trend figure for annual sales might be obtained. Then this figure might be allocated among months based on an appropriate set of seasonal indices. Of course, the basic underlying assumption in this procedure is the persistence of the historical pattern of trend and seasonal variations of the sales of this company into the next year. A more complete forecast might involve superimposing a cyclical prediction as well. Thus, for example, again the first step may involve a projection of trend to obtain an annual sales figure. Then, an adjustment of this estimate may be made based on judgment with respect to recent cyclical growth rates. Suppose that the past few years represented the expansion phase of a business cycle and the cyclical growth rate for the economy during the next year was predicted at about 4% by a group of economists. Assume further that the company in question had found these forecasts in the past were quite accurate and applicable to the company's own cyclical growth rate—over and above its own forecast of trend levels. Then the company might increase its trend forecast by this 4% figure to obtain a trend-cycle prediction. Again, if predictions by months were required, a monthly average could be obtained from the trend-cycle forecast and seasonal indices could be applied to yield the monthly allocation. Ordinarily, no attempt would be made to predict the irregular movements.

Cyclical forecasting and business indicators

Cyclical movements are more difficult to forecast than trend and seasonal elements. These cyclical fluctuations in a specific time series are strongly influenced by the general business cycle movements characteristic of large sectors of the overall economy. However, since there is considerable variability in the timing and amplitude with which many individual economic series trace out their cyclical swings, there is no simple mechanical method of projecting these movements.

Relatively "naive methods" such as the extension of the same percentage rate of increase or decrease in, say, sales as occurred last year or during the past few years are often made. These may be quite accurate, particularly if the period for which the forecast is made occurs during the same phase of the business cycle as the time periods from which the projections are made. However, the most difficult and most important items to forecast are the cyclical turning points at which reversals in direction occur. Obviously, managerial planning and implementation that has presupposed a continuation of a cyclical expansion phase can give rise to serious problems if an unpredicted cyclical downturn occurs during the planning period.

Many statistical series produced by governmental and private sources have been extensively used as business indicators. Some of these series represent activity in specific areas of the economy such as employment in non-agricultural establishments or average hours worked per week in manufacturing. Others are very broad measures of aggregate activity pertaining to the economy as a whole, as for example, gross national product and personal income. We have noted earlier that economic series, while exhibiting a certain amount of commonality in business cycle fluctuations, nevertheless display differences in timing and amplitude. The National Bureau of Economic Research has studied these differences carefully and has specified a number of time series as statistical indicators of cyclical revivals and recessions.

These time series have been classified into three groups. The first group consists of the so-called "*leading series.*" These series have usually reached their cyclical turning points prior to the analogous turns in general economic activity. The group includes series such as the layoff rate in manufacturing; value of new orders, durable goods industries; and the common stock price index, industrials, rails, and utilities. The second group are series whose cyclical turns have roughly *coincided* with those of the general business cycle. Included are such series as the unemployment rate, the industrial production index, gross national product, and dollar sales of retail stores. Finally, the third group consists of the "*lagging series,*" those whose arrivals at cyclical peaks and troughs usually lag behind those of the general business cycle. This group includes series such as plant and equipment expenditures, consumer installment debt, and bank interest on short-term business loans. It is worth noting that rational explanations stemming from economic theory can be given for the logic of the placement of the various series into the respective groups, in addition to the empirical observations themselves. These statistical indicators are adjusted for seasonal movements. They are published monthly in *Business Conditions Digest,* by the Bureau of the Census. Another publication which carries the National Bureau statistical indicators, as well as other time series with accompanying analyses, is *Economic Indicators,* published by the Council of Economic Advisors.

Probably the most widespread application of these cycle indicators is as an aid in the prediction of the timing of *cyclical turning points.* If, for example, most of the leading indicators move in an opposite direction from the prevailing phase of the cyclical activity, this is taken to be a possible harbinger of a cyclical turning point. A subsequent similar movement by a majority of the roughly coincident indices would be considered a confirmation of the fact that a cyclical turn was in progress. These cyclical indicators, like all other statistical tools, have their limitations and must be used carefully. They are not completely consistent in their timing, and leading indicators sometimes give incorrect signals of forthcoming turning points because of erratic fluctuations in individual series. Furthermore, it is not possible to predict, with any high degree of assurance, the length of time between a signal given by the leading series group of an impending cyclical turning point and the turning point itself. There has been considerable variation in this lead time during past cycles of business activity.

Other forecasting methods

Most individuals and companies engaged in forecasting do not depend upon any single method, but rather utilize a variety of different approaches. It stands to reason that if there is substantial agreement among a number of forecasts arrived at by relatively independent methods, greater reliance would be placed on this consensus than would have been on the results of any single technique.

Other methods of prediction range from very informal judgmental techniques to highly sophisticated mathematical models. At the informal end of this scale, for example, sales forecasts are sometimes derived from the combined outlooks of the sales force of a company, from panels of executive opinion, or from a composite of both of these. At the other end of the scale are the more formal mathematical models such as regression equations or complex econometric models. There is widespread usage of various types of regression equations by which firms attempt to predict the movements of their own company's or industry's activity on the basis of relationships to other economic and demographic factors. Often, for example, a company's sales are predicted on the basis of relationships with other series whose movements precede those of the sales series to be forecasted.

Among the most formal and mathematically sophisticated methods of forecasting in current use are econometric models. An *econometric model* is a set of two or more simultaneous mathematical equations that describe the interrelationships among the variables in the system. Some of the more complex models in current use for prediction of movements in overall economic activity include dozens of individual equations. Special methods of solution for the parameters of these equation systems have been developed, since in many instances ordinary least squares techniques are not appropriate. These econometric models have been primarily used for prediction at the level of the economy as a whole, and for industries, but are coming into increasing use for prediction at the company level as well.

Management uses forecasts as an important ingredient of its planning, operational, and control functions. Invariably, no single method is relied upon but judgment is applied to the results of various forecasting methods. Often, formal prediction techniques make their greatest contribution by narrowing considerably the area within which intuitive judgment is applied.

EXERCISES

1. Given the following data from the ABC Shirt Company:

$$Y = \text{Actual Sales, June } 1973 = 46{,}500$$
$$Y_t = \text{Trend Value, June } 1973 = 50{,}000$$
$$SI = \text{June Seasonal Index} = 90$$

(a) Express seasonally adjusted sales as a percent of trend. What general factors account for the difference between your calculated value and 100%?

(b) What is the meaning of the trend value?

2. A certain department store experiences marked seasonal variations in sales. The July seasonal index is 70, and the trend value for sales in July 1973 was $28,000. Do you think sales in July 1973 were closer to $28,000 or $19,600 ($28,000 × 0.7)? Discuss.

3. (a) In the percent-of-moving average method (ratio-to-moving average method) of determining seasonal indices, how is each of the non-seasonal elements removed? Explain.

(b) Given the following data on milk production in a section of the United States during June, July, and August of 1973 and relevant seasonal indices on milk production:

MONTH	MILK PRODUCTION (THOUSANDS OF LBS)	SEASONAL INDEX
June 1973	13,178	126.54
July 1973	12,663	117.90
August 1973	11,625	105.99

(1) Would you attribute the decline in milk production during this period merely to seasonal variations? Specify the calculations you would make to answer this question, but do not perform these calculations.

(2) Explain specifically how you would determine the effect of business cycle fluctuations on milk production. Specify any information that would be required in addition to that given above.

4. The trend equation for sales of the Expo Corporation is as follows:

$$Y_t = 190 + 0.24x$$
$$x = 0 \text{ in July 1963}$$

x is in monthly intervals

Y is monthly sales in millions of dollars

Actual figures for certain months in 1967 are given below:

MONTH	ACTUAL SALES (MILLIONS OF DOLLARS)	SEASONAL INDEX
September	167	88
October	198	104
November	210	114
December	391	200

(a) What does the seasonal index of 88 for September mean?

(b) Isolate for December the effect of each component of a time series (trend, cycle and random, seasonal).

5. The following data pertain to the number of automobiles sold by the Easy Wheels Corporation:

YEAR	QUARTER	NO. OF AUTOS SOLD
1965	1st	152
	2nd	277
	3rd	203
	4th	174
1966	1st	205
	2nd	363
	3rd	255
	4th	182
1967	1st	171
	2nd	325
	3rd	233
	4th	180
1968	1st	202
	2nd	396
	3rd	274
	4th	238
1969	1st	212
	2nd	350
	3rd	246
	4th	208
1970	1st	241
	2nd	453
	3rd	362
	4th	355

(a) Using the ratio-to-moving average method, determine constant seasonal indices for each of the four quarters.

(b) Do you think constant seasonal indices should be employed in this problem? Why or why not?

(c) Assuming the constant seasonal indices are appropriate, adjust the quarterly sales figures between 1965 and 1970 for seasonal variations.

(d) Assume the trend in sales for the Easy Wheels Corporation can be described by the following equation:

$$Y_t = 200 + 25x$$
$$x = 0 \text{ in 2nd quarter of 1965}$$
$$x \text{ is in one-year intervals}$$
$$Y \text{ is the number of autos sold}$$

Do you see any evidence of cycles in data between 1965 and 1970? What is the basis of your answer?

6. The following table presents the consumption of electric power in the United States:

YEAR	CONSUMPTION (BILLION KILOWATT-HOURS)
1910	20
1920	57
1930	116
1940	182
1950	396
1960	832

(a) Fit a linear trend line to the above data by the method of least squares.

(b) Interpret the meaning of the constants of the trend equation to the natural numbers specifically in terms of this problem.

(c) In 1932, consumption of electric power was 100 billion kilowatt-hours. Compute and interpret the relative cyclical residual for that year.

7. Given the following values for a particular series: actual value, December 1972 = 320 units; seasonal index, December = 200. $Y_t = 150 + 5.0x - 0.2x^2$, with $x = 0$ at June 15, 1965; x is in one-year units.

(a) What is the meaning of the seasonal index, 200?

(b) What is the relative cyclical residual for December 1972 adjusted for seasonal variation?

(c) What does the relative cyclical residual mean?

8. The following sentences refer to the ratio-to-moving average method of measuring seasonal variation when applied to United States monthly gasoline sales from 1950 to 1973. Insert in the blank space of each sentence the letter corresponding to the phrase that will complete the sentence most appropriately.

(a) A twelve-month moving total was computed because _____.
   (1) Trend is thus eliminated.
   (2) This will give column totals equal to 1200.
   (3) Seasonal variation cancels out over a period of twelve months.

(b) A two-item total is then taken of the twelve-month totals in order to _____.
   (1) Obtain moving average figures for the first and last six months.
   (2) Center the moving average properly.
   (3) Eliminate the rest of the random movements.

(c) This two-item total of a twelve-month moving total is divided by _____ to get the centered twelve-month moving average.
   (1) 14
   (2) 2

        (3) 12
        (4) 24
(d) This moving average contains _____.
        (1) All of the trend, most of the cycle, all of the seasonal variation, and some irregular (random) variation.
        (2) All of the trend, most of the cycle, and possibly some irregular (random) variation.
        (3) Most of the trend, most of the cycle, and all of the irregular (random) variation.
(e) The original data are then divided by the moving average figures. These ratio-to-moving average figures, Y/MA values, contain ____.
        (1) All of the seasonal, possibly some of the cycle, and practically all of the irregular.
        (2) Seasonal only.
        (3) All of the trend, most of the cycle, and none of the irregular.
(f) Modified means are taken of the specific seasonal relatives, Y/MA in order to _____.
        (1) Eliminate the nonseasonal elements from the Y/MA values.
        (2) Get rid of seasonal elements.
        (3) Eliminate the trend in the Y/MA values.
        (4) Compensate for a changing seasonal pattern.
(g) To adjust the original data for seasonal variation, one computes _____.
        (1) Original data times seasonal index.
        (2) Seasonal index divided by original data.
        (3) Original data divided by seasonal index.

14 *Index Numbers*

14.1   THE NEED FOR AND USE OF INDEX NUMBERS

In our daily lives, we often make judgments that involve summarizing the differences that have taken place in economic variables over time or differences that exist among two or more places. For example, a family's income may have increased by 20 percent over a five-year period. Suppose this was a period of inflation, or generally rising prices. Has the family's "real income" increased? That is, can the family now purchase more goods and services with its income than five years earlier? If the general price level of items has increased more than the 20 percent figure, the family clearly cannot purchase as much. On the other hand, if prices have increased less than 20 percent, the family's real income has increased.

As a second example, let us consider a company that wishes to transfer an executive from St. Louis to New York City. What should his minimum salary increase be to allow for the cost of living in New York?

Both these cases require measurements of general price levels. The prices of the various items purchased by the family have doubtlessly increased at different rates, a few may even have decreased. Similarly, although some prices in New York City are much higher than in St. Louis, some may be lower. How can we summarize in a single composite figure the average differences that exist between the two time periods or the two cities? Index numbers serve to answer questions of this type. The term *index number* refers to a summary measure that states a relative comparison between groups of related items.

In its simplest form, an index number is nothing more than a percentage relative that expresses the relationship between two figures with one of the figures used as the base. For example, in a time series of prices of a particular commodity, the prices may be expressed as price relatives by dividing each figure by the price in the base period. In the calculation of economic indices, it is conventional to state the relative numbers as percentages, where the base period is 100 (%). For example, suppose the prices of a pound of a certain brand of coffee in three different years were 1970, $0.80; 1971, $0.90; and 1972, $1.00. Then the price relatives of the three figures, with 1970 as a base (written, 1970 = 100), are

YEAR	PRICE RELATIVES (1970 = 100)
1970	$0.80/$0.80 = 1.000 or 100.0
1971	$0.90/$0.80 = 1.125 or 112.5
1972	$1.00/$0.80 = 1.250 or 125.0

As indicated, the price relative for any given year is obtained by dividing the price for that year by the figure for the base period. The resulting figure is multiplied by 100 to express the price relative in percentage form.

The price relatives may be interpreted as follows. In 1971 it would have cost 112.5% of the price in 1970 to purchase a pound of this brand of coffee, or there was a 12.5% increase from 1970 to 1971 in the price. Similarly, in 1972 the price was 125% of the 1970 price, or the price had risen 25% from 1970 to 1972. Of course, usually we are interested in changes of prices of more than one item. For example, in the cases of the family and the transferred executive cited earlier, we were interested in the prices of all commodities and services included in the cost of living. For such purposes, we are interested in combining the price relatives for many different items into a single summary figure for each time period. As other examples, we may wish to compute a food price index, a clothing price index, or an index of medical costs. Such summary figures constitute composite index number series. Consequently, our discussion will pertain solely to composite indices. In keeping with general practice, we will ordinarily use the term "index number" to mean "composite index number."

Series of index numbers are extremely useful in the study and analysis of economic activity. Every economy, regardless of the political and social structure of the environment within which it operates is engaged in the production, distribution, and consumption of goods and services. Convenient methods of aggregation, averaging, and approximation are required to summarize the myriad of individual activities and transactions that take place. Index numbers have proved to be very useful tools in this connection. Thus, we find indices of industrial production, agricultural production, stock market prices, wholesale prices, consumer prices, prices of exports and imports, incomes of various types, and so forth in common use. A convenient classification for economic indices is in terms of indices of price, quantity, or value. The present discussion concentrates primarily on price indices because most of the problems of construction, interpretation, and use of indices may be illustrated in terms of such measures. At first we will deal with methods of construction of index numbers, using the illustrative data of a simple example. Then we will consider some of the general problems of index number construction.

14.2   AGGREGATIVE PRICE INDICES

In order to illustrate the construction and interpretation of price indices, we will consider an artificial problem of constructing a price index for a list of only four food commodities. A more realistic counterpart of this problem is represented by the Consumer Price Index produced by the Bureau of Labor Statistics (BLS) of the U.S. Department of Labor. This index is used in many ways and provides the basis for many economic decisions. For example, fluctuations in the wages of over 3 million workers are partially based on changes that occur in the index figures. The series is also closely watched by monetary authorities as an indicator of inflationary movements of prices.

Unweighted aggregates index

In our simplified illustration, we will use a base period of 1970, and we will be interested in the change in these prices from 1970 to 1973 for a typical family of four that purchased these products at retail prices in a certain city. The universe and other basic elements of the problem should be very carefully defined. However, we will purposely leave these matters very indefinite and will concentrate on the methods of construction and interpretation of the various indices. Hence, these indices will deliver different answers to our vaguely worded problem. In Table 14-1 are shown the basic data of the problem and the calculation of the unweighted aggregates index. As indicated at the bottom of Table 14-1, the prices per unit are summed (or aggregated) for each year. Then, one year is selected as a base, in our case, 1970. The price index for any given year is obtained by dividing the sum of prices for that year by the similar sum for the base period. The resulting figure is multiplied by 100 to express the index in percentage form. Hence, the index takes the value 100 in the base period. If the symbol P_0 is used to denote the price in a base period and P_n the price in a nonbase period, the general formula for the unweighted aggregates index may be expressed as follows:

UNWEIGHTED AGGREGATES PRICE INDEX

(14.1)
$$\frac{\Sigma P_n}{\Sigma P_0} \cdot 100$$

Let us interpret the index figure of 117.7 for 1973. It would have cost $3.00 in 1970 to have purchased one pound of coffee, one loaf of bread, one dozen eggs, and one pound of hamburger. The corresponding cost in 1973 was $3.53. Expressing $3.53 as a percentage of $3.00, we find that in 1973 it would have cost 117.7% of the cost in 1970 to have purchased one unit each of the specified commodities. Stated in terms of percentage change, it would have cost 17.7% *more* in 1973 than in 1970 to have purchased the stated bill of goods.

The interpretation of the unweighted relative of aggregates index is very straightforward. However, this type of index suffers from the serious limitation that it is unduly influenced by high priced commodities. The total of prices in

TABLE 14-1
Calculation of the Unweighted Aggregates
Index for Food Prices, 1970 and 1973.

| | UNIT PRICE | |
| | 1970 | 1973 |
FOOD COMMODITY	P_{70}	P_{73}
Coffee (pound)	$1.00	$1.15
Bread (loaf)	0.35	0.43
Eggs (dozen)	0.65	0.70
Hamburger (pound)	1.00	1.25
	$3.00	$3.53

**UNWEIGHTED AGGREGATES INDEX
FOR 1973, ON 1970 BASE**

$$\frac{\Sigma P_{73}}{\Sigma P_{70}} \cdot 100 = \frac{\$3.53}{\$3.00} \times 100 = 117.7$$

1970 and 1973, respectively, were $3.00 and $3.53, an increase of $0.53. If we added to the list of commodities one which declined from $4.00 to $3.00 per unit from 1970 to 1973, the totals for 1970 and 1973 would then become $7.00 and $6.53. Hence the price index figure for 1973 would be 93.3, indicating a decline in prices of 6.7%. Although the prices of four commodities increased and only one decreased, the overall index shows a decline, because of the dominance of the one high priced commodity. Furthermore, this high priced commodity may be a relatively unimportant one in the consumption pattern of the group to which the index pertains. Clearly, this type of so-called "un-weighted index" is one which has an inherent haphazard weighting scheme, as indicated above.

Another deficiency of this type of index is the arbitrary nature of its calculation because of the quoted units for which the prices are stated. For example, if the price of eggs were stated per half-dozen rather than per dozen or if any of the other prices were stated on a different basis, the calculated price index figure would change. However, even if all of the prices were stated for the same quoted unit of each commodity, say, per pound, the problems concerning the inherent haphazard weighting scheme would still remain. In this case, the index would be dominated by the commodities that happened to have high prices on a per pound basis. These may be the very commodities that are purchased least, because of their expensive nature. Because of the difficulties of converting a simple aggregative index into an economically meaningful measure, the need for applying explicit weights is apparent. We now turn to weighted aggregative price indices.

Weighted aggregates indices

In order to attribute the appropriate importance to each of the items included in an aggregative index, some reasonable weighting plan must be used. The weights to be used depend on the purposes of the index calculation, that is, on the economic question that the index attempts to answer. In the case of a consumer food price index such as the one we have been discussing, reasonable weights would be given by the amounts of the individual food commodities purchased by the consumer units to whom the indices pertain. These would constitute so-called "quantity weights," since they represent quantities of commodities purchased. The specific types of quantities to be used in an aggregative index would depend, of course, on the economic nature of the index computed. Hence, an aggregative index of export prices would use quantities of commodities and services exported, an index of import prices would use quantities imported, and so forth.

Table 14-2 shows the prices of the same food commodities given in Table 14-1, but also quantities consumed during the base period, 1970. Specifically, these figures given in the column Q_{70} (the symbol Q denotes quantity) represent average quantities consumed per week in 1970 by the consumer units to which the index pertains. Hence, they indicate an average consumption of one pound of coffee, three loaves of bread, etc. The figures given under the column labeled $\dot{P}_{70}Q_{70}$ indicate the dollar expenditures for the quantities purchased in 1970. Correspondingly, the numbers under the column headed $P_{73}Q_{70}$ specify what it would have cost to purchase these amounts of food in 1973. Hence the sums, $\Sigma P_{70}Q_{70} = \$4.35$ and $\Sigma P_{73}Q_{70} = \$5.09$, indicate what it could have cost to purchase the specified quantities of food commodities in 1970 and 1973, respectively. The index number for 1973 on a 1970 base is given by expressing the figure for $\Sigma P_{73}Q_{70}$ as a percentage of the $\Sigma P_{70}Q_{70}$ figure, yielding in this case a figure of 117.0, as shown at the bottom of Table 14-2. Of course, the index number for the base period 1970 would be 100.0.

What this type of index measures is the change in the total cost of a fixed bill of goods. For example, in this case, the 117.0 figure indicates that in 1973 it would have cost 117.0% of what it cost in 1970 to purchase the weekly market basket of commodities representing an average consumption pattern in 1970.

TABLE 14-2
Calculation of the Weighted Aggregates Index for Food Prices,
Using Base Period Quantities Consumed as Weights (Laspeyres Method).

| | | | QUANTITY | | |
| | 1970 | 1973 | 1970 | | |
FOOD COMMODITY	P_{70}	P_{73}	Q_{70}	$P_{70}Q_{70}$	$P_{73}Q_{70}$
Coffee (pound)	$1.00	$1.15	1	$1.00	$1.15
Bread (loaf)	0.35	0.43	3	1.05	1.29
Eggs (dozen)	0.65	0.70	2	1.30	1.40
Hamburger (pound)	1.00	1.25	1	1.00	1.25
				$4.35	$5.09

WEIGHTED RELATIVE OF AGGREGATES INDEX, WITH BASE PERIOD WEIGHTS:
FOR 1973, ON 1970 BASE

$$\frac{\Sigma P_{73}Q_{70}}{\Sigma P_{70}Q_{70}} \cdot 100 = \frac{\$5.09}{\$4.35} \times 100 = 117.0$$

Roughly speaking, this indicates an average price rise of 17.0% for this food market basket from 1970 to 1973. Referring back to the corresponding index figure for the simple or unweighted index of 117.7%, we see that it is quite close to the 117.0 figure for the weighted index. The reason for the closeness of the two figures is that in our example we have assumed that the prices of all four commodities have moved in the same direction with the percentage changes all falling between about 8 to 25%. On the other hand, if there had been more dispersion in price movements, for example, if some prices increased while some decreased, as is often actually the case, the weighted index would have tended to differ more from the unweighted one.

The weighted aggregative index using base period weights is also known as the Laspeyres index. The general formula for this type of index may be expressed as follows:

WEIGHTED AGGREGATES PRICE INDEX,
BASE PERIOD WEIGHTS (LASPEYRES METHOD)

(14.2)
$$\frac{\Sigma P_n Q_0}{\Sigma P_0 Q_0} \cdot 100$$

where P_0 = price in a base period
P_n = price in a non-base period
Q_0 = quantity in a base period

The basic dilemma posed by the use of any weighting system is clearly illustrated by a consideration of the Laspeyres index. Since an aggregative price index attempts to measure price changes and contains data on both prices and quantities, it appears logical to hold the quantity factor constant in order to isolate change attributable to price movements. If both prices and quantities were permitted to vary, their changes would be entangled and it would not be possible to ascertain that part of the movement due to price changes. However, by keeping quantities fixed as of the base period in a consumer price index, the Laspeyres index assumes a frozen consumption pattern. As time goes on, this becomes a more and more unrealistic and untenable assumption. The consumption pattern of the current period would seem to represent a more realistic set of weights from the economic viewpoint.

However, let us consider the implications of the use of an aggregative index using current period (non-base period) weights. This type of index is known as the Paasche method. The general formula for a Paasche index is

WEIGHTED AGGREGATES PRICE INDEX
CURRENT PERIOD WEIGHTS (PAASCHE METHOD)

(14.3)
$$\frac{\Sigma P_n Q_n}{\Sigma P_0 Q_n} \cdot 100$$

Hence, if such an index is prepared on an annual basis, the weights would have to change each year, since they would consist of current year quantity figures. The 1971 Paasche index would be computed by the formula $\Sigma P_{71} Q_{71} / \Sigma P_{70} Q_{71}$, the 1972 index would be $\Sigma P_{72} Q_{72} / \Sigma P_{70} Q_{72}$, and so forth. The interpretation of any one of the resulting figures in terms of price change from the base period, assuming the consumption pattern of the current period, is clear. However, the use of changing current period weights destroys the possibility of obtaining unequivocal measures of year-to-year price change. For example, if the Paasche formulas for the 1971 and 1972 indices given above are compared, it will be noted that both prices and quantities have changed. Therefore, no clear statement can be made about price movements from 1971 to 1972. Thus, the use of current year weights makes year-to-year comparisons of price changes impossible.

Another practical disadvantage of using current period weights is the necessity of obtaining a new set of weights in each period. Let us consider the U.S. Bureau of Labor Statistics Consumer Price Index as an example of the difficulty of obtaining such weights. In order to obtain an appropriate set of weights for this index, the BLS conducts a massive sample survey of the expenditure patterns of families in a large number of cities. Such surveys have been carried out at about ten-year intervals. They are large-scale, expensive undertakings. From a practical standpoint, it would be simply infeasible for such surveys to be conducted at, say one-year or more frequent time intervals. Because of these disadvantages, the current period weighted aggregative method is not used in any well-known price index number series.

In summary, because of the above-mentioned considerations, and other factors as well, probably the most generally satisfactory type of price index is the weighted relative of aggregates index, using a fixed set of weights. The term "fixed set of weights" rather than "base period weights" is used here, because the weights may pertain to a period which is somewhat different from the time period which represents the base for measuring price changes. For example, one of the base periods for the Consumer Price Index was 1957–1959, whereas the weights were derived from a 1960–1961 survey of consumer expenditures.[1] The BLS revises its weighting system about every ten years and also changes the reference base period for the measurement of price changes with about the same frequency. This procedure constitutes a workable solution to the dilemma of needing to retain constant weights in order to isolate price change, and requiring up-to-date weights in order to have a recent realistic description of consumption patterns.

The weighted relative of aggregates index using a fixed set of weights described in the preceding paragraph is referred to as the *fixed-weight aggregative index* and is defined by the formula

WEIGHTED RELATIVE OF AGGREGATES PRICE INDEX WITH FIXED WEIGHTS

(14.4)
$$\frac{\Sigma P_n Q_f}{\Sigma P_0 Q_f}$$

where Q_f denotes a fixed set of quantity weights. The Laspeyres method may be viewed as a special case of this index in which the period to which the

[1] The most recent reference base for prices is 1967.

weights refer is the same as the base period for prices. In order to clarify discussion of the two different time periods, the term "weight base" is used for the period to which the quantity weights pertain, whereas the term "reference base" is used to designate the time period from which price changes are measured. Of course, a distinct advantage of a fixed-weight aggregative index is that the reference base period for measuring price changes may be changed without a corresponding change in the weight base. In certain instances, this is a useful and practical procedure, particularly in the case of some U.S. government indices that utilize data from censuses or large-scale sample surveys for changes in weights.

AVERAGE OF RELATIVES INDICES   14.3

A second basic method of construction of price indices is the *average of relatives* procedures. In an average of relatives index, the first step involves the calculation of a price relative for each commodity by dividing its price in a non-base period by the price in a base period. Then an average of these price relatives is calculated. Just as in the case of aggregative indices, averages of relatives indices may be either unweighted or weighted. We consider first the unweighted indices, using the same data on prices as were used in the preceding section.

Unweighted arithmetic mean of relatives index

The price data previously shown in Tables 14-1 and 14-2 are given in Table 14-3. The first step in the calculation of any average of relatives price index is the calculation of price relatives in which the price of each commodity is expressed as a percentage of the price in the base period. These price relatives for 1973 on a 1970 base, denoted $(P_{73}/P_{70}) \times 100$ are shown in Column (4) of Table 14-3. Theoretically, once the price relatives are obtained, any average, including the arithmetic mean, median, mode, etc., could conceivably be used as a measure of their central tendency. The arithmetic mean has been most frequently used, doubtless because of its simplicity and familiarity. The method of computation and the interpretation of the unweighted arithmetic mean of relatives are given below. At the bottom of Table 14-3 is shown the calculation of the unweighted arithmetic mean of relatives for 1973 on a 1970 base. The formula is Equation (3.1), with the price relatives representing the items to be averaged. Stated in general form, the formula for this unweighted arithmetic mean of relatives is:

UNWEIGHTED ARITHMETIC MEAN OF RELATIVES INDEX

(14.5)
$$\frac{\Sigma\left(\dfrac{P_n}{P_0} \cdot 100\right)}{n}$$

TABLE 14-3

Calculation of the Unweighted Arithmetic Mean
of Relatives Index of Food Prices for 1973
on a 1970 Base.

FOOD COMMODITY (1)	UNIT PRICE 1970 P_{70} (2)	UNIT PRICE 1973 P_{73} (3)	PRICE RELATIVE $\dfrac{P_{73}}{P_{70}} = 100$ (4)
Coffee (pound)	$1.00	$1.15	115.0
Bread (loaf)	0.35	0.43	122.9
Eggs (dozen)	0.65	0.70	107.7
Hamburger (pound)	1.00	1.25	125.0
			470.6

**UNWEIGHTED ARITHMETIC MEAN OF RELATIVES INDEX
FOR 1973, ON A 1970 BASE**

$$\frac{\Sigma \left(\frac{P_{73}}{P_{70}} \cdot 100 \right)}{4} = \frac{470.6}{4} = 117.7$$

where

$$\frac{P_n}{P_0} \cdot 100 = \text{the price relative for a commodity or service}$$

$$n = \text{the number of commodities and services}$$

Although this is an unweighted index, just as in the case of the unweighted aggregative index, there is, in fact, an inherent weighting pattern present. It is useful to consider the implications of this inherent weighting system. In the unweighted arithmetic mean of relatives, percentage increases are balanced off against equal percentage decreases. For example, if we consider two commodities, one whose price increased by 10% and one whose price declined by 10% from 1970 to 1973, the respective price relatives for 1973 on a 1970 base would be 110 and 90. The unweighted arithmetic mean of these two figures is 100, indicating that, on the average, prices have remained unchanged.

In summary, the unweighted arithmetic mean attaches the same weight to equal percentage changes in opposite directions. However, this method does not provide for an explicit weighting in terms of the importance of the commodities whose prices have changed. Since it is widely recognized that explicit weighting is required to permit the individual items in an index to exert their proper influence, virtually none of the important governmental or private organization price indices are of the "unweighted" variety. We now turn to a consideration of weighted average of relatives indices.

EXERCISES

1. Assume the following is an index number series for department store prices in a certain city:

CITY DEPARTMENT STORE INDEX

1961	100
1970	270
1971	230
1972	290
1973	310

As the analyst for a large department store you are asked to construct your own index for this store using a survey taken in 1965 which showed that the average store customer spent $250 on clothing, $50 on furniture, and $100 on all other items in that year. You also know the following:

AVERAGE PRICES (IN DOLLARS)

	1961	1965	1972
Clothing	20	25	40
Furniture	40	50	60
All Others	8	10	11

Using the weighted relative of aggregate price index with base period weights
(a) Calculate the index for 1961, 1965, and 1972 if 1965 is the base year.
(b) Compare the increase in price levels for your store from 1961 to 1972 with the corresponding increase as determined from the city department store index.

2. The Jones Metal Company uses three raw materials in its business. Given below are the average prices and the quantities consumed of these three products in 1967 and 1973.

	1967		1973	
PRODUCT	PRICE	QUANTITY	PRICE	QUANTITY
A	$20	20	$25	30
B	1	100	2	120
C	5	50	6	70

(a) Compute an appropriate weighted aggregates price index for 1973 on a 1967 base.
(b) A competitor reported that his 1973 index for the same products (1967 base year) was 120. Would you conclude that Jones paid more per unit in 1973 than his competitor? Why?

3. A price index of two commodities is to be constructed from the following data:

	UNIT PRICE		QUANTITIES CONSUMED	
COMMODITY	1972	1973	1972	1973
A	$1.00	$.50	3	2
B	.30	.60	7	8

A simple unweighted arithmetic mean of the two price relatives for 1973 on a 1972 base indicates that prices in 1973 were, on the average, 25% higher than in 1972. A simple unweighted arithmetic mean of the two price relatives for 1972 on a 1973 base indicates that prices in 1972 were, on the average, 25% higher than in 1973.

(a) How do you explain these paradoxical results?

(b) Compute what you consider to be the most generally satisfactory price index for 1973 using 1972 as a base year. You may use any of the above data you deem appropriate.

(c) Explain precisely the meaning of the answer obtained from your calculation in (b).

Weighted arithmetic mean of relatives indices

Although several averages can theoretically be used for calculating weighted averages of relatives, in fact, only the weighted arithmetic mean is ordinarily employed.

The general formula for a weighted arithmetic mean of price relatives is

WEIGHTED ARITHMETIC MEAN OF RELATIVES, GENERAL FORM

$$\frac{\Sigma\left(\frac{P_n}{P_0} \cdot 100\right)w}{\Sigma w}$$

(14.6)

where $w =$ the weight applied to the price relatives.

Customarily, the weights used in this type of index are values, such as values consumed, produced, purchased, or sold. For example, in the type of food price index we have used as our illustrative problem, the weights would be values consumed, that is, the dollar expenditures on the individual food commodities by the typical family to whom the index pertains. It seems reasonable that the importance attached to the price change for each commodity be indicated by the amounts spent on these commodities. In the field of index number construction, value = price × quantity. For example, if a commodity has a price of $.10 and the quantity consumed is three units, then the *value* of the commodity consumed is $.10 × 3 = $.30. Since prices and quantities can pertain to either a base period or a current period, the following systems of weights are all possibilities: P_0Q_0, P_0Q_n, P_nQ_0, and P_nQ_n. The weights P_0Q_0 and P_nQ_n are, respectively, base period values and current period values, the other two are mixtures of base and current period prices and quantities. Interestingly, the weighting systems P_0Q_0 and P_0Q_n, when used in the weighted arithmetic mean of relatives, result in indices which are algebraically identical to the Laspeyres and Paasche aggregative indices, respectively. This point is illustrated in (14.7), where base period weights P_0Q_0 are used in the weighted arithmetic mean of relatives.

WEIGHTED ARITHMETIC MEAN OF RELATIVES, WITH BASE PERIOD VALUE WEIGHTS

$$\frac{\Sigma\left(\frac{P_n}{P_0}\right)P_0Q_0}{\Sigma P_0Q_0} \cdot 100 = \frac{\Sigma P_nQ_0}{\Sigma P_0Q_0} \cdot 100$$

(14.7)

As is clear from (14.7), the P_0's in the numerator cancel, yielding the

Laspeyres index. The calculation of the weighted arithmetic mean of relatives, using 1970 base period value weights is given in Table 14-4, for the data of our illustrative problem. The numerical value of the index is, of course, exactly the same as that obtained previously for the weighted aggregative index with base period quantity weights (Laspeyres Method) in Table 14-2.

TABLE 14-4

Calculation of the Weighted Arithmetic Mean of Relatives Index of Food Prices
for 1973 on a 1970 Base, Using Base Period Weights

FOOD COMMODITY (1)	PRICES		PRICE RELATIVES $\frac{P_{73}}{P_{70}} \cdot 100$ (4)	QUANTITY 1970 Q_{70} (5)	$P_{70}Q_{70}$ (6)	WEIGHTED PRICE RELATIVES COL. (4) × COL. (6) $\left(\frac{P_{73}}{P_{70}} \cdot 100\right)\left(P_{70}Q_{70}\right)$
	1970 P_{70} (2)	1973 P_{73} (3)				
Coffee (pound)	$1.00	1.15	115.0	1	$1.00	$115.00
Bread (loaf)	0.35	0.43	122.9	3	1.05	129.05
Eggs (dozen)	0.65	0.70	107.7	2	1.30	140.01
Hamburger (pound)	1.00	1.25	125.0	1	1.00	125.00
					$4.35	$509.06

WEIGHTED ARITHMETIC MEAN OF RELATIVES FOR 1973, ON 1970 BASE, USING BASE PERIOD VALUE WEIGHTS

$$\frac{\Sigma\left(\frac{P_{73}}{P_{70}} \cdot 100\right)(P_{70}Q_{70})}{\Sigma P_{70}Q_{70}} = \frac{\$509.06}{4.35} = 117.0$$

Since the two indices in (14.7) are algebraically identical, it would seem immaterial which is used, but there are instances when it is more feasible to compute one rather than the other. For example, it is more convenient to use the weighted average of relatives than the Laspeyres index when value weights are easier to obtain than quantity weights; when the basic price data are more easily obtainable in the form of relatives than absolute values, and when an overall index is broken down into a number of component indices and there is a desire for comparison of the individual components in the form of relatives. As an illustration of the first of these situations, it is usually easier for manufacturing firms to furnish value of production weights in the form of "value added by manufacturing" (sales minus cost of raw materials) than to provide detailed data on quantities produced.

As indicated earlier, the Paasche index and the weighted arithmetic mean of relatives with a P_0Q_n weighting system are algebraically identical. The reasons given for the wider use of the Laspeyres than the Paasche index analogously apply to a similarly wider usage of weighted means of relatives with a P_0Q_0 weighting system than with a P_0Q_n scheme. The other two possible value weighting systems P_nQ_0 and P_nQ_n create interpretational difficulties, and therefore are not utilized in any of the important indices.

EXERCISES

1. A small electrical company produces three models of household exhaust fans. Average unit selling prices and quantities sold in 1970 and 1973 were as follows.

	1970		1973	
MODEL	PRICE	QUANTITY (IN 1000's)	PRICE	QUANTITY (IN 1000's)
Economy	$20	12	$22	15
Model *B*	30	4	36	4
Model *A*	35	8	40	9

(a) Calculate the index of fan prices for 1970 on a base year of 1973. Use the arithmetic mean of relatives method with base period weights.

(b) Explain, in words understandable to a layman, precisely what the value of your index means.

2. (a) Compute an index of apple prices for the data below by the weighted arithmetic mean of relatives method with 1966 = 100, using base year weights.

(b) Compute a price index by the weighted aggregate method, using base year weights and the same base year.

	PRICE (DOLLARS PER BUSHEL)		PRODUCTION (MILLIONS OF BUSHELS)	
	WINESAP	MACINTOSH	WINESAP	MACINTOSH
1966	$2.45	$2.15	1.28	1.40
1967	2.52	2.23	1.32	1.47
1968	2.60	2.41	1.31	1.52

3. (a) Using the data of Problem 2, compute an index of apple *production* by the weighted arithmetic mean of relatives method, with 1966 = 100 using base year weights.

(b) Compute an index by the weighted aggregate method, on the same base.

(c) Compare your results in (a) and (b).

14.4 GENERAL PROBLEMS OF INDEX NUMBER CONSTRUCTION

In a brief treatment, it is not feasible to discuss all of the relevant problems of index number construction. However, many of the important matters are subsumed under the following categories: (1) selection of items to be included, and (2) choice of a base period.

Selection of items to be included

In the construction of price indices as in other problems involving statistical methods, the definition of the problem and the statistical universe to be investigated are of paramount importance. Most of the widely used price index number series are produced by governmental agencies or sizable private organizations and are used in a large variety of ways. Hence, it is not feasible to state a simple purpose for each price index from which a clear definition of the problem and statistical universe might follow. However, every index attempts to answer meaningful questions, and it is these general purposes of an index that determine the specific items to be included. As an example, let us consider the Consumer Price Index produced by the Bureau of Labor Statistics of the U.S. Department of Labor. This index attempts to answer a question concerning the average movement of certain prices over time. The specific nature of this question about price movements determines the items to be included in the index. Similarly, many of the limitations of the use of the index for the aforementioned widely different purposes stem from what the index does and does not attempt to measure.

Let us pursue the illustration of the Consumer Price Index. Essentially, what this index attempts to measure is how much it would cost at retail to purchase a particular combination of goods and services compared to what it would have cost in a base period. More specifically, the combination of goods and services consists of items selected to represent a typical "market basket" of purchases by city wage earners and city clerical workers and their families. These families are considered to have "moderate incomes." The relevant universe comprises about 40% of the U.S. population. Hence, the index does not attempt to describe changes in prices of purchases by low income families, high income families, farm families, or the families of business men or professional people. By means of periodic consumer surveys, the Bureau determines the goods and services purchased by the specified families and how these families spread their spending among these items. In summary, the general question the index purports to answer determines the items to be included. Obviously, if the indices have other purposes, as for example, indices of export prices or agricultural prices, very different lists of items would be included.

However, even when the general purpose of an index is clearly defined, many problems remain concerning the choice of items to be included. In the case of the Consumer Price Index, the BLS has determined that there are about 2000 items that moderate income city wage earner families purchase. However, the BLS includes only about 400 of these goods and services, having found that these few hundred accurately reflect the average change in the cost of the entire market basket. The choice of the commodities to be included in a price index is ordinarily not determined by usual sampling procedures. Each good and service cannot be considered a random sampling unit equally as representative as any other unit. Rather, an attempt is made to include practically all of the most important items, and by pricing these, to obtain a representative portrayal of the movement of the entire population of prices. If subgroup indices are required, for example, indices of food, housing, medical care, etc., as well as an overall consumers' price index, more items must be included than if only the overall index were desired. After the decisions have been made concerning the commodities to be included, sophisticated sampling procedures are often utilized to determine the specific prices that will be included.

Choice of a base period

A second problem in the construction of a price index is the choice of a base period, that is, a period whose level of prices represents the base from which changes in prices are measured. As indicated earlier, the level of prices in the base period is taken as 100%. Price levels in non-base periods are stated as percentages of the base period level. The base period may be a conventional calendar time interval such as a month or a year, or even a period of years. It is usually considered advisable to use a time period which is "normal" as regards levels of prices. Of course, it is virtually impossible to devise a meaningful definition of what constitutes "normality" in almost any area of economic experience. However, the criterion of normality of prices in the choice of a base period implies operationally that the time period selected should not be one which is at or near the peaks or troughs of price fluctuations. Actually, there is nothing mathematically incorrect about using as a base a period when price levels were unusually low or high. The point is that the use of such time intervals as bases tends to produce distorted concepts, since comparisons are made with atypical periods.

The use of a period of years as a base provides an averaging effect on year-to-year variations. Any particular year may have relatively unique influences present, but if, say, a three- to five-year base period is used, these will tend to be evened out. Most of the United States governmental indices have used such time intervals as base periods, as for example, 1935–1939, 1947–1949, and 1957–1959.[2]

Another point in the choice of a base time interval is suggested by the aforementioned three time periods. That is, it is desirable that the base period be not too distant from the present. The farther away we move from the base period the dimmer are our recollections of economic conditions prevailing at that time. Consequently, comparisons with these remote periods tend to lose significance and to become rather tenuous in meaning. Therefore, producers of index number series, such as United States governmental agencies, shift their base periods every decade or so, in order that comparisons may be made with a base time interval in the recent past. Furthermore, it is desirable to shift the base from time to time because a period that previously may have been thought of as normal or average may no longer be so considered after a long lapse of time.

Other considerations may also be involved in choosing a base period from an index. If a number of important existing indices have a certain base period, it is desirable for purposes of ease of comparability for newly constructed indices to use the same time periods. Also, as new commodities are developed

---

[2] A recent exception is the use of 1967 as a base. This was a year that was neither a peak nor a trough in business activity.

and indices are revised to include them, it becomes desirable to shift the base period to a time interval that reflects the newer economic environment.

The discussion in the preceding sections has referred to price indices. Another important group of summary measures of economic change is represented by *quantity indices.* Such indices measure changes in physical *quantities* such as the volume of industrial production, physical volume of imports and exports, quantities of goods and services consumed, volume of stock transactions, etc. In virtually all currently used *quantity indices,* what is actually measured is the change in the *value* of a set of goods from the base period to the current period attributed to changes in *quantities* only, prices being held constant. This corresponds to the interpretation of what is measured in weighted price indices as being the change in the *value* of a set of goods from the base period to the current period and is attributed to changes in *prices* only, quantities being held constant. The same types of procedures used for the calculation of price indices are also employed to obtain quantity indices. Except for the case of the unweighted aggregate index, where its calculation for a quantity index would not be meaningful, corresponding quantity indices may be obtained by interchanging P's and Q's in the formulas given earlier in this chapter.

An unweighted average of relatives quantity index can be determined by establishing quantity relatives $Q_n/Q_0 \cdot 100$ and calculating the arithmetic mean of these figures. As indicated in the preceding paragraph, an unweighted aggregative quantity index would not be meaningful. The reason is that it does not make sense to add up quantities which are stated in different units.

As was true for price indices, weighted quantity indices are preferable to unweighted ones. A weighted aggregative index of the Laspeyres type is given by the following formula:

**WEIGHTED RELATIVE OF AGGREGATES QUANTITY INDEX,
BASE PERIOD WEIGHTS (LASPEYRES METHOD)**

(14.8)
$$\frac{\Sigma Q_n P_0}{\Sigma Q_0 P_0} \cdot 100$$

Just as the corresponding Laspeyres price index measures the change in price levels from a base period assuming a fixed set of quantities produced or consumed in the base period, etc., this quantity index measures change in quantities produced or consumed, etc., assuming a fixed set of prices which existed in the base period. Paralleling the corresponding situation for price indices, the weighted average of relatives quantity index, using base period value weights given in (14.9), is algebraically identical to this Laspeyres index.

WEIGHTED ARITHMETIC MEAN OF RELATIVES QUANTITY INDEX, BASE PERIOD VALUE WEIGHTS

(14.9)

$$\frac{\Sigma\left(\dfrac{Q_n}{Q_0} \cdot 100\right)Q_0 P_0}{\Sigma Q_0 P_0}$$

Let us interpret the meaning of these two equivalent weighted indices by considering the Laspeyres version, given in (14.9). Also, we continue with the assumption that the raw data refer to quantities of food items consumed (during a week) and prices paid by a typical family in an urban area. The numerator of the index shows the value of the specified food items consumed in year n at base year prices. The denominator refers to the value of the food items consumed in the base year. Suppose a figure of 125 resulted from such an index. Since prices were kept constant, the increase would be solely attributable to an average increase of 25% in the quantity of these food items consumed.

FRB index of industrial production

Probably the most widely used and best known quantity index in the United States is the Federal Reserve Board (FRB) Index of Industrial Production. This index measures changes in the physical volume of output of manufacturing, mining, and utilities. In addition to the overall index of industrial production, component indices are published by industry groupings such as Manufactures and Minerals, and by sub-components such as Durable Manufactures and Nondurable Manufactures. Separate indices are reported for the output of consumer goods, output of equipment for business and government use, and of materials. Following the groupings used by the Standard Industrial Classification of the U.S. Bureau of the Budget, indices are also prepared for major industrial groups and subgroups. The indices are issued monthly, utilizing 1967 both as a reference base and weight base. The Index of Industrial Production is closely watched by businessmen, economists, financial analysts, and others as a major indicator of the physical output of the economy.

The method of construction is the weighted arithmetic mean of relatives, using the aforementioned base periods. Numerous problems have had to be resolved concerning both the quantity relatives and value weights. Many industries cannot easily provide physical output data for the quantity relatives. Therefore, related data are sometimes used instead which tend to move more or less parallel to output, such as shipments and man-hours worked. The weights used are value-added data, which at the individual company level represent the sales of the firm minus all purchases of materials and services from other business firms. The reason that value-added rather than value of final production weights are used is to avoid the problem of double counting. For example, if the value of final product were used for a steel company which sells its steel to an automobile company, and the value of the final product of the automobile company were also used, there would be double-counting of the steel which went into the making of the automobile. Hence, the weights used follow the value-added approach in which the value of so-called "intermediate products" that are produced at all stages prior to the final product are excluded. From the viewpoint of the economist, the value-added of a firm is conceptually equivalent to the total of its factor of production payments — wages, interest, rent, and profits. (See footnote to Table 13-1.)

DEFLATION OF VALUE SERIES BY PRICE INDICES   14.6

One of the most useful applications of price indices is to adjust series of dollar figures for changes in levels of prices. The result of this adjustment procedure, known as "deflation," is to restate the original dollar figures in terms of so-called "constant dollars." The rationale of the procedure is illustrated in terms of the simple example given in Table 14-5. In Column (2) are shown average (arithmetic mean) weekly wage figures for factory workers in a large city in 1970 and 1973. Such unadjusted dollar figures are usually referred to as stated in "current dollars." In Column (3) is shown a consumer price index for the given city, on a reference base period of 1970. The notation (1970 = 100) is a conventional method of specifying the base period. For simplicity of interpretation, let us assume the consumer price index was computed by the Laspeyres method. As we note from Column (2), average weekly wages of the given workers has increased from $110.00 in 1970 to $140.00 in 1973, a gain of 27.3%. However, can these workers purchase 27.3% more goods and services with this increased income? If prices of all of these goods and services had remained unchanged between 1970 and 1973, all other things being equal, the answer would be "yes." But, as can be seen in Column (3), prices rose 15% over the period. To determine what average weekly wages are in terms of 1970 constant dollars (dollars with 1970 purchasing power) we carry out the division $140.00/1.15 = $121.74. That is, we divide the 1973 weekly wage figure in current dollars by the 1973 consumer price index stated as a decimal figure (around a base of 1.00 rather than 100) to obtain the $121.74 figure for average weekly wages in 1970 constant dollars. As indicated in the heading of Column (4), the result of this adjustment for price change is referred to as "real wages," in this case "real average weekly wages." The impli-

TABLE 14-5
Calculation of Average Weekly Wages in 1970 Constant Dollars
for Factory Workers in a Large City, 1970 and 1973.

YEAR (1)	AVERAGE WEEKLY WAGES (2)	CONSUMER PRICE INDEX (1970 = 100) (3)	"REAL" AVERAGE WEEKLY WAGES (1970 CONSTANT DOLLARS) (4)
1970	$110.00	100	$110.00
1973	140.00	115	121.74

cation of the term "real" is that a portion of the increase in wages in dollars is absorbed by the increase in prices. The adjustment attempts to isolate the "real change," in terms of the volume of goods and services which the weekly wages can purchase at base year prices. In summary, the dollar value figures in Column (2) are divided by the price index figures in Column (3) (stated on a base of 1.00) to obtain real value figures in Column (4). The same procedure would have been followed if there had been a series of figures in Columns (2) and (3), say annually, rather than just the current and base period figures of the example. This process of dividing a dollar figure by a price index is referred to as a deflation of the current dollar value series, whether a decrease or an increase occurs in going from figures in current dollars to constant dollars.

The rationale of the deflation procedure stems from the basic relationship of value = price × quantity. The weekly wages in current dollars are value figures. They may be viewed as value aggregates composed of a sum of prices of labor times quantities of such labor. By dividing such a figure by a price index, an attempt is made to isolate the change attributable to the concept of quantity or physical volume. Hence, we may think of the real average weekly wage figures as reflecting the changes in quantities of goods over which the wage figures have command.

This deflation procedure is very widely used in business and economics. One interesting application is in connection with attempts to measure economic well-being and economic growth. For example, in comparing growth rates among countries, frequently one of the most important indicators used is per capita growth in real gross national product. The division of gross national product by population to yield per capita figures may be viewed as an adjustment for differences in population size. The division of the gross national product figures by a relevant price index to obtain real gross national product is an adjustment for change in price levels. The resultant figures for per capita real gross national product are extremely useful measures of physical volume of production.

Of course, there are numerous limitations to the use of the deflation procedure. For example, in the weekly wages illustration, the "market basket" of commodities and services implicit in the consumer price index may not refer specifically to the factory workers to whom the weekly wages pertain. Secondly, even if the index had been constructed for this specific group of factory workers, the index is, after all, only an average. Hence, it is subject to all the interpretational problems of any such measure of central tendency. Furthermore, inferences from such data must be used with care. For example, even if there has been an increase in real average weekly wages from one period to another, we obviously cannot immediately infer an increase in economic welfare for the factory worker group in question. In the later period there may be a less equitable distribution of this income, taxes may be higher leading to lower disposable income, and so forth. Nevertheless, despite such limitations and caveats, the deflation procedure is a very useful, practical, and widely utilized tool of business and economic analysis.

In the above illustration, we have seen how a price index may be used to

remove from a value aggregate the change that is attributable to price movements. Another way to view the deflation procedure is as a method of adjustment of a value figure for changes in the purchasing power of money. In this
connection, it is important to note that a *purchasing power index* is conceptually the *reciprocal of a price index.* For example, assume you have $20 to
purchase shoes in a certain year when a pair of shoes costs $10. The $20
enables you to purchase two pairs of shoes. Suppose, in a later year the price
of shoes has risen to $20. You now can only purchase one pair of shoes. Let
us imagine a price index composed solely of the price of this pair of shoes.
If the earlier year is the base period, the base period price index is 100 and
the later period figure is 200. On the other hand, if the price of these shoes has
doubled, the purchasing power of the dollar relative to shoes has halved.
Hence, a purchasing power index which was at a level of 100 in the earlier
base year should stand at a level of 50 in the later period. If the indices are
stated around 1.00 rather than around 100 in the base period, the reciprocal
relationship can be expressed as $2 \times \frac{1}{2} = 1$. That is, the doubling in the price
index and the halving in the purchasing power index are reciprocals. It is this
relationship between price and purchasing power indices which is implied in
comparative popular statements of the nature that a dollar today is worth only
(say) fifty cents in terms of the dollar in some earlier period.

EXERCISES

1. A company's gross sales in 1970 were $10,000,000 and in 1973 were $15,000,000. It
uses the following price index as a price deflator.

PRICE INDEX	
1967	100
1970	125
1973	150

By what percentage did "real gross sales" increase from 1970 to 1973?

2. Assume the figures given below represent the Disposable Income and Consumer
Price Index of a certain country.

	DISPOSABLE INCOME (BILLIONS OF DOLLARS)	CPI (1960 = 100)
1955	$100	90
1960	160	100
1965	180	120
1970	220	125

Adjust the above series on disposable income so that it reflects changes in disposable
income in 1965 constant-dollars.

3. Data on mean weekly earnings of factory workers in a certain city and values of the Consumer Price Index (CPI) for that city for 1972 and 1973 are presented.

YEAR	MEAN WEEKLY EARNINGS	CONSUMER PRICE INDEX (1967 = 100)
1972	$150.10	120
1973	154.20	125

Did average weekly earnings, when deflated by the CPI, increase from 1972 to 1973? Show your calculations and briefly explain your results.

14.7 SELECTED CONSIDERATIONS IN THE USE OF INDEX NUMBERS

Numerous problems arise in connection with the use of index numbers for analysis and decision purposes. A couple of these are discussed below.

Quality changes

In the construction of an index such as the Consumer Price Index, the basic data on prices are collected by well trained investigators who price goods for which detailed specifications have been made. Rigidity is essential with respect to pricing the same items in the same stores. However, over time, as a result of technological and other improvements, there often is a corresponding improvement in the quality of many commodities. It is very difficult and in many cases impossible to make suitable adjustments in a price index for quality changes. The artificial but practical procedure adopted by the Bureau of Labor Statistics is to consider a product's quality improved only if changes have occurred that increase the cost of producing the product. Hence, for example, an automobile tire is not considered to have been improved if it delivers increased mileage at the same cost of production. Because of such actual improvements in product quality, many analysts feel that over reasonably long periods indices such as the Consumer Price Index which have shown steady rises in price levels overstate actual price increases in terms of a fixed market basket of goods.

Uses of indices

Index number series are widely used in connection with business and governmental decision making and analysis. One of the best known applications of a price index is the use of the Consumer Price Index as an escalator in collective bargaining contracts. In this connection, over three million workers are covered by contracts which specify periodic changes in wage rates depending upon the amount by which the Consumer Price Index moves up or down. The Bureau of Labor Statistics Wholesale Price Index is similarly used for escalation clauses in contracts between business firms.

A great deal of use of index numbers is made at the company, industry, and overall economy level. In certain industries it is standard practice to key changes in selling prices to changes in indices of prices of raw materials and wage earnings. Assessments of past trends and current status and projection of future economic activity are made on the basis of appropriate indices. Economists follow many of the various indices for purposes of appraisal of performance of the economy and for analyzing its structure and behavior.

Bibliography

1. Probability

Feller, W. *An Introduction to Probability Theory and Its Applications.* Vols. 1 and 2. New York: John Wiley & Sons, Inc., 1957, 1966.

Goldberg, S. *Probability: An Introduction.* Englewood Cliffs, N.J.: Prentice-Hall, Inc., 1960.

Hodges, J. and E. Lehman. *Basic Concepts of Probability and Statistics.* San Francisco: Holden-Day, Inc., 1964.

Mosteller, F., R. Rourke, and G. Thomas, Jr. *Probability and Statistics.* Reading, Mass.: Addison-Wesley Publishing Co., Inc., 1961.

Parzen, E. *Modern Probability and Its Applications.* New York: John Wiley & Sons, Inc., 1960.

2. General Statistics

Clelland, R. C., J. S. de Cani, and F. E. Brown. *Basic Statistics with Business Applications,* 2nd ed. New York: John Wiley & Sons, Inc., 1973.

Dixon, W. and F. Massey, Jr. *Introduction to Statistical Analysis,* 3rd ed. New York: McGraw-Hill Book Co., 1969.

Ehrenfeld, S., and S. Littauer. *Introduction to Statistical Method.* New York: McGraw-Hill Book Co., 1964.

Ezekiel, M., and K. Fox. *Methods of Correlation and Regression Analysis,* 3rd ed. New York: John Wiley & Sons, Inc., 1959.

Fox, K. *Intermediate Economic Statistics.* New York: John Wiley & Sons, Inc., 1968.

Freund, J. and F. Williams. *Modern Business Statistics,* 2nd ed. Englewood Cliffs, N.J.: Prentice-Hall, Inc., 1969.

Huff, D. *How to Lie with Statistics.* New York: W. W. Norton, 1954.

Lapin, L. L. *Statistics for Modern Business Decisions.* New York: Harcourt Brace Jovanovich, Inc., 1973.

Mood, A. and F. Graybill. *Introduction to the Theory of Statistics.* New York: McGraw-Hill Book Co., 1963.

Moroney, M. J. *Facts from Figures.* Baltimore: Penguin Books, Inc., 1957.

Peters, W. S. *Readings in Applied Statistics.* Englewood Cliffs, N.J.: Prentice-Hall, Inc., 1969.

Summers, G. and W. Peters. *Statistical Analysis for Decision Making.* Englewood Cliffs, N.J.: Prentice-Hall, Inc., 1968.

Tanur, J. M., et al. *Statistics: A Guide to the Unknown.* San Francisco: Holden-Day, Inc., 1972.

Wonnacott, T. H. and R. J. Wonnacott. *Introductory Statistics for Business and Economics.* New York: John Wiley & Sons, Inc., 1972.

Yule, G., and M. Kendall. *An Introduction to the Theory of Statistics,* 14th ed. New York: Hafner Publishing Co., 1950.

3. Decision Theory

Chernoff, H. and L. Moses. *Elementary Decision Theory.* New York: John Wiley & Sons, Inc., 1959.

DeGroot, M. H. *Optimal Statistical Decisions.* New York: McGraw-Hill Book Co., 1970.

Forester, J. *Statistical Selection of Business Strategies.* Homewood, Ill.: Richard D. Irwin, Inc., 1968.

Hadley, G. *Introduction to Probability and Statistical Decision Theory.* San Francisco: Holden-Day, Inc., 1967.

Lindley, D. V. *Introduction to Probability and Statistics from a Bayesian Viewpoint,* Part 2 "Inference." New York: Cambridge University Press, 1965.

Luce, R. D. and H. Raiffa. *Games and Decisions.* New York: John Wiley & Sons, Inc., 1957.

Morris, W. T. *Management Science — A Bayesian Introduction.* Englewood Cliffs, N.J.: Prentice-Hall, Inc., 1968.

Pratt, J., H. Raiffa, and R. Schlaifer. *Introduction to Statistical Decision Theory.* New York: McGraw-Hill Book Co., 1965.

Raiffa, H. *Decision Analysis, Introductory Lectures on Choices Under Uncertainty.* Reading, Mass.: Addison-Wesley Publishing Co., Inc., 1968.

Raiffa, H. and R. Schlaifer. *Applied Statistical Decision Theory.* Cambridge, Mass.: Division of Research, Graduate School of Business Administration, Harvard University, 1961.

Schlaifer, R. *Probability and Statistics for Business Decisions.* New York: McGraw-Hill Book Co., Inc., 1959.

Wald, A. *Statistical Decision Functions.* New York: John Wiley & Sons, Inc., 1950.

Winkler, R. L. *Introduction to Bayesian Inference and Decision.* New York: Holt, Rinehart and Winston, Inc., 1972.

4. Sample Survey Methods

Cochran, W. G. *Sampling Techniques,* 2nd ed. New York: John Wiley & Sons, Inc., 1963.

Deming, W. E. *Sample Designs in Business Research.* New York: John Wiley & Sons, Inc., 1960.

Hansen, M. H., W. N. Hurwitz, and W. G. Madow. *Sample Survey Methods and Theory,* Vol. I: *Methods and Applications;* Vol. II: *Theory.* New York: John Wiley & Sons, Inc., 1953.

Kish, L. *Survey Sampling.* New York: John Wiley & Sons, Inc., 1965.

Namias, J. *Handbook of Selected Sample Surveys in the Federal Government* (with Annotated Bibliography). New York: St. John's University Press, 1969.

5. Nonparametric Statistics

Bradley, J. V. *Distribution-Free Statistical Tests.* Englewood Cliffs, N.J.: Prentice-Hall, Inc., 1968.

Fisz, M. *Probability Theory and Mathematical Statistics,* 3rd ed. New York: John Wiley & Sons, Inc., 1963.

Gibbons, J. D. *Nonparametric Statistical Inference.* New York: McGraw-Hill Book Co., 1971.

Kraft, C. H. and C. van Eeden. *A Nonparametric Introduction to Statistics..* New York: The Macmillan Co., 1968.

Siegel, S. *Nonparametric Statistics for the Behavioral Sciences.* New York: McGraw-Hill Book Co., 1956.

6. Sources of Statistical Data

Coman, E. T. *Sources of Business Information,* rev. ed. New York: Prentice-Hall, Inc., 1964.

National Referral Center for Science and Technology. *A Directory of Information Resources in the United States, Social Sciences.* Library of Congress, Washington, D.C.: U.S. Government Printing Office, October 1967.

Silk, L. S., and M. L. Curley. *A Primer on Business Forecasting with a Guide to Sources of Business Data.* New York: Random House, 1970.

Wasserman, P., E. Allen, A. Kruzas, and C. Georgi. *Statistics Sources,* 4th ed. Detroit: Gale Research Co., 1971.

7. Statistical Tables

Burington, R. S., and D. C. May. *Handbook of Probability and Statistics with Tables,* 2nd ed. New York: McGraw-Hill Book Co., 1970.

Hald, A. *Statistical Tables and Formulas.* New York: John Wiley & Sons, Inc., 1952.

Owen, D. *Handbook of Statistical Tables.* Reading, Mass.: Addison-Wesley Publishing Co., Inc., 1962.

Pearson, E. S., and H. O. Hartley. *Biometrika Tables for Statisticians,* 2nd ed. Cambridge, Eng.: Cambridge University Press, 1962.

RAND Corporation. *A Million Random Digits with 100,000 Normal Deviates.* New York: Free Press of Glencoe, 1955.

8. Dictionary of Statistical Terms

Freund, J. and F. Williams. *Dictionary/Outline of Basic Statistics.* New York: McGraw-Hill Book Co., 1966.

Appendix A
Statistical Tables

TABLE A-1 SELECTED VALUES OF THE BINOMIAL PROBABILITY DISTRIBUTION

$$P(x) = \binom{n}{x}(1-p)^{n-x}p^x$$

Example: If $p = 0.15$, $n = 4$, $x = 3$, then $P(3) = 0.0115$. When $p > 0.5$, the value of $P(x)$ for a given n, x, and p is obtained by finding the tabular entry for the given n, with $n - x$ in place of the given x and $1 - p$ in place of the given p.

n	x	.05	.10	.15	.20	.25	.30	.35	.40	.45	.50
1	0	.9500	.9000	.8500	.8000	.7500	.7000	.6500	.6000	.5500	.5000
	1	.0500	.1000	.1500	.2000	.2500	.3000	.3500	.4000	.4500	.5000
2	0	.9025	.8100	.7225	.6400	.5625	.4900	.4225	.3600	.3025	.2500
	1	.0950	.1800	.2550	.3200	.3750	.4200	.4550	.4800	.4950	.5000
	2	.0025	.0100	.0225	.0400	.0625	.0900	.1225	.1600	.2025	.2500
3	0	.8574	.7290	.6141	.5120	.4219	.3430	.2746	.2160	.1664	.1250
	1	.1354	.2430	.3251	.3840	.4219	.4410	.4436	.4320	.4084	.3750
	2	.0071	.0270	.0574	.0960	.1406	.1890	.2389	.2880	.3341	.3750
	3	.0001	.0010	.0034	.0080	.0156	.0270	.0429	.0640	.0911	.1250
4	0	.8145	.6561	.5220	.4096	.3164	.2401	.1785	.1296	.0915	.0625
	1	.1715	.2916	.3685	.4096	.4219	.4116	.3845	.3456	.2995	.2500
	2	.0135	.0486	.0975	.1536	.2109	.2646	.3105	.3456	.3675	.3750
	3	.0005	.0036	.0115	.0256	.0469	.0756	.1115	.1536	.2005	.2500
	4	.0000	.0001	.0005	.0016	.0039	.0081	.0150	.0256	.0410	.0625
5	0	.7738	.5905	.4437	.3277	.2373	.1681	.1160	.0778	.0503	.0312
	1	.2036	.3280	.3915	.4096	.3955	.3602	.3124	.2592	.2059	.1562
	2	.0214	.0729	.1382	.2048	.2637	.3087	.3364	.3456	.3369	.3125
	3	.0011	.0081	.0244	.0512	.0879	.1323	.1811	.2304	.2757	.3125
	4	.0000	.0004	.0022	.0064	.0146	.0284	.0488	.0768	.1128	.1562
	5	.0000	.0000	.0001	.0003	.0010	.0024	.0053	.0102	.0185	.0312
6	0	.7351	.5314	.3771	.2621	.1780	.1176	.0754	.0467	.0277	.0156
	1	.2321	.3543	.3993	.3932	.3560	.3025	.2437	.1866	.1359	.0938
	2	.0305	.0984	.1762	.2458	.2966	.3241	.3280	.3110	.2780	.2344
	3	.0021	.0146	.0415	.0819	.1318	.1852	.2355	.2765	.3032	.3125
	4	.0001	.0012	.0055	.0154	.0330	.0595	.0951	.1382	.1861	.2344
	5	.0000	.0001	.0004	.0015	.0044	.0102	.0205	.0369	.0609	.0938
	6	.0000	.0000	.0000	.0001	.0002	.0007	.0018	.0041	.0083	.0156
7	0	.6983	.4783	.3206	.2097	.1335	.0824	.0490	.0280	.0152	.0078
	1	.2573	.3720	.3960	.3670	.3115	.2471	.1848	.1306	.0872	.0547
	2	.0406	.1240	.2097	.2753	.3115	.3177	.2985	.2613	.2140	.1641
	3	.0036	.0230	.0617	.1147	.1730	.2269	.2679	.2903	.2918	.2734
	4	.0002	.0026	.0109	.0287	.0577	.0972	.1442	.1935	.2388	.2734
	5	.0000	.0002	.0012	.0043	.0115	.0250	.0466	.0774	.1172	.1641
	6	.0000	.0000	.0001	.0004	.0013	.0036	.0084	.0172	.0320	.0547
	7	.0000	.0000	.0000	.0000	.0001	.0002	.0006	.0016	.0037	.0078
8	0	.6634	.4305	.2725	.1678	.1001	.0576	.0319	.0168	.0084	.0039
	1	.2793	.3826	.3847	.3355	.2670	.1977	.1373	.0896	.0548	.0312
	2	.0515	.1488	.2376	.2936	.3115	.2965	.2587	.2090	.1569	.1094
	3	.0054	.0331	.0839	.1468	.2076	.2541	.2786	.2787	.2568	.2188
	4	.0004	.0046	.0185	.0459	.0865	.1361	.1875	.2322	.2627	.2734
	5	.0000	.0004	.0026	.0092	.0231	.0467	.0808	.1239	.1719	.2188
	6	.0000	.0000	.0002	.0011	.0038	.0100	.0217	.0413	.0703	.1094
	7	.0000	.0000	.0000	.0001	.0004	.0012	.0033	.0079	.0164	.0312
	8	.0000	.0000	.0000	.0000	.0000	.0001	.0002	.0007	.0017	.0039

Source: From *Handbook of Probability and Statistics with Tables* by Burington and May. Second Edition, Copyright© 1970 by McGraw-Hill, Inc. Used with permission of McGraw-Hill Book Company.

TABLE A-1 (continued)

n	x	.05	.10	.15	.20	.25	.30	.35	.40	.45	.50
							p				
9	0	.6302	.3874	.2316	.1342	.0751	.0404	.0207	.0101	.0046	.0020
	1	.2985	.3874	.3679	.3020	.2253	.1556	.1004	.0605	.0339	.0176
	2	.0629	.1722	.2597	.3020	.3003	.2668	.2162	.1612	.1110	.0703
	3	.0077	.0446	.1069	.1762	.2336	.2668	.2716	.2508	.2119	.1641
	4	.0006	.0074	.0283	.0661	.1168	.1715	.2194	.2508	.2600	.2461
	5	.0000	.0008	.0050	.0165	.0389	.0735	.1181	.1672	.2128	.2461
	6	.0000	.0001	.0006	.0028	.0087	.0210	.0424	.0743	.1160	.1641
	7	.0000	.0000	.0000	.0003	.0012	.0039	.0098	.0212	.0407	.0703
	8	.0000	.0000	.0000	.0000	.0001	.0004	.0013	.0035	.0083	.0176
	9	.0000	.0000	.0000	.0000	.0000	.0000	.0001	.0003	.0008	.0020
10	0	.5987	.3487	.1969	.1074	.0563	.0282	.0135	.0060	.0025	.0010
	1	.3151	.3874	.3474	.2684	.1877	.1211	.0725	.0403	.0207	.0098
	2	.0746	.1937	.2759	.3020	.2816	.2335	.1757	.1209	.0763	.0439
	3	.0105	.0574	.1298	.2013	.2503	.2668	.2522	.2150	.1665	.1172
	4	.0010	.0112	.0401	.0881	.1460	.2001	.2377	.2508	.2384	.2051
	5	.0001	.0015	.0085	.0264	.0584	.1029	.1536	.2007	.2340	.2461
	6	.0000	.0001	.0012	.0055	.0162	.0368	.0689	.1115	.1596	.2051
	7	.0000	.0000	.0001	.0008	.0031	.0090	.0212	.0425	.0746	.1172
	8	.0000	.0000	.0000	.0001	.0004	.0014	.0043	.0106	.0229	.0439
	9	.0000	.0000	.0000	.0000	.0000	.0001	.0005	.0016	.0042	.0098
	10	.0000	.0000	.0000	.0000	.0000	.0000	.0000	.0001	.0003	.0010
11	0	.5688	.3138	.1673	.0859	.0422	.0198	.0088	.0036	.0014	.0005
	1	.3293	.3835	.3248	.2362	.1549	.0932	.0518	.0266	.0125	.0054
	2	.0867	.2131	.2866	.2953	.2581	.1998	.1395	.0887	.0513	.0269
	3	.0137	.0710	.1517	.2215	.2581	.2568	.2254	.1774	.1259	.0806
	4	.0014	.0158	.0536	.1107	.1721	.2201	.2428	.2365	.2060	.1611
	5	.0001	.0025	.0132	.0388	.0803	.1321	.1830	.2207	.2360	.2256
	6	.0000	.0003	.0023	.0097	.0268	.0566	.0985	.1471	.1931	.2256
	7	.0000	.0000	.0003	.0017	.0064	.0173	.0379	.0701	.1128	.1611
	8	.0000	.0000	.0000	.0002	.0011	.0037	.0102	.0234	.0462	.0806
	9	.0000	.0000	.0000	.0000	.0001	.0005	.0018	.0052	.0126	.0269
	10	.0000	.0000	.0000	.0000	.0000	.0000	.0002	.0007	.0021	.0054
	11	.0000	.0000	.0000	.0000	.0000	.0000	.0000	.0000	.0002	.0005
12	0	.5404	.2824	.1422	.0687	.0317	.0138	.0057	.0022	.0008	.0002
	1	.3413	.3766	.3012	.2062	.1267	.0712	.0368	.0174	.0075	.0029
	2	.0988	.2301	.2924	.2835	.2323	.1678	.1088	.0639	.0339	.0161
	3	.0173	.0852	.1720	.2362	.2581	.2397	.1954	.1419	.0923	.0537
	4	.0021	.0213	.0683	.1329	.1936	.2311	.2367	.2128	.1700	.1208
	5	.0002	.0038	.0193	.0532	.1032	.1585	.2039	.2270	.2225	.1934
	6	.0000	.0005	.0040	.0155	.0401	.0792	.1281	.1766	.2124	.2256
	7	.0000	.0000	.0006	.0033	.0115	.0291	.0591	.1009	.1489	.1934
	8	.0000	.0000	.0001	.0005	.0024	.0078	.0199	.0420	.0762	.1208
	9	.0000	.0000	.0000	.0001	.0004	.0015	.0048	.0125	.0277	.0537
	10	.0000	.0000	.0000	.0000	.0000	.0002	.0008	.0025	.0068	.0161
	11	.0000	.0000	.0000	.0000	.0000	.0000	.0001	.0003	.0010	.0029
	12	.0000	.0000	.0000	.0000	.0000	.0000	.0000	.0000	.0001	.0002
13	0	.5133	.2542	.1209	.0550	.0238	.0097	.0037	.0013	.0004	.0001
	1	.3512	.3672	.2774	.1787	.1029	.0540	.0259	.0113	.0045	.0016
	2	.1109	.2448	.2937	.2680	.2059	.1388	.0836	.0453	.0220	.0095
	3	.0214	.0997	.1900	.2457	.2517	.2181	.1651	.1107	.0660	.0349
	4	.0028	.0277	.0838	.1535	.2097	.2337	.2222	.1845	.1350	.0873
	5	.0003	.0055	.0266	.0691	.1258	.1803	.2154	.2214	.1989	.1571
	6	.0000	.0008	.0063	.0230	.0559	.1030	.1546	.1968	.2169	.2095
	7	.0000	.0001	.0011	.0058	.0186	.0442	.0833	.1312	.1775	.2095
	8	.0000	.0000	.0001	.0011	.0047	.0142	.0336	.0656	.1089	.1571
	9	.0000	.0000	.0000	.0001	.0009	.0034	.0101	.0243	.0495	.0873
	10	.0000	.0000	.0000	.0000	.0001	.0006	.0022	.0065	.0162	.0349
	11	.0000	.0000	.0000	.0000	.0000	.0001	.0003	.0012	.0036	.0095
	12	.0000	.0000	.0000	.0000	.0000	.0000	.0000	.0001	.0005	.0016
	13	.0000	.0000	.0000	.0000	.0000	.0000	.0000	.0000	.0000	.0001

TABLE A-1 (continued)

n	x	.05	.10	.15	.20	.25	.30	.35	.40	.45	.50
14	0	.4877	.2288	.1028	.0440	.0178	.0068	.0024	.0008	.0002	.0001
	1	.3593	.3559	.2539	.1539	.0832	.0407	.0181	.0073	.0027	.0009
	2	.1229	.2570	.2912	.2501	.1802	.1134	.0634	.0317	.0141	.0056
	3	.0259	.1142	.2056	.2501	.2402	.1943	.1366	.0845	.0462	.0222
	4	.0037	.0349	.0998	.1720	.2202	.2290	.2022	.1549	.1040	.0611
	5	.0004	.0078	.0352	.0860	.1468	.1963	.2178	.2066	.1701	.1222
	6	.0000	.0013	.0093	.0322	.0734	.1262	.1759	.2066	.2088	.1833
	7	.0000	.0002	.0019	.0092	.0280	.0618	.1082	.1574	.1952	.2095
	8	.0000	.0000	.0003	.0020	.0082	.0232	.0510	.0918	.1398	.1833
	9	.0000	.0000	.0000	.0003	.0018	.0066	.0183	.0408	.0762	.1222
	10	.0000	.0000	.0000	.0000	.0003	.0014	.0049	.0136	.0312	.0611
	11	.0000	.0000	.0000	.0000	.0000	.0002	.0010	.0033	.0093	.0222
	12	.0000	.0000	.0000	.0000	.0000	.0000	.0001	.0005	.0019	.0056
	13	.0000	.0000	.0000	.0000	.0000	.0000	.0000	.0001	.0002	.0009
	14	.0000	.0000	.0000	.0000	.0000	.0000	.0000	.0000	.0000	.0001
15	0	.4633	.2059	.0874	.0352	.0134	.0047	.0016	.0005	.0001	.0000
	1	.3658	.3432	.2312	.1319	.0668	.0305	.0126	.0047	.0016	.0005
	2	.1348	.2669	.2856	.2309	.1559	.0916	.0476	.0219	.0090	.0032
	3	.0307	.1285	.2184	.2501	.2252	.1700	.1110	.0634	.0318	.0139
	4	.0049	.0428	.1156	.1876	.2252	.2186	.1792	.1268	.0780	.0417
	5	.0006	.0105	.0449	.1032	.1651	.2061	.2123	.1859	.1404	.0916
	6	.0000	.0019	.0132	.0430	.0917	.1472	.1906	.2066	.1914	.1527
	7	.0000	.0003	.0030	.0138	.0393	.0811	.1319	.1771	.2013	.1964
	8	.0000	.0000	.0005	.0035	.0131	.0348	.0710	.1181	.1647	.1964
	9	.0000	.0000	.0001	.0007	.0034	.0116	.0298	.0612	.1048	.1527
	10	.0000	.0000	.0000	.0001	.0007	.0030	.0096	.0245	.0515	.0916
	11	.0000	.0000	.0000	.0000	.0001	.0006	.0024	.0074	.0191	.0417
	12	.0000	.0000	.0000	.0000	.0000	.0001	.0004	.0016	.0052	.0139
	13	.0000	.0000	.0000	.0000	.0000	.0000	.0001	.0003	.0010	.0032
	14	.0000	.0000	.0000	.0000	.0000	.0000	.0000	.0000	.0001	.0005
	15	.0000	.0000	.0000	.0000	.0000	.0000	.0000	.0000	.0000	.0000
16	0	.4401	.1853	.0743	.0281	.0100	.0033	.0010	.0003	.0001	.0000
	1	.3706	.3294	.2097	.1126	.0535	.0228	.0087	.0030	.0009	.0002
	2	.1463	.2745	.2775	.2111	.1336	.0732	.0353	.0150	.0056	.0018
	3	.0359	.1423	.2285	.2463	.2079	.1465	.0888	.0468	.0215	.0085
	4	.0061	.0514	.1311	.2001	.2252	.2040	.1553	.1014	.0572	.0278
	5	.0008	.0137	.0555	.1201	.1802	.2099	.2008	.1623	.1123	.0667
	6	.0001	.0028	.0180	.0550	.1101	.1649	.1982	.1983	.1684	.1222
	7	.0000	.0004	.0045	.0197	.0524	.1010	.1524	.1889	.1969	.1746
	8	.0000	.0001	.0009	.0055	.0197	.0487	.0923	.1417	.1812	.1964
	9	.0000	.0000	.0001	.0012	.0058	.0185	.0442	.0840	.1318	.1746
	10	.0000	.0000	.0000	.0002	.0014	.0056	.0167	.0392	.0755	.1222
	11	.0000	.0000	.0000	.0000	.0002	.0013	.0049	.0142	.0337	.0667
	12	.0000	.0000	.0000	.0000	.0000	.0002	.0011	.0040	.0115	.0278
	13	.0000	.0000	.0000	.0000	.0000	.0000	.0002	.0008	.0029	.0085
	14	.0000	.0000	.0000	.0000	.0000	.0000	.0000	.0001	.0005	.0018
	15	.0000	.0000	.0000	.0000	.0000	.0000	.0000	.0000	.0001	.0002
	16	.0000	.0000	.0000	.0000	.0000	.0000	.0000	.0000	.0000	.0000
17	0	.4181	.1668	.0631	.0225	.0075	.0023	.0007	.0002	.0000	.0000
	1	.3741	.3150	.1893	.0957	.0426	.0169	.0060	.0019	.0005	.0001
	2	.1575	.2800	.2673	.1914	.1136	.0581	.0260	.0102	.0035	.0010
	3	.0415	.1556	.2359	.2393	.1893	.1245	.0701	.0341	.0144	.0052
	4	.0076	.0605	.1457	.2093	.2209	.1868	.1320	.0796	.0411	.0182
	5	.0010	.0175	.0668	.1361	.1914	.2081	.1849	.1379	.0875	.0472
	6	.0061	.0039	.0236	.0680	.1276	.1784	.1991	.1839	.1432	.0944
	7	.0000	.0007	.0065	.0267	.0668	.1201	.1685	.1927	.1841	.1484
	8	.0000	.0001	.0014	.0084	.0279	.0644	.1134	.1606	.1883	.1855
	9	.0000	.0000	.0003	.0021	.0093	.0276	.0611	.1070	.1540	.1855
	10	.0000	.0000	.0000	.0004	.0025	.0095	.0263	.0571	.1008	.1484
	11	.0000	.0000	.0000	.0001	.0005	.0026	.0090	.0242	.0525	.0944
	12	.0000	.0000	.0000	.0000	.0001	.0006	.0024	.0081	.0215	.0472
	13	.0000	.0000	.0000	.0000	.0000	.0001	.0005	.0021	.0068	.0182
	14	.0000	.0000	.0000	.0000	.0000	.0000	.0001	.0004	.0016	.0052

TABLE A-1 (continued)

n	x	.05	.10	.15	.20	.25	.30	.35	.40	.45	.50
17	15	.0000	.0000	.0000	.0000	.0000	.0000	.0000	.0001	.0003	.0010
	16	.0000	.0000	.0000	.0000	.0000	.0000	.0000	.0000	.0000	.0001
	17	.0000	.0000	.0000	.0000	.0000	.0000	.0000	.0000	.0000	.0000
18	0	.3972	.1501	.0536	.0180	.0056	.0016	.0004	.0001	.0000	.0000
	1	.3763	.3002	.1704	.0811	.0338	.0126	.0042	.0012	.0003	.0001
	2	.1683	.2835	.2556	.1723	.0958	.0458	.0190	.0069	.0022	.0006
	3	.0473	.1680	.2406	.2297	.1704	.1046	.0547	.0246	.0095	.0031
	4	.0093	.0700	.1592	.2153	.2130	.1681	.1104	.0614	.0291	.0117
	5	.0014	.0218	.0787	.1507	.1988	.2017	.1664	.1146	.0666	.0327
	6	.0002	.0052	.0301	.0816	.1436	.1873	.1941	.1655	.1181	.0708
	7	.0000	.0010	.0091	.0350	.0820	.1376	.1792	.1892	.1657	.1214
	8	.0000	.0002	.0022	.0120	.0376	.0811	.1327	.1734	.1864	.1669
	9	.0000	.0000	.0004	.0033	.0139	.0386	.0794	.1284	.1694	.1855
	10	.0000	.0000	.0001	.0008	.0042	.0149	.0385	.0771	.1248	.1669
	11	.0000	.0000	.0000	.0001	.0010	.0046	.0151	.0374	.0742	.1214
	12	.0000	.0000	.0000	.0000	.0002	.0012	.0047	.0145	.0354	.0708
	13	.0000	.0000	.0000	.0000	.0000	.0002	.0012	.0045	.0134	.0327
	14	.0000	.0000	.0000	.0000	.0000	.0000	.0002	.0011	.0039	.0117
	15	.0000	.0000	.0000	.0000	.0000	.0000	.0000	.0002	.0009	.0031
	16	.0000	.0000	.0000	.0000	.0000	.0000	.0000	.0000	.0001	.0006
	17	.0000	.0000	.0000	.0000	.0000	.0000	.0000	.0000	.0000	.0001
	18	.0000	.0000	.0000	.0000	.0000	.0000	.0000	.0000	.0000	.0000
19	0	.3774	.1351	.0456	.0144	.0042	.0011	.0003	.0001	.0000	.0000
	1	.3774	.2852	.1529	.0685	.0268	.0093	.0029	.0008	.0002	.0000
	2	.1787	.2852	.2428	.1540	.0803	.0358	.0138	.0046	.0013	.0003
	3	.0533	.1796	.2428	.2182	.1517	.0869	.0422	.0175	.0062	.0018
	4	.0112	.0798	.1714	.2182	.2023	.1491	.0909	.0467	.0203	.0074
	5	.0018	.0266	.0907	.1636	.2023	.1916	.1468	.0933	.0497	.0222
	6	.0002	.0069	.0374	.0955	.1574	.1916	.1844	.1451	.0949	.0518
	7	.0000	.0014	.0122	.0443	.0974	.1525	.1844	.1797	.1443	.0961
	8	.0000	.0002	.0032	.0166	.0487	.0981	.1489	.1797	.1771	.1442
	9	.0000	.0000	.0007	.0051	.0198	.0514	.0980	.1464	.1771	.1762
	10	.0000	.0000	.0001	.0013	.0066	.0220	.0528	.0976	.1449	.1762
	11	.0000	.0000	.0000	.0003	.0018	.0077	.0233	.0532	.0970	.1442
	12	.0000	.0000	.0000	.0000	.0004	.0022	.0083	.0237	.0529	.0961
	13	.0000	.0000	.0000	.0000	.0001	.0005	.0024	.0085	.0233	.0518
	14	.0000	.0000	.0000	.0000	.0000	.0001	.0006	.0024	.0082	.0222
	15	.0000	.0000	.0000	.0000	.0000	.0000	.0001	.0005	.0022	.0074
	16	.0000	.0000	.0000	.0000	.0000	.0000	.0000	.0001	.0005	.0018
	17	.0000	.0000	.0000	.0000	.0000	.0000	.0000	.0000	.0001	.0003
	18	.0000	.0000	.0000	.0000	.0000	.0000	.0000	.0000	.0000	.0000
	19	.0000	.0000	.0000	.0000	.0000	.0000	.0000	.0000	.0000	.0000
20	0	.3585	.1216	.0388	.0115	.0032	.0008	.0002	.0000	.0000	.0000
	1	.3774	.2702	.1368	.0576	.0211	.0068	.0020	.0005	.0001	.0000
	2	.1887	.2852	.2293	.1369	.0669	.0278	.0100	.0031	.0008	.0002
	3	.0596	.1901	.2428	.2054	.1339	.0716	.0323	.0123	.0040	.0011
	4	.0133	.0898	.1821	.2182	.1897	.1304	.0738	.0350	.0139	.0046
	5	.0022	.0319	.1028	.1746	.2023	.1789	.1272	.0746	.0365	.0148
	6	.0003	.0089	.0454	.1091	.1686	.1916	.1712	.1244	.0746	.0370
	7	.0000	.0020	.0160	.0545	.1124	.1643	.1844	.1659	.1221	.0739
	8	.0000	.0004	.0046	.0222	.0609	.1144	.1614	.1797	.1623	.1201
	9	.0000	.0001	.0011	.0074	.0271	.0654	.1158	.1597	.1771	.1602
	10	.0000	.0000	.0002	.0020	.0099	.0308	.0686	.1171	.1593	.1762
	11	.0000	.0000	.0000	.0005	.0030	.0120	.0336	.0710	.1185	.1602
	12	.0000	.0000	.0000	.0001	.0008	.0039	.0136	.0355	.0727	.1201
	13	.0000	.0000	.0000	.0000	.0002	.0010	.0045	.0146	.0366	.0739
	14	.0000	.0000	.0000	.0000	.0000	.0002	.0012	.0049	.0150	.0370
	15	.0000	.0000	.0000	.0000	.0000	.0000	.0003	.0013	.0049	.0148
	16	.0000	.0000	.0000	.0000	.0000	.0000	.0000	.0003	.0013	.0046
	17	.0000	.0000	.0000	.0000	.0000	.0000	.0000	.0000	.0002	.0011
	18	.0000	.0000	.0000	.0000	.0000	.0000	.0000	.0000	.0000	.0002
	19	.0000	.0000	.0000	.0000	.0000	.0000	.0000	.0000	.0000	.0000
	20	.0000	.0000	.0000	.0000	.0000	.0000	.0000	.0000	.0000	.0000

TABLE A-2 COEFFICIENTS OF THE BINOMIAL DISTRIBUTION

Example: If $n = 8$ and $x = 6$, $\binom{8}{6} = 28$.

This table gives the value of $\binom{n}{x}$ in $\binom{n}{x}q^{n-x}p^x$, the general term of $(q + p)^n$.

n	$\binom{n}{0}$	$\binom{n}{1}$	$\binom{n}{2}$	$\binom{n}{3}$	$\binom{n}{4}$	$\binom{n}{5}$	$\binom{n}{6}$	$\binom{n}{7}$	$\binom{n}{8}$	$\binom{n}{9}$	$\binom{n}{10}$
0	1										
1	1	1									
2	1	2	1								
3	1	3	3	1							
4	1	4	6	4	1						
5	1	5	10	10	5	1					
6	1	6	15	20	15	6	1				
7	1	7	21	35	35	21	7	1			
8	1	8	28	56	70	56	28	8	1		
9	1	9	36	84	126	126	84	36	9	1	
10	1	10	45	120	210	252	210	120	45	10	1
11	1	11	55	165	330	462	462	330	165	55	11
12	1	12	66	220	495	792	924	792	495	220	66
13	1	13	78	286	715	1287	1716	1716	1287	715	286
14	1	14	91	364	1001	2002	3003	3432	3003	2002	1001
15	1	15	105	455	1365	3003	5005	6435	6435	5005	3003
16	1	16	120	560	1820	4368	8008	11440	12870	11440	8008
17	1	17	136	680	2380	6188	12376	19448	24310	24310	19448
18	1	18	153	816	3060	8568	18564	31824	43758	48620	43758
19	1	19	171	969	3876	11628	27132	50388	75582	92378	92378
20	1	20	190	1140	4845	15504	38760	77520	125970	167960	184756

Source: From *Statistical Analysis for Decision Making* by Morris Hamburg, © 1970 by Harcourt Brace Jovanovich, Inc.

TABLE A-3 SELECTED VALUES OF THE POISSON PROBABILITY DISTRIBUTION

$$P(x) = \frac{\mu^x e^{-\mu}}{x!}$$

Example: If $\mu = 1$, $x = 2$, then $P(2) = 0.1839$.

x	0.1	0.2	0.3	0.4	0.5	0.6	0.7	0.8	0.9	1.0
0	.9048	.8187	.7408	.6703	.6065	.5488	.4966	.4493	.4066	.3679
1	.0905	.1637	.2222	.2681	.3033	.3293	.3476	.3595	.3659	.3679
2	.0045	.0164	.0333	.0536	.0758	.0988	.1217	.1438	.1647	.1839
3	.0002	.0011	.0033	.0072	.0126	.0198	.0284	.0383	.0494	.0613
4	.0000	.0001	.0002	.0007	.0016	.0030	.0050	.0077	.0111	.0153
5	.0000	.0000	.0000	.0001	.0002	.0004	.0007	.0012	.0020	.0031
6	.0000	.0000	.0000	.0000	.0000	.0000	.0001	.0002	.0003	.0005
7	.0000	.0000	.0000	.0000	.0000	.0000	.0000	.0000	.0000	.0001

x	1.1	1.2	1.3	1.4	1.5	1.6	1.7	1.8	1.9	2.0
0	.3329	.3012	.2725	.2466	.2231	.2019	.1827	.1653	.1496	.1353
1	.3662	.3614	.3543	.3452	.3347	.3230	.3106	.2975	.2842	.2707
2	.2014	.2169	.2303	.2417	.2510	.2584	.2640	.2678	.2700	.2707
3	.0738	.0867	.0998	.1128	.1255	.1378	.1496	.1607	.1710	.1804
4	.0203	.0260	.0324	.0395	.0471	.0551	.0636	.0723	.0812	.0902
5	.0045	.0062	.0084	.0111	.0141	.0176	.0216	.0260	.0309	.0361
6	.0008	.0012	.0018	.0026	.0035	.0047	.0061	.0078	.0098	.0120
7	.0001	.0002	.0003	.0005	.0008	.0011	.0015	.0020	.0027	.0034
8	.0000	.0000	.0001	.0001	.0001	.0002	.0003	.0005	.0006	.0009
9	.0000	.0000	.0000	.0000	.0000	.0000	.0001	.0001	.0001	.0002

x	2.1	2.2	2.3	2.4	2.5	2.6	2.7	2.8	2.9	3.0
0	.1225	.1108	.1003	.0907	.0821	.0743	.0672	.0608	.0550	.0498
1	.2572	.2438	.2306	.2177	.2052	.1931	.1815	.1703	.1596	.1494
2	.2700	.2681	.2652	.2613	.2565	.2510	.2450	.2384	.2314	.2240
3	.1890	.1966	.2033	.2090	.2138	.2176	.2205	.2225	.2237	.2240
4	.0992	.1082	.1169	.1254	.1336	.1414	.1488	.1557	.1622	.1680
5	.0417	.0476	.0538	.0602	.0668	.0735	.0804	.0872	.0940	.1008
6	.0146	.0174	.0206	.0241	.0278	.0319	.0362	.0407	.0455	.0504
7	.0044	.0055	.0068	.0083	.0099	.0118	.0139	.0163	.0188	.0216
8	.0011	.0015	.0019	.0025	.0031	.0038	.0047	.0057	.0068	.0081
9	.0003	.0004	.0005	.0007	.0009	.0011	.0014	.0018	.0022	.0027
10	.0001	.0001	.0001	.0002	.0002	.0003	.0004	.0005	.0006	.0008
11	.0000	.0000	.0000	.0000	.0000	.0001	.0001	.0001	.0002	.0002
12	.0000	.0000	.0000	.0000	.0000	.0000	.0000	.0000	.0000	.0001

x	3.1	3.2	3.3	3.4	3.5	3.6	3.7	3.8	3.9	4.0
0	.0450	.0408	.0369	.0334	.0302	.0273	.0247	.0224	.0202	.0183
1	.1397	.1304	.1217	.1135	.1057	.0984	.0915	.0850	.0789	.0733
2	.2165	.2087	.2008	.1929	.1850	.1771	.1692	.1615	.1539	.1465
3	.2237	.2226	.2209	.2186	.2158	.2125	.2087	.2046	.2001	.1954
4	.1734	.1781	.1823	.1858	.1888	.1912	.1931	.1944	.1951	.1954
5	.1075	.1140	.1203	.1264	.1322	.1377	.1429	.1477	.1522	.1563
6	.0555	.0608	.0662	.0716	.0771	.0826	.0881	.0936	.0989	.1042
7	.0246	.0278	.0312	.0348	.0385	.0425	.0466	.0508	.0551	.0595
8	.0095	.0111	.0129	.0148	.0169	.0191	.0215	.0241	.0269	.0298
9	.0033	.0040	.0047	.0056	.0066	.0076	.0089	.0102	.0116	.0132
10	.0010	.0013	.0016	.0019	.0023	.0028	.0033	.0039	.0045	.0053
11	.0003	.0004	.0005	.0006	.0007	.0009	.0011	.0013	.0016	.0019
12	.0001	.0001	.0001	.0002	.0002	.0003	.0003	.0004	.0005	.0006
13	.0000	.0000	.0000	.0000	.0001	.0001	.0001	.0001	.0002	.0002
14	.0000	.0000	.0000	.0000	.0000	.0000	.0000	.0000	.0000	.0001

Source: From *Handbook of Probability and Statistics with Tables* by Burington and May. Second Edition, Copyright © 1970 by McGraw-Hill, Inc. Used with permission of McGraw-Hill Book Company.

TABLE A-3 (continued)

x	4.1	4.2	4.3	4.4	4.5	4.6	4.7	4.8	4.9	5.0
0	.0166	.0150	.0136	.0123	.0111	.0101	.0091	.0082	.0074	.0067
1	.0679	.0630	.0583	.0540	.0500	.0462	.0427	.0395	.0365	.0337
2	.1393	.1323	.1254	.1188	.1125	.1063	.1005	.0948	.0894	.0842
3	.1904	.1852	.1798	.1743	.1687	.1631	.1574	.1517	.1460	.1404
4	.1951	.1944	.1933	.1917	.1898	.1875	.1849	.1820	.1789	.1755
5	.1600	.1633	.1662	.1687	.1708	.1725	.1738	.1747	.1753	.1755
6	.1093	.1143	.1191	.1237	.1281	.1323	.1362	.1398	.1432	.1462
7	.0640	.0686	.0732	.0778	.0824	.0869	.0914	.0959	.1002	.1044
8	.0328	.0360	.0393	.0428	.0463	.0500	.0537	.0575	.0614	.0653
9	.0150	.0168	.0188	.0209	.0232	.0255	.0280	.0307	.0334	.0363
10	.0061	.0071	.0081	.0092	.0104	.0118	.0132	.0147	.0164	.0181
11	.0023	.0027	.0032	.0037	.0043	.0049	.0056	.0064	.0073	.0082
12	.0008	.0009	.0011	.0014	.0016	.0019	.0022	.0026	.0030	.0034
13	.0002	.0003	.0004	.0005	.0006	.0007	.0008	.0009	.0011	.0013
14	.0001	.0001	.0001	.0001	.0002	.0002	.0003	.0003	.0004	.0005
15	.0000	.0000	.0000	.0000	.0001	.0001	.0001	.0001	.0001	.0002

x	5.1	5.2	5.3	5.4	5.5	5.6	5.7	5.8	5.9	6.0
0	.0061	.0055	.0050	.0045	.0041	.0037	.0033	.0030	.0027	.0025
1	.0311	.0287	.0265	.0244	.0225	.0207	.0191	.0176	.0162	.0149
2	.0793	.0746	.0701	.0659	.0618	.0580	.0544	.0509	.0477	.0446
3	.1348	.1293	.1239	.1185	.1133	.1082	.1033	.0985	.0938	.0892
4	.1719	.1681	.1641	.1600	.1558	.1515	.1472	.1428	.1383	.1339
5	.1753	.1748	.1740	.1728	.1714	.1697	.1678	.1656	.1632	.1606
6	.1490	.1515	.1537	.1555	.1571	.1584	.1594	.1601	.1605	.1606
7	.1086	.1125	.1163	.1200	.1234	.1267	.1298	.1326	.1353	.1377
8	.0692	.0731	.0771	.0810	.0849	.0887	.0925	.0962	.0998	.1033
9	.0392	.0423	.0454	.0486	.0519	.0552	.0586	.0620	.0654	.0688
10	.0200	.0220	.0241	.0262	.0285	.0309	.0334	.0359	.0386	.0413
11	.0093	.0104	.0116	.0129	.0143	.0157	.0173	.0190	.0207	.0225
12	.0039	.0045	.0051	.0058	.0065	.0073	.0082	.0092	.0102	.0113
13	.0015	.0018	.0021	.0024	.0028	.0032	.0036	.0041	.0046	.0052
14	.0006	.0007	.0008	.0009	.0011	.0013	.0015	.0017	.0019	.0022
15	.0002	.0002	.0003	.0003	.0004	.0005	.0006	.0007	.0008	.0009
16	.0001	.0001	.0001	.0001	.0001	.0002	.0002	.0002	.0003	.0003
17	.0000	.0000	.0000	.0000	.0000	.0001	.0001	.0001	.0001	.0001

x	6.1	6.2	6.3	6.4	6.5	6.6	6.7	6 8	6.9	7.0
0	.0022	.0020	.0018	.0017	.0015	.0014	.0012	.0011	.0010	.0009
1	.0137	.0126	.0116	.0106	.0098	.0090	.0082	.0076	.0070	.0064
2	.0417	.0390	.0364	.0340	.0318	.0296	.0276	.0258	.0240	.0223
3	.0848	.0806	.0765	.0726	.0688	.0652	.0617	.0584	.0552	.0521
4	.1294	.1249	.1205	.1162	.1118	.1076	.1034	.0992	.0952	.0912
5	.1579	.1549	.1519	.1487	.1454	.1420	.1385	.1349	.1314	.1277
6	.1605	.1601	.1595	.1586	.1575	.1562	.1546	.1529	.1511	.1490
7	.1399	.1418	.1435	.1450	.1462	.1472	.1480	.1486	.1489	.1490
8	.1066	.1099	.1130	.1160	.1188	.1215	.1240	.1263	.1284	.1304
9	.0723	.0757	.0791	.0825	.0858	.0891	.0923	.0954	.0985	.1014
10	.0441	.0469	.0498	.0528	.0558	.0588	.0618	.0649	.0679	.0710
11	.0245	.0265	.0285	.0307	.0330	.0353	.0377	.0401	.0426	.0452
12	.0124	.0137	.0150	.0164	.0179	.0194	.0210	.0227	.0245	.0264
13	.0058	.0065	.0073	.0081	.0089	.0098	.0108	.0119	.0130	.0142
14	.0025	.0029	.0033	.0037	.0041	.0046	.0052	.0058	.0064	.0071
15	.0010	.0012	.0014	.0016	.0018	.0020	.0023	.0026	.0029	.0033
16	.0004	.0005	.0005	.0006	.0007	.0008	.0010	.0011	.0013	.0014
17	.0001	.0002	.0002	.0002	.0003	.0003	.0004	.0004	.0005	.0006
18	.0000	.0001	.0001	.0001	.0001	.0001	.0001	.0002	.0002	.0002
19	.0000	.0000	.0000	.0000	.0000	.0000	.0000	.0001	.0001	.0001

TABLE A-3 (continued)

	μ									
x	7.1	7.2	7.3	7.4	7.5	7.6	7.7	7.8	7.9	8.0
0	.0008	.0007	.0007	.0006	.0006	.0005	.0005	.0004	.0004	.0003
1	.0059	.0054	.0049	.0045	.0041	.0038	.0035	.0032	.0029	.0027
2	.0208	.0194	.0180	.0167	.0156	.0145	.0134	.0125	.0116	.0107
3	.0492	.0464	.0438	.0413	.0389	.0366	.0345	.0324	.0305	.0286
4	.0874	.0836	.0799	.0764	.0729	.0696	.0663	.0632	.0602	.0573
5	.1241	.1204	.1167	.1130	.1094	.1057	.1021	.0986	.0951	.0916
6	.1468	.1445	.1420	.1394	.1367	.1339	.1311	.1282	.1252	.1221
7	.1489	.1486	.1481	.1474	.1465	.1454	.1442	.1428	.1413	.1396
8	.1321	.1337	.1351	.1363	.1373	.1382	.1388	.1392	.1395	.1396
9	.1042	.1070	.1096	.1121	.1144	.1167	.1187	.1207	.1224	.1241
10	.0740	.0770	.0800	.0829	.0858	.0887	.0914	.0941	.0967	.0993
11	.0478	.0504	.0531	.0558	.0585	.0613	.0640	.0667	.0695	.0722
12	.0283	.0303	.0323	.0344	.0366	.0388	.0411	.0434	.0457	.0481
13	.0154	.0168	.0181	.0196	.0211	.0227	.0243	.0260	.0278	.0296
14	.0078	.0086	.0095	.0104	.0113	.0123	.0134	.0145	.0157	.0169
15	.0037	.0041	.0046	.0051	.0057	.0062	.0069	.0075	.0083	.0090
16	.0016	.0019	.0021	.0024	.0026	.0030	.0033	.0037	.0041	.0045
17	.0007	.0008	.0009	.0010	.0012	.0013	.0015	.0017	.0019	.0021
18	.0003	.0003	.0004	.0004	.0005	.0006	.0006	.0007	.0008	.0009
19	.0001	.0001	.0001	.0002	.0002	.0002	.0003	.0003	.0003	.0004
20	.0000	.0000	.0001	.0001	.0001	.0001	.0001	.0001	.0001	.0002
21	.0000	.0000	.0000	.0000	.0000	.0000	.0000	.0000	.0001	.0001

	μ									
x	8.1	8.2	8.3	8.4	8.5	8.6	8.7	8.8	8.9	9.0
0	.0003	.0003	.0002	.0002	.0002	.0002	.0002	.0002	.0001	.0001
1	.0025	.0023	.0021	.0019	.0017	.0016	.0014	.0013	.0012	.0011
2	.0100	.0092	.0086	.0079	.0074	.0068	.0063	.0058	.0054	.0050
3	.0269	.0252	.0237	.0222	.0208	.0195	.0183	.0171	.0160	.0150
4	.0544	.0517	.0491	.0466	.0443	.0420	.0398	.0377	.0357	.0337
5	.0882	.0849	.0816	.0784	.0752	.0722	.0692	.0663	.0635	.0607
6	.1191	.1160	.1128	.1097	.1066	.1034	.1003	.0972	.0941	.0911
7	.1378	.1358	.1338	.1317	.1294	.1271	.1247	.1222	.1197	.1171
8	.1395	.1392	.1388	.1382	.1375	.1366	.1356	.1344	.1332	.1318
9	.1256	.1269	.1280	.1290	.1299	.1306	.1311	.1315	.1317	.1318
10	.1017	.1040	.1063	.1084	.1104	.1123	.1140	.1157	.1172	.1186
11	.0749	.0776	.0802	.0828	.0853	.0878	.0902	.0925	.0948	.0970
12	.0505	.0530	.0555	.0579	.0604	.0629	.0654	.0679	.0703	.0728
13	.0315	.0334	.0354	.0374	.0395	.0416	.0438	.0459	.0481	.0504
14	.0182	.0196	.0210	.0225	.0240	.0256	.0272	.0289	.0306	.0324
15	.0098	.0107	.0116	.0126	.0136	.0147	.0158	.0169	.0182	.0194
16	.0050	.0055	.0060	.0066	.0072	.0079	.0086	.0093	.0101	.0109
17	.0024	.0026	.0029	.0033	.0036	.0040	.0044	.0048	.0053	.0058
18	.0011	.0012	.0014	.0015	.0017	.0019	.0021	.0024	.0026	.0029
19	.0005	.0005	.0006	.0007	.0008	.0009	.0010	.0011	.0012	.0014
20	.0002	.0002	.0002	.0003	.0003	.0004	.0004	.0005	.0005	.0006
21	.0001	.0001	.0001	.0001	.0001	.0002	.0002	.0002	.0002	.0003
22	.0000	.0000	.0000	.0000	.0001	.0001	.0001	.0001	.0001	.0001

	μ									
x	9.1	9.2	9.3	9.4	9.5	9.6	9.7	9.8	9.9	10
0	.0001	.0001	.0001	.0001	.0001	.0001	.0001	.0001	.0001	.0000
1	.0010	.0009	.0009	.0008	.0007	.0007	.0006	.0005	.0005	.0005
2	.0046	.0043	.0040	.0037	.0034	.0031	.0029	.0027	.0025	.0023
3	.0140	.0131	.0123	.0115	.0107	.0100	.0093	.0087	.0081	.0076
4	.0319	.0302	.0285	.0269	.0254	.0240	.0226	.0213	.0201	.0189
5	.0581	.0555	.0530	.0506	.0483	.0460	.0439	.0418	.0398	.0378
6	.0881	.0851	.0822	.0793	.0764	.0736	.0709	.0682	.0656	.0631
7	.1145	.1118	.1091	.1064	.1037	.1010	.0982	.0955	.0928	.0901
8	.1302	.1286	.1269	.1251	.1232	.1212	.1191	.1170	.1148	.1126
9	.1317	.1315	.1311	.1306	.1300	.1293	.1284	.1274	.1263	.1251

TABLE A-3 (continued)

x	9.1	9.2	9.3	9.4	9.5	9.6	9.7	9.8	9.9	10
10	.1198	.1210	.1219	.1228	.1235	.1241	.1245	.1249	.1250	.1251
11	.0991	.1012	.1031	.1049	.1067	.1083	.1098	.1112	.1125	.1137
12	.0752	.0776	.0799	.0822	.0844	.0866	.0888	.0908	.0928	.0948
13	.0526	.0549	.0572	.0594	.0617	.0640	.0662	.0685	.0707	.0729
14	.0342	.0361	.0380	.0399	.0419	.0439	.0459	.0479	.0500	.0521
15	.0208	.0221	.0235	.0250	.0265	.0281	.0297	.0313	.0330	.0347
16	.0118	.0127	.0137	.0147	.0157	.0168	.0180	.0192	.0204	.0217
17	.0063	.0069	.0075	.0081	.0088	.0095	.0103	.0111	.0119	.0128
18	.0032	.0035	.0039	.0042	.0046	.0051	.0055	.0060	.0065	.0071
19	.0015	.0017	.0019	.0021	.0023	.0026	.0028	.0031	.0034	.0037
20	.0007	.0008	.0009	.0010	.0011	.0012	.0014	.0015	.0017	.0019
21	.0003	.0003	.0004	.0004	.0005	.0006	.0006	.0007	.0008	.0009
22	.0001	.0001	.0002	.0002	.0002	.0002	.0003	.0003	.0004	.0004
23	.0000	.0001	.0001	.0001	.0001	.0001	.0001	.0001	.0002	.0002
24	.0000	.0000	.0000	.0000	.0000	.0000	.0000	.0001	.0001	.0001

μ

x	11	12	13	14	15	16	17	18	19	20
0	.0000	.0000	.0000	.0000	.0000	.0000	.0000	.0000	.0000	.0000
1	.0002	.0001	.0000	.0000	.0000	.0000	.0000	.0000	.0000	.0000
2	.0010	.0004	.0002	.0001	.0000	.0000	.0000	.0000	.0000	.0000
3	.0037	.0018	.0008	.0004	.0002	.0001	.0000	.0000	.0000	.0000
4	.0102	.0053	.0027	.0013	.0006	.0003	.0001	.0001	.0000	.0000
5	.0224	.0127	.0070	.0037	.0019	.0010	.0005	.0002	.0001	.0001
6	.0411	.0255	.0152	.0087	.0048	.0026	.0014	.0007	.0004	.0002
7	.0646	.0437	.0281	.0174	.0104	.0060	.0034	.0018	.0010	.0005
8	.0888	.0655	.0457	.0304	.0194	.0120	.0072	.0042	.0024	.0013
9	.1085	.0874	.0661	.0473	.0324	.0213	.0135	.0083	.0050	.0029
10	.1194	.1048	.0859	.0663	.0486	.0341	.0230	.0150	.0095	.0058
11	.1194	.1144	.1015	.0844	.0663	.0496	.0355	.0245	.0164	.0106
12	.1094	.1144	.1099	.0984	.0829	.0661	.0504	.0368	.0259	.0176
13	.0926	.1056	.1099	.1060	.0956	.0814	.0658	.0509	.0378	.0271
14	.0728	.0905	.1021	.1060	.1024	.0930	.0800	.0655	.0514	.0387
15	.0534	.0724	.0885	.0989	.1024	.0992	.0906	.0786	.0650	.0516
16	.0367	.0543	.0719	.0866	.0960	.0992	.0963	.0884	.0772	.0646
17	.0237	.0383	.0550	.0713	.0847	.0934	.0963	.0936	.0863	.0760
18	.0145	.0256	.0397	.0554	.0706	.0830	.0909	.0936	.0911	.0844
19	.0084	.0161	.0272	.0409	.0557	.0699	.0814	.0887	.0911	.0888
20	.0046	.0097	.0177	.0286	.0418	.0559	.0692	.0798	.0866	.0888
21	.0024	.0055	.0109	.0191	.0299	.0426	.0560	.0684	.0783	.0846
22	.0012	.0030	.0065	.0121	.0204	.0310	.0433	.0560	.0676	.0769
23	.0006	.0016	.0037	.0074	.0133	.0216	.0320	.0438	.0559	.0669
24	.0003	.0008	.0020	.0043	.0083	.0144	.0226	.0328	.0442	.0557
25	.0001	.0004	.0010	.0024	.0050	.0092	.0154	.0237	.0336	.0446
26	.0000	.0002	.0005	.0013	.0029	.0057	.0101	.0164	.0246	.0343
27	.0000	.0001	.0002	.0007	.0016	.0034	.0063	.0109	.0173	.0254
28	.0000	.0000	.0001	.0003	.0009	.0019	.0038	.0070	.0117	.0181
29	.0000	.0000	.0001	.0002	.0004	.0011	.0023	.0044	.0077	.0125
30	.0000	.0000	.0000	.0001	.0002	.0006	.0013	.0026	.0049	.0083
31	.0000	.0000	.0000	.0000	.0001	.0003	.0007	.0015	.0030	.0054
32	.0000	.0000	.0000	.0000	.0001	.0001	.0004	.0009	.0018	.0034
33	.0000	.0000	.0000	.0000	.0000	.0001	.0002	.0005	.0010	.0020
34	.0000	.0000	.0000	.0000	.0000	.0000	.0001	.0002	.0006	.0012
35	.0000	.0000	.0000	.0000	.0000	.0000	.0000	.0001	.0003	.0007
36	.0000	.0000	.0000	.0000	.0000	.0000	.0000	.0001	.0002	.0004
37	.0000	.0000	.0000	.0000	.0000	.0000	.0000	.0000	.0001	.0002
38	.0000	.0000	.0000	.0000	.0000	.0000	.0000	.0000	.0000	.0001
39	.0000	.0000	.0000	.0000	.0000	.0000	.0000	.0000	.0000	.0001

TABLE A-4 FOUR-PLACE COMMON LOGARITHMS

N	0	1	2	3	4	5	6	7	8	9	1	2	3	4	5	6	7	8	9
														Proportional Parts					
10	0000	0043	0086	0128	0170	0212	0253	0294	0334	0374	4	8	12	17	21	25	29	33	37
11	0414	0453	0492	0531	0569	0607	0645	0682	0719	0755	4	8	11	15	19	23	26	30	34
12	0792	0828	0864	0899	0934	0969	1004	1038	1072	1106	3	7	10	14	17	21	24	28	31
13	1139	1173	1206	1239	1271	1303	1335	1367	1399	1430	3	6	10	13	16	19	23	26	29
14	1461	1492	1523	1553	1584	1614	1644	1673	1703	1732	3	6	9	12	15	18	21	24	27
15	1761	1790	1818	1847	1875	1903	1931	1959	1987	2014	3	6	8	11	14	17	20	22	25
16	2041	2068	2095	2122	2148	2175	2201	2227	2253	2279	3	5	8	11	13	16	18	21	24
17	2304	2330	2355	2380	2405	2430	2455	2480	2504	2529	2	5	7	10	12	15	17	20	22
18	2553	2577	2601	2625	2648	2672	2695	2718	2742	2765	2	5	7	9	12	14	16	19	21
19	2788	2810	2833	2856	2878	2900	2923	2945	2967	2989	2	4	7	9	11	13	16	18	20
20	3010	3032	3054	3075	3096	3118	3139	3160	3181	3201	2	4	6	8	11	13	15	17	19
21	3222	3243	3263	3284	3304	3324	3345	3365	3385	3404	2	4	6	8	10	12	14	16	18
22	3424	3444	3464	3483	3502	3522	3541	3560	3579	3598	2	4	6	8	10	12	14	15	17
23	3617	3636	3655	3674	3692	3711	3729	3747	3766	3784	2	4	6	7	9	11	13	15	17
24	3802	3820	3838	3856	3874	3892	3909	3927	3945	3962	2	4	5	7	9	11	12	14	16
25	3979	3997	4014	4031	4048	4065	4082	4099	4116	4133	2	3	5	7	9	10	12	14	15
26	4150	4166	4183	4200	4216	4232	4249	4265	4281	4298	2	3	5	7	8	10	11	13	15
27	4314	4330	4346	4362	4378	4393	4409	4425	4440	4456	2	3	5	6	8	9	11	13	14
28	4472	4487	4502	4518	4533	4548	4564	4579	4594	4609	2	3	5	6	8	9	11	12	14
29	4624	4639	4654	4669	4683	4698	4713	4728	4742	4757	1	3	4	6	7	9	10	12	13
30	4771	4786	4800	4814	4829	4843	4857	4871	4886	4900	1	3	4	6	7	9	10	11	13
31	4914	4928	4942	4955	4969	4983	4997	5011	5024	5038	1	3	4	6	7	8	10	11	12
32	5051	5065	5079	5092	5105	5119	5132	5145	5159	5172	1	3	4	5	7	8	9	11	12
33	5185	5198	5211	5224	5237	5250	5263	5276	5289	5302	1	3	4	5	6	8	9	10	12
34	5315	5328	5340	5353	5366	5378	5391	5403	5416	5428	1	3	4	5	6	8	9	10	11
35	5441	5453	5465	5478	5490	5502	5514	5527	5539	5551	1	2	4	5	6	7	9	10	11
36	5563	5575	5587	5599	5611	5623	5635	5647	5658	5670	1	2	4	5	6	7	8	10	11
37	5682	5694	5705	5717	5729	5740	5752	5763	5775	5786	1	2	3	5	6	7	8	9	10
38	5798	5809	5821	5832	5843	5855	5866	5877	5888	5899	1	2	3	5	6	7	8	9	10
39	5911	5922	5933	5944	5955	5966	5977	5988	5999	6010	1	2	3	4	5	7	8	9	10
40	6021	6031	6042	6053	6064	6075	6085	6096	6107	6117	1	2	3	4	5	6	8	9	10
41	6128	6138	6149	6160	6170	6180	6191	6201	6212	6222	1	2	3	4	5	6	7	8	9
42	6232	6243	6253	6263	6274	6284	6294	6304	6314	6325	1	2	3	4	5	6	7	8	9
43	6335	6345	6355	6365	6375	6385	6395	6405	6415	6425	1	2	3	4	5	6	7	8	9
44	6435	6444	6454	6464	6474	6484	6493	6503	6513	6522	1	2	3	4	5	6	7	8	9
45	6532	6542	6551	6561	6571	6580	6590	6599	6609	6618	1	2	3	4	5	6	7	8	9
46	6628	6637	6646	6656	6665	6675	6684	6693	6702	6712	1	2	3	4	5	6	7	7	8
47	6721	6730	6739	6749	6758	6767	6776	6785	6794	6803	1	2	3	4	5	5	6	7	8
48	6812	6821	6830	6839	6848	6857	6866	6875	6884	6893	1	2	3	4	4	5	6	7	8
49	6902	6911	6920	6928	6937	6946	6955	6964	6972	6981	1	2	3	4	4	5	6	7	8
50	6990	6998	7007	7016	7024	7033	7042	7050	7059	7067	1	2	3	3	4	5	6	7	8
51	7076	7084	7093	7101	7110	7118	7126	7135	7143	7152	1	2	3	3	4	5	6	7	8
52	7160	7168	7177	7185	7193	7202	7210	7218	7226	7235	1	2	2	3	4	5	6	7	7
53	7243	7251	7259	7267	7275	7284	7292	7300	7308	7316	1	2	2	3	4	5	6	6	7
54	7324	7332	7340	7348	7356	7364	7372	7380	7388	7396	1	2	2	3	4	5	6	6	7
N	0	1	2	3	4	5	6	7	8	9	1	2	3	4	5	6	7	8	9

Source: From *Statistical Analysis for Decision Making* by Morris Hamburg, © 1970 by Harcourt Brace Jovanovich, Inc.

N	0	1	2	3	4	5	6	7	8	9	Proportional Parts								
											1	2	3	4	5	6	7	8	9
55	7404	7412	7419	7427	7435	7443	7451	7459	7466	7474	1	2	2	3	4	5	5	6	7
56	7482	7490	7497	7505	7513	7520	7528	7536	7543	7551	1	2	2	3	4	5	5	6	7
57	7559	7566	7574	7582	7589	7597	7604	7612	7619	7627	1	2	2	3	4	5	5	6	7
58	7634	7642	7649	7657	7664	7672	7679	7686	7694	7701	1	1	2	3	4	4	5	6	7
59	7709	7716	7723	7731	7738	7745	7752	7760	7767	7774	1	1	2	3	4	4	5	6	7
60	7782	7789	7796	7803	7810	7818	7825	7832	7839	7846	1	1	2	3	4	4	5	6	6
61	7853	7860	7868	7875	7882	7889	7896	7903	7910	7917	1	1	2	3	4	4	5	6	6
62	7924	7931	7938	7945	7952	7959	7966	7973	7980	7987	1	1	2	3	3	4	5	6	6
63	7993	8000	8007	8014	8021	8028	8035	8041	8048	8055	1	1	2	3	3	4	5	5	6
64	8062	8069	8075	8082	8089	8096	8102	8109	8116	8122	1	1	2	3	3	4	5	5	6
65	8129	8136	8142	8149	8156	8162	8169	8176	8182	8189	1	1	2	3	3	4	5	5	6
66	8195	8202	8209	8215	8222	8228	8235	8241	8248	8254	1	1	2	3	3	4	5	5	6
67	8261	8267	8274	8280	8287	8293	8299	8306	8312	8319	1	1	2	3	3	4	5	5	6
68	8325	8331	8338	8344	8351	8357	8363	8370	8376	8382	1	1	2	3	3	4	4	5	6
69	8388	8395	8401	8407	8414	8420	8426	8432	8439	8445	1	1	2	2	3	4	4	5	6
70	8451	8457	8463	8470	8476	8482	8488	8494	8500	8506	1	1	2	2	3	4	4	5	6
71	8513	8519	8525	8531	8537	8543	8549	8555	8561	8567	1	1	2	2	3	4	4	5	5
72	8573	8579	8585	8591	8597	8603	8609	8615	8621	8627	1	1	2	2	3	4	4	5	5
73	8633	8639	8645	8651	8657	8663	8669	8675	8681	8686	1	1	2	2	3	4	4	5	5
74	8692	8698	8704	8710	8716	8722	8727	8733	8739	8745	1	1	2	2	3	4	4	5	5
75	8751	8756	8762	8768	8774	8779	8785	8791	8797	8802	1	1	2	2	3	3	4	5	5
76	8808	8814	8820	8825	8831	8837	8842	8848	8854	8859	1	1	2	2	3	3	4	5	5
77	8865	8871	8876	8882	8887	8893	8899	8904	8910	8915	1	1	2	2	3	3	4	4	5
78	8921	8927	8932	8938	8943	8949	8954	8960	8965	8971	1	1	2	2	3	3	4	4	5
79	8976	8982	8987	8993	8998	9004	9009	9015	9020	9025	1	1	2	2	3	3	4	4	5
80	9031	9036	9042	9047	9053	9058	9063	9069	9074	9079	1	1	2	2	3	3	4	4	5
81	9085	9090	9096	9101	9106	9112	9117	9122	9128	9133	1	1	2	2	3	3	4	4	5
82	9138	9143	9149	9154	9159	9165	9170	9175	9180	9186	1	1	2	2	3	3	4	4	5
83	9191	9196	9201	9206	9212	9217	9222	9227	9232	9238	1	1	2	2	3	3	4	4	5
84	9243	9248	9253	9258	9263	9269	9274	9279	9284	9289	1	1	2	2	3	3	4	4	5
85	9294	9299	9304	9309	9315	9320	9325	9330	9335	9340	1	1	2	2	3	3	4	4	5
86	9345	9350	9355	9360	9365	9370	9375	9380	9385	9390	1	1	2	2	3	3	4	4	5
87	9395	9400	9405	9410	9415	9420	9425	9430	9435	9440	0	1	1	2	2	3	3	4	4
88	9445	9450	9455	9460	9465	9469	9474	9479	9484	9489	0	1	1	2	2	3	3	4	4
89	9494	9499	9504	9509	9513	9518	9523	9528	9533	9538	0	1	1	2	2	3	3	4	4
90	9542	9547	9552	9557	9562	9566	9571	9576	9581	9586	0	1	1	2	2	3	3	4	4
91	9590	9595	9600	9605	9609	9614	9619	9624	9628	9633	0	1	1	2	2	3	3	4	4
92	9638	9643	9647	9652	9657	9661	9666	9671	9675	9680	0	1	1	2	2	3	3	4	4
93	9685	9689	9694	9699	9703	9708	9713	9717	9722	9727	0	1	1	2	2	3	3	4	4
94	9731	9736	9741	9745	9750	9754	9759	9763	9768	9773	0	1	1	2	2	3	3	4	4
95	9777	9782	9786	9791	9795	9800	9805	9809	9814	9818	0	1	1	2	2	3	3	4	4
96	9823	9827	9832	9836	9841	9845	9850	9854	9859	9863	0	1	1	2	2	3	3	4	4
97	9868	9872	9877	9881	9886	9890	9894	9899	9903	9908	0	1	1	2	2	3	3	4	4
98	9912	9917	9921	9926	9930	9934	9939	9943	9948	9952	0	1	1	2	2	3	3	4	4
99	9956	9961	9965	9969	9974	9978	9983	9987	9991	9996	0	1	1	2	2	3	3	3	4
N	0	1	2	3	4	5	6	7	8	9	1	2	3	4	5	6	7	8	9

TABLE A-5 AREAS UNDER THE STANDARD NORMAL PROBABILITY DISTRIBUTION BETWEEN THE MEAN AND SUCCESSIVE VALUES OF z.

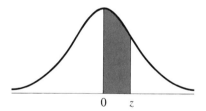

Example: If z = 1.00, then the area between the mean and this value of z is 0.3413.

z	.00	.01	.02	.03	.04	.05	.06	.07	.08	.09
0.0	.0000	.0040	.0080	.0120	.0160	.0199	.0239	.0279	.0319	.0359
0.1	.0398	.0438	.0478	.0517	.0557	.0596	.0636	.0675	.0714	.0753
0.2	.0793	.0832	.0871	.0910	.0948	.0987	.1026	.1064	.1103	.1141
0.3	.1179	.1217	.1255	.1293	.1331	.1368	.1406	.1443	.1480	.1517
0.4	.1554	.1591	.1628	.1664	.1700	.1736	.1772	.1808	.1844	.1879
0.5	.1915	.1950	.1985	.2019	.2054	.2088	.2123	.2157	.2190	.2224
0.6	.2257	.2291	.2324	.2357	.2389	.2422	.2454	.2486	.2518	.2549
0.7	.2580	.2612	.2642	.2673	.2704	.2734	.2764	.2794	.2823	.2852
0.8	.2881	.2910	.2939	.2967	.2995	.3023	.3051	.3078	.3106	.3133
0.9	.3159	.3186	.3212	.3238	.3264	.3289	.3315	.3340	.3365	.3389
1.0	.3413	.3438	.3461	.3485	.3508	.3531	.3554	.3577	.3599	.3621
1.1	.3643	.3665	.3686	.3708	.3729	.3749	.3770	.3790	.3810	.3830
1.2	.3849	.3869	.3888	.3907	.3925	.3944	.3962	.3980	.3997	.4015
1.3	.4032	.4049	.4066	.4082	.4099	.4115	.4131	.4147	.4162	.4177
1.4	.4192	.4207	.4222	.4236	.4251	.4265	.4279	.4292	.4306	.4319
1.5	.4332	.4345	.4357	.4370	.4382	.4394	.4406	.4418	.4429	.4441
1.6	.4452	.4463	.4474	.4484	.4495	.4505	.4515	.4525	.4535	.4545
1.7	.4554	.4564	.4573	.4582	.4591	.4599	.4608	.4616	.4625	.4633
1.8	.4641	.4649	.4656	.4664	.4671	.4678	.4686	.4693	.4699	.4706
1.9	.4713	.4719	.4726	.4732	.4738	.4744	.4750	.4756	.4761	.4767
2.0	.4772	.4778	.4783	.4788	.4793	.4798	.4803	.4808	.4812	.4817
2.1	.4821	.4826	.4830	.4834	.4838	.4842	.4846	.4850	.4854	.4857
2.2	.4861	.4864	.4868	.4871	.4875	.4878	.4881	.4884	.4887	.4890
2.3	.4893	.4896	.4898	.4901	.4904	.4906	.4909	.4911	.4913	.4916
2.4	.4918	.4920	.4922	.4925	.4927	.4929	.4931	.4932	.4934	.4936
2.5	.4938	.4940	.4941	.4943	.4945	.4946	.4948	.4949	.4951	.4952
2.6	.4953	.4955	.4956	.4957	.4959	.4960	.4961	.4962	.4963	.4964
2.7	.4965	.4966	.4967	.4968	.4969	.4970	.4971	.4972	.4973	.4974
2.8	.4974	.4975	.4976	.4977	.4977	.4978	.4979	.4979	.4980	.4981
2.9	.4981	.4982	.4982	.4983	.4984	.4984	.4985	.4985	.4986	.4986
3.0	.49865	.4987	.4987	.4988	.4988	.4989	.4989	.4989	.4990	.4990
4.0	.49997									

Source: From *Statistical Analysis for Decision Making* by Morris Hamburg, © 1970 by Harcourt Brace Jovanovich, Inc.

TABLE A-6   STUDENT'S *t*-DISTRIBUTION

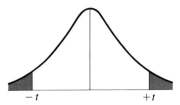

$-t$ $+t$

Example: For 15 degrees of freedom, the *t*-value which corresponds to an area of 0.5 in both tails combined is 2.131.

Degrees of Freedom	Area in Both Tails Combined			
	.10	.05	.02	.01
1	6.314	12.706	31.821	63.657
2	2.920	4.303	6.965	9.925
3	2.353	3.182	4.541	5.841
4	2.132	2.776	3.747	4.604
5	2.015	2.571	3.365	4.032
6	1.943	2.447	3.143	3.707
7	1.895	2.365	2.998	3.499
8	1.860	2.306	2.896	3.355
9	1.833	2.262	2.821	3.250
10	1.812	2.228	2.764	3.169
11	1.796	2.201	2.718	3.106
12	1.782	2.179	2.681	3.055
13	1.771	2.160	2.650	3.012
14	1.761	2.145	2.624	2.977
15	1.753	2.131	2.602	2.947
16	1.746	2.120	2.583	2.921
17	1.740	2.110	2.567	2.898
18	1.734	2.101	2.552	2.878
19	1.729	2.093	2.539	2.861
20	1.725	2.086	2.528	2.845
21	1.721	2.080	2.518	2.831
22	1.717	2.074	2.508	2.819
23	1.714	2.069	2.500	2.807
24	1.711	2.064	2.492	2.797
25	1.708	2.060	2.485	2.787
26	1.706	2.056	2.479	2.779
27	1.703	2.052	2.473	2.771
28	1.701	2.048	2.467	2.763
29	1.699	2.045	2.462	2.756
30	1.697	2.042	2.457	2.750
40	1.684	2.021	2.423	2.704
60	1.671	2.000	2.390	2.660
120	1.658	1.980	2.358	2.617
Normal Distribution	1.645	1.960	2.326	2.576

Source: Table A-6 is taken from Table III of Fisher and Yates: *Statistical Tables for Biological, Agricultural and Medical Research*, published by Longman Group Ltd., London (previously published by Oliver & Boyd, Edinburgh), and by permission of the authors and publishers.

TABLE A-7   CHI SQUARE (χ^2) DISTRIBUTION

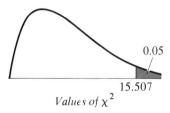

Values of χ^2

Example: In a chi square distribution with $df = 8$ degrees of freedom, the area to the right of a chi square value of 15.507 is 0.05.

Degrees of Freedom df	Area in Right Tail				
	.20	.10	.05	.02	.01
1	1.642	2.706	3.841	5.412	6.635
2	3.219	4.605	5.991	7.824	9.210
3	4.642	6.251	7.815	9.837	11.345
4	5.989	7.779	9.488	11.668	13.277
5	7.289	9.236	11.070	13.388	15.086
6	8.558	10.645	12.592	15.033	16.812
7	9.803	12.017	14.067	16.622	18.475
8	11.030	13.362	15.507	18.168	20.090
9	12.242	14.684	16.919	19.679	21.666
10	13.442	15.987	18.307	21.161	23.209
11	14.631	17.275	19.675	22.618	24.725
12	15.812	18.549	21.026	24.054	26.217
13	16.985	19.812	22.362	25.472	27.688
14	18.151	21.064	23.685	26.873	29.141
15	19.311	22.307	24.996	28.259	30.578
16	20.465	23.542	26.296	29.633	32.000
17	21.615	24.769	27.587	30.995	33.409
18	22.760	25.989	28.869	32.346	34.805
19	23.900	27.204	30.144	33.687	36.191
20	25.038	28.412	31.410	35.020	37.566
21	26.171	29.615	32.671	36.343	38.932
22	27.301	30.813	33.924	37.659	40.289
23	28.429	32.007	35.172	38.968	41.638
24	29.553	33.196	36.415	40.270	42.980
25	30.675	34.382	37.652	41.566	44.314
26	31.795	35.563	38.885	42.856	45.642
27	32.912	36.741	40.113	44.140	46.963
28	34.027	37.916	41.337	45.419	48.278
29	35.139	39.087	42.557	46.693	49.588
30	36.250	40.256	43.773	47.962	50.892

Source: Table A-7 is taken from Table IV of Fisher and Yates: *Statistical Tables for Biological, Agricultural and Medical Research*, published by Longman Group Ltd., London (previously published by Oliver & Boyd, Edinburgh), and by permission of the authors and publishers.

TABLE A-8 F-DISTRIBUTION

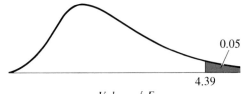

0.05

4.39

Values of F

Example: In an F-distribution with $df_1 = 5$ and $df_2 = 6$ degrees of freedom, the area to the right of an F value of 4.39 is 0.05. The value on the F-scale to the right of which lies .05 of the area is in lightface type. The value of the F-scale to the right of which lies .01 of the area is in boldface type. df_1 = number of degrees of freedom for numerator; df_2 = number of degrees of freedom for denominator.

df_2 \ df_1	1	2	3	4	5	6	7	8	9	10	20	30	40	50	100	200	∞	df_2
1	161 **4,052**	200 **4,999**	216 **5,403**	225 **5,625**	230 **5,764**	234 **5,859**	237 **5,928**	239 **5,981**	241 **6,022**	242 **6,056**	248 **6,208**	250 **6,261**	251 **6,286**	252 **6,302**	253 **6,334**	254 **6,352**	254 **6,366**	1
2	18.51 **98.49**	19.00 **99.00**	19.16 **99.17**	19.25 **99.25**	19.30 **99.30**	19.33 **99.33**	19.36 **99.36**	19.37 **99.37**	19.38 **99.39**	19.39 **99.40**	19.44 **99.45**	19.46 **99.47**	19.47 **99.48**	19.47 **99.48**	19.49 **99.49**	19.49 **99.49**	19.50 **99.50**	2
3	10.13 **34.12**	9.55 **30.82**	9.28 **29.46**	9.12 **28.71**	9.01 **28.24**	8.94 **27.91**	8.88 **27.67**	8.84 **27.49**	8.81 **27.34**	8.78 **27.23**	8.66 **26.69**	8.62 **26.50**	8.60 **26.41**	8.58 **26.35**	8.56 **26.23**	8.54 **26.18**	8.53 **26.12**	3
4	7.71 **21.20**	6.94 **18.00**	6.59 **16.69**	6.39 **15.98**	6.26 **15.52**	6.16 **15.21**	6.09 **14.98**	6.04 **14.80**	6.00 **14.66**	5.96 **14.54**	5.80 **14.02**	5.74 **13.83**	5.71 **13.74**	5.70 **13.69**	5.66 **13.57**	5.65 **13.52**	5.63 **13.46**	4
5	6.61 **16.26**	5.79 **13.27**	5.41 **12.06**	5.19 **11.39**	5.05 **10.97**	4.95 **10.67**	4.88 **10.45**	4.82 **10.29**	4.78 **10.15**	4.74 **10.05**	4.56 **9.55**	4.50 **9.38**	4.46 **9.29**	4.44 **9.24**	4.40 **9.13**	4.38 **9.07**	4.36 **9.02**	5
6	5.99 **13.74**	5.14 **10.92**	4.76 **9.78**	4.53 **9.15**	4.39 **8.75**	4.28 **8.47**	4.21 **8.26**	4.15 **8.10**	4.10 **7.98**	4.06 **7.87**	3.87 **7.39**	3.81 **7.23**	3.77 **7.14**	3.75 **7.09**	3.71 **6.99**	3.69 **6.94**	3.67 **6.88**	6
7	5.59 **12.25**	4.74 **9.55**	4.35 **8.45**	4.12 **7.85**	3.97 **7.46**	3.87 **7.19**	3.79 **7.00**	3.73 **6.84**	3.68 **6.71**	3.63 **6.62**	3.44 **6.15**	3.38 **5.98**	3.34 **5.90**	3.32 **5.85**	3.28 **5.75**	3.25 **5.70**	3.23 **5.65**	7
8	5.32 **11.26**	4.46 **8.65**	4.07 **7.59**	3.84 **7.01**	3.69 **6.63**	3.58 **6.37**	3.50 **6.19**	3.44 **6.03**	3.39 **5.91**	3.34 **5.82**	3.15 **5.36**	3.08 **5.20**	3.05 **5.11**	3.03 **5.06**	2.98 **4.96**	2.96 **4.91**	2.93 **4.86**	8
9	5.12 **10.56**	4.26 **8.02**	3.86 **6.99**	3.63 **6.42**	3.48 **6.06**	3.37 **5.80**	3.29 **5.62**	3.23 **5.47**	3.18 **5.35**	3.13 **5.26**	2.93 **4.80**	2.86 **4.64**	2.82 **4.56**	2.80 **4.51**	2.76 **4.41**	2.73 **4.36**	2.71 **4.31**	9
10	4.96 **10.04**	4.10 **7.56**	3.71 **6.55**	3.48 **5.99**	3.33 **5.64**	3.22 **5.39**	3.14 **5.21**	3.07 **5.06**	3.02 **4.95**	2.97 **4.85**	2.77 **4.41**	2.70 **4.25**	2.67 **4.17**	2.64 **4.12**	2.59 **4.01**	2.56 **3.96**	2.54 **3.91**	10
20	4.35 **8.10**	3.49 **5.85**	3.10 **4.94**	2.87 **4.43**	2.71 **4.10**	2.60 **3.87**	2.52 **3.71**	2.45 **3.56**	2.40 **3.45**	2.35 **3.37**	2.12 **2.94**	2.04 **2.77**	1.99 **2.69**	1.96 **2.63**	1.90 **2.53**	1.87 **2.47**	1.84 **2.42**	20
30	4.17 **7.56**	3.32 **5.39**	2.92 **4.51**	2.69 **4.02**	2.53 **3.70**	2.42 **3.47**	2.34 **3.30**	2.27 **3.17**	2.21 **3.06**	2.16 **2.98**	1.93 **2.55**	1.84 **2.38**	1.79 **2.29**	1.76 **2.24**	1.69 **2.13**	1.66 **2.07**	1.62 **2.01**	30
40	4.08 **7.31**	3.23 **5.18**	2.84 **4.31**	2.61 **3.83**	2.45 **3.51**	2.34 **3.29**	2.25 **3.12**	2.18 **2.99**	2.12 **2.88**	2.07 **2.80**	1.84 **2.37**	1.74 **2.20**	1.69 **2.11**	1.66 **2.05**	1.59 **1.94**	1.55 **1.88**	1.51 **1.81**	40
50	4.03 **7.17**	3.18 **5.06**	2.79 **4.20**	2.56 **3.72**	2.40 **3.41**	2.29 **3.18**	2.20 **3.02**	2.13 **2.88**	2.07 **2.78**	2.02 **2.70**	1.78 **2.26**	1.69 **2.10**	1.63 **2.00**	1.60 **1.94**	1.52 **1.82**	1.48 **1.76**	1.44 **1.68**	50
100	3.94 **6.90**	3.09 **4.82**	2.70 **3.98**	2.46 **3.51**	2.30 **3.20**	2.19 **2.99**	2.10 **2.82**	2.03 **2.69**	1.97 **2.59**	1.92 **2.51**	1.68 **2.06**	1.57 **1.89**	1.51 **1.79**	1.48 **1.73**	1.39 **1.59**	1.34 **1.51**	1.28 **1.43**	100
200	3.89 **6.76**	3.04 **4.71**	2.65 **3.88**	2.41 **3.41**	2.26 **3.11**	2.14 **2.90**	2.05 **2.73**	1.98 **2.60**	1.92 **2.50**	1.87 **2.41**	1.62 **1.97**	1.52 **1.79**	1.45 **1.69**	1.42 **1.62**	1.32 **1.48**	1.26 **1.39**	1.19 **1.28**	200
∞	3.84 **6.64**	2.99 **4.60**	2.60 **3.78**	2.37 **3.32**	2.21 **3.02**	2.09 **2.80**	2.01 **2.64**	1.94 **2.51**	1.88 **2.41**	1.83 **2.32**	1.57 **1.87**	1.46 **1.69**	1.40 **1.59**	1.35 **1.52**	1.24 **1.36**	1.17 **1.25**	1.00 **1.00**	∞

Source: Reprinted by permission from *Statistical Methods*, 5th ed., by George W. Snedecor, © 1956, Iowa State University Press, Ames, Iowa.

TABLE A-9 TABLE OF EXPONENTIAL FUNCTIONS

x	e^x	e^{-x}	x	e^x	e^{-x}
0.00	1.000	1.000	3.00	20.086	0.050
0.10	1.105	0.905	3.10	22.198	0.045
0.20	1.221	0.819	3.20	24.533	0.041
0.30	1.350	0.741	3.30	27.113	0.037
0.40	1.492	0.670	3.40	29.964	0.033
0.50	1.649	0.607	3.50	33.115	0.030
0.60	1.822	0.549	3.60	36.598	0.027
0.70	2.014	0.497	3.70	40.447	0.025
0.80	2.226	0.449	3.80	44.701	0.022
0.90	2.460	0.407	3.90	49.402	0.020
1.00	2.718	0.368	4.00	54.598	0.018
1.10	3.004	0.333	4.10	60.340	0.017
1.20	3.320	0.301	4.20	66.686	0.015
1.30	3.669	0.273	4.30	73.700	0.014
1.40	4.055	0.247	4.40	81.451	0.012
1.50	4.482	0.223	4.50	90.017	0.011
1.60	4.953	0.202	4.60	99.484	0.010
1.70	5.474	0.183	4.70	109.95	0.009
1.80	6.050	0.165	4.80	121.51	0.008
1.90	6.686	0.150	4.90	134.29	0.007
2.00	7.389	0.135	5.00	148.41	0.007
2.10	8.166	0.122	5.10	164.02	0.006
2.20	9.025	0.111	5.20	181.27	0.006
2.30	9.974	0.100	5.30	200.34	0.005
2.40	11.023	0.091	5.40	221.41	0.005
2.50	12.182	0.082	5.50	244.69	0.004
2.60	13.464	0.074	5.60	270.43	0.004
2.70	14.880	0.067	5.70	298.87	0.003
2.80	16.445	0.061	5.80	330.30	0.003
2.90	18.174	0.055	5.90	365.04	0.003
3.00	20.086	0.050	6.00	403.43	0.002

Source: From *Statistical Analysis for Decision Making* by Morris Hamburg, © 1970 by Harcourt Brace Jovanovich, Inc.

TABLE A-10 SQUARES, SQUARE ROOTS, AND RECIPROCALS

N	$\sqrt{N}$	N^2	$\sqrt{10N}$	$1000/N$	N	$\sqrt{N}$	N^2	$\sqrt{10N}$	$1000/N$
					50	7.07107	2500	22.36068	20.00000
1	1.00000	1	3.16228	1000.00000	51	7.14143	2601	22.58318	19.60784
2	1.41421	4	4.47214	500.00000	52	7.21110	2704	22.80351	19.23077
3	1.73205	9	5.47723	333.33333	53	7.28011	2809	23.02173	18.86792
4	2.00000	16	6.32456	250.00000	54	7.34847	2916	23.23790	18.51852
5	2.23607	25	7.07107	200.00000	55	7.41620	3025	23.45208	18.18182
6	2.44949	36	7.74597	166.66667	56	7.48331	3136	23.66432	17.85714
7	2.64575	49	8.36660	142.85714	57	7.54983	3249	23.87467	17.54386
8	2.82843	64	8.94427	125.00000	58	7.61577	3364	24.08319	17.24138
9	3.00000	81	9.48683	111.11111	59	7.68115	3481	24.28992	16.94915
10	3.16228	100	10.00000	100.00000	60	7.74597	3600	24.49490	16.66667
11	3.31662	121	10.48809	90.90909	61	7.81025	3721	24.69818	16.39344
12	3.46410	144	10.95445	83.33333	62	7.87401	3844	24.89980	16.12903
13	3.60555	169	11.40175	76.92308	63	7.93725	3969	25.09980	15.87302
14	3.74166	196	11.83216	71.42857	64	8.00000	4096	25.29822	15.62500
15	3.87298	225	12.24745	66.66667	65	8.06226	4225	25.49510	15.38462
16	4.00000	256	12.64911	62.50000	66	8.12404	4356	25.69047	15.15152
17	4.12311	289	13.03840	58.82353	67	8.18535	4489	25.88436	14.92537
18	4.24264	324	13.41641	55.55556	68	8.24621	4624	26.07681	14.70588
19	4.35890	361	13.78405	52.63158	69	8.30662	4761	26.26785	14.49275
20	4.47214	400	14.14214	50.00000	70	8.36660	4900	26.45751	14.28571
21	4.58258	441	14.49138	47.61905	71	8.42615	5041	26.64583	14.08451
22	4.69042	484	14.83240	45.45455	72	8.48528	5184	26.83282	13.88889
23	4.79583	529	15.16575	43.47826	73	8.54400	5329	27.01851	13.69863
24	4.89898	576	15.49193	41.66667	74	8.60233	5476	27.20294	13.51351
25	5.00000	625	15.81139	40.00000	75	8.66025	5625	27.38613	13.33333
26	5.09902	676	16.12452	38.46154	76	8.71780	5776	27.56810	13.15789
27	5.19615	729	16.43168	37.03704	77	8.77496	5929	27.74887	12.98701
28	5.29150	784	16.73320	35.71429	78	8.83176	6084	27.92848	12.82051
29	5.38516	841	17.02939	34.48276	79	8.88819	6241	28.10694	12.65823
30	5.47723	900	17.32051	33.33333	80	8.94427	6400	28.28427	12.50000
31	5.56776	961	17.60682	32.25806	81	9.00000	6561	28.46050	12.34568
32	5.65685	1024	17.88854	31.25000	82	9.05539	6724	28.63564	12.19512
33	5.74456	1089	18.16590	30.30303	83	9.11043	6889	28.80972	12.04819
34	5.83095	1156	18.43909	29.41176	84	9.16515	7056	28.98275	11.90476
35	5.91608	1225	18.70829	28.57143	85	9.21954	7225	29.15476	11.76471
36	6.00000	1296	18.97367	27.77778	86	9.27362	7396	29.32576	11.62791
37	6.08276	1369	19.23538	27.02703	87	9.32738	7569	29.49576	11.49425
38	6.16441	1444	19.49359	26.31579	88	9.38083	7744	29.66479	11.36364
39	6.24500	1521	19.74842	25.64103	89	9.43398	7921	29.83287	11.23596
40	6.32456	1600	20.00000	25.00000	90	9.48683	8100	30.00000	11.11111
41	6.40312	1681	20.24846	24.39024	91	9.53939	8281	30.16621	10.98901
42	6.48074	1764	20.49390	23.80952	92	9.59166	8464	30.33150	10.86957
43	6.55744	1849	20.73644	23.25581	93	9.64365	8649	30.49590	10.75269
44	6.63325	1936	20.97618	22.72727	94	9.69536	8836	30.65942	10.63830
45	6.70820	2025	21.21320	22.22222	95	9.74679	9025	30.82207	10.52632
46	6.78233	2116	21.44761	21.73913	96	9.79796	9216	30.98387	10.41667
47	6.85565	2209	21.67948	21.27660	97	9.84886	9409	31.14482	10.30928
48	6.92820	2304	21.90890	20.83333	98	9.89949	9604	31.30495	10.20408
49	7.00000	2401	22.13594	20.40816	99	9.94987	9801	31.46427	10.10101
50	7.07107	2500	22.36068	20.00000	100	10.00000	10000	31.62278	10.00000

Source: From *Statistics for Modern Business Decisions* by Lawrence L. Lapin, © 1973 by Harcourt Brace Jovanovich, Inc. and reproduced with their permission.

N	$\sqrt{N}$	N^2	$\sqrt{10N}$	$1000/N$	N	$\sqrt{N}$	N^2	$\sqrt{10N}$	$1000/N$
100	10.00000	10000	31.62278	10.00000	150	12.24745	22500	38.72983	6.66667
101	10.04988	10201	31.78050	9.90099	151	12.28821	22801	38.85872	6.62252
102	10.09950	10404	31.93744	9.80392	152	12.32883	23104	38.98718	6.57895
103	10.14889	10609	32.09361	9.70874	153	12.36932	23409	39.11521	6.53595
104	10.19804	10816	32.24903	9.61538	154	12.40967	23716	39.24283	6.49351
105	10.24695	11025	32.40370	9.52381	155	12.44990	24025	39.37004	6.45161
106	10.29563	11236	32.55764	9.43396	156	12.49000	24336	39.49684	6.41026
107	10.34408	11449	32.71085	9.34579	157	12.52996	24649	39.62323	6.36943
108	10.39230	11664	32.86335	9.25926	158	12.56981	24964	39.74921	6.32911
109	10.44031	11881	33.01515	9.17431	159	12.60952	25281	39.87480	6.28931
110	10.48809	12100	33.16625	9.09091	160	12.64911	25600	40.00000	6.25000
111	10.53565	12321	33.31666	9.00901	161	12.68858	25921	40.12481	6.21118
112	10.58301	12544	33.46640	8.92857	162	12.72792	26244	40.24922	6.17284
113	10.63015	12769	33.61547	8.84956	163	12.76715	26569	40.37326	6.13497
114	10.67708	12996	33.76389	8.77193	164	12.80625	26896	40.49691	6.09756
115	10.72381	13225	33.91165	8.69565	165	12.84523	27225	40.62019	6.06061
116	10.77033	13456	34.05877	8.62069	166	12.88410	27556	40.74310	6.02410
117	10.81665	13689	34.20526	8.54701	167	12.92285	27889	40.86563	5.98802
118	10.86278	13924	34.35113	8.47458	168	12.96148	28224	40.98780	5.95238
119	10.90871	14161	34.49638	8.40336	169	13.00000	28561	41.10961	5.91716
120	10.95445	14400	34.64102	8.33333	170	13.03840	28900	41.23106	5.88235
121	11.00000	14641	34.78505	8.26446	171	13.07670	29241	41.35215	5.84795
122	11.04536	14884	34.92850	8.19672	172	13.11488	29584	41.47288	5.81395
123	11.09054	15129	35.07136	8.13008	173	13.15295	29929	41.59327	5.78035
124	11.13553	15376	35.21363	8.06452	174	13.19091	30276	41.71331	5.74713
125	11.18034	15625	35.35534	8.00000	175	13.22876	30625	41.83300	5.71429
126	11.22497	15876	35.49648	7.93651	176	13.26650	30976	41.95235	5.68182
127	11.26943	16129	35.63706	7.87402	177	13.30413	31329	42.07137	5.64972
128	11.31371	16384	35.77709	7.81250	178	13.34166	31684	42.19005	5.61798
129	11.35782	16641	35.91657	7.75194	179	13.37909	32041	42.30839	5.58659
130	11.40175	16900	36.05551	7.69231	180	13.41641	32400	42.42641	5.55556
131	11.44552	17161	36.19392	7.63359	181	13.45362	32761	42.54409	5.52486
132	11.48913	17424	36.33180	7.57576	182	13.49074	33124	42.66146	5.49451
133	11.53256	17689	36.46917	7.51880	183	13.52775	33489	42.77850	5.46448
134	11.57584	17956	36.60601	7.46269	184	13.56466	33856	42.89522	5.43478
135	11.61895	18225	36.74235	7.40741	185	13.60147	34225	43.01163	5.40541
136	11.66190	18496	36.87818	7.35294	186	13.63818	34596	43.12772	5.37634
137	11.70470	18769	37.01351	7.29927	187	13.67479	34969	43.24350	5.34759
138	11.74734	19044	37.14835	7.24638	188	13.71131	35344	43.35897	5.31915
139	11.78983	19321	37.28270	7.19424	189	13.74773	35721	43.47413	5.29101
140	11.83216	19600	37.41657	7.14286	190	13.78405	36100	43.58899	5.26316
141	11.87434	19881	37.54997	7.09220	191	13.82027	36481	43.70355	5.23560
142	11.91638	20164	37.68289	7.04225	192	13.85641	36864	43.81780	5.20833
143	11.95826	20449	37.81534	6.99301	193	13.89244	37249	43.93177	5.18135
144	12.00000	20736	37.94733	6.94444	194	13.92839	37636	44.04543	5.15464
145	12.04159	21025	38.07887	6.89655	195	13.96424	38025	44.15880	5.12821
146	12.08305	21316	38.20995	6.84932	196	14.00000	38416	44.27189	5.10204
147	12.12436	21609	38.34058	6.80272	197	14.03567	38809	44.38468	5.07614
148	12.16553	21904	38.47077	6.75676	198	14.07125	39204	44.49719	5.05051
149	12.20656	22201	38.60052	6.71141	199	14.10674	39601	44.60942	5.02513
150	12.24745	22500	38.72983	6.66667	200	14.14214	40000	44.72136	5.00000

N	$\sqrt{N}$	N^2	$\sqrt{10N}$	$1000/N$	N	$\sqrt{N}$	N^2	$\sqrt{10N}$	$1000/N$
200	14.14214	40000	44.72136	5.00000	250	15.81139	62500	50.00000	4.00000
201	14.17745	40401	44.83302	4.97512	251	15.84298	63001	50.09990	3.98406
202	14.21267	40804	44.94441	4.95050	252	15.87451	63504	50.19960	3.96825
203	14.24781	41209	45.05552	4.92611	253	15.90597	64009	50.29911	3.95257
204	14.28286	41616	45.16636	4.90196	254	15.93738	64516	50.39841	3.93701
205	14.31782	42025	45.27693	4.87805	255	15.96872	65025	50.49752	3.92157
206	14.35270	42436	45.38722	4.85437	256	16.00000	65536	50.59644	3.90625
207	14.38749	42849	45.49725	4.83092	257	16.03122	66049	50.69517	3.89105
208	14.42221	43264	45.60702	4.80769	258	16.06238	66564	50.79370	3.87597
209	14.45683	43681	45.71652	4.78469	259	16.09348	67081	50.89204	3.86100
210	14.49138	44100	45.82576	4.76190	260	16.12452	67600	50.99020	3.84615
211	14.52584	44521	45.93474	4.73934	261	16.15549	68121	51.08816	3.83142
212	14.56022	44944	46.04346	4.71698	262	16.18641	68644	51.18594	3.81679
213	14.59452	45369	46.15192	4.69484	263	16.21727	69169	51.28353	3.80228
214	14.62874	45796	46.26013	4.67290	264	16.24808	69696	51.38093	3.78788
215	14.66288	46225	46.36809	4.65116	265	16.27882	70225	51.47815	3.77358
216	14.69694	46656	46.47580	4.62963	266	16.30951	70756	51.57519	3.75940
217	14.73092	47089	46.58326	4.60829	267	16.34013	71289	51.67204	3.74532
218	14.76482	47524	46.69047	4.58716	268	16.37071	71824	51.76872	3.73134
219	14.79865	47961	46.79744	4.56621	269	16.40122	72361	51.86521	3.71747
220	14.83240	48400	46.90416	4.54545	270	16.43168	72900	51.96152	3.70370
221	14.86607	48841	47.01064	4.52489	271	16.46208	73441	52.05766	3.69004
222	14.89966	49284	47.11688	4.50450	272	16.49242	73984	52.15362	3.67647
223	14.93318	49729	47.22288	4.48430	273	16.52271	74529	52.24940	3.66300
224	14.96663	50176	47.32864	4.46429	274	16.55295	75076	52.34501	3.64964
225	15.00000	50625	47.43416	4.44444	275	16.58312	75625	52.44044	3.63636
226	15.03330	51076	47.53946	4.42478	276	16.61325	76176	52.53570	3.62319
227	15.06652	51529	47.64452	4.40529	277	16.64332	76729	52.63079	3.61011
228	15.09967	51984	47.74935	4.38596	278	16.67333	77284	52.72571	3.59712
229	15.13275	52441	47.85394	4.36681	279	16.70329	77841	52.82045	3.58423
230	15.16575	52900	47.95832	4.34783	280	16.73320	78400	52.91503	3.57143
231	15.19868	53361	48.06246	4.32900	281	16.76305	78961	53.00943	3.55872
232	15.23155	53824	48.16638	4.31034	282	16.79286	79524	53.10367	3.54610
233	15.26434	54289	48.27007	4.29185	283	16.82260	80089	53.19774	3.53357
234	15.29706	54756	48.37355	4.27350	284	16.85230	80656	53.29165	3.52113
235	15.32971	55225	48.47680	4.25532	285	16.88194	81225	53.38539	3.50877
236	15.36229	55696	48.57983	4.23729	286	16.91153	81796	53.47897	3.49650
237	15.39480	56169	48.68265	4.21941	287	16.94107	82369	53.57238	3.48432
238	15.42725	56644	48.78524	4.20168	288	16.97056	82944	53.66563	3.47222
239	15.45962	57121	48.88763	4.18410	289	17.00000	83521	53.75872	3.46021
240	15.49193	57600	48.98979	4.16667	290	17.02939	84100	53.85165	3.44828
241	15.52417	58081	49.09175	4.14938	291	17.05872	84681	53.94442	3.43643
242	15.55635	58564	49.19350	4.13223	292	17.08801	85264	54.03702	3.42466
243	15.58846	59049	49.29503	4.11523	293	17.11724	85849	54.12947	3.41297
244	15.62050	59536	49.39636	4.09836	294	17.14643	86436	54.22177	3.40136
245	15.65248	60025	49.49747	4.08163	295	17.17556	87025	54.31390	3.38983
246	15.68439	60516	49.59839	4.06504	296	17.20465	87616	54.40588	3.37838
247	15.71623	61009	49.69909	4.04858	297	17.23369	88209	54.49771	3.36700
248	15.74802	61504	49.79960	4.03226	298	17.26268	88804	54.58938	3.35570
249	15.77973	62001	49.89990	4.01606	299	17.29162	89401	54.68089	3.34448
250	15.81139	62500	50.00000	4.00000	300	17.32051	90000	54.77226	3.33333

N	$\sqrt{N}$	N^2	$\sqrt{10N}$	$1000/N$	N	$\sqrt{N}$	N^2	$\sqrt{10N}$	$1000/N$
300	17.32051	90000	54.77226	3.33333	350	18.70829	122500	59.16080	2.85714
301	17.34935	90601	54.86347	3.32226	351	18.73499	123201	59.24525	2.84900
302	17.37815	91204	54.95453	3.31126	352	18.76166	123904	59.32959	2.84091
303	17.40690	91809	55.04544	3.30033	353	18.78829	124609	59.41380	2.83286
304	17.43560	92416	55.13620	3.28947	354	18.81489	125316	59.49790	2.82486
305	17.46425	93025	55.22681	3.27869	355	18.84144	126025	59.58188	2.81690
306	17.49286	93636	55.31727	3.26797	356	18.86796	126736	59.66574	2.80899
307	17.52142	94249	55.40758	3.25733	357	18.89444	127449	59.74948	2.80112
308	17.54993	94864	55.49775	3.24675	358	18.92089	128164	59.83310	2.79330
309	17.57840	95481	55.58777	3.23625	359	18.94730	128881	59.91661	2.78552
310	17.60682	96100	55.67764	3.22581	360	18.97367	129600	60.00000	2.77778
311	17.63519	96721	55.76737	3.21543	361	19.00000	130321	60.08328	2.77008
312	17.66352	97344	55.85696	3.20513	362	19.02630	131044	60.16644	2.76243
313	17.69181	97969	55.94640	3.19489	363	19.05256	131769	60.24948	2.75482
314	17.72005	98596	56.03570	3.18471	364	19.07878	132496	60.33241	2.74725
315	17.74824	99225	56.12486	3.17460	365	19.10497	133225	60.41523	2.73973
316	17.77639	99856	56.21388	3.16456	366	19.13113	133956	60.49793	2.73224
317	17.80449	100489	56.30275	3.15457	367	19.15724	134689	60.58052	2.72480
318	17.83255	101124	56.39149	3.14465	368	19.18333	135424	60.66300	2.71739
319	17.86057	101761	56.48008	3.13480	369	19.20937	136161	60.74537	2.71003
320	17.88854	102400	56.56854	3.12500	370	19.23538	136900	60.82763	2.70270
321	17.91647	103041	56.65686	3.11526	371	19.26136	137641	60.90977	2.69542
322	17.94436	103684	56.74504	3.10559	372	19.28730	138384	60.99180	2.68817
323	17.97220	104329	56.83309	3.09598	373	19.31321	139129	61.07373	2.68097
324	18.00000	104976	56.92100	3.08642	374	19.33908	139876	61.15554	2.67380
325	18.02776	105625	57.00877	3.07692	375	19.36492	140625	61.23724	2.66667
326	18.05547	106276	57.09641	3.06748	376	19.39072	141376	61.31884	2.65957
327	18.08314	106929	57.18391	3.05810	377	19.41649	142129	61.40033	2.65252
328	18.11077	107584	57.27128	3.04878	378	19.44222	142884	61.48170	2.64550
329	18.13836	108241	57.35852	3.03951	379	19.46792	143641	61.56298	2.63852
330	18.16590	108900	57.44563	3.03030	380	19.49359	144400	61.64414	2.63158
331	18.19341	109561	57.53260	3.02115	381	19.51922	145161	61.72520	2.62467
332	18.22087	110224	57.61944	3.01205	382	19.54482	145924	61.80615	2.61780
333	18.24829	110889	57.70615	3.00300	383	19.57039	146689	61.88699	2.61097
334	18.27567	111556	57.79273	2.99401	384	19.59592	147456	61.96773	2.60417
335	18.30301	112225	57.87918	2.98507	385	19.62142	148225	62.04837	2.59740
336	18.33030	112896	57.96551	2.97619	386	19.64688	148996	62.12890	2.59067
337	18.35756	113569	58.05170	2.96736	387	19.67232	149769	62.20932	2.58398
338	18.38478	114244	58.13777	2.95858	388	19.69772	150544	62.28965	2.57732
339	18.41195	114921	58.22371	2.94985	389	19.72308	151321	62.36986	2.57069
340	18.43909	115600	58.30952	2.94118	390	19.74842	152100	62.44998	2.56410
341	18.46619	116281	58.39521	2.93255	391	19.77372	152881	62.52999	2.55754
342	18.49324	116964	58.48077	2.92398	392	19.79899	153664	62.60990	2.55102
343	18.52026	117649	58.56620	2.91545	393	19.82423	154449	62.68971	2.54453
344	18.54724	118336	58.65151	2.90698	394	19.84943	155236	62.76942	2.53807
345	18.57418	119025	58.73670	2.89855	395	19.87461	156025	62.84903	2.53165
346	18.60108	119716	58.82176	2.89017	396	19.89975	156816	62.92853	2.52525
347	18.62794	120409	58.90671	2.88184	397	19.92486	157609	63.00794	2.51889
348	18.65476	121104	58.99194	2.87356	398	19.94994	158404	63.08724	2.51256
349	18.68154	121801	59.07622	2.86533	399	19.97498	159201	63.16645	2.50627
350	18.70829	122500	59.16080	2.85714	400	20.00000	160000	63.24555	2.50000

N	$\sqrt{N}$	N^2	$\sqrt{10N}$	$1000/N$	N	$\sqrt{N}$	N^2	$\sqrt{10N}$	$1000/N$
400	20.00000	160000	63.24555	2.50000	450	21.21320	202500	67.08204	2.22222
401	20.02498	160801	63.32456	2.49377	451	21.23676	203401	67.15653	2.21729
402	20.04994	161604	63.40347	2.48756	452	21.26029	204304	67.23095	2.21239
403	20.07486	162409	63.48228	2.48139	453	21.28380	205209	67.30527	2.20751
404	20.09975	163216	63.56099	2.47525	454	21.30728	206116	67.37952	2.20264
405	20.12461	164025	63.63961	2.46914	455	21.33073	207025	67.45369	2.19780
406	20.14944	164836	63.71813	2.46305	456	21.35416	207936	67.52777	2.19298
407	20.17424	165649	63.79655	2.45700	457	21.37756	208849	67.60178	2.18818
408	20.19901	166464	63.87488	2.45098	458	21.40093	209764	67.67570	2.18341
409	20.22375	167281	63.95311	2.44499	459	21.42429	210681	67.74954	2.17865
410	20.24846	168100	64.03124	2.43902	460	21.44761	211600	67.82330	2.17391
411	20.27313	168921	64.10928	2.43309	461	21.47091	212521	67.89698	2.16920
412	20.29778	169744	64.18723	2.42718	462	21.49419	213444	67.97058	2.16450
413	20.32240	170569	64.26508	2.42131	463	21.51743	214369	68.04410	2.15983
414	20.34699	171396	64.34283	2.41546	464	21.54066	215296	68.11755	2.15517
415	20.37155	172225	64.42049	2.40964	465	21.56386	216225	68.19091	2.15054
416	20.39608	173056	64.49806	2.40385	466	21.58703	217156	68.26419	2.14592
417	20.42058	173889	64.57554	2.39808	467	21.61018	218089	68.33740	2.14133
418	20.44505	174724	64.65292	2.39234	468	21.63331	219024	68.41053	2.13675
419	20.46949	175561	64.73021	2.38663	469	21.65641	219961	68.48357	2.13220
420	20.49390	176400	64.80741	2.38095	470	21.67948	220900	68.55655	2.12766
421	20.51828	177241	64.88451	2.37530	471	21.70253	221841	68.62944	2.12314
422	20.54264	178084	64.96153	2.36967	472	21.72556	222784	68.70226	2.11864
423	20.56696	178929	65.03845	2.36407	473	21.74856	223729	68.77500	2.11416
424	20.59126	179776	65.11528	2.35849	474	21.77154	224676	68.84766	2.10970
425	20.61553	180625	65.19202	2.35294	475	21.79449	225625	68.92024	2.10526
426	20.63977	181476	65.26868	2.34742	476	21.81742	226576	68.99275	2.10084
427	20.66398	182329	65.34524	2.34192	477	21.84033	227529	69.06519	2.09644
428	20.68816	183184	65.42171	2.33645	478	21.86321	228484	69.13754	2.09205
429	20.71232	184041	65.49809	2.33100	479	21.88607	229441	69.20983	2.08768
430	20.73644	184900	65.57439	2.32558	480	21.90890	230400	69.28203	2.08333
431	20.76054	185761	65.65059	2.32019	481	21.93171	231361	69.35416	2.07900
432	20.78461	186624	65.72671	2.31481	482	21.95450	232324	69.42622	2.07469
433	20.80865	187489	65.80274	2.30947	483	21.97726	233289	69.49820	2.07039
434	20.83267	188356	65.87868	2.30415	484	22.00000	234256	69.57011	2.06612
435	20.85665	189225	65.95453	2.29885	485	22.02272	235225	69.64194	2.06186
436	20.88061	190096	66.03030	2.29358	486	22.04541	236196	69.71370	2.05761
437	20.90454	190969	66.10598	2.28833	487	22.06808	237169	69.78539	2.05339
438	20.92845	191844	66.18157	2.28311	488	22.09072	238144	69.85700	2.04918
439	20.95233	192721	66.25708	2.27790	489	22.11334	239121	69.92853	2.04499
440	20.97618	193600	66.33250	2.27273	490	22.13594	240100	70.00000	2.04082
441	21.00000	194481	66.40783	2.26757	491	22.15852	241081	70.07139	2.03666
442	21.02380	195364	66.48308	2.26244	492	22.18107	242064	70.14271	2.03252
443	21.04757	196249	66.55825	2.25734	493	22.20360	243049	70.21396	2.02840
444	21.07131	197136	66.63332	2.25225	494	22.22611	244036	70.28513	2.02429
445	21.09502	198025	66.70832	2.24719	495	22.24860	245025	70.35624	2.02020
446	21.11871	198916	66.78323	2.24215	496	22.27106	246016	70.42727	2.01613
447	21.14237	199809	66.85806	2.23714	497	22.29350	247009	70.49823	2.01207
448	21.16601	200704	66.93280	2.23214	498	22.31591	248004	70.56912	2.00803
449	21.18962	201601	67.00746	2.22717	499	22.33831	249001	70.63993	2.00401
450	21.21320	202500	67.08204	2.22222	500	22.36068	250000	70.71068	2.00000

N	$\sqrt{N}$	N^2	$\sqrt{10N}$	$1000/N$	N	$\sqrt{N}$	N^2	$\sqrt{10N}$	$1000/N$
500	22.36068	250000	70.71068	2.00000	550	23.45208	302500	74.16198	1.81818
501	22.38303	251001	70.78135	1.99601	551	23.47339	303601	74.22937	1.81488
502	22.40536	252004	70.85196	1.99203	552	23.49468	304704	74.29670	1.81159
503	22.42766	253009	70.92249	1.98807	553	23.51595	305809	74.36397	1.80832
504	22.44994	254016	70.99296	1.98413	554	23.53720	306916	74.43118	1.80505
505	22.47221	255025	71.06335	1.98020	555	23.55844	308025	74.49832	1.80180
506	22.49444	256036	71.13368	1.97628	556	23.57965	309136	74.56541	1.79856
507	22.51666	257049	71.20393	1.97239	557	23.60085	310249	74.63243	1.79533
508	22.53886	258064	71.27412	1.96850	558	23.62202	311364	74.69940	1.79211
509	22.56103	259081	71.34424	1.96464	559	23.64318	312481	74.76630	1.78891
510	22.58318	260100	71.41428	1.96078	560	23.66432	313600	74.83315	1.78571
511	22.60531	261121	71.48426	1.95695	561	23.68544	314721	74.89993	1.78253
512	22.62742	262144	71.55418	1.95313	562	23.70654	315844	74.96666	1.77936
513	22.64950	263169	71.62402	1.94932	563	23.72762	316969	75.03333	1.77620
514	22.67157	264196	71.69379	1.94553	564	23.74868	318096	75.09993	1.77305
515	22.69361	265225	71.76350	1.94175	565	23.76973	319225	75.16648	1.76991
516	22.71563	266256	71.83314	1.93798	566	23.79075	320356	75.23297	1.76678
517	22.73763	267289	71.90271	1.93424	567	23.81176	321489	75.29940	1.76367
518	22.75961	268324	71.97222	1.93050	568	23.83275	322624	75.36577	1.76056
519	22.78157	269361	72.04165	1.92678	569	23.85372	323761	75.43209	1.75747
520	22.80351	270400	72.11103	1.92308	570	23.87467	324900	75.49834	1.75439
521	22.82542	271441	72.18033	1.91939	571	23.89561	326041	75.56454	1.75131
522	22.84732	272484	72.24957	1.91571	572	23.91652	327184	75.63068	1.74825
523	22.86919	273529	72.31874	1.91205	573	23.93742	328329	75.69676	1.74520
524	22.89105	274576	72.38784	1.90840	574	23.95830	329476	75.76279	1.74216
525	22.91288	275625	72.45688	1.90476	575	23.97916	330625	75.82875	1.73913
526	22.93469	276676	72.52586	1.90114	576	24.00000	331776	75.89466	1.73611
527	22.95648	277729	72.59477	1.89753	577	24.02082	332929	75.96052	1.73310
528	22.97825	278784	72.66361	1.89394	578	24.04163	334084	76.02631	1.73010
529	23.00000	279841	72.73239	1.89036	579	24.06242	335241	76.09205	1.72712
530	23.02173	280900	72.80110	1.88679	580	24.08319	336400	76.15773	1.72414
531	23.04344	281961	72.86975	1.88324	581	24.10394	337561	76.22336	1.72117
532	23.06513	283024	72.93833	1.87970	582	24.12468	338724	76.28892	1.71821
533	23.08679	284089	73.00685	1.87617	583	24.14539	339889	76.35444	1.71527
534	23.10844	285156	73.07530	1.87266	584	24.16609	341056	76.41989	1.71233
535	23.13007	286225	73.14369	1.86916	585	24.18677	342225	76.48529	1.70940
536	23.15167	287296	73.21202	1.86567	586	24.20744	343396	76.55064	1.70648
537	23.17326	288369	73.28028	1.86220	587	24.22808	344569	76.61593	1.70358
538	23.19483	289444	73.34848	1.85874	588	24.24871	345744	76.68116	1.70068
539	23.21637	290521	73.41662	1.85529	589	24.26932	346921	76.74634	1.69779
540	23.23790	291600	73.48469	1.85185	590	24.28992	348100	76.81146	1.69492
541	23.25941	292681	73.55270	1.84843	591	24.31049	349281	76.87652	1.69205
542	23.28089	293764	73.62065	1.84502	592	24.33105	350464	76.94154	1.68919
543	23.30236	294849	73.68853	1.84162	593	24.35159	351649	77.00649	1.68634
544	23.32381	295936	73.75636	1.83824	594	24.37212	352836	77.07140	1.68350
545	23.34524	297025	73.82412	1.83486	595	24.39262	354025	77.13624	1.68067
546	23.36664	298116	73.89181	1.83150	596	24.41311	355216	77.20104	1.67785
547	23.38803	299209	73.95945	1.82815	597	24.43358	356409	77.26578	1.67504
548	23.40940	300304	74.02702	1.82482	598	24.45404	357604	77.33046	1.67224
549	23.43075	301401	74.09453	1.82149	599	24.47448	358801	77.39509	1.66945
550	23.45208	302500	74.16198	1.81818	600	24.49490	360000	77.45967	1.66667

N	$\sqrt{N}$	N^2	$\sqrt{10N}$	$1000/N$	N	$\sqrt{N}$	N^2	$\sqrt{10N}$	$1000/N$
600	24.49490	360000	77.45967	1.66667	650	25.49510	422500	80.62258	1.53846
601	24.51530	361201	77.52419	1.66389	651	25.51470	423801	80.68457	1.53610
602	24.53569	362404	77.58866	1.66113	652	25.53429	425104	80.74652	1.53374
603	24.55606	363609	77.65307	1.65837	653	25.55386	426409	80.80842	1.53139
604	24.57641	364816	77.71744	1.65563	654	25.57342	427716	80.87027	1.52905
605	24.59675	366025	77.78175	1.65289	655	25.59297	429025	80.93207	1.52672
606	24.61707	367236	77.84600	1.65017	656	25.61250	430336	80.99383	1.52439
607	24.63737	368449	77.91020	1.64745	657	25.63201	431649	81.05554	1.52207
608	24.65766	369664	77.97435	1.64474	658	25.65151	432964	81.11720	1.51976
609	24.67793	370881	78.03845	1.64204	659	25.67100	434281	81.17881	1.51745
610	24.69818	372100	78.10250	1.63934	660	25.69047	435600	81.24038	1.51515
611	24.71841	373321	78.16649	1.63666	661	25.70992	436921	81.30191	1.51286
612	24.73863	374544	78.23043	1.63399	662	25.72936	438244	81.36338	1.51057
613	24.75884	375769	78.29432	1.63132	663	25.74879	439569	81.42481	1.50830
614	24.77902	376996	78.35815	1.62866	664	25.76820	440896	81.48620	1.50602
615	24.79919	378225	78.42194	1.62602	665	25.78759	442225	81.54753	1.50376
616	24.81935	379456	78.48567	1.62338	666	25.80698	443556	81.60882	1.50150
617	24.83948	380689	78.54935	1.62075	667	25.82634	444889	81.67007	1.49925
618	24.85961	381924	78.61298	1.61812	668	25.84570	446224	81.73127	1.49701
619	24.87971	383161	78.67655	1.61551	669	25.86503	447561	81.79242	1.49477
620	24.89980	384400	78.74008	1.61290	670	25.88436	448900	81.85353	1.49254
621	24.91987	385641	78.80355	1.61031	671	25.90367	450241	81.91459	1.49031
622	24.93993	386884	78.86698	1.60772	672	25.92296	451584	81.97561	1.48810
623	24.95997	388129	78.93035	1.60514	673	25.94224	452929	82.03658	1.48588
624	24.97999	389376	78.99367	1.60256	674	25.96151	454276	82.09750	1.48368
625	25.00000	390625	79.05694	1.60000	675	25.98076	455625	82.15838	1.48148
626	25.01999	391876	79.12016	1.59744	676	26.00000	456976	82.21922	1.47929
627	25.03997	393129	79.18333	1.59490	677	26.01922	458329	82.28001	1.47710
628	25.05993	394384	79.24645	1.59236	678	26.03843	459684	82.34076	1.47493
629	25.07987	395641	79.30952	1.58983	679	26.05763	461041	82.40146	1.47275
630	25.09980	396900	79.37254	1.58730	680	26.07681	462400	82.46211	1.47059
631	25.11971	398161	79.43551	1.58479	681	26.09598	463761	82.52272	1.46843
632	25.13961	399424	79.49843	1.58228	682	26.11513	465124	82.58329	1.46628
633	25.15949	400689	79.56130	1.57978	683	26.13427	466489	82.64381	1.46413
634	25.17936	401956	79.62412	1.57729	684	26.15339	467856	82.70429	1.46199
635	25.19921	403225	79.68689	1.57480	685	26.17250	469225	82.76473	1.45985
636	25.21904	404496	79.74961	1.57233	686	26.19160	470596	82.82512	1.45773
637	25.23886	405769	79.81228	1.56986	687	26.21068	471969	82.88546	1.45560
638	25.25866	407044	79.87490	1.56740	688	26.22975	473344	82.94577	1.45349
639	25.27845	408321	79.93748	1.56495	689	26.24881	474721	83.00602	1.45138
640	25.29822	409600	80.00000	1.56250	690	26.26785	476100	83.06624	1.44928
641	25.31798	410881	80.06248	1.56006	691	26.28688	477481	83.12641	1.44718
642	25.33772	412164	80.12490	1.55763	692	26.30589	478864	83.18654	1.44509
643	25.35744	413449	80.18728	1.55521	693	26.32489	480249	83.24662	1.44300
644	25.37716	414736	80.24961	1.55280	694	26.34388	481636	83.30666	1.44092
645	25.39685	416025	80.31189	1.55039	695	26.36285	483025	83.36666	1.43885
646	25.41653	417316	80.37413	1.54799	696	26.38181	484416	83.42661	1.43678
647	25.43619	418609	80.43631	1.54560	697	26.40076	485809	83.48653	1.43472
648	25.45584	419904	80.49845	1.54321	698	26.41969	487204	83.54639	1.43266
649	25.47548	421201	80.56054	1.54083	699	26.43861	488601	83.60622	1.43062
650	25.49510	422500	80.62258	1.53846	700	26.45751	490000	83.66600	1.42857

N	$\sqrt{N}$	N^2	$\sqrt{10N}$	$1000/N$	N	$\sqrt{N}$	N^2	$\sqrt{10N}$	$1000/N$
70C	26.45751	490000	83.66600	1.42857	750	27.38613	562500	86.60254	1.33333
701	26.47640	4914C1	83.72574	1.42653	751	27.40438	564001	86.66026	1.33156
702	26.49528	4928C4	83.78544	1.42450	752	27.42262	565504	86.71793	1.32979
703	26.51415	494209	83.84510	1.42248	753	27.44085	567009	86.77557	1.32802
704	26.5330C	495616	83.90471	1.42045	754	27.45906	568516	86.83317	1.32626
705	26.55184	497025	83.96428	1.41844	755	27.47726	570025	86.89074	1.32450
706	26.57066	498436	84.02381	1.41643	756	27.49545	571536	86.94826	1.32275
707	26.58947	499849	84.08329	1.41443	757	27.51363	573049	87.00575	1.32100
708	26.60827	501264	84.14274	1.41243	758	27.53180	574564	87.06320	1.31926
709	26.62705	502681	84.20214	1.41C44	759	27.54995	576081	87.12061	1.31752
710	26.64583	504100	84.26150	1.40845	760	27.56810	577600	87.17798	1.31579
711	26.66458	505521	84.32082	1.40647	761	27.58623	579121	87.23531	1.31406
712	26.68333	506944	84.38009	1.40449	762	27.60435	580644	87.29261	1.31234
713	26.70206	5C8369	84.43933	1.40252	763	27.62245	582169	87.34987	1.31C62
714	26.72078	509796	84.49852	1.40056	764	27.64055	583696	87.40709	1.30890
715	26.73948	511225	84.55767	1.39860	765	27.65863	585225	87.46428	1.30719
716	26.75818	512656	84.61678	1.39665	766	27.67671	586756	87.52143	1.30548
717	26.77686	514089	84.67585	1.39470	767	27.69476	588289	87.57854	1.30378
718	26.79552	515524	84.73488	1.39276	768	27.71281	589824	87.63561	1.30208
719	26.81418	516961	84.79387	1.39082	769	27.73085	591361	87.69265	1.30039
720	26.83282	518400	84.85281	1.38889	770	27.74887	592900	87.74964	1.29870
721	26.85144	519841	84.91172	1.38696	771	27.76689	594441	87.80661	1.29702
722	26.87006	521284	84.97058	1.38504	772	27.78489	595984	87.86353	1.29534
723	26.88866	522729	85.02941	1.38313	773	27.8C288	597529	87.92042	1.29366
724	26.90725	524176	85.08819	1.38122	774	27.82086	599076	87.97727	1.29199
725	26.92582	525625	85.14693	1.37931	775	27.83882	600625	88.03408	1.29032
726	26.94439	527076	85.20563	1.37741	776	27.85678	602176	88.09C86	1.28866
727	26.96294	528529	85.26429	1.37552	777	27.87472	603729	88.14760	1.28700
728	26.98148	529984	85.32292	1.37363	778	27.89265	605284	88.20431	1.28535
729	27.00000	531441	85.38150	1.37174	779	27.91057	606841	88.26098	1.28370
730	27.01851	532900	85.44004	1.36986	780	27.92848	608400	88.31761	1.282C5
731	27.03701	534361	85.49854	1.36799	781	27.94638	609961	88.37420	1.28C41
732	27.05550	535824	85.55700	1.36612	782	27.96426	611524	88.43C76	1.27877
733	27.07397	537289	85.61542	1.36426	783	27.98214	613089	88.48729	1.27714
734	27.09243	538756	85.67380	1.36240	784	28.00000	614656	88.54377	1.27551
735	27.11088	540225	85.73214	1.36054	785	28.01785	616225	88.60023	1.27389
736	27.12932	541696	85.79044	1.35870	786	28.03569	617796	88.65664	1.27226
737	27.14774	543169	85.84870	1.35685	787	28.05352	619369	88.71302	1.27065
738	27.16616	544644	85.90693	1.35501	788	28.07134	62C944	88.76936	1.26904
739	27.18455	546121	85.96511	1.35318	789	28.08914	622521	88.82567	1.26743
740	27.20294	547600	86.02325	1.35135	790	28.10694	624100	88.88194	1.26582
741	27.22132	549081	86.08136	1.34953	791	28.12472	625681	88.93818	1.26422
742	27.23968	550564	86.13942	1.34771	792	28.14249	627264	88.99438	1.26263
743	27.25803	552049	86.19745	1.34590	793	28.16026	628849	89.05C55	1.26103
744	27.27636	553536	86.25543	1.34409	794	28.17801	630436	89.10668	1.25945
745	27.29469	555025	86.31338	1.34228	795	28.19574	632025	89.16277	1.25786
746	27.31300	556516	86.37129	1.34048	796	28.21347	633C16	89.21883	1.25628
747	27.33130	558009	86.42916	1.33869	797	28.23119	635209	89.27486	1.25471
748	27.34959	559504	86.48699	1.33690	798	28.24889	636804	89.33085	1.25313
749	27.36786	561001	86.54479	1.33511	799	28.26659	638401	89.38680	1.25156
750	27.38613	562500	86.60254	1.33333	800	28.28427	640000	89.44272	1.25000

N	$\sqrt{N}$	N^2	$\sqrt{10N}$	$1000/N$	N	$\sqrt{N}$	N^2	$\sqrt{10N}$	$1000/N$
800	28.28427	640000	89.44272	1.25000	850	29.15476	722500	92.19544	1.17647
801	28.30194	641601	89.49860	1.24844	851	29.17190	724201	92.24966	1.17509
802	28.31960	643204	89.55445	1.24688	852	29.18904	725904	92.30385	1.17371
803	28.33725	644809	89.61027	1.24533	853	29.20616	727609	92.35800	1.17233
804	28.35489	646416	89.66605	1.24378	854	29.22328	729316	92.41212	1.17096
805	28.37252	648025	89.72179	1.24224	855	29.24038	731025	92.46621	1.16959
806	28.39014	649636	89.77750	1.24069	856	29.25748	732736	92.52027	1.16822
807	28.40775	651249	89.83318	1.23916	857	29.27456	734449	92.57429	1.16686
808	28.42534	652864	89.88882	1.23762	858	29.29164	736164	92.62829	1.16550
809	28.44293	654481	89.94443	1.23609	859	29.30870	737881	92.68225	1.16414
810	28.46050	656100	90.00000	1.23457	860	29.32576	739600	92.73618	1.16279
811	28.47806	657721	90.05554	1.23305	861	29.34280	741321	92.79009	1.16144
812	28.49561	659344	90.11104	1.23153	862	29.35984	743044	92.84396	1.16009
813	28.51315	660969	90.16651	1.23001	863	29.37686	744769	92.89779	1.15875
814	28.53069	662596	90.22195	1.22850	864	29.39388	746496	92.95160	1.15741
815	28.54820	664225	90.27735	1.22699	865	29.41088	748225	93.00538	1.15607
816	28.56571	665856	90.33272	1.22549	866	29.42788	749956	93.05912	1.15473
817	28.58321	667489	90.38805	1.22399	867	29.44486	751689	93.11283	1.15340
818	28.60070	669124	90.44335	1.22249	868	29.46184	753424	93.16652	1.15207
819	28.61818	670761	90.49862	1.22100	869	29.47881	755161	93.22017	1.15075
820	28.63564	672400	90.55385	1.21951	870	29.49576	756900	93.27379	1.14943
821	28.65310	674041	90.60905	1.21803	871	29.51271	758641	93.32738	1.14811
822	28.67054	675684	90.66422	1.21655	872	29.52965	760384	93.38094	1.14679
823	28.68798	677329	90.71935	1.21507	873	29.54657	762129	93.43447	1.14548
824	28.70540	678976	90.77445	1.21359	874	29.56349	763876	93.48797	1.14416
825	28.72281	680625	90.82951	1.21212	875	29.58040	765625	93.54143	1.14286
826	28.74022	682276	90.88454	1.21065	876	29.59730	767376	93.59487	1.14155
827	28.75761	683929	90.93954	1.20919	877	29.61419	769129	93.64828	1.14025
828	28.77499	685584	90.99451	1.20773	878	29.63106	770884	93.70165	1.13895
829	28.79236	687241	91.04944	1.20627	879	29.64793	772641	93.75500	1.13766
830	28.80972	688900	91.10434	1.20482	880	29.66479	774400	93.80832	1.13636
831	28.82707	690561	91.15920	1.20337	881	29.68164	776161	93.86160	1.13507
832	28.84441	692224	91.21403	1.20192	882	29.69848	777924	93.91486	1.13379
833	28.86174	693889	91.26883	1.20048	883	29.71532	779689	93.96808	1.13250
834	28.87906	695556	91.32360	1.19904	884	29.73214	781456	94.02127	1.13122
835	28.89637	697225	91.37833	1.19760	885	29.74895	783225	94.07444	1.12994
836	28.91366	698896	91.43304	1.19617	886	29.76575	784996	94.12757	1.12867
837	28.93095	700569	91.48770	1.19474	887	29.78255	786769	94.18068	1.12740
838	28.94823	702244	91.54234	1.19332	888	29.79933	788544	94.23375	1.12613
839	28.96550	703921	91.59694	1.19190	889	29.81610	790321	94.28680	1.12486
840	28.98275	705600	91.65151	1.19048	890	29.83287	792100	94.33981	1.12360
841	29.00000	707281	91.70605	1.18906	891	29.84962	793881	94.39280	1.12233
842	29.01724	708964	91.76056	1.18765	892	29.86637	795664	94.44575	1.12108
843	29.03446	710649	91.81503	1.18624	893	29.88311	797449	94.49868	1.11982
844	29.05168	712336	91.86947	1.18483	894	29.89983	799236	94.55157	1.11857
845	29.06888	714025	91.92388	1.18343	895	29.91655	801025	94.60444	1.11732
846	29.08608	715716	91.97826	1.18203	896	29.93326	802816	94.65728	1.11607
847	29.10326	717409	92.03260	1.18064	897	29.94996	804609	94.71008	1.11483
848	29.12044	719104	92.08692	1.17925	898	29.96665	806404	94.76286	1.11359
849	29.13760	720801	92.14120	1.17786	899	29.98333	808201	94.81561	1.11235
850	29.15476	722500	92.19544	1.17647	900	30.00000	810000	94.86833	1.11111

TABLE A-10 (continued)

N	$\sqrt{N}$	N^2	$\sqrt{10N}$	$1000/N$	N	$\sqrt{N}$	N^2	$\sqrt{10N}$	$1000/N$
900	30.00000	810000	94.86833	1.11111	950	30.82207	902500	97.46794	1.05263
901	30.01666	811801	94.92102	1.10988	951	30.83829	904401	97.51923	1.05152
902	30.03331	813604	94.97368	1.10865	952	30.85450	906304	97.57049	1.05042
903	30.04996	815409	95.02631	1.10742	953	30.87070	908209	97.62172	1.04932
904	30.06659	817216	95.07891	1.10619	954	30.88689	910116	97.67292	1.04822
905	30.08322	819025	95.13149	1.10497	955	30.90307	912025	97.72410	1.04712
906	30.09983	820836	95.18403	1.10375	956	30.91925	913936	97.77525	1.04603
907	30.11644	822649	95.23655	1.10254	957	30.93542	915849	97.82638	1.04493
908	30.13304	824464	95.28903	1.10132	958	30.95158	917764	97.87747	1.04384
909	30.14963	826281	95.34149	1.10011	959	30.96773	919681	97.92855	1.04275
910	30.16621	828100	95.39392	1.09890	960	30.98387	921600	97.97959	1.04167
911	30.18278	829921	95.44632	1.09769	961	31.00000	923521	98.03061	1.04058
912	30.19934	831744	95.49869	1.09649	962	31.01612	925444	98.08160	1.03950
913	30.21589	833569	95.55103	1.09529	963	31.03224	927369	98.13256	1.03842
914	30.23243	835396	95.60335	1.09409	964	31.04835	929296	98.18350	1.03734
915	30.24897	837225	95.65563	1.09290	965	31.06445	931225	98.23441	1.03627
916	30.26549	839056	95.70789	1.09170	966	31.08054	933156	98.28530	1.03520
917	30.28201	840889	95.76012	1.09051	967	31.09662	935089	98.33616	1.03413
918	30.29851	842724	95.81232	1.08932	968	31.11270	937024	98.38699	1.03306
919	30.31501	844561	95.86449	1.08814	969	31.12876	938961	98.43780	1.03199
920	30.33150	846400	95.91663	1.08696	970	31.14482	940900	98.48858	1.03093
921	30.34798	848241	95.96874	1.08578	971	31.16087	942841	98.53933	1.02987
922	30.36445	850084	96.02083	1.08460	972	31.17691	944784	98.59006	1.02881
923	30.38092	851929	96.07289	1.08342	973	31.19295	946729	98.64076	1.02775
924	30.39737	853776	96.12492	1.08225	974	31.20897	948676	98.69144	1.02669
925	30.41381	855625	96.17692	1.08108	975	31.22499	950625	98.74209	1.02564
926	30.43025	857476	96.22889	1.07991	976	31.24100	952576	98.79271	1.02459
927	30.44667	859329	96.28084	1.07875	977	31.25700	954529	98.84331	1.02354
928	30.46309	861184	96.33276	1.07759	978	31.27299	956484	98.89388	1.02249
929	30.47950	863041	96.38465	1.07643	979	31.28898	958441	98.94443	1.02145
930	30.49590	864900	96.43651	1.07527	980	31.30495	960400	98.99495	1.02041
931	30.51229	866761	96.48834	1.07411	981	31.32092	962361	99.04544	1.01937
932	30.52868	868624	96.54015	1.07296	982	31.33688	964324	99.09591	1.01833
933	30.54505	870489	96.59193	1.07181	983	31.35283	966289	99.14636	1.01729
934	30.56141	872356	96.64368	1.07066	984	31.36877	968256	99.19677	1.01626
935	30.57777	874225	96.69540	1.06952	985	31.38471	970225	99.24717	1.01523
936	30.59412	876096	96.74709	1.06838	986	31.40064	972196	99.29753	1.01420
937	30.61046	877969	96.79876	1.06724	987	31.41656	974169	99.34787	1.01317
938	30.62679	879844	96.85040	1.06610	988	31.43247	976144	99.39819	1.01215
939	30.64311	881721	96.90201	1.06496	989	31.44837	978121	99.44848	1.01112
940	30.65942	883600	96.95360	1.06383	990	31.46427	980100	99.49874	1.01010
941	30.67572	885481	97.00515	1.06270	991	31.48015	982081	99.54898	1.00908
942	30.69202	887364	97.05668	1.06157	992	31.49603	984064	99.59920	1.00806
943	30.70831	889249	97.10819	1.06045	993	31.51190	986049	99.64939	1.00705
944	30.72458	891136	97.15966	1.05932	994	31.52777	988036	99.69955	1.00604
945	30.74085	893025	97.21111	1.05820	995	31.54362	990025	99.74969	1.00503
946	30.75711	894916	97.26253	1.05708	996	31.55947	992016	99.79980	1.00402
947	30.77337	896809	97.31393	1.05597	997	31.57531	994009	99.84989	1.00301
948	30.78961	898704	97.36529	1.05485	998	31.59114	996004	99.89995	1.00200
949	30.80584	900601	97.41663	1.05374	999	31.60696	998001	99.94999	1.00100
950	30.82207	902500	97.46794	1.05263	1000	31.62278	1000000	100.00000	1.00000

Symbols, Subscripts, and Summations

In statistics, *symbols* such as X, Y, and Z are used to represent different sets of data. Hence, if we have data for five families, we might let

X = family income
Y = family clothing expenditures
Z = family savings

Subscripts are used to represent individual observations within these sets of data. Thus, we write X_i to represent the income of the ith family, where i takes on the values 1, 2, 3, 4, and 5. In this notation X_1, X_2, X_3, X_4, and X_5 stand for the incomes of the first family, the second family, etc. The data are arranged in some order, such as by size of income, the order in which the data were gathered, or any other way suitable to the purposes or convenience of the investigator. The subscript i is a variable used to index the individual data observations. Continuing with the example, X_i, Y_i, and Z_i represent the income, clothing expenditures, and savings of the ith family. For example, X_2 represents the income of the second family, Y_2 clothing expenditures of the second family (same family), and Z_5 the savings of the fifth family.

Now, let us suppose that we have data for two different samples, say the net worths of 100 corporations and the test scores of 20 students. To refer to individual observations in these samples, we can let X_i denote the net worth of the ith corporation, where i assumes values from 1 to 100. This latter idea is indicated by the notation $i = 1, 2, 3, \ldots, 100$. Also we can let Y_j denote the test score of the jth student, where $j = 1, 2, 3, \ldots, 20$. Thus, the different subscript letters make it clear that different samples are involved. Letters such as X, Y, and Z are generally used to represent the different variables or types of measurements involved, whereas subscripts such as i, j, k, and l are used to designate individual observations.

We now turn to the method of expressing summations of sets of data. Suppose we want to add a set of four observations, denoted X_1, X_2, X_3, and X_4. A convenient way of designating this addition is

$$\sum_{i=1}^{4} X_i = X_1 + X_2 + X_3 + X_4$$

where the symbol Σ (Greek capital "sigma") means "the sum of." Hence, the symbol

$$\sum_{i=1}^{4} X_i$$

is read "the sum of the X_i's, i going from 1 to 4." For example, if $X_1 = 3$, $X_2 = 1$, $X_3 = 10$, and $X_4 = 5$,

$$\sum_{i=1}^{4} X_i = 3 + 1 + 10 + 5 = 19$$

In general, if there are n observations, we write

$$\sum_{i=1}^{n} X_i = X_1 + X_2 + \cdots + X_n$$

EXAMPLE 1

Let $X_1 = -2$, $X_2 = 3$, $X_3 = 5$. Find

(a)

$$\sum_{i=1}^{3} X_i$$

(b)

$$\sum_{j=1}^{3} X_j^2$$

(c)

$$\sum_{j=1}^{3} (2X_j + 3)$$

SOLUTION

(a)

$$\sum_{i=1}^{3} X_i = X_1 + X_2 + X_3$$
$$= -2 + 3 + 5 = 6$$

(b)

$$\sum_{j=1}^{3} X_j^2 = X_1^2 + X_2^2 + X_3^2$$
$$= (-2)^2 + (3)^2 + (5)^2 = 38$$

(c)

$$\sum_{j=1}^{3} (2X_j + 3) = (2X_1 + 3) + (2X_2 + 3) + (2X_3 + 3)$$
$$= (-4 + 3) + (6 + 3) + (10 + 3)$$
$$= -1 + 9 + 13 = 21$$

EXAMPLE 2

Prove

(a)
$$\sum_{i=1}^{n} aX_i = a \sum_{i=1}^{n} X_i$$

(b)
$$\sum_{i=1}^{n} a = na$$

(c)
$$\sum_{i=1}^{n} (X_i + Y_i) = \sum_{i=1}^{n} X_i + \sum_{i=1}^{n} Y_i$$

where a is a constant.

SOLUTION

(a)
$$\sum_{i=1}^{n} aX_i = aX_1 + aX_2 \cdots + aX_n$$
$$= a(X_1 + X_2 \cdots + X_n)$$
$$= a \sum_{i=1}^{n} X_i$$

(b)
$$\sum_{i=1}^{n} a = a \sum_{i=1}^{n} 1$$
$$= a\underbrace{(1 + 1 + \cdots + 1)}_{n \text{ terms}}$$
$$= na$$

(c)
$$\sum_{i=1}^{n} (X_i + Y_i) = X_1 + Y_1 + X_2 + Y_2 + \cdots + X_n + Y_n$$
$$= (X_1 + X_2 + \cdots + X_n) + (Y_1 + Y_2 + \cdots + Y_n)$$
$$= \sum_{i=1}^{n} X_i + \sum_{i=1}^{n} Y_i$$

These three summation properties have been indicated as Rules 1, 2, and 3 at the end of this appendix.

Double summations are used to indicate summations of more than one variable, where different subscript indexes are involved. For example, the symbol

$$\sum_{j=1}^{3} \sum_{i=1}^{2} X_i Y_j$$

means "the sum of the products of X_i and Y_j where $i = 1, 2$ and $j = 1, 2, 3$." Thus, we can write

$$\sum_{j=1}^{3} \sum_{i=1}^{2} X_i Y_i = X_1 Y_1 + X_2 Y_1 + X_1 Y_2 + X_2 Y_2 + X_1 Y_3 + X_2 Y_3$$

SIMPLIFIED SUMMATION NOTATIONS

In this text, simplified summation notations are often used in which subscripts are eliminated. Thus, for example, ΣX, ΣX^2, and ΣY^2 are used instead of

$$\sum_{i=1}^{n} X_i \qquad \sum_{i=1}^{n} X_i^2 \qquad \text{and} \qquad \sum_{i=1}^{n} Y_i^2 \text{ respectively.}$$

SUMMATION PROPERTIES

Rule 1
$$\sum_{i=1}^{n} aX_i = a\sum_{i=1}^{n} X_i$$

Rule 2
$$\sum_{i=1}^{n} a = \underbrace{a + a + \cdots + a}_{n \text{ terms}} = na$$

Rule 3
$$\sum_{i=1}^{n} (X_i + Y_i) = \Sigma X_i + \Sigma Y_i$$

APPENDIX C

Shortcut Formulas

SHORTCUT CALCULATION OF THE MEAN — STEP-DEVIATION METHOD

A shortcut calculation known as the step-deviation method is useful when class intervals in a frequency distribution are of equal size. The method results in simpler arithmetic than the direct definitional formula, particularly if the class intervals and frequencies involve a large number of digits.

The step-deviation method of computing the mean involves three basic steps:

1. The selection of an assumed (or arbitrary) mean.
2. Calculation of an average deviation from this assumed mean.
3. The addition of this average deviation as a correction factor to the assumed mean to obtain the true mean. This correction factor is positive if the assumed mean lies below the true mean, negative if above.

To accomplish step 2, a midpoint of a class (preferably near the center of the distribution) is taken as the assumed mean. Then deviations of the midpoints of the other classes are taken from the assumed mean in class interval units. These deviations are denoted d. After the d values are averaged, the result must be multiplied by the size of the class interval to return to the units of the original data.

The formula for the step-deviation method is given in equation (C.1).

STEP-DEVIATION METHOD FOR THE ARITHMETIC MEAN — GROUPED DATA

(C.1)
$$\bar{X} = \bar{X}_a + \left(\frac{\Sigma fd}{n}\right)(i)$$

where $\overline{X}$ = the arithmetic mean
$\overline{X}_a$ = the assumed arithmetic mean
f = frequencies
d = deviations of midpoints from the assumed mean in class interval units
n = the number of observations
i = the size of a class interval

The step-deviation method is illustrated in Table C-1 for the same frequency distribution of earnings of 100 semi-skilled workers given in Table 3-5 on page 49, where the arithmetic mean was computed by the direct definitional method. In Table C-1, the assumed arithmetic mean is $195. As may be noted in the d column, these values indicate the number of class intervals below or above the one in which the assumed mean is taken.

TABLE C-1
Calculation of the Arithmetic Mean for Grouped Data by the
Step-Deviation Method: Weekly Earnings Data.

WEEKLY EARNINGS	NUMBER OF EMPLOYEES f	d	fd
$160.00 and under 170.00	4	−3	−12
170.00 and under 180.00	14	−2	−28
180.00 and under 190.00	18	−1	−18
190.00 and under 200.00	28	0	0
200.00 and under 210.00	20	1	20
210.00 and under 220.00	12	2	24
220.00 and under 230.00	4	3	12
	100		−2

$\overline{X}_a = \$195$

$$\overline{X} = \overline{X}_a + \left(\frac{\Sigma fd}{n}\right)(i) = \$195 + \left(\frac{-2}{100}\right)(\$10) = \$194.80$$

SHORTCUT CALCULATION OF THE STANDARD DEVIATION— STEP-DEVIATION METHOD

Just as in the case of the arithmetic mean, the step-deviation method of calculating the standard deviation is useful when class intervals in a frequency distribution are of the same size. The saving in computational effort accomplished by the use of the step-deviation method is illustrated for the case of the weekly earnings data given in Table C-1.

As in the case of the calculation for the arithmetic mean, the procedure involves taking deviations of midpoints of classes from an assumed mean and stating them in class interval units. Only one additional column of values, fd^2, is required to compute the standard deviation by the step-deviation method as compared to the corresponding arithmetic mean computation given in Table C-1. The formula for the step-deviation method is given in equation (C.2). All of the symbols have the same meaning as in equation (C.1) for the arithmetic mean and the computation in Table C-1.

STEP-DEVIATION METHOD FOR THE SAMPLE STANDARD DEVIATION – GROUPED DATA

(C.2)
$$s = (i)\sqrt{\frac{\Sigma fd^2 - \frac{(\Sigma fd)^2}{n}}{n-1}}$$

The use of equation (C.2) is illustrated in Table C-2.

TABLE C-2

Calculation of the Standard Deviation for Grouped Data by
the Step-Deviation Method: Weekly Earnings Data.

WEEKLY EARNINGS	NUMBER OF EMPLOYEES f	d	fd	fd²
$160.00 and under $170.00	4	−3	−12	36
170.00 and under 180.00	14	−2	−28	56
180.00 and under 190.00	18	−1	−18	18
190.00 and under 200.00	28	0	0	0
200.00 and under 210.00	20	1	20	20
210.00 and under 220.00	12	2	24	48
220.00 and under 230.00	4	3	12	36
	100		−2	214

$\overline{X}_a = \$195$

$$s = (i)\sqrt{\frac{\Sigma fd^2 - \frac{(\Sigma fd)^2}{n}}{n-1}} = (\$10)\sqrt{\frac{214 - \frac{(-2)^2}{100}}{99}} = \$10 \, (1.47)$$

$s = \$14.70$

The same assumed mean, $\overline{X}_a = \$195$ was used in Table C-2, as in the calculation of the arithmetic mean given in Table C-1.

Answers To Even-Numbered Exercises

2. **(a)** A census of the students in the class would appear to be appropriate. Any type of sample would deprive some students of the opportunity to express an opinion. From the standpoint of fairness, such exclusions would be inadvisable.
   (b) A sample of the subscribers would be advisable. The greater accuracy of a census would probably not warrant the additional cost.
   (c) The total number of customers would be an important factor. If this number were relatively small, a census would be preferable. If the number were very large, a sample would doubtless be a better solution.
   (d) A census would be preferable. Since there are only 25 firms, the trade association should obtain the employment data from every company.

4. The director used as his universe those employees who had attended the party the previous year. Since employees who thought husbands and wives should be included probably had a lower attendance rate than other employees, such as unmarrieds or married persons who did not prefer the inclusion of husbands and wives, the director's sample was biased. The subgroup of employees who attended the Christmas party last year undoubtedly was non-homogeneous with all other employees with respect to preference for an "employees only" party.

6. Conceptually, the population of interest is all male consumers in the New York City area. We are interested in the proportion of this population who possess the characteristic "would wear a certain type of man's suit." As a practical matter, the population might be defined as those male consumers who shop in the stores and departments of stores that sell this type of man's suit. This is a dynamic population, since men enter and leave this population primarily because of shifting shopping patterns and changing tastes.

2. (a) If less than 6% of all persons in the United States in the given year came from families with $10,000 or more in income, this would indicate a *higher* incidence of polio for this group than for lower income groups. A direct comparison could be made between the incidence of polio in families with incomes over $10,000 (treatment group) and the corresponding incidence of polio in families with incomes of $10,000 and under (control group).

(b) If women constituted less than 20% of the student body, this suggests that women were worse students than men in this university. A direct comparison could be made between the failure rate for women (treatment group) and the corresponding failure rate for men (control group).

(c) Assume that we define a control group consisting of cold sufferers who use no cold remedy at all. Suppose that more than 95% of this control group were free of their colds within a one week period. This would constitute evidence that the cold preparation was ineffective as regards shortening the duration of the cold.

4. See page 22.

2. (a) Systematic and random error. The systematic portion arises from the fact that since the student lounge is the only area in the building where smoking is permitted, there would tend to be an overrepresentation of smokers as compared to the percentage in the general student population. The random portion of the error arises from the difference in the percentage of smokers among all students who use the lounge.

(b) Random error. As sample size increases, the average life of tubes in the sample would tend to be closer to the "true" average life. On the other hand, if for any reason, the tubes in the particular stores included in the sample had a lower average life than was true for all vacuum tubes of this type, then clearly there might also be a systematic error component as well.

(c) Systematic and random error. The systematic error component is due to incorrect scale measurement. If only this component were present, we would expect the average weight to be 1.1 pounds. We attribute the residual error to chance fluctuations in the sampling and measurement of the contents of 50 cans.

(d) Systematic and random. A systematic error component is introduced because people who feel strongly one way or the other on an issue, (in this case, those who would rate the services bad or good), tend to have higher response rates than others. A random component is also introduced because of sample size.

4. Yes, there are probably both sampling and systematic errors present. There would be sampling error in the random sample of 10,000 as compared to the population of 200,000 subscribers. A systematic error component is probably introduced because of the differences between the attitudes of the 1500 respondents and the corresponding attitudes of the total sample of 10,000.

2. (a) We would have to know the denominators of the savings-income ratios, that is, the actual amounts of income attributable to the families in the two income classes in 1967 and 1972. Alternatively, we could use a percentage breakdown of income for each of the two years.

(b) Assume that the proportionate breakdowns of incomes were as given below for 1967 and 1972. Then the average savings-income ratios for 1967 and 1972 are calculated as follows:

1967

FAMILY INCOME	SAVINGS INCOME RATIO	PERCENTAGE OF TOTAL INCOME	WEIGHTED SAVINGS-INCOME RATIO
COL(1)	COL(2)	COL(3)	COL(2) × COL(3)
Under $10,000	0.03	0.90	0.027
$10,000 and over	0.30	0.10 ‾‾‾‾ 1.00	0.030 ‾‾‾‾ 0.057

Combined Ratio $= \dfrac{0.057}{1.00} = 0.057 = 5.7\%$

1972

FAMILY INCOME	SAVINGS INCOME RATIO	PERCENTAGE OF TOTAL INCOME	WEIGHTED SAVINGS-INCOME RATIO
COL(1)	COL(2)	COL(3)	COL(2) × COL(3)
Under $10,000	0.02	0.40	0.008
$10,000 and over	0.25	0.60 ‾‾‾‾ 1.00	0.150 ‾‾‾‾ 0.158

Combined Ratio $= \dfrac{0.158}{1.00} = 0.158 = 15.8\%$

The savings-income ratio for all families combined is higher in 1972 (15.8%) than in 1967 (5.7%) because in 1972 the higher savings-income ratio (0.25) received greater weight (0.60) than was true in 1967. In 1967, the higher ratio (0.30) received a weight of only 0.10. Similar demonstrations can be made by assuming other appropriate proportionate breakdowns of income (weights).

CHAPTER 3

Page 45

2. (a) A possible table is shown below:

DAILY SALES OF A FIRM	
SALES	FREQUENCY
$ 80.00 and under $100.00	5
$100.00 and under $120.00	6
$120.00 and under $140.00	12
$140.00 and under $160.00	4
$160.00 and under $180.00	3
TOTAL	30

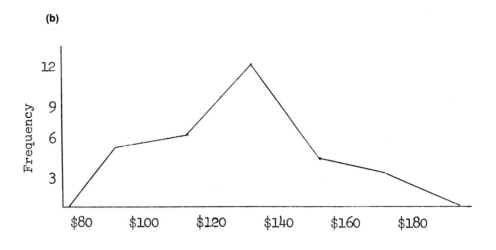

(b)

DAILY SALES

4.

ANALYSIS OF ORDINARY LIFE INSURANCE

POLICY SIZE	NUMBER OF ORDINARY LIFE INSURANCE POLICIES IN FORCE PER 1,000 POLICIES
Less than $1,000	31
Less than $2,500	212
Less than $5,000	320
Less than $10,000	533
Less than maximum	1000

---

6. (a)

NUMBER OF CASES OF BEER PER EIGHT-HOUR SHIFT

NUMBER OF CASES	(A) TOTAL FREQUENCY	(B) LINE 1 FREQUENCY	(B) LINE 2 FREQUENCY
5500 and under 6000	2	1	1
6000 and under 6500	3	2	1
6500 and under 7000	9	7	2
7000 and under 7500	4	2	2
7500 and under 8000	9	2	7
8000 and under 8500	3	1	2
	30	15	15

(b)

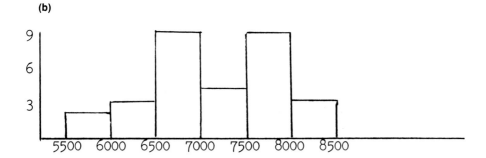

(b)

Line 1

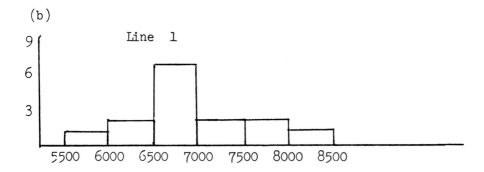

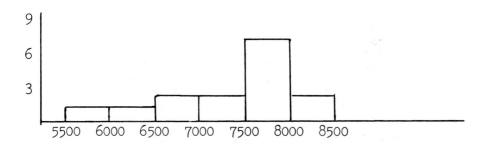

(c) The answer in part (b) is preferable since a bimodal distribution results if the data for the two production lines are merged.

Page 51

2. No. For example, suppose 99 percent of Center City's total dollar credit extended consists of personal loans and 99 percent of Neighborhood Bank's business is industrial loans, then the overall debt ratios are:

$$\text{Center City} = (0.99)(0.04) + (0.01)(0.02) = 0.0398$$
$$\text{Neighborhood} = (0.01)(0.05) + (0.99)(0.03) = 0.0302$$

Page 60

2. Raw data:

$$\bar{X} = \frac{\Sigma X}{n} = \frac{3767.97}{30} = \$125.60$$

Median = mean of fifteenth and sixteenth observations $= \dfrac{\$125.27 + \$125.41}{2}$

$$= \$125.34$$

$$\bar{X} = \$130 + \left(\frac{-6}{30}\right)\$20 = \$126.00$$

$$\text{Median} = \$120 + \left(\frac{15 - 11}{12}\right)\$20 = \$126.67$$

4. (a)

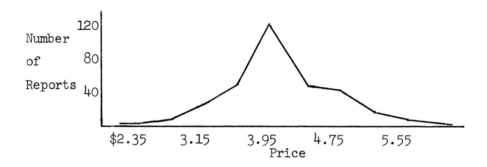

Composition Roofing

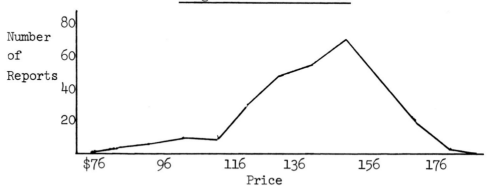

Douglas and Inland Firs

(b) Composition Roofing:
  Median Class is $3.95 and under $4.35
  Modal   Class is $3.95 and under $4.35
  Douglas and Inland Firs:
  Median Class is $136–$145
  Modal   Class is $146–$155

(c)

COMPOSITION ROOFING		FIRS	
PRICE RANGE	NUMBER OF REPORTS	PRICE RANGE	NUMBER OF REPORTS
Less than $2.35	0	Less than 76	0
Less than 2.75	1	Less than 86	2
Less than 3.15	7	Less than 96	6
Less than 3.55	40	Less than 106	14
Less than 3.95	91	Less than 116	21
Less than 4.35	212	Less than 126	49
Less than 4.75	262	Less than 136	97
Less than 5.15	306	Less than 146	154
Less than 5.55	319	Less than 156	225
Less than 5.95	324	Less than 166	270
		Less than 176	290
		Less than 186	291

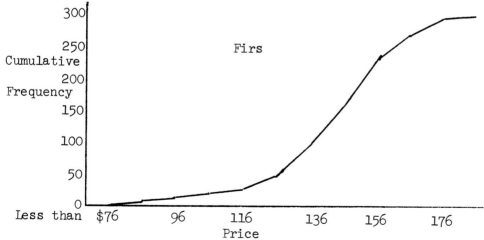

(d) The distribution of composition roofing prices is nearly symmetric while the distribution of fir prices is skewed to the left.

6. (a) $\bar{X} = \dfrac{\$5,680.00}{50} = \$113.60$

(b) No. The ratio gives the exact mean while part (a) gives only a close approximation, since a frequency table was used.

(c) Median class is "$110.00 and under $117.50"

Median observation is the $\dfrac{50 + 1}{2} = 25.5$th observation

Median $= \$110 + \left(\dfrac{25 - 18}{14}\right) \$7.50 = \$113.75$

(d) So slightly skewed to left, that for practical purposes, it is a symmetrical distribution.

(e) No, since the Σfx is $5,680. This figure is an *estimate* of the total payroll which is close to the true figure of $5675.18.

(f) Yes, the mean here seems typical since there is little skewness to distort it.

Page 70

2. No. It would be necessary to compare the respective coefficients of variation. As it stands, we have no basis on which to judge the variation of sales relative to the average levels of sales of the two companies over the past five years.

4. Using the shortcut formulas of Appendix C:

$$\bar{X} = 350 + \left(\dfrac{18}{100}\right)(100) = 368 \text{ hours}$$

$$s = 100 \sqrt{\dfrac{222 - \dfrac{222}{100}}{99}} = 149.0 \text{ hours}$$

$$CV = \dfrac{149.0}{368} = 40.5\%$$

(b) $Md = 300 + \left(\dfrac{50 - 40}{20}\right)(100) = 350 \text{ hours}$

Approximately fifty percent of the vacuum tubes lasted longer than 350 hours.

(c) For part (a), we would have to recalculate the mean on the basis of an estimate of the mean for the open interval. For part (b) there would be no effect.

CHAPTER 4

Page 79

2. $\{D, GD, GGD, \ldots\}$

4.

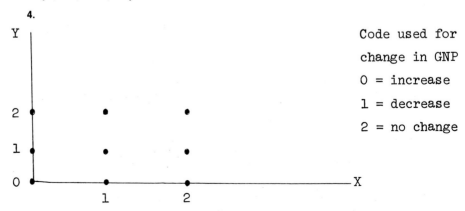

Code used for change in GNP

0 = increase

1 = decrease

2 = no change

6.

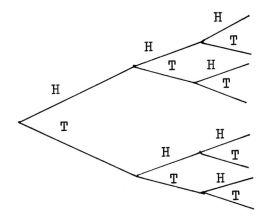

2. $\frac{2}{5}$;

	MALE	FEMALE	TOTAL
Republican	5	4	9
Democrat	3	8	11
TOTAL	8	12	20

4. (a) $P(\bar{A} \text{ and } \bar{B}) = 1 - \frac{1}{6} - \frac{2}{9} - \frac{1}{3} = \frac{5}{18}$

(b) $P(A) = \frac{1}{6} + \frac{2}{9} = \frac{7}{18}$ $P(B) = \frac{1}{6} + \frac{1}{3} = \frac{9}{18} = \frac{1}{2}$

$P(A \text{ and } B) = \frac{1}{6} \neq P(A)\ P(B)$

Hence A and B are *not* independent.

6. (a) $\frac{18}{38} = \frac{9}{19}$ **(d)** $18:20 = 9:10$

  (e) $1:37$

(b) $\frac{1}{38}$

  (f) Win: $\frac{18}{38} = \frac{9}{19}$

(c) $\frac{18}{38} = \frac{9}{19}$

  Lose: $\frac{20}{38} = \frac{10}{19}$

8. Let S_1 = sells first client $P(S_1) = 0.3$
    S_2 = sells second client $P(S_2) = 0.3$
  (a) $P(S_1)\ P(S_2) = 0.3 \times 0.3 = 0.09$
  (b) $P(S_1\bar{S_2} \text{ and } \bar{S_1}S_2) = (0.3)(0.7) + (0.7)(0.3) = 0.21 + 0.21 = 0.42$
  (c) $P(S_1 \text{ and } S_2) = P(S_1) + P(S_2) - P(S_1 \text{ and } S_2) = 0.3 + 0.3 - (0.3)(0.3) = .51$

10. Probability of all accepting position is $(0.6)^6 = .0467$
  Thus, the probability of not over-hiring is 0.9533.

12.

OPINION	JOB TYPE WHITE-COLLAR	BLUE-COLLAR	TOTAL
For	125	250	375
Against	25	100	125
TOTAL	150	350	500

(a) $\dfrac{100}{500} = \dfrac{1}{5}$ (b) $\dfrac{375}{500} = \dfrac{3}{4}$ (c) $\dfrac{250}{350} = \dfrac{5}{7}$

(d) Dependent. Let $A_1 = $ For $A_2 = $ Against
$\qquad\qquad\qquad B_1 = $ White-Collar $B_2 = $ Blue-Collar
$P(A_1 \text{ and } B_1) \neq P(A_1)\,P(B_1)$
$\dfrac{125}{500} \neq \left(\dfrac{375}{500}\right)\left(\dfrac{150}{500}\right)$
Thus, an employee's opinion concerning the proposal is *not* independent of whether he is a white-collar or blue-collar worker.

14. (a) $\dfrac{20}{500} = 0.04$ (b) $\dfrac{100}{500} = 0.2$ (c) $\dfrac{15}{50} = 0.3$ (d) No.

OPINION	REGION EAST	MIDDLE WEST	PACIFIC COAST	TOTAL
Opposed	10	10	30	50
Not opposed	90	90	270	450
TOTAL	100	100	300	500

Page 95

2. $\dfrac{(0.7)(0.2)}{(0.7)(0.2) + (0.3)(0.8)} = 0.37$

4. $\dfrac{(0.5)(0.7)}{(0.5)(0.7) + (0.5)(0.3)} = 0.70$

CHAPTER 5

Page 105

2. $E(X) = (0)(0.10) + (1)(0.90) = 0.9$
On the average, 90% of the items are not defective.

4. $E(X) = (.01)(10) + (0.05)(11) + (0.13)(12) + (0.18)(13)$
$\qquad\qquad + (0.26)(14) + (0.18)(15) + (0.13)(16)$
$\qquad\qquad + (0.05)(17) + (0.01)(18) = 14.00$

6. $X = +\$2$ if numbers 1–12 occur on a given roll
$\qquad = -\$1$ if numbers other than 1–12 occur on a given roll
$E(X) = (\$2)(12/38) + (-\$1)(26/38) = -\$1/19 = \$ -0.0526$
Expected profit for 1000 rolls $= 1000(\$ -0.0526) = -\$52.60$

2. Binomial distribution; $n = 6$, $p = .4$
   (a) P (3 decline) $= 0.2765$
   (b) P (3 or more decline) $= 0.2765 + 0.1382 + 0.0369 + 0.0041 = 0.4557$
   (c) No, stock price movements cannot ordinarily be considered independent. Hence, the binomial is not the appropriate probability distribution.

4. Binomial distribution; $n = 10$, $p = 0.05$
   P (2 or more will not develop) $= 1 - (0.5987 + 0.3151) = 0.0862$

6. Binomial distribution; $n = 8$, $p = 0.25$
   P (at least 1 sale) $= 1 - 0.1002 = .8998$

8. Binomial distribution; $n = 4$, $p = .15$
   P (zero contracts) $= 0.5220$

10. $p = 0.2$ $p = 0.5$ $p = 0.8$
   Symmetric when $p = 0.5$
   Skewed to the right for $p = 0.2$
   Skewed to the left for $p = 0.8$
   Skewness increases as p departs from 0.5 in either direction.

2. (a) Binomial distribution; $n = 20$, $p = 0.10$
   P (1 or less) $= 0.1216 + 0.2702 = 0.3918$
   (b) Poisson distribution; $\mu = np = (20)(0.10) = 2$
   P (1 or less) $= 0.1353 + 0.2707 = 0.4060$

2. μ	σ
4	$\sqrt{(10)(0.4)(0.6)} = 1.55$
8	$\sqrt{(20)(0.4)(0.6)} = 2.19$
16	$\sqrt{(40)(0.4)(0.6)} = 3.10$

2. (a) 0.5
   (b) 0.1056
   (c) 0.0062
   (d) 0.9876
   (e) 0.9938
   (f) 0.0124

4. $P\left(z > \dfrac{150,000 - 120,000}{10,000}\right) = 0.00135$

6. (a) 0.5
   (b) $P\left(z < \dfrac{12.75 - 12}{0.5}\right) = 0.9332$

(c) $P\left(\dfrac{11.5 - 12}{0.5} < z < \dfrac{12.75 - 12}{0.5}\right) = 0.7745$

Page 142

2. $P\left(z > \dfrac{370 - 356}{55/\sqrt{121}}\right) = 0.0026$

4. (a) $P\left(z > \dfrac{20,000 - 22,000}{1500}\right) = 0.9082$

(b) $P\left(z > \dfrac{25,000 - 22,000}{1500}\right) = 0.0228$

(c) $P\left(z > \dfrac{20,000 - 22,000}{1500/\sqrt{4}}\right) = 0.9962$

6. $\$1,200 \pm 3(\$200) = (\$600, \$1,800)$

$\$1,200 \pm 3\left(\dfrac{\$200}{\sqrt{25}}\right) = (\$1,080, \$1,320)$

$\$1,200 \pm 3\left(\dfrac{\$200}{\sqrt{100}}\right) = (\$1,140, \$1,260)$

CHAPTER 7

Page 153

2. $\bar{x}$ is a random variable since it can take on different values depending upon which sample is drawn from a population with mean μ. As a random variable it has a probability distribution, and the standard deviation of $\bar{x}$ is a measure of the dispersion of that probability distribution. This standard deviation is a measure of the spread of $\bar{x}$ values around μ, the value to be estimated. It can thus be interpreted as a measure of errors due to sampling. From this viewpoint, $\sigma_{\bar{x}}$ is a measure of the error involved in using $\bar{x}$ as an estimator of μ.

4. $\$550 \pm 2.17 \dfrac{\$120}{\sqrt{100}} = (\$523.96, \$576.04)$

Page 156

2. $0.56 \pm 2.33\sqrt{\dfrac{(0.56)(0.44)}{100}} = (0.444, 0.676)$ of the time

Page 162

2. $1.96\sqrt{\dfrac{(0.10)(0.90)}{n}} = 0.02$

$n = 865$

Page 164

2. (a) 95.5% of all the possible samples he could draw will yield a correct interval estimate, while 4.5% of the possible samples will lead to incorrect interval estimates.

(b) $(2)\dfrac{\$.25}{\sqrt{n}} = \$.02$

$n = 625$

2. The statement is incorrect, since there are two types of errors that can occur, Type I and Type II. If α, the probability of a Type I error, is made extremely small, the Type II error probability in most cases, with a fixed n, will become quite large. The two errors, for a fixed n, move inversely to one another. As an extreme case, for example, if we always accept H_0 regardless of test results, the α error is 0 but the β error is equal to 1.

4. Power is defined as the probability of rejecting H_0 when H_1 is the true state of nature. Power $= 1 - \beta$.

6. (a) Disagree. The operating characteristic curve has the probability of accepting the null hypothesis on the vertical axis and the possible values of the parameter on the horizontal axis.
(b) Disagree. β represents the probability of incorrectly accepting the null hypothesis.
(c) Agree.

8. (a) $H_0 : \mu \geq 101$ millimeters
$H_1 : \mu < 101$ millimeters
Assume $\alpha = 0.05$
DECISION RULE
  1. Reject H_0 if $\bar{x} < 99.52$ millimeters
  2. Accept H_0, otherwise
Decision: Reject H_0

Critical value $= 101 - 1.65\left(\dfrac{9}{\sqrt{100}}\right)$ millimeters

(Note: if $\alpha = 0.01$, the critical value is 98.9 millimeters, and the correct decision is to accept.)

2. $H_0 : p \geq 0.667$
$H_1 : p < 0.667$
$\alpha = 0.05$
DECISION RULE
  1. Reject H_0 if $\bar{p} < 0.585$
  2. Accept H_0, otherwise
Decision: Accept H_0

Critical value $= 0.667 - 1.65\sqrt{\dfrac{(0.667)(0.333)}{90}}$

4. $H_0 : p = 0.10$
$H_1 : p \neq 0.10$
DECISION RULE
  1. Reject H_0 if $0.0743 > \bar{p}$ or $\bar{p} > 0.1257$
  2. Accept H_0 if $0.0743 \leq \bar{p} \leq 0.1257$

Critical value $= 0.10 \pm (2.57)\left(\sqrt{\dfrac{(0.10)(0.90)}{900}}\right)$

Page 192

2. (a) $H_0 : \mu_1 = \mu_2$
$H_1 : \mu_1 \neq \mu_2$
$\alpha = 0.05$
DECISION RULE
  1. Reject H_0 if $\bar{x}_1 - \bar{x}_2 < -\$34.20$ or $> \$34.20$
  2. Accept H_0 if $-\$34.20 \leq \bar{x}_1 - \bar{x}_2 \leq \$34.20$
Decision: Reject H_0 since $\$1282 - \$1208 > \$34.20$

Critical value $= 0 \pm 1.96\sqrt{\dfrac{(\$80)^2}{50} + \dfrac{(\$94)^2}{50}}$

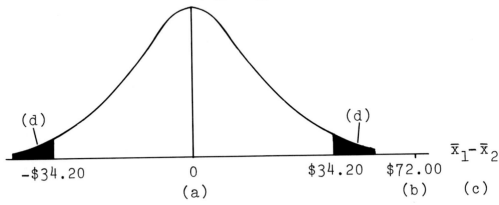

(d)   (d)

$\bar{x}_1 - \bar{x}_2$

$-\$34.20$ 0 $\$34.20$ $\$72.00$

(a)   (b)   (c)

Page 198

2.

$H_0 : p_1 = p_2$
$H_1 : p_1 \neq p_2$
$\alpha = 0.05$
$\bar{p} = \dfrac{398}{900} = 0.442$
DECISION RULE
  1. Reject H_0 if $|\bar{p}_1 - \bar{p}_2| > 0.065$
  2. Accept H_0 if $|\bar{p}_1 - \bar{p}_2| \leq 0.065$
Decision: Accept H_0 since $|0.47 - 0.42| < 0.065$

Critical value: $0 \pm 1.96\sqrt{(0.442)(0.558)\left[\dfrac{1}{400} + \dfrac{1}{500}\right]}$

Page 203

2. (a) $s^2 = \dfrac{9(10)^2 + 16(8)^2}{10 + 17 - 2} = 76.96$

$s_{\bar{x}_1 - \bar{x}_2} = 3.50$

$t = \dfrac{30 - 28}{3.50} = 0.57$

For 25 degrees of freedom and $\alpha = 0.10$, the critical value of $t = 1.708$. Since $0.57 < 1.708$, we conclude that there was not a statistically significant difference between the sample averages.

(b) $H_0 : \mu_1 - \mu_2 = 0$
$H_1 : \mu_1 - \mu_2 \neq 0$
DECISION RULE
  1. Reject H_0 if $t < -1.708$ or $t > 1.708$
  2. Accept H_0 if $-1.708 \leq t \leq 1.708$

2. H_0: Preferences for different magazines are independent of geographical location.
 H_1: Preferences for different magazines are not independent of geographical location.

f_0	f_t
75	100.0
50	81.7
175	118.3
120	100.0
85	81.7
95	118.3
105	100.0
110	81.7
85	118.3
900	900.0

$\chi^2 = 73.875$
$df = 4$
$\chi^2_{.05} = 9.488$
Hence, reject H_0 at $\alpha = 0.05$, and conclude that preferences for different magazines are not independent of geographical location.

4. H_0: Choice of newspaper and social class are independent.
 H_1: Choice of newspaper and social class are not independent.

f_0	f_t
80	73.3
70	76.7
90	73.3
60	76.7
50	73.3
100	76.7
450	450.0

$\chi^2 = 23.122$
$df = 2$
$\chi^2_{.05} = 5.991$
Hence, reject H_0, and conclude that choice of newspaper and social class are not independent.

6. H_0: City type and geographical region are independent.
 H_1: City type and geographical region are not independent.

f_0	f_t
35	35.6
10	11.9
25	23.8
25	23.8
15	20.6
10	6.9
15	13.8
15	13.8
25	18.8
5	6.3
10	12.5
10	12.5
200	200.3

$\chi^2 = 6.871$
$df = 6$
$\chi^2_{.05} = 12.592$
Hence, accept H_0. The data are consistent with the hypothesis that city type and geographical region are independent.

Page 229

2. H_0: There is no real difference among subcontractors in average number of defectives per batch.

H_1: There is a real difference among subcontractors in average number of defectives per batch.

ANALYSIS OF VARIANCE TABLE

SOURCE OF VARIATION	SUM OF SQUARES	DEGREES OF FREEDOM	MEAN SQUARE
Between Columns	430	2	215
Between Rows	510	12	42.5
TOTAL	940	14	

$$F(2,12) = \frac{215}{42.5} = 5.06$$

$F_{.05}(2,12) \approx 3.98$ (by linear interpolation in Appendix A, Table A-8)

Since $5.06 > 3.98$, reject H_0.

4. H_0: The mean lives of the four brands are equal.

H_1: The mean lives of the four brands are not equal.

ANALYSIS OF VARIANCE TABLE

SOURCE OF VARIATION	SUM OF SQUARES	DEGREES OF FREEDOM	MEAN SQUARE
Between Columns	15	3	5
Between Rows	24	8	3
TOTAL	39	11	

$$F(3,8) = \frac{5}{3} = 1.67$$

$$F_{.05}(3,8) = 4.07$$

Since $1.66 < 4.07$ accept H_0.

CHAPTER 10

Page 243

2. (a) $b = \dfrac{15,000 - (10)(20)(4.2)}{10,000 - (10)(20)^2} = 2.36$

$a = 4.2 - (2.36)(20) = -43$

(b) The "method of least squares" is a method of fitting a regression line that imposes the following requirement: the sum of the squares of the deviations of the observed values of the dependent variable from the corresponding computed values on the regression line must be a minimum.

Page 262

2. This result is possible if the research workers used different significance levels in testing the hypothesis of no correlation.

4. (a) $r_c^2 = 1 - \dfrac{0.16}{0.36} = 0.56$

Fifty-six percent of the variation in percentage of income spent for medical care was explained or accounted for by the regression line relating percentage of income spent for medical care and family income.

(b) $t = \dfrac{-0.76}{\sqrt{\dfrac{0.44}{100}}} = -11.5$

The critical t value at the 5 percent level is -1.98. Since -11.5 is less than -1.98 we reject H_0 and assume that a linear relationship exists between family income and percentage of income spent for medical care. If the null hypothesis had been accepted, we would have concluded that the data were consistent with the hypothesis that the two variables are not linearly related.

(c) $z = \dfrac{Y - Y_c}{s_{Y.X}} = \dfrac{10\% - 6\%}{0.4\%} = 10$

Since the z value is so large, the medical care expenditure level would be considered unusual. Note that this is an equivalent type of answer to that given in problem 3(d).

6. (a) $t = \dfrac{0.8}{\sqrt{0.36/48}} = 9.2$

The critical t value at the 5% level of significance for 48 degrees of freedom is approximately equal to 2.01. Since the computed t value is 9.2, far in excess of the critical value, we reject H_0 and conclude that a linear relationship exists between test scores and output for *all* workers in the assembly department.

(b) $Y_c = 19.7 + 0.6(50) = 49.7$ units per hour

* *95% Confidence Interval*

$Y_c \pm 1.96 s_{Y.X} = 49.7 \pm 1.96(6)$

8. (a) $b = \dfrac{1608 - (10)(8)(18.6)}{686 - 10(8)^2} = 2.608$

$a = 18.6 - (2.608)(8) = -2.264$

$Y_c = -2.264 + 2.608\ X$

(b) The regression coefficient $b = 2.608$ means that for two families whose income differs by \$1000, the estimated difference in the amount of life insurance held by the heads of these families is \$2,608.

(c)

$s_{Y.X} = \sqrt{\dfrac{3828 - (-2.264)(186) - (2.608)(1608)}{10 - 2}} = 2.63$ (thousand dollars)

$s_b = \dfrac{2.63}{\sqrt{46}} = 0.39$

$t = \dfrac{2.608 - 0}{0.39} = 6.7$

Since $6.7 > 2.306$ and $6.7 > 3.355$, we reject the hypothesis that the population regression coefficient is equal to zero at $\alpha = 0.05$ and $\alpha = 0.01$.

(d) $Y_c = -2.264 + 2.608(10) = 23.816$ (thousand dollars)

$Y_c \pm 2 s_{Y.X} = 23.816 \pm 2(2.63) = (18.556, 29.076)$ thousand dollars

CHAPTER 11

Page 280

2. $H_0 : p = 0.50$

$H_1 : p \neq 0.50$

$\alpha = 0.05$

$$\sigma_{\bar{p}} = \sqrt{\frac{(0.50)(0.50)}{200}} = 0.035$$

$$\bar{p} = \frac{110}{200} = 0.55$$

$$z = \frac{0.55 - 0.50}{0.035} = 1.43$$

Since $1.43 < 1.96$, we accept the null hypothesis that the proportion of components weighing in excess of 14.5 pounds is 0.50.

4. $H_0: p \leq 0.50$
   $H_1: p > 0.50$
   $\alpha = 0.01$

$$\sigma_{\bar{p}} = \sqrt{\frac{(0.50)(0.50)}{298}} = 0.029$$

$$\bar{p} = \frac{178}{298} = 0.60$$

$$z = \frac{0.60 - 0.50}{0.029} = 3.45$$

Since $3.45 > 2.33$, we reject the null hypothesis that $p \leq 0.50$. Hence we conclude that more than 50 percent of television viewers felt that there had been a decrease in the amount of violence depicted on television programs.

Page 285

2. $R_1 = 214$ $n_1 = 15$
   $R_2 = 251$ $n_2 = 15$
   $U = 131$
   $\mu_U = 112.5$ $\alpha = 0.01$

$$\sigma_U = \sqrt{\frac{(15)(15)(31)}{12}} = 24.1$$

$$z = \frac{131 - 112.5}{24.1} = 0.77$$

Since $0.77 < 2.58$, we conclude that there is not a significant difference in the average level of sales in the two samples.

Page 287

2. $n_1 = 24$
   $n_2 = 26$
   $r = 20$

$$\mu_r = \frac{2(24)(26)}{24 + 26} + 1 = 25.96$$

$$\sigma_r = \sqrt{\frac{2(24)(26)\,[2(24)(26) - 24 - 26]}{(24 + 26)^2(24 + 26 - 1)}} = 3.49$$

$$z = \frac{20 - 25.96}{3.49} = -1.71$$

Since $-1.71 > -1.96$ and $-1.71 > -2.58$, we accept the hypothesis of randomness of runs above and below the median 4.5 at the 0.05 and 0.01 levels of significance, respectively.

4. $n_1 = 12$
   $n_2 = 12$
   $r = 6$

$$\mu_r = \frac{2(12)(12)}{12 + 12} + 1 = 13$$

$$\sigma_r = \sqrt{\frac{2(12)(12)\,[2(12)(12) - 12 - 12]}{(12 + 12)^2(12 + 12 - 1)}} = 2.40$$

$$z = \frac{6 - 13}{2.40} = \frac{-7}{2.40} = -2.92$$

Since $-2.92 < -2.58$, we reject the hypothesis of randomness of runs above and below the median. Since runs below the median tend to occur in the earlier part of the series, and runs above the median in the later part, there is evidence of an increasing trend in the monthly number of on-the-job accidents.

Page 291

2. $r_r = 1 - \frac{6(46)}{12(12^2 - 1)} = 0.84$

Page 297

2. H_0: The distribution is binomial with $p = \frac{1}{2}$.
H_1: The distribution is not binomial with $p = \frac{1}{2}$.

NUMBER OF HEADS	OBSERVED FREQUENCY f_0	EXPECTED FREQUENCY f_t
0	36	31.3
1	138	156.3
2	348	312.5
3	287	312.5
4	165	156.3
5	26	31.3
TOTAL	1000	1000.2

$$\chi^2 = 10.344$$
$$df = 5$$
$$\chi^2_{.05} = 11.070$$

Hence, accept H_0, and conclude that the binomial distribution with $p = \frac{1}{2}$ is a good fit.

CHAPTER 12

Page 305

2. P (Major appliance) $= 0.11$
P (Home appliance) $= 0.22$
P (Small appliance) $= 0.67$

4. Expected Profit (Preparing bid) $= (0.10)(175,000)$
$+ (0.9)(- 10,000) = \$8,500$
Therefore, R.B.A., Inc. should prepare the bid.

6. Expected Profit $(0) = 0$

Expected Profit $(1) = (2/20)(-\$1) + (18/20)(\$4) = \$3.50$

Expected Profit $(2) = (2/20)(-\$2) + (3/20)(\$3)$
$+ (15/20)(\$8) = \$6.25$

Expected Profit $(3) = (2/20)(-\$3) + (3/20)(\$2)$
$+ (5/20)(\$7) + (10/20)(\$12) = \$7.75$

Expected Profit $(4) = (2/20)(-\$4) + (3/20)(\$1)$
$+ (5/20)(\$6) + (5/20)(\$11)$
$+ (5/20)(\$16) = \$8.00$

Expected Profit $(5) = (2/20)(-\$5) + (3/20)(\$0)$
$+ (5/20)(\$5) + (5/20)(\$10)$
$+ (4/20)(\$15) + (1/20)(\$20) = \$7.25$

The optimal stocking level is 4 units.

The expected profit is $8.00.

8. The gain from depositing the money in the bank is $4,000.

If the two investments are to be equally attractive

$P(\text{Recession}) \$2,000 + [(1 - P)(\text{Recession})] \$10,000 = \$4,000$

$P(\text{Recession}) = 0.75$

Page 312

2. The expected value of perfect information is the expected opportunity loss of the optimal act under uncertainty.

4. (a) (In units of $1000)

$EOL(A_1) = (0.3)(0) + (0.2)(3) + (0.4)(0) + (0.1)(6) = 1.2$

$EOL(A_2) = (0.3)(4) + (0.2)(0) + (0.4)(3) + (0.1)(4) = 2.8$

$EOL(A_3) = (0.3)(7) + (0.2)(4) + (0.4)(4) + (0.1)(2) = 4.7$

$EOL(A_4) = (0.3)(9) + (0.2)(2) + (0.4)(7) + (0.1)(0) = 5.9$

Therefore, EVPI = minimum EOL = 1.2

(b) Under certainty, the expected payoff is

$(0.3)(10) + (0.2)(8) + (0.4)(12) + (0.1)(14) = 10.8$

Under uncertainty, the expected payoff of the best act A_1 is

$(0.3)(10) + (0.2)(5) + (0.4)(12) + (0.1)(8) = 9.6$

$EVPI = 10.8 - 9.6 = 1.2$

(In units of $1000)

6.

OPPORTUNITY LOSS TABLE

OUTCOME	A_1 PUBLISH	A_2 DO NOT PUBLISH
Success	0	8
Failure	4	0

$EOL(A_1) = (1/3)(0) + (2/3)(4) = \$2.67$ million

$EOL(A_2) = (1/3)(8) + (2/3)(0) = \$2.67$ million

$EVPI = \$2.67$ million

The publisher should be indifferent as to publishing or not publishing the book.

8. *Under certainty*

Expected return $= (0.7)(\$500) + (0.3)(\$800) = \$590$

Under uncertainty
Expected return (Guess A) = (0.7)($500) + (0.3)(−$300) = $260
Expected return (Guess B) = (0.7)(−$40) + (0.3)($800) = $212
EVPI = $590 − $260 = $330

Alternatively,
  EOL (Guess A) = (0.7)($0) + (0.3)($1100) = $330
  EOL (Guess B) = (0.7)($540) + (0.3)($0) = $378
  Hence, EVPI = $330.

10. (a) Expected gain (Open new region) = (0.3)($100,000)
    + (0.4)($10,000) + (0.3)(−$80,000) = $10,000
    Yes, open the new region.
(b) Under certainty, the expected gain is
    (0.3)($100,000) + (0.4)($10,000) + (0.3)(0) = $34,000
    EVPI = $34,000 − $10,000 = $24,000

12. (a) (in $10,000 units)
    $EOL(A_1)$ = (0.5)(1) + (0.2)(0) + (0.1)(3) + (0.2)(3) = 1.4
    $EOL(A_2)$ = (0.5)(2) + (0.2)(1) + (0.1)(0) + (0.2)(2) = 1.6
    $EOL(A_3)$ = (0.5)(3) + (0.2)(2) + (0.1)(1) + (0.2)(0) = 2.0
    $EOL(A_4)$ = (0.5)(0) + (0.2)(4) + (0.1)(3) + (0.2)(3) = 1.7
(b) EVPI = 1.4
(c) Since A_1 has the lowest EOL, the optimal decision would be A_1.

Page 318

2.

STATE OF NATURE	$P(S)$	$P(X \mid S)$	$P(S)P(X \mid S)$	REVISED PROBABILITIES $P(S \mid X)$
S_1	0.1	0.8	0.08	8/43
S_2	0.2	0.6	0.12	12/43
S_3	0.3	0.5	0.15	15/43
S_4	0.4	0.2	0.08	8/43
			0.43	

4.

p	PRIOR PROBABILITIES $P_0(p)$	CONDITIONAL PROBABILITIES $P(X = 1 \mid n = 10, p)$	JOINT PROBABILITIES	POSTERIOR PROBABILITIES $P_1(p)$
0.10	0.8	0.3874	0.310	155/182
0.20	0.2	0.2684	0.054	27/182

6. (a) EOL (accept) = (0.5)(0) + (0.3)(100) + (0.2)(200) = 70
(b)

$P(X \mid \theta)$	$P_1(p)$
0.1937	0.41
0.3020	0.39
0.2335	0.20

EOL (accept) = (0.41)(0) + (0.39)(100) + (0.20)(200)
         = 79

Page 328

2. **(a)** $(0.4)(10) + (0.6)(2) > 5$ utiles. Thus, prefer *AE* combination.
   (b) $(0.5)(8) + (0.5)(2) = 5$ utiles. Thus, indifferent.
   (c) $(0.3)(10) + (0.7)(3) > 5$ utiles. Thus, prefer *AD* combination.
   (d) $(0.4)(8) + (0.6)(2) < 5$ utiles. Thus, prefer *C*.

4. **(a)** Expected Gain (100% interest) $= (0.8)(-\$50,000)$
       $+ (0.1)(\$130,000) + (0.1)(\$430,000) = \$16,000$
       Expected Gain (50% interest) $= (0.8)(-\$25,000)$
       $+ (0.1)(\$65,000) + (0.1)(\$215,000) = \$8,000$
       Expected Gain (Don't drill) $= 0$
       The best act is to drill with 100% interest.
   (b) Expected $[U(100\%)] = (0.8)(-30) + (0.1)(60) + (0.1)(200)$
       $= 2$ utiles
       Expected $[U(50\%)] = (0.8)(-10) + (0.1)(25) + (0.1)(120)$
       $= 6.5$ utiles
       The best act is to drill with 50% interest.

6. Expected Utility [Market drug] $= (0.99)(9000) + (0.01)(-900,000) = -90$ utiles
   Therefore, the FDA should not accept the drug.

CHAPTER 13

Page 346

2. **(a)** The labor force of this county is increasing by decreasing amounts and at a decreasing percentage rate.
   (b) $Y_t = 49.17 + 4.23(10) - 0.19(10)^2 = 72.47$ thousands
       The deviation of 95,185 from the computed trend figure of 72,470 does not necessarily imply a poor fit. It is conceivable that the fit is adequate, and that the deviation represents an unusually strong cyclical influence.

Page 360

2. Sales in July would tend to be closer to $19,600 because the trend value of $28,000 does not take into account seasonal variations. Thus, with a July seasonal index of 70, sales in that month would tend to be below the trend figure, in the absence of cyclical and random factors.

4. **(a)** Roughly, the seasonal index of 88 means that sales in September are typically 88% of average monthly sales for Expo Corporation. More precisely, the effect of seasonality in September is to reduce the sales of that month by 12% below what they would have been in the absence of seasonal influence.
   (b) In the multiplicative model of time series analysis the original monthly figure is viewed as consisting of the product of the trend level, the cyclical relative (combined effect of cyclical and irregular factor), and the seasonal index. In symbols,

$$Y = (Y_t)\left(\frac{Y/SI}{Y_t}\right)(SI)$$

For December, 1967
$$Y_t = 190 + (.24)(53) = 202.72$$
$$Y/SI = \frac{391}{2.00} = 195.5$$

and $\dfrac{Y/SI}{Y_t} = \dfrac{195.5}{202.72} = 0.9643$

$SI = 2.00$

Hence:

$391 = (202.72)(0.9643)(2.00)$

represents the breakdown of components of the December 1967 figure.

6. (a) $a = \dfrac{\Sigma Y}{n} = \dfrac{1603}{6} \doteq 267.17$

$b = \dfrac{\Sigma xY}{\Sigma x^2} = \dfrac{5143}{70} = 73.47$

$Y_t = 267.17 + 73.47x$

$x = 0$ in 1935

x is in five year intervals

y is in billions of kilowatt hours

(b) a or 267.17 billion kilowatt hours is the computed trend figure for 1935.
b or 73.47 is the slope of the trend line. That is, there was an estimated change of 73.47 billion kilowatt hours per five years in the trend level of the consumption of electric power in the U.S. from 1910 to 1960.

(c) $Y_t = 267.17 + 73.47 \,(-.6) = 223.09$

Relative cyclical residual $= \dfrac{100 - 223.09}{223.09} \cdot 100 = -55.2\%$

The consumption of electric power in 1932 was 55.2% below the trend level because of cyclical and irregular factors.

8. (a) 3   **(e)** 1

(b) 2   **(f)** 1

(c) 4   **(g)** 3

(d) 2

CHAPTER 14

Page 372

2. (a) *Weighted Aggregates Index, with 1967 Weights*

$$\dfrac{\Sigma P_{73}Q_{67}}{\Sigma P_{67}Q_{67}} \cdot 100 = \dfrac{\$1000}{\$750} \cdot 100 = 133.3$$

(b) We cannot conclude that Jones Metal Company paid more per unit in 1973 than its competitor. All that we know is that in 1973 it costs Jones 133.3% of what it did in 1967, whereas in 1973 it costs its competitor 120% of the cost to purchase the 1967 quantities. However, the absolute level of prices as measured by price per unit may very well be lower for Jones than for its competitor. The prices paid by Jones have risen more, on the average, than its competitor's, but its prices per unit in 1967 may have been at a much lower level than its competitor's.

Page 376

2. (a) *For 1967 on 1966 base*

$$\dfrac{\Sigma \left(\dfrac{P_{67}}{P_{66}} \right)(P_{66}Q_{66})}{\Sigma (P_{66}Q_{66})} \cdot 100 = \dfrac{\Sigma P_{67}Q_{66}}{\Sigma P_{66}Q_{66}} \cdot 100$$

$$= \dfrac{(\$2.52)(1.28) + (\$2.23)(1.40)}{(\$2.45)(1.28) + (\$2.15)(1.40)} \cdot 100$$

$$= \dfrac{\$6.3476}{\$6.1460} \cdot 100 = 103.3$$

For 1968 on 1966 base

$$\frac{\Sigma\left(\frac{P_{68}}{P_{66}}\right)(P_{66}Q_{66})}{\Sigma(P_{66}Q_{66})} \cdot 100 = \frac{\Sigma P_{68}Q_{66}}{\Sigma P_{66}Q_{66}} \cdot 100 = \frac{(\$2.60)(1.28) + (\$2.41)(1.4)}{\$6.1460}$$

$$= \frac{\$6.702}{\$6.146} \cdot 100 = 109.0$$

(b) *For 1967 on 1966 base*

$$\frac{\Sigma P_{67}Q_{66}}{\Sigma P_{66}Q_{66}} \cdot 100 = \frac{\Sigma\left(\frac{P_{67}}{P_{66}}\right)(P_{66}Q_{66})}{\Sigma(P_{66}Q_{66})} \cdot 100 = 103.3$$

For 1968 on 1966 base

$$\frac{\Sigma P_{68}Q_{66}}{\Sigma P_{66}Q_{66}} \cdot 100 = \frac{\Sigma\left(\frac{P_{68}}{P_{66}}\right)(P_{66}Q_{66})}{\Sigma(P_{66}Q_{66})} \cdot 100 = 109.0$$

Page 383

2.

	DISPOSABLE INCOME (BILLIONS OF DOLLARS)	DISPOSABLE INCOME IN 1965 CONSTANT-DOLLARS (BILLIONS OF DOLLARS)
1955	$100	$100 \div \dfrac{90}{120} = \$133.33$
1960	160	$160 \div \dfrac{100}{120} = \$192.00$
1965	180	$180 \div \dfrac{120}{120} = \$180.00$
1970	220	$220 \div \dfrac{125}{120} = \$211.20$

$Index$

TABLE A-6   STUDENT'S *t*-DISTRIBUTION

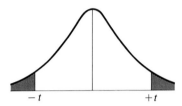

$-t$ $+t$

Example: For 15 degrees of freedom, the *t*-value which corresponds to an area of 0.5 in both tails combined is 2.131.

Degrees of Freedom	Area in Both Tails Combined			
	.10	.05	.02	.01
1	6.314	12.706	31.821	63.657
2	2.920	4.303	6.965	9.925
3	2.353	3.182	4.541	5.841
4	2.132	2.776	3.747	4.604
5	2.015	2.571	3.365	4.032
6	1.943	2.447	3.143	3.707
7	1.895	2.365	2.998	3.499
8	1.860	2.306	2.896	3.355
9	1.833	2.262	2.821	3.250
10	1.812	2.228	2.764	3.169
11	1.796	2.201	2.718	3.106
12	1.782	2.179	2.681	3.055
13	1.771	2.160	2.650	3.012
14	1.761	2.145	2.624	2.977
15	1.753	2.131	2.602	2.947
16	1.746	2.120	2.583	2.921
17	1.740	2.110	2.567	2.898
18	1.734	2.101	2.552	2.878
19	1.729	2.093	2.539	2.861
20	1.725	2.086	2.528	2.845
21	1.721	2.080	2.518	2.831
22	1.717	2.074	2.508	2.819
23	1.714	2.069	2.500	2.807
24	1.711	2.064	2.492	2.797
25	1.708	2.060	2.485	2.787
26	1.706	2.056	2.479	2.779
27	1.703	2.052	2.473	2.771
28	1.701	2.048	2.467	2.763
29	1.699	2.045	2.462	2.756
30	1.697	2.042	2.457	2.750
40	1.684	2.021	2.423	2.704
60	1.671	2.000	2.390	2.660
120	1.658	1.980	2.358	2.617
Normal Distribution	1.645	1.960	2.326	2.576

Source: Table A-6 is taken from Table III of Fisher and Yates: *Statistical Tables for Biological, Agricultural and Medical Research*, published by Longman Group Ltd., London (previously published by Oliver & Boyd, Edinburgh), and by permission of the authors and publishers.